While self-contained and entirely the responsibility of the authors, this study was prepared within the context of a wider project concerning East-West Technology Transfer conducted under the auspices of the Committee for Scientific and Technological Policy.

It was composed of two distinct, though interrelated, phases. The first was exploratory in nature and delineated the field of inquiry. It was concluded with the publication entitled *Technology Transfer between East and West*.

The second seeks to analyse and assess two broad issues: the factors determining the assimilative capacity of the Eastern countries with regard to technology in general and Western technology in particular and the impact of Eastern imports of Western technology on East-West trade in technology.

The method of analysis consisted of country studies, analyses based on East-West and intra-CMEA trade flows, and on econometric and sectoral analyses.

Publication of this report has been authorised by the Secretary-General.

## *Also available*

EAST-WEST TECHNOLOGY TRANSFER. Study of Hungary 1968-1984 by Paul Marer (February 1986)
(92 86 01 1) ISBN 92-64-12800-X 244 pages £14.00 US$28.00 F140.00 DM62.00

EAST-WEST TECHNOLOGY TRANSFER. The Transfer of Western Technology to the USSR by Morris Bornstein (December 1985)
(92 85 04 1) ISBN 92-64-12779-8 190 pages £14.00 US$28.00 F140.00 DM62.00

EAST-WEST TECHNOLOGY TRANSFER. Study of Czechoslovakia by Friedrich Levcik and Jiri Skolka (August 1984)
(92 84 03 1) ISBN 92-64-12600-7 102 pages £4.50 US$9.00 F45.00 DM22.00

EAST-WEST TECHNOLOGY TRANSFER. I. Contribution to Eastern Growth: An Econometric Evaluation by Stanislaw Gomulka and Alec Nove. II. Survey of Sectoral Case Studies by George D. Holliday (June 1984)
(92 84 02 1) ISBN 92-64-12565-5 94 pages £6.20 US$12.50 F62.00 DM28.00

EAST-WEST TECHNOLOGY TRANSFER. Study of Poland 1971-1980 by Zbigniew Fallenbuchl (November 1983)
(92 83 01 1) ISBN 92-64-12484-5 200 pages £11.00 US$22.00 F110.00 DM49.00

TECHNOLOGY TRANSFER BETWEEN EAST AND WEST by Eugène Zaleski and Helgard Wienert (September 1980)
(92 80 02 1) ISBN 92-64-12125-0 436 pages £22.00 US$50.00 F200.00 DM100.00

*Prices charged at the OECD Bookshop.*

*THE OECD CATALOGUE OF PUBLICATIONS and supplements will be sent free of charge on request addressed either to OECD Publications Service, Sales and Distribution Division, 2, rue André-Pascal, 75775 PARIS CEDEX 16, or to the OECD Sales Agent in your country.*

# EAST-WEST TECHNOLOGY TRANSFER

# THE TRADE AND ECONOMIC ASPECTS

by
Helgard Wienert and John Slater



ORGANISATION FOR ECONOMIC CO-OPERATION AND DEVELOPMENT

Pursuant to article 1 of the Convention signed in Paris on 14th December, 1960, and which came into force on 30th September, 1961, the Organisation for Economic Co-operation and Development (OECD) shall promote policies designed:

- to achieve the highest sustainable economic growth and employment and a rising standard of living in Member countries, while maintaining financial stability, and thus to contribute to the development of the world economy;
- to contribute to sound economic expansion in Member as well as non-member countries in the process of economic development; and
- to contribute to the expansion of world trade on a multilateral, non-discriminatory basis in accordance with international obligations.

The Signatories of the Convention on the OECD are Austria, Belgium, Canada, Denmark, France, the Federal Republic of Germany, Greece, Iceland, Ireland, Italy, Luxembourg, the Netherlands, Norway, Portugal, Spain, Sweden, Switzerland, Turkey, the United Kingdom and the United States. The following countries acceded subsequently to this Convention (the dates are those on which the instruments of accession were deposited): Japan (28th April, 1964), Finland (28th January, 1969), Australia (7th June, 1971) and New Zealand (29th May, 1973).

The Socialist Federal Republic of Yugoslavia takes part in certain work of the OECD (agreement of 28th October, 1961).

Publié en français sous le titre:

TRANSFERT DE TECHNOLOGIE
ENTRE L'EST ET L'OUEST
LES ASPECTS COMMERCIAUX
ET ÉCONOMIQUES

# TABLE OF CONTENTS

*Part II*

**EMBODIED TECHNOLOGY TRANSFER AND TOTAL TRADE**

## *Part III*

## EASTERN POLICIES AND THE ACQUISITION OF WESTERN TECHNOLOGY



## *Part IV*

## THE IMPACT OF EAST-WEST TECHNOLOGY TRANSFER

## NOTES AND REFERENCES



## LIST OF TEXT TABLES

*Chapter 1*



*Chapter 2*



*Chapter 3*



*Chapter 4*

*Chapter 5*



*Chapter 6*

*Chapter 7*



*Chapter 8*

## *LIST OF ANNEX TABLES*

### *Chapter 4*



### *Chapter 7*



## *SUMMARY GRAPHICS*

## ACKNOWLEDGEMENTS

The authors wish to thank Ronald Amann, Morris Bornstein, Josef C. Brada, John P. Hardt, Ed.A. Hewett and Harriet Matejka for helpful comments and advice on an earlier version of this study.

The authors also wish to express their particular thanks to John A. Martens for having contributed Chapter 3 of this study, as well as to Israel Borenstein and Zdenek Drabek for their help in reclassifying the OECD and Eastern countries' capital goods trade data and to Peter Elkan and Dario Tosi for their research contributions on which this study draws.


**The text of this study incorporates data available up to July 1985**

# PREFACE

## I. PURPOSE OF THE STUDY

This study, a component part of the OECD's programme of work on East-West technology transfer, attempts to analyse the main factors which determine East-West trade in technology and technology-based products. The desirability of undertaking an analysis of the technology transfer phenomenon originates in the political/economic relationship between East and West – a relationship which adds a military/strategic dimension to general concerns about the effects of large-scale transfers of technology on the economic strength of the Eastern countries.

During the 1970s, these countries expanded their imports of Western technology at substantially higher rates than their export performance alone justified. This was made possible by the extension of large-scale credits by private financial institutions (and some Western governments) on very favourable terms. The period of expanded – and essentially credit-financed – imports of Western capital goods by the Eastern countries was accompanied by an expansion of industrial co-operation activities between Western firms and Eastern enterprises and a growth in such phenomena as countertrade and compensation arrangements.

It was felt that these developments should be comprehensively reviewed and that, in particular, an effort should be made to gauge the possible consequences of transfers of productive technology and know-how on Eastern output and export potential. Clearly, this can best be undertaken within the context of a review of the past and likely future evolution of overall East-West trade. The study was also expected to analyse the Eastern countries' capacity for assimilating Western technology and, more generally, to provide essential elements for assessing the presumed advantages and disadvantages of East-West trade as a whole to the OECD countries.

## II. TERMINOLOGY

### 1. East-West

While this study deals primarily with the transfer of technology from West to East, the term "East-West" has been retained since it is standard usage in the literature. The term "West" refers, by and large, to the OECD area, though when applied to Eastern reported import figures, it may include insignificant amounts of technology imported from elsewhere. "East", "Eastern countries" or "CMEA" (Council for Mutual Economic Assistance) pertains only to the six East European countries (Bulgaria, Czechoslovakia, the German Democratic Republic, Hungary, Poland and Romania) and the Soviet Union – also referred to in this study as the CMEA countries.

## 2. Technology Transfer

In selecting an approach to the work, it was necessary to establish a working definition of technology transfer. For evaluation purposes, there is no single, totally satisfactory, definition of this term. An earlier OECD study employed the following definition: "technology transfer is a process whereby innovations (new products or know-how) obtained in one country are then transmitted for use to another...."[1].

This is comprehensive, but may not be suitable in practice for several reasons. A basic problem is that definitions may vary according to the questions asked (are we interested in all transfers of technology, "high technology" only, productive technology only, or all products with a significant technology content?) Moreover, the scope for answering questions about different classes of technology depends on whether appropriate information is available.

The definition of technology transfer retained in the present study is thus conditioned by the two general bodies of information used. The first includes published data on co-operation arrangements between Western firms and Eastern entities, as well as a compendium of information on those elements of intra-CMEA co-operation likely to be relevant to the process of East-West technology transfer. This information has been used to make qualitative assessments of the mechanisms involved in technology transfer and to provide, where possible, an assessment of their importance in Eastern countries' trade, and notably, East-West trade. No specific definition of technology transfer is involved here. The second body of information is the published statistical data on foreign trade in capital goods and in certain commodities incorporating a significant level of technology (mainly intermediate products). These data have been used as a focus for the quantitative work contained in the study. By virtue of this, the definition retained in the quantitative assessments covers only "embodied" technology and trade in such technology. It is evident that "disembodied" technology (know-how, training and so forth) is excluded from the analysis.

# III. THE APPROACH AND CONTENT OF THE STUDY

Apart from the definitional problems described above, the approach to the study was conditioned by the impossibility of relating a given Eastern import of technology directly to specific items of output or exports.

When the work on which this document is based was beginning, it was suggested that work at whole economy, sectoral and branch levels (embodied in trade statistics, the indicators of domestic development, etc.) should be supplemented where possible by a direct approach. This would take the form of case studies covering inter-firm technology transfers, followed up by descriptions of such issues as lead times, working up-times, manning levels and manpower savings, costs and benefits per unit of technology transferred, disposal of final product (for export or the domestic market, etc.). In practice, due to their high cost, such case study work could not be carried out[2].

The approach has thus been two-pronged. First a major effort has been devoted to a detailed scrutiny of published sources in East and West in order to collect various items of information on the facts and policy attitudes relating to the technology transfer issue. Second, efforts have been made to extract and reclassify in an appropriate form trade and other data to clarify the factual basis for discussion of the main technology transfer issues. Given the difficulties of defining technology transfer already noted, these operations were carried out to overcome a fundamental problem: since technology transfer is not a simple concept, and since essential information for a direct approach is lacking even in some of the central areas which might preoccupy those responsible for it, it can inevitably be examined only obliquely in many of its aspects.

The study thus attempts to examine the technology transfer issue from as many angles and viewpoints as possible. The results are presented in four parts. Part I is designed to provide the background for the detailed statistical analysis undertaken in the second part. Consequently, Chapters 1 and 2 detail the changes in the macro-economic conditions and performance which characterised the Eastern economies during the 1970s, and into the 1980s, and the response to these changes as reflected in the Five-Year Plans. In Chapter 3, the Eastern countries' R&D systems are examined. Chapter 4 describes the mutual economic relations of the CMEA countries and the possible consequences of intra-CMEA co-operation for East-West technology transfer.

As noted, Part II is devoted to a statistical analysis and assessment of what the trade data suggest about the possible long-range attitudes of Eastern policy-makers towards the purchase of Western technology. Chapter 5 describes the statistical sources used and reviews the statistical facts of East-West technology transfer, while Chapter 6 assesses the role of embodied technology transfer in the Eastern countries' trade with the world. Emphasis is placed on trade with the OECD countries, but developments in intra-Eastern area technology trade are also reviewed.

Part III looks at technology transfer from the viewpoint of industrial co-operation. The basic purpose of Chapter 7 is to highlight the individual Eastern countries' approach to technology transfer.

Finally, Part IV is devoted to the issue of impact. Chapter 8 attempts to gather together the main conclusions which can be drawn on the past and likely future impact of West-to-East technology transfers on the foreign trade (primarily East-West) of the Eastern countries and also on domestic output and efficiency developments.

## IV. PROBLEMS AND LIMITATIONS OF THE APPROACH

The limitations of the approach are clear. First of all, it was not possible (given time constraints and the scope as well as the complexity of the subject matter) to exploit systematically all the available literature. Secondly, the quantitative analysis is essentially macro-economic in nature and, therefore, deals with the various aspects of technology trade at a relatively high level of aggregation. As noted, it was not possible to conduct the independent case studies originally foreseen. Consequently, the conclusions drawn from the study are limited primarily to the impact on trade – as reflected in trade flows between the OECD and the CMEA countries and in intra-CMEA trade. This precluded assessments of the impact of technology transfer at the level of the Western firm or Eastern enterprise, of the effectiveness of particular technologies transferred, or comprehensive treatment of such issues as the impact of Western technology on the overall production capacity of the Eastern countries. Nevertheless, these gaps have been partially filled by disaggregating the statistical information available to illustrate developments at sector and industrial branch levels. This enables some relationships between the capital goods imports of a particular sector or industrial branch and developments in that branch's efficiency (labour or capital productivity) or export performance to be shown.

The approach cannot avoid a number of inherent methodological limitations. For instance, the rate of transmission of know-how is acknowledged to be an important element of technology transfer in some cases. But data to quantify such sales are unsystematic. They may be hard to pin down in monetary terms (patents), or subject to commercial secrecy (cost of licence acquisitions), or simply not available (such as data having a bearing on royalties or other intangibles which are not published by governments in sufficient detail in balance of payments statistics). There is often a gap which is impossible to bridge between the observed facts of a given technology import and its impact; outcomes are rarely simply related with a particular input of technology. For instance, at the level of the economy as a whole, or of sectors and industrial branches, it is not possible to say how far a given technology import (however defined) has resulted in a given increase in exports by the recipient country, or even to show how much of the increase in output was directly due to the technology import. The same is true at sector or branch level.

A statistical examination may indicate a relationship, but it is possible that an observed rise in exports apparently relating to a transfer of technology stems wholly or partly from indigenous capacity and know-how. It is even less possible to establish a direct connection between an import of technology and exports to a given market: an increase in export capacity might show itself in East-South or intra-CMEA, rather than in East-West trade. A transfer of technology may also show up in the form of import saving.

In conclusion, it should again be noted that the focus of the quantitative analysis is on the transfer of investment (capital goods) and intermediate goods with a significant technological content. The study does not deal with matters related to export controls (considered to fall outside its scope and purpose) nor does it address the "resource-releasing" issue, an aspect often evoked in the Western literature. Consequently, no conclusions can be drawn as to the impact of Western technology transfer on the Eastern countries' military potential.

## V. THE OUTCOME OF THE STUDY

Notwithstanding these real limitations, it is felt that the information presented – both statistical and institutional – on the development of East-West technology transfer in the 1970s and early 1980s constitutes a more detailed body of information on the subject than has previously been available. It provides a framework for assessing fluctuations in East-West technology trade. It thus illustrates some of the determining factors likely to influence East-West technology transfer in the future. Second, direct comparisons of technology imports in aggregate, by type and by end-use sectors and branches have been made with relevant domestic indicators and exports, enabling some idea of Eastern technology import strategies to be deduced. This again suggests some of the factors likely to influence future Eastern technology import strategies. In addition, trade data have been reclassified and used to show the level of sophistication of Western technology exports to the Eastern area.

In sum, and though none of the approaches used can completely overcome the data, methodological and definitional difficulties mentioned earlier, the various elements brought together in the findings and conclusions which follow, do add up to a reasonably consistent set of ansers to the questions posed. While many areas of uncertainty remain, this empirical approach provides, it is hoped, enough global and detailed information for this study to be used to quantify the limits within which decisions related to East-West technology transfer can be made. It may also suggest some further areas of study appropriate for those most concerned at individual firm or national government level.

## NOTES AND REFERENCES

1. *Technology Transfer and Scientific Co-Operation Between the United States and the Soviet Union: A Review*, prepared for the Sub-Committee on International Security and Scientific Affairs of the Committee on International Relations by the Congressional Research Service, Library of Congress, Washington, D.C., May 26, 1977. See also E. Zaleski and H. Wienert, *Technology Transfer Between East and West,* OECD, Paris, 1980.
2. However, for a summary of some recent case-study work, see George D. Holliday, *East-West Technology Transfer, Survey of Sectoral Case Studies,* Book II, OECD, Paris, 1984.

# FINDINGS AND CONCLUSIONS

## I. EAST-WEST TECHNOLOGY TRADE 1970-1984: THE BACKGROUND

### 1. Principal Developments 1970-1984

The first half of the 1970s saw an unprecedented opening of the Eastern countries' market to Western goods and, in particular, Western embodied technology exports. Nevertheless, the period during which expansion of Eastern imports of Western technology took place was relatively short. Moreover, Bulgaria, Poland and the Soviet Union were together responsible for the bulk of the increase recorded by the seven countries as a whole. Czechoslovakia, Hungary and Romania, and to a much lesser extent the German Democratic Republic, were already heavy importers of Western technology relative to their share in the Eastern countries' aggregate output in 1970. Up to the mid-1970s therefore, differences in the growth of the various countries' technology imports had the effect of levelling off previous differences between the Eastern countries on this score. The exception was Poland, whose share of the Eastern countries' total imports of Western technology rose by the mid-1970s to levels much higher than its share of the region's output would suggest. By 1983, the total value of the OECD area's capital goods exports to the seven countries together was over $10 billion, and of intermediate goods with a significant technology content, nearly $11 billion – according to Western-reported trade data.

Eastern imports of Western technology have always been an important component of East-West trade: not only in the post-war trade of Eastern Europe, but also in the Soviet Union's pre-war trading relations with the West. A new element, which appeared in the early 1970s, was the Eastern countries' recognition that previous patterns of factor intensive growth was jeopardised by the exhaustion of the labour reserves on which such growth depended. This reflected the diminishing size of the farm population and the slowdown in the growth of the active population. The rate of capital-labour substitution could not be increased sufficiently to offset this.

For the seven countries together, big increases in imports as a whole were registered up to the mid-1970s, when the Eastern countries' rates of economic and investment growth were high. Technology imports over the period rose much faster, doubling in volume over 1970-1975. They were the main elements in the widening Eastern deficits on trade with the West – the cumulative totals for which amounted to $68 billion for capital goods and $49 billion for intermediate goods with a significant technology content over the ten years 1971-1980. The massive rise in Eastern imports of agricultural goods during the 1970s as a whole and the failure to expand exports of almost all classes of goods (fuel and a few other items excepted) also contributed to the growth of Eastern deficits on overall East-West trade.

The deficits were largely financed by recycled petro-dollars which had greatly expanded the liquidity of Western banks following the oil price rises of 1973. Furthermore, the political thaw culminating in the 1975 Helsinki agreements, was accompanied by a greater willingness on the part of the Eastern countries to extend their economic ties with the West to cover large-scale borrowing. The evident belief that greater access to Western technology would be beneficial to economic growth was a major motivation for this shift in policy. Large-scale borrowing was also made possible by the willingness of financial institutions and some Western governments to extend credit on very favourable terms. Expectations of higher export earnings, gaining access to new markets and maintaining employment levels were among the primary Western motivations.

As growth slowed down under the impact of domestic and external growth constraints (and in particular the need to redress growing trade imbalances), the Eastern countries were unable to maintain the pace of investment. Hence their demand for imports of capital goods from the West slackened, rising in fact by only about 10 per cent in volume between 1975 and 1983.

The redressment efforts undertaken in 1975/1976, and more severely in 1979/80, took the form of import restraints within which technology imports (and domestic investment expenditures) bore the brunt. This may also have reflected a certain disenchantment with Western technology, and explains, in part,the emphasis in the 1981-1985 Five-Year Plans on basing economic growth more on the development of indigenous technologies. This intention was explicitly reiterated at the 37th session of the CMEA which took place in October 1983 and at the subsequent CMEA Summit meeting held in June 1984. The proposal made at the Summit meeting to the effect that a "Complex Programme for Scientific-Technical Progress for 15-20 Years" be elaborated also mirrors the pre-occupation with accelerating the development of domestic S&T and with reducing thereby, dependence on imports of Western technology.

The emphasis on developing indigenous technology may, of course, also reflect failure to absorb and diffuse the imported technology more generally throughout the economy. This would confirm, on a macro-economic level, the findings of a number of Western studies which have drawn attention to the long time lags between the development and application of innovations in individual industries. It also explains the preoccupation with the organisation of research and the implementation of research results which have featured in recent Five-Year Plans, as well as the institutional changes which have been promulgated in several countries of the Eastern region.

In sum, although the period of fast growth of Western embodied technology imports was rather short-lived and their rates of growth, in some countries, relatively modest, their contribution to the build-up of Eastern indebtedness to the Western countries was very big. In spite of investment slowdowns and even absolute declines, the share of imported capital goods in total investment by 1983 was lower or only marginally higher, than in 1970 in most Eastern countries. The two exceptions are Hungary (where the import build-up had been much slower than average and the investment decline particularly big) and the Soviet Union (due to the sudden recovery in capital goods imports in 1982). The principal reason for the cut-backs was the growth of debt. Fears of over-exposure on the part of Western institutions lending to the Eastern countries gradually restricted new credits – particularly after the Polish crisis, which became acute in 1980. Eastern willingness to borrow was also affected by the manifest failure of export growth to provide the funds necessary to cover imports, and of the Eastern economies to achieve import savings [Chapters 1, 2, 4, 5 (Section 2) and 6].

## 2. The Technology Gap

The reason for the growth of Eastern demand for Western technology up to the mid-1970s and its gradual slow-down, in real terms thereafter, is a basic question addressed in this study. It can be asserted that the underlying potential Eastern demand for Western technology is strong simply because the level of technology produced in the West is higher than in the East. The large imbalance between Western exports of technology to the Eastern countries and imports from them, is the most obvious indication of the substantial lead of the OECD countries in technology applications relative to the Eastern countries. For capital goods this is about 9:1 in the West's favour. Another indicator from foreign trade is the loss of Western market shares by Eastern (non-fuel) industrial commodities which has occurred since 1970 (Chapters 5 and 8).

The technological gap can be illustrated by differences in productivity per worker in East and West. Bearing in mind that it is difficult because of pricing anomalies and for other reasons to make direct comparisons of productivity levels between Eastern and Western countries, available Western estimates indicate that Eastern productivity levels are barely half of those in the West. For instance, it was estimated that the German Democratic Republic's level was some 60 per cent of the level in the Federal Republic of Germany in the mid-1970s. In Czechoslovakia (also in the mid-1970s) it was 50-60 per cent of the West European average, while in the Soviet Union it was estimated that labour productivity was only one-third of the US level in 1977[1]. Moreover, no significant narrowing of the East-West technology gap has occurred over the last 15-20 years (Chapter 3). Judging by labour productivity growth deceleration and absolute capital productivity declines, it may even be widening.

Nonetheless, the Eastern countries have made and are making great efforts to attack the problem. The RDI (research-development-innovation) complex employs a high share of the work force and R&D expenditures account for a substantial proportion of national income. However, forces promoting technological progress are often thwarted by obstacles to innovation arising from certain features of the planning system at both national and intra-CMEA level. While advanced technology has been developed and applied in some sectors of the Eastern economies – the space and defence efforts spring instantly to mind – large lags in most other areas bear heavily on the civilian economy. The size of the technology gap is not directly addressed in this study. Nonetheless, the work takes as its starting point the frequently – if only obliquely – acknowledged existence of such a gap by Eastern policy makers.

Eastern technological inadequacy reflects, in the first place, the effects of late industrial development in most of the countries of the area. Bulgaria, Hungary, Poland, Romania and the Soviet Union, in particular remained predominantly agrarian economies long after the industrial take-off occurred in the most developed countries of the OECD area. The emergence of industry as the predominant sector of the economy took place as late as the 1930s in the Soviet Union and not until the 1950s and 1960s in the other countries mentioned.

A further element in the technological lag stems from certain features of the planning system. The lack of direct incentives – indeed the presence of disincentives – at the level of production enterprises as well as an over-centralised R&D system restricts the economy's responsiveness to innovative needs. A related element is the low level of efficiency with which resources are used. Rigidities in planning, which are not the result of technological shortcomings, lead, among other things, to below-capacity use of production equipment and uneconomic resource allocation decisions. The latter, in particular, have hindered the majority of the Eastern countries from exporting those commodities that would be appropriate for Western markets.

As early as 1970, the imperatives for a wide-ranging improvement in efficiency had become urgent. This was reflected in the much stronger emphasis placed on "intensification" in the Five-Year Plans of all the countries for 1970-1975 and subsequently, and much more explicitly in the Five-Year Plans for 1976-1980 and 1981-1985. However, the trade deficits built up in the mid- to late-1970s dictated a shift of resources from domestic demand and, in particular, from investment demand, to the foreign balances in order to reverse the continuing rise in deficits and to service and ultimately repay accumulated debt. Thus while the level of Eastern demand for Western technologies may have remained high, basically determined by the technology gap, the domestic resources to pay for them were lacking. Of course, whether access to unlimited supplies of Western technology would have significantly accelerated the intensification process is far from certain (Chapters 2, 3 and 4).

## II. EASTERN TECHNOLOGY IMPORT STRATEGIES

A part of the study is devoted to assessing Eastern embodied technology import strategies in terms of main regional suppliers, the corresponding institutional mechanisms, levels of sophistication of Western technologies acquired by the Eastern countries and their allocation by final and by sector/branch end-uses.

### 1. Acquisition Strategies

#### a) *Regional Suppliers*

Capital goods from all suppliers accounted for around one third of total imports in the 1970s. As balance of payments readjustment policies were progressively tightened from the late 1970s, mainly by cutting back the growth of domestic investment and imports, the share of capital goods in total imports declined throughout Eastern Europe. A similar process in the Soviet Union was reversed after 1981. As will be noted later, intermediate goods with a significant technology content (mainly sophisticated industrial inputs) which also accounted for about a third of total imports from the West, were generally not cut back so sharply except in the Soviet Union after 1980.

The Western countries are important but minority suppliers of capital goods to the Eastern countries. Eastern-reported trade data indicate that the Western share was just under one-quarter in the early 1970s and also in the 1980s, though it rose briefly to nearly 40 per cent in the mid-1970s. The fall thereafter was mainly concentrated in Eastern Europe, where it was very steep. By 1983, the share of Western suppliers in total East European capital goods imports from all sources had fallen to 10-20 per cent, but was still about 40 per cent in the Soviet case. The actual shares quoted above are based on dollar conversions of official Eastern reported figures. They may be inaccurate because the relationship between intra- CMEA trade prices and those on world markets is tenuous, though the *changes* in these shares should be reasonably reliable. However, due to inadequacies in the coverage of Eastern-reported trade statistics, about one-quarter of Eastern capital goods imports could not be identified by region of origin. Most of this probably originates within the Eastern area itself, but it is possible that part represents re-exports within Eastern Europe from unspecified regions of origin (Chapter 6).

Such statistical problems complicate a judgement on whether or not the Eastern countries have "turned inward" with regard to supplies of technology since 1980. In current prices, the available data suggest that they have and to a considerable extent. But when account is taken of a fall in the prices of Western exports of machinery and equipment to the East, and, in contrast, a rise in such prices in intra-Eastern area trade, the statistical evidence suggests a more nuanced conclusion. ECE data, based on Eastern-country reported trade in machinery and equipment with "socialist" and "non-socialist" countries, indicate first that rises in the volume, as well as in the value, of Soviet machinery and equipment imports from other socialist countries far exceeded the rise in such imports from the "non-socialist" countries between 1980 and 1984. The volume of East European imports of these items from the non-socialist countries declined slightly.

For Eastern Europe, the smallness of the decline suggests that falls in the prices of machinery and equipment imports from the West enabled such purchases to be maintained. The much bigger rise in intra-area trade was thus almost exclusively due to the rise in East European machinery and equipment exports to the Soviet Union. This can be viewed as part of the effort made to limit the recent growth of East European trade deficits with the Soviet Union. In neither the Soviet Union nor in Eastern Europe is there evidence of a fundamental reconsideration of the West's role as a supplier of technology (Chapter 6).

b) *Mechanisms*

i) *East-West*

For most of the Eastern countries, the hard currency constraints which developed from the mid-1970s explain their interest in pursuing new forms of industrial co-operation which provide access to Western technology, make provision for financial support and ensure some form of market sharing.

East-West industrial co-operation (EWIC) developed rapidly in the 1960s, and especially during the first half of the 1970s. The United Nations estimates that some 1 000 industrial co-operation agreements existed at the end of 1975. While EWIC continued to expand during 1975-1980 (albeit at a slower rate), available data suggest that since 1980 a downward trend in its growth (at least as measured by the number of agreements concluded) has occurred for all countries except the Soviet Union. Though it is difficult to quantify industrial co-operation, it has been estimated that it represents a marginal share in East-West trade as a whole – not more than 5-7 per cent. The share of Eastern imports of Western techology covered by industrial co-operation agreements is not known.

Gaining access to Western technology (inclusive of production know-how) and export promotion are the primary motivations for the Eastern countries to seek industrial co-operation with Western firms. While Eastern interest in pursuing and expanding the actual scope of industrial co-operation remains high (notably such complex forms as co-production and joint ventures), these objectives have either not been, or are only partially met.

Recent developments suggest that the Eastern countries will continue to make a considerable effort to pursue co-operative arrangements to help them meet their economic development objectives and to extend co-operation to third-country markets. For some of them (notably the German Democratic Republic, Romania and the Soviet Union) the compensatory aspect of co-operation is clearly favoured and can be expected to be pursued vigorously (Chapter 7).

ii) *East-East*

The renewed emphasis placed on furthering intra-CMEA co-operation reflects the problems and developments which have made the pursuit of "intensive growth" the central theme of the development strategies adopted by all Eastern countries. More effective co-operation, notably in the sphere of science and technology, is clearly seen as one means of reversing the long-term trend of decelerating economic growth in the region. Furthermore, some countries (notably the Soviet Union) are advocating closer and more effective intra-CMEA co-operation as a means of decreasing dependence on Western technology. Particular emphasis is placed on the need to expand the mutual exchange of machinery and equipment. Some Soviet commentators have stressed that the Eastern countries have become too reliant on Western suppliers for such items – particularly in such branches as chemicals, mining and metallurgy and in the petroleum industry.

For various reasons, all the Eastern countries support the effort to be made to strengthen those mechanisms which are regarded as particularly appropriate for improving their production structures and raising factor productivity (for instance, the specialisation and production co-operation agreements which are themselves linked to programmes for joint RDI work). The aim is thereby to enhance not only the Eastern countries' export potential, but also to provide a stimulus for expanding mutual exchanges of high quality and technically sophisticated new products.

Nevertheless, it appears that the efforts made so far to raise the technical level of existing production capacities have not been successful. The share of advanced technological products in mutual deliveries also remains low – some 10-15 per cent of the total value of trade arising for instance, from specialisation and co-operation agreements. While the figure cited is from the Eastern literature, some Western analysts have likewise concluded that intra-CMEA co-operation, particularly in the S&T sphere (and despite the fact that intra-CMEA co-operation in S&T concerns mainly high and moderately R&D-intensive products or groups of products), has not fostered intra-CMEA exchanges of high technology products.

Thus, whether – in the interest of promoting more rapid production and wider diffusion of more sophisticated technologies – the changes undertaken (or proposed) to improve general economic management and the RDI systems (both in the individual Eastern countries and on the level of the CMEA) will be effective, remains to be seen. Current evidence suggests that this is not the case and a number of Western specialists maintain that continuing inadequacies in RDI performance will call for sustained imports of Western technology. Technology cannot solve problems related to inefficient planning of output. However, without reforms of management systems to encourage innovation at the enterprise level, the full effect of new (imported or domestically produced) technology on output may not be realised (Chapters 4 and 6).

## 2. Level of Sophistication of Eastern Technology Imports from the West

Within Eastern total capital and intermediate goods imports from the West, it is possible to separate out sub-groups classified according to their level of technological sophistication on the basis of the R&D intensity of the industrial branches producing them. This exercise shows that low R&D-itensive items accounted for the bulk of Eastern embodied technology imports over the entire period reviewed, although their share fell quite sharply in the early 1980s.

The Soviet Union in fact imported low R&D-intensive items to a substantially greater extent (80-86 per cent) than Eastern Europe (71-77 per cent) in the years 1970 to 1983.

Embodied technology with a high R&D content accounted for a very much smaller share in the total: less than 3 per cent in all years 1970-1983. Medium R&D-intensive goods rose from one-fifth to about a quarter of the total during the same period. The increase in both the small share of high R&D-intensive and also of medium R&D-intensive items in recent years does, however, indicate a pattern of increasing selectivity which appears to have been accompanied, if not provoked by, the cut back in the growth of technology imports in the late 1970s. This increasing selectivity is more apparent in those countries where technology import cut backs have been biggest (Poland and Romania).

The relative stability of the overall pattern of embodied technology imports of different R&D-intensities throughout the 1970s and early 1980s, and notably the still preponderant shares of low-to-medium R&D-intensive technology, suggests the following comments. First, the general strategy towards the level of sophistication of technology acquisitions does not seem inappropriate for countries in which the general level of technology is considered to be lower than in OECD countries. Second, under these conditions, the lower the level of technology imported, the less it is likely to be disruptive. Third, though the generalised decline in the share of low R&D-intensive imports in the total is not dramatic, it is consistent with the hypothesis that more sophisticated technology items are currently accorded a higher priority than the less sophisticated. This is not very surprising, if only because more sophisticated technology items are likely to be harder to replace from domestic or other CMEA country suppliers.

Nevertheless, given the existence and possible widening of the technology gap, it is still worthy of note that high and even medium R&D-intensive items account for such a low share in the total. The mere fact of importing larger quantities of the more highly sophisticated items would not, of course, mean that they would necessarily be effectively absorbed. It may be that present patterns are realistic under the circumstances of central planning, where flexibility (for instance, in ensuring inputs of the requisite type and quality needed for the more complex production processes for which high technology items are designed) is in short supply. However, the fact remains, as noted below, that neither export competitiveness nor domestic efficiency (total factor productivity) has noticeably improved in the Eastern countries since the beginning of the 1970s. This must at least raise questions about the appropriateness of a low R&D-intensive technology import strategy (Chapter 6).

## 3. The Use of Imported Technology

The study also examines the distribution of Eastern imports of Western capital goods by end-users. In the first instance, their distribution between broad categories of final demand (exports as distinct from domestic consumption) is assessed. The distribution of capital goods imports from West and East by Eastern end-use sectors and industrial branches is also scrutinised in some detail.

### a) *Capital Goods Imports by Broad Demand Categories*

The breakdown by broad end-use categories shows that in the East European countries a fairly steady 40 per cent of their total capital goods imports from the West has consisted of items for the production of tradeable goods (engineering and metal working machinery, agricultural and metallurgical equipment and various other types of electrical and non-electrical machinery and equipment). In the Soviet Union this share was also about 40 per cent in 1970s, but by 1983 it had fallen by half. The decline was largely offset by a rise in the share of a oil and gas distribution equipment.

This immediately raises questions concerning this aspect of Eastern strategy towards technology imports. First, it is highly likely that the debts incurred in relation to such imports were expected to be repaid by increased exports. However, as noted, less than half of East European capital goods imports and an even smaller share in the case of the Soviet Union, were purchased directly for the manufacture of tradeable goods. Over half consisted of equipment for the production of non-tradeable goods (for example, power and transport machinery) and for the production-assisting services which are, by definition, not traded. Moreover, not all production from capital goods which are capable of producing tradeable goods was actually exported (Chapter 6). This leads to the conclusion that the technology import effort of all the Eastern countries, was only partially designed to stimulate the export industries directly. This is so even when Soviet emphasis on importing pipeline equipment of which only part was used to export oil and gas to the West, is taken into account.

b) *Capital Goods Imports by Sector and Branch End-Users*

The productive (non-service) sectors of the economy – the material sphere according to Eastern terminology – accounted for at least 90 per cent of all Eastern capital goods imports from the West in 1975-1983. Between half and three-quarters of Eastern capital goods imports were allocated to the industry sector; together with transport, including pipeline and related imports on a 2:1 basis. This made up by far the greater part of the total allocated to the material sphere. Only in Hungary was the share of agriculture important (as much as 10 per cent in some years). Construction equipment rose as high as 5-10 per cent in most countries, but only in some years. Eastern reported data show that, within industry, the energy and fuel branches together accounted for 15-40 per cent of the total in 1983. Their share doubled in some East European countries and in the Soviet Union since 1975. The intermediate and investment goods supply industries – notably chemicals and engineering – usually accounted for between half and three-quarters of industrial capital goods imported from the West, but their share fell in most countries since 1975.

Perhaps surprising, given the need to restructure and improve the efficiency of output, is the fact that the share of goods for the engineering branch in total industrial capital goods imports from the West is generally very low relative to its share in industrial investment. At best, it was about the same in some countries. The quality of engineering products is, after all, an important pre-condition for technical progress throughout the economy. Moreover, the share of the engineering branch in industrial capital goods imports has fallen in the Soviet Union since 1975, though it has risen in most East European countries. In contrast, the corresponding share of chemicals has been very high relative to its share in investment. It is, with fuel, one of the few successful Eastern export branches.

These patterns contrast quite sharply with the pattern of capital goods trade within the Eastern area, where the share of industry in the material sphere total is much lower. Within industry (but not within the total), the share of engineering was a good deal higher than in capital goods imports from the West both in 1983 and in the 1970s. This is also the case for fuel and energy, but not for chemicals or the "supply" industries in general.

As noted earlier, if possible pricing discrepancies between East-West and intra-Eastern trade are discounted, the West supplied more than half the Eastern area's total import of industrial capital goods in 1975. This share fell subsequently, but was still above 40 per cent in 1983.

Despite the fall in their share of total Eastern capital goods imports, Western countries apparently remained important, or even the principal, suppliers of capital goods for the oil extraction, metallurgical, chemical and wood and paper branches. However, by 1983, some of the branches most heavily dependent on Western suppliers in 1975 – notably oil extraction and mining, engineering, the light and food industries – show a considerable rise in the share of Eastern suppliers. In several of these branches, the "substitution" which apparently occurred was accompanied by falls or slow growth in overall imports of capital goods for the branches concerned. This was the case particularly in chemicals and metallurgy. Moreover, it should be recalled that, since 1980, there was little change in the value of Eastern Europe's imports of capital goods from all sources. For the Soviet Union, where growth in imports from both East and West was very fast, substantial dependence on Western suppliers still remains with regard to equipment for oil extraction equipment, engineering, chemicals and the wood and paper branches (Chapter 6).

## III. IMPACT OF WESTERN TECHNOLOGY ON THE EASTERN ECONOMIES

As has been indicated above, Eastern technology import strategies contained a number of different elements. Hence their impact was, at the outset, likely to be diffused among several different outcomes (for instance, between efforts on exports on the one hand, and domestic output and efficiency indicators on the other). The question of impact is thus discussed under these two broad headings.

### a) *Exports*

For most of the 1970s, Eastern exports to the West failed to match import growth. The Eastern export failure did not only reflect sluggish demand in the West. The competitiveness of Eastern exports did, to some extent, increase between 1970 and 1977 but, as indicated by changes in their share of OECD's total imports, it did not rise spectacularly. Even these modest gains were wiped out after 1977. By 1983, the Eastern share of total OECD imports of non-fuel commodities, had fallen below the 1970 level. Most of this was due to intense and successful competition by the newly industrialised countries. Thus while market conditions in the Western countries certainly contributed to the growth of Eastern deficits, the main cause of the growth of trade deficits was the Eastern failure to export goods of the kind, quality and in the quantities necessary even to maintain their previous shares of a more slowly growing Western market.

These failures point up the different elements of Eastern strategy towards imports of Western technology. There are clearly elements of export promotion – notably towards the chemical industry (Soviet Union, the German Democratic Republic and perhaps also Hungary), motor cars (Poland), and clothing and processed food (Hungary). But the industries producing tradeable goods, let alone enterprises producing mainly for export, have not – as far as can be traced – been the main recipients of Western technology. This apparent lack of emphasis on exports, despite (or because of) a degree of success in individual cases noted in the study, is the key to the lack of impact of Western technology on Eastern export performance in Western markets (Chapter 8).

### b) *The Domestic Impact*

The most recent econometric approaches to measuring the impact of Eastern technology imports from the West on the economy as a whole, though not always consistent with earlier work, show some measure of agreement that differences between the productivity of imported Western and indigenous capital goods are rather small. The rationale underlying such imports is, therefore, to be found partly in improvements in domestic performance which they bring about, but also in an educative effect which cannot be measured.

These somewhat imprecise findings are echoed in attempts made in the study to find correlations between the growth rates of imports of Western capital goods and factor productivity changes. While no completely conclusive results have emerged, some of the relationships among the data are suggestive. A positive, though not a strong, correlation was found between sector and branch rates of growth of capital goods imports as a whole and labour productivity growth in most Eastern countries. A number of individual sectors and branches also gave a similar result.

Such findings suggest that imports of Western capital goods cannot provide the sole or even the main contribution to low or only slowly rising factor productivity. Nevertheless, the findings are consistent with what the continuing import of these items already suggests. Western capital goods do make a contribution to Eastern economic performance, though it is not always possible to pinpoint it.

In general, it has to be said that the study's attempts to quantify the impact of Western technology on Eastern output and efficiency is inconclusive. The period of rapid growth of Eastern purchases of Western technology was, in fact, followed by a period of declining growth and decelerating factor productivity performance. It may, of course, be that performance with less Western capital goods would have been worse. It can also be argued that the cut-backs in technology imports in the late 1970s and, especially in the early 1980s, themselves contributed to disruptions in the production process, notably a faster decline of capital productivity due, apparently, to the under-utilisation of capacity. But there is no hard evidence to show that the mere expansion of Western imported technology offers a short-cut for "closing the technology gap" (Chapter 8).

## IV. PROSPECTS FOR EAST-WEST TECHNOLOGY TRANSFER

Over the next few years, the formulation of Eastern and Western policies towards East-West technology transfer is likely to be cautious, inspired by the unpredictability of developments during the years 1970-1984. Reservations on the Western side will focus on the consequences of possible over-exposure by Western financial institutions if Eastern debt appears to be rising too fast. On the Eastern side, a more considered and realistic appraisal of the likely performance of Western technology in an Eastern production context is likely. The failure of Eastern exports to penetrate Western markets may reinforce this caution, but could also result in a strategy more highly concentrated on export development.

A more reserved attitude on the Eastern side could be reinforced by recent efforts made to improve the co-ordination of the Eastern countries' own S&T policies within the CMEA as well as within the national framework. However, as noted above, intra-CMEA S&T co-operation and co-operation in the production of high technology products during the 1970s indicates that the implementation of relevant co-operative programmes has not met expectations. Intra-CMEA trade of technology-embodying products shows that the share of high-technology products in this trade has been, and continues to be, limited. The Eastern countries have not, by and large, met their needs for advanced technologies by means of intra-CMEA co-operation.

Latest data show that considerable changes in the level and structure of import demand have been brought about in Eastern Europe and the Soviet Union since the beginning of this decade. The Soviet Union, despite the recent partial reversal of terms of trade gains made in the 1970s, succeeded in maintaining relative freedom of action via volume rises in its fuel exports. Compared with Eastern Europe, its import policies needed to be somewhat less restrictive. Although, and in sharp contrast with Eastern Europe, Soviet imports of Western technology items rose fast in 1982 and were maintained in 1983 and 1984, the Soviet Union also recorded surpluses in both years. While in 1983 and 1984, the East European countries (taken as a whole) also achieved a surplus in their East-West trade, it mainly reflected, as in the earlier years of adjustment, import cuts rather than real export growth. The striking exception was the German Democratic Republic, though the exports of other East European countries also turned up in 1984.

Concerning Eastern Europe, it may be noted that the cut-backs in Western technology imports (notably, capital goods) have clearly been a defensive measure and one response to the inability to increase exports. As far as the Soviet Union is concerned, the principal factor determining future imports may well be how far the volume of oil exports can be maintained or increased since domestic output of oil (though not yet gas) is slowing down. Soviet oil output has virtually stabilised since 1982, but exports to the OECD area nevertheless rose steadily over 1980-1984. It is unlikely – given the large movements which have already occurred – that the future will bring windfall oil price gains on the same scale as hitherto. The Soviet Union is thus likely to be much more dependent on import saving, as well as on increasing exports of non-oil products, than in the past. The present explicit emphasis laid on import substitution in the annual and Five-Year Plans of several Eastern countries suggests that this class of problems is currently receiving priority attention throughout the area.

The East European countries' relationships with the Soviet Union will continue to limit their margin of manœuvre for importing Western technology – at least in the immediate future. This is because they are faced with the tasks of reducing Western debt, maintaining recently achieved trade surpluses with the Western countries and redressing trade imbalances with the Soviet Union. The increasing unwillingness of the Soviet Union to continue to carry the sizeable trade deficits incurred by its Eastern partners has been made explicit. That is, the Soviet Union has clearly indicated that it will continue to supply energy and raw materials at acceptable levels only if the Eastern countries change their production structures in such a manner that they can deliver more and higher quality goods – principally machinery and equipment. If this shows export efforts on Western markets, it will mean that less hard currency will be available for the purchase of Western goods of any kind. It is worthy of note, that trade between the Soviet Union and its six East European partners is more important than trade among the Six. Furthermore, while the East European countries' share in total Soviet trade turnover has been declining since the 1970s, the Soviet Union's share in the Six's total trade has been increasing. By 1983, some two-thirds of the East European countries' intra area trade was accounted for by the Soviet Union. In sum, the volume of exchanges *among* the East European countries has been declining.

There are now indications that, to a greater or lesser extent, the Eastern countries have overcome the crisis situation of the early 1980s; one that was pointed up in an extreme form by Poland. The disruptive effects of import stagnation on economic activity in general, which brought about abrupt decelerations and even declines in the absolute level of NMP (Net Material Product) in the first three years of the current plan period, thus eased somewhat. By 1984, this provided a breathing space for further reassessment of policies towards future growth, to investment and towards imports of Western technology, the first results of which can already be traced.

Given the share of industrial goods other than fuels in total exports (low in the Soviet Union – though likely now to increase – and considerably higher in Eastern Europe), the choice of new projects will, in future, need to be geared more closely to projects and technology imports capable of achieving high rates of market penetration. Any further sustained recovery in Western demand would certainly facilitate this task. Nevertheless, difficulties in increasing exports to Western markets (Western market restrictions, inability or difficulty in gaining access to advanced Western technologies) are always compounded by the Eastern countries' lack of modern, attractive products and the lack of a large output of goods for the domestic market to achieve economies of scale. Large-scale domestic production of industrial commodities for the potentially big Soviet market is, in practice, restricted to basic products, not easily saleable in the West.

With regard to investment, a principal determining factor of capital goods imports in the past, both the 1981-1985 Five-Year Plans and recent performance indicate a fall in investment ratios to the relatively low average levels at the end of the 1960s. The Soviet Union, where investment growth has been more buoyant in recent years, has already begun to accelerate its purchases of Western capital goods. It is unlikely that the momentum of technology imports can be maintained while investment stagnates or declines. Thus a key question is: will the pace of investment be accelerated in the 1986-1990 Five-Year Plans? Little information was available on the Five-Year Plans for 1986-1990 at the time of writing. However, the investment acceleration in the Soviet Union and Bulgaria in recent years, as well as a continuing fall in capital productivity levels associated, in part, with ageing capital stocks throughout the area, all suggest that some upturn in purchases of Western technology is likely for the seven countries up to 1990.

Little detail on specific investment projects planned by the Eastern countries is known for the years beyond 1985. The possible development of a new Siberian natural gas pipeline, with a capacity of 20-22 million cu.m. annually, could, on past precedent, maintain Soviet and East European demand for Western pipeline and related equipment. It is, however, possible that part of this will consist of smaller gauge pipe and lower capacity turbines than used in previous projects. These can be produced domestically. Demand for Western pipeline equipment will still remain for projects not yet completed, but probably at relatively lower levels than in the past. The bulk of the oil pipeline network, to Western Europe and to the CMEA countries, which accounted for the very high share of pipeline in total investment goods imports from the West in the 1970s, is already in place. The demand for fuel extraction equipment, particularly to cope with deposits in permafrost and offshore areas, will likely remain high. For the East European countries, manufactures can be expected to remain the priority area.

## V. CONCLUDING OBSERVATIONS

The recent bottoming out of Eastern imports of Western technology suggests that most of the Eastern countries have reached a tolerable lower limit of technology imports, below which certain real gains from the principle of comparative advantage would be sacrificed.

Eastern Europe's scope for increasing purchases of Western technology and of capital goods in particular, will remain rather limited in the immediate future, but could increase during the next Five-Year Plan period. In the Soviet case, the upsurge in capital goods imports in 1982/1983 reflected, above all, the maintenance of hard currency exports from oil deliveries. In the medium term, prospects for further increases in earnings from fuel exports are restricted by falling oil prices. An as yet unknown factor is how Western attitudes to Eastern imports of strategic but non-military high technology goods develop.

The experience of the 1970s suggests that the import of Western technologies is unlikely to move up significantly faster than domestic resource availabilities, and especially investment resources and export earnings, will allow. Consequently, this will necessitate a greater selectivity in purchases: notably, a continuation of recent trends to upgrade the level of sophistication of technology acquisitions, a better choice of technologies permitting export expansion and also, a more controlled demand for imports of intermediate (technology) goods. To the extent that these requirements are met, the prospects are for a period of much harder bargaining among Western suppliers for a much more slowly expanding market.

As for export development prospects, the Soviet Union has been the best placed of all the seven countries, given its ability to increase fuel exports for hard currency. Its future ability to do so, and the prospects for the Eastern European countries to increase sales of their principal manufactured export commodities, will be one determining factor in the future development of East-West technology trade. But unless this is accompanied by favourable, or at least neutral, terms of trade developments which would end unrequited transfers of real resources abroad, investment restrictions will continue to limit the prospects for an upturn in Western capital good exports to the area.

These constraints suggest the likelihood that a more concerted effort will, generally, be made to gain access to Western technology by means other than outright purchasing arrangements: that is, by means of more sophisticated (and effective) forms of industrial co-operation. Demands for various forms of counter-trade – on the part of some of the Eastern countries (particularly the Soviet Union, Romania and the German Democratic Republic) – can also be expected to increase.

The Eastern countries have made explicit their interest in maintaining economic ties with the West and the desirability (if not necessity) of having continued access to Western technology. Despite economic and political difficulties, the share of the West in the Eastern countries' total imports of capital goods has not diminished more than could be expected at a time of stringent balance of trade re-adjustments. Indeed, it has risen since 1980 in real terms in the Soviet Union and for the seven countries as a whole. Continuation of these trends will greatly depend on the manner in which East-West trade develops. Perhaps the most crucial factor here is judgements made – in the East and in the West – as to the economic and political desirability and/or profitability of forging closer links.

## NOTES AND REFERENCES

1. See respectively: P. Knirsch, *The Special Role of the German Democratic Republic in the Transfer of Technology Between East and West* (working Paper) OECD, 1981; F. Levcik and J. Skolka, *East-West Technology Transfer, Study of Czechoslovakia*, OECD Paris, 1984, p. 30 and H. Block, "Soviet Economic Performance in a Global Context", in *Soviet Economy in a Time of Change*, Vol. 1, Joint Economic Committee, Congress of the United States, Washington D.C., October 10, 1979, p. 131.

*Part I*

# THE SETTING FOR TECHNOLOGY TRANSFER

*Chapter 1*

# THE ECONOMIC SITUATION IN THE EASTERN COUNTRIES

## 1. Slowdown in Economic Growth: General Considerations

Since the early 1960s, growth rates of net material product[1], industrial production and labour and capital productivity began to decline in the CMEA countries compared with the rates of growth that had been achieved throughout much of the previous decade[2]. Until the first half of the 1970s the slowdown was not uniform. Indeed, the slowdown in growth rates of net material product (NMP), industrial production and labour productivity were reversed in some countries at various times, notably during the last half of the 1960s and the first half of the 1970s (see Tables 1, 2 and 5 of Chapter 2). Capital productivity growth rates were already decelerating in some countries during the 1960s. This movement intensified in the 1970s, and absolute levels of capital productivity declined in all countries of the region over the decade and into the 1980s.

The lower output growth targets retained in the 1976-1980 Five-year Plans show that important constraints were clearly anticipated. However, even the lower objectives for 1976-1980 were not met and a further reduction of overall growth objectives was included for the 1981-1985 plan period as well. The main exception was the German Democratic Republic where pre-1975 NMP and industrial growth performance, relatively weaker than in the other countries of the region, were maintained at almost unchanged rates in the second half of the decade. A combination of internal developments (some beyond the control of Eastern planners), systemic weaknesses and a number of exogenous factors contributed to the phenomenon of decelerating growth. Most were common to all countries of the area, but not all countries were affected to the same extent or at the same time.

### a) *Internal Developments*

With regard to internal developments, the post-war deceleration in the growth of the total population and the almost continuous slowdown since 1966 in the growth of the active population in most Eastern countries had an adverse effect on economic growth. Demographic shifts have exacerbated this situation. In the Soviet Union, in particular, labour shortages have become concentrated in the European areas of the country, the main centres of industry and hence of labour demand. They have also been felt acutely in the newly developing areas of the country – notably Siberia. Similar, though less marked, regional patterns can also be discerned in the other countries of the region. Furthermore, the extent of rural industrialisation, coupled with diminished rural-to-urban area migration and the decline in agricultural work forces, means that the agricultural sector is no longer a plentiful reservoir of manpower. In all countries, labour participation rates are now as high as practicable, with a very large proportion of women working. Thus, apart from the declining natural addition to the working-age population, labour reserves have been almost fully mobilized in virtually every East European country and in the Soviet Union.

A second constraint on economic growth pertains to developments in the energy sector. In the Soviet Union, easily accessible oil and gas reserves have been subject to gradual depletion. The need to exploit and develop new fields has led to higher energy extraction costs in the Soviet Union itself, and to higher import bills for Eastern Europe. The East European countries are, for the most part, poorly endowed with energy resources and hence greatly reliant on the Soviet Union for energy supplies. Economic expansion has been conditioned by this and by related developments. The new CMEA pricing forumula adopted in 1975 (in response to increases in world market prices of crude oil), provided for annual revisions of intra-CMEA trade prices according to a moving average of world market prices over the previous five years. While this limited the growth of oil prices to rates lower than on world markets while the latter were weakening, the marked increase in the price of Soviet oil nevertheless resulted in a deterioration of the East European countries' terms of trade with the Soviet Union. More recently (1984-1985), the fall in world oil prices has not yet been reflected in the price paid by Eastern Europe to the Soviet Union.

In 1982, it was reported that the volume of Soviet oil deliveries to all of its East European partners (except Poland) would be reduced by 10 per cent[3]. Soviet oil output figures suggest that oil deliveries to Eastern Europe are unlikely to rise in the foreseeable future. This means that the East European countries will not only be faced with increasing long-term fuel costs but with a squeeze on supplies. This explains the great attention currently being paid to reducing energy consumption levels. These evident constraints on the economic development of the region cannot be solved quickly.

A third constraint on economic growth has been poor agricultural performance, particularly in the Soviet Union[4]. Recurrent bouts of bad weather have contributed to declining agricultural output. However, inefficient management and inadequate reforms (notwithstanding the variety of measures undertaken to increase agricultural production) are the root cause of poor agricultural performance. The increase in farm and other inputs in the Soviet Union has not been paralleled by improvements in planning and management on the scale necessary to cope with the greater complexity of modern farming methods. Poland has suffered from the grudging attitude towards small-scale private farming which provides the bulk of agricultural output.

A more relaxed attitude to "auxiliary" small-scale private farming, and organisational changes in the state and collective farm sectors in other East European countries, have yet to prove their worth. Higher per capita output and consumption levels have not been accompanied by the creation of a sufficient margin for food reserves or export. Taking the area as a whole, a trend towards increasing capital intensity and decelerating total productivity became apparent in the agricultural sector over the 1970s – a pattern of development similar to, but more marked than, developments in the economy as a whole.

Variable and only slowly improving agricultural performance, in the Soviet Union especially, has led to increased costs per unit of output. Equally important, it has given rise to reliance on large scale imports of grain and other food and feed products from the West and elsewhere. This has pre-empted a large and growing share of hard currency expenditures over the last 15 years and particularly, since 1979, in the Soviet Union.

Finally, future economic development will be constrained by the emergence of resource allocation problems in recent years. These have become more acute as a result of the slowdown in the growth of investment resources. Not only was the total increment of resources slated for "consumption and capital accumulation" reduced relative to already slower-growing ouput during the 1976-1980 period and the early 1980s (because of the need to divert resources to reduce imports and raise exports), but the political and social necessity of maintaining or preventing further declines in living standards also reduced the share of that total available for investment. This occurred at a time when labour shortages increased the need to substitute capital for employment, and rising energy and other material input costs called for increased investment in materials-saving production technologies.

These constraints on growth potential intensified very markedly in the late 1970s and in the 1980s, and have been reinforced by partly consequential decelerations and even absolute declines in factor productivity. Some of the reasons for this are taken up in the following section.

b) *Systemic Considerations*

As noted, a major problem, closely linked with the supply-side constraints mentioned above, has been the failure of the Eastern countries to improve the efficiency of resource use in the production process. Despite the increasing emphasis on "intensification" in successive Five-Year and annual plans, there has, in fact, been virtually no shift from the previous "extensive" pattern of growth on the scale needed to offset the reduced supply of labour and capital.

Essentially, intensive growth implies a more efficient exploitation of all inputs – labour, capital stocks and material inputs (energy, raw materials and materials for further processing) in order to reduce the dependence of the growth of output on quantitative increases of such inputs.

The call for a shift to the intensive growth path was an implicit feature of the 1971-1975 period, though the emphasis placed on it in Five-Year Plans was not uniform throughout the area. It became an explicit feature of all countries' plans for the 1976-1980 period. For 1981-1985, intensification moved to the forefront of planners' concerns and occupied a central position in the annual and medium-term plans adopted by all seven countries. However, none of the main partial efficiency indicators – labour and capital productivity or costs per unit of output – improved at the rates required. Indeed, their evolution has either slowed down (labour productivity) and even, as in the case of capital productivity, declined significantly in absolute terms during the 1970s. These two latter tendencies became even more marked during 1976-1980, and again in the early 1980s. Only very modest success has been achieved in reducing material costs per unit of output. These adverse developments account very largely for the fact that the provisions of the Five-Year Plans adopted in most of the CMEA countries were not achieved in most countries in either 1976-1980 or in 1981-1985.

Inability to realise the shift from extensive to intensive growth is intimately related to the failure to carry out economic reforms. While this can be judged the major systemic reason for the slowdown in economic growth, its negative influence cannot be quantified. This is not the place to analyse in detail the workings of centrally planned economic systems, but some general observations on their characteristics and development are pertinent here.

The negative effects of the overcentralised system of planning and management have long been apparent to the Eastern authorities. That they were large is evident from the critical discussions reported in the professional literature in virtually all countries of the region at various times. The debates initiated by Liberman in the Soviet Union in the early 1960s, Sik in Czechoslovakia, as well as by economists in Poland and Hungary, were followed by modifications to the traditional models of central planning. Thus, as early as the final years of the 1960s, attempts had been made, contemplated, or even, as in the case of Czechoslovakia initiated and then abandoned, to institute economic reforms throughout the region[5]. These aimed at reducing the degree of economic centralisation and essentially foresaw measures designed: to improve the system of economic incentives; to introduce a measure of indicative planning ("spontaneous" adjustments of the system, to be achieved by the use of "economic levers"); to introduce plan-related bonuses for the attainment of profitability, cost reduction or other efficiency targets as opposed to a rigid command type of planning; to shift emphasis in general away from gross output towards sales (and profits) as a means for gauging economic performance; to improve the organisation and performance of the S&T system in order to stimulate innovation and, to ensure the more rapid and effective application of the results of research and development.

However, interest in fundamental economic reforms had largely faded away by the early 1970s. Despite a good deal of systemic tinkering, no fundamental changes in economic planning and management have in fact been undertaken in the past decade (except in Hungary). These seem insufficient to meet the increasing complexity of modern production processes which place a premium on the timeliness of supplies, the attainment of high quality specifications and the fulfillment of precise orders from clients.

The lack of basic improvements in the economic systems of the Eastern countries has another dimension which is closely related to technology transfer policy. It is the view of some Western commentators, for instance, that technology imports can be substitutes, at least to some extent, for potentially politically hazardous organisational reforms. However, it is more generally argued that the absorption of non-indigenous technologies, particularly advanced technologies, cannot be successfully accomplished on a large scale *without* such reforms. It is argued, *inter alia*, that tighter quality specifications for inputs and timely deliveries of final products – perennial sources of weakness and critical comment in the Eastern press and the professional literature – are essential if expensive imported machinery is to be exploited to the full.

The recent experience in Poland provides a further, and related, example of the negative effects of rigid centralised planning on economic development in general, and on technology absorption, in particular. The initial decision to rely on credit-financed imported technology on a massive scale was a high-level one, not subject to, or based on, informed debate on a case-by-case basis. As a result, to take one documented example, traditional and high-potential exporting sectors were *not* the primary recipients of Western technology. Furthermore, the all out effort made from the late 1970s to reduce imports while increasing exports (the basic measure chosen to combat the serious macro-economic disequilibrium which arose from these policies) were extremely costly because decisions relating to the allocation of investment priorities and hard currency were, again, made "from above" by planners far removed from the operational level[6].

Changes made in the planning and management systems have, nonetheless, tended towards a degree of decentralisation in most of the Eastern countries. Nevertheless, even enterprise level decision-making is still carried out for the most part according to criteria which are centrally determined. These may in individual cases be contradictory, do not encourage rational resource use nor are they flexible enough to allow rapid adjustment to changed circumstances. These criteria are, at best, adjusted to changing circumstances retrospectively. In particular, price formation – even in Hungary – does not provide an independent framework for decision-making. Hence, the vital element of "spontaneous" adjustment to changes in the economic climate is still largely absent.

### c) *Exogenous Factors*

The domestic problems which became apparent in the second half of the 1970s were compounded, especially in Eastern Europe but, to a smaller extent also, in the Soviet Union, by external developments. During the 1971-1975 period, national income for domestic use ("distributed" NMP) grew considerably faster than "produced" national income. Thus the growth of resources available for domestic use was bigger than the growth of output. To fill this gap, the CMEA countries were able to turn to an import-led investment strategy. There was a shift in priority towards the accelerated import of disembodied and embodied technology (primarily capital goods – machinery and equipment – and technology-based intermediate goods), in order to modernise and restructure productive capacities. The rapid expansion of Eastern imports of Western technology was made possible by the extension to the East of Western credit on a large scale and at real interest rates which were often, in fact, negative.

These financial transfers undoubtedly contributed to the Eastern countries' economic growth during the first half of the 1970s by making available resources over and above the level of domestic output – as seen from the higher growth rates of domestically-used NMP vis-à-vis produced NMP. These credits were used in some countries to finance the import of investment and intermediate goods. The rise in these credits closely paralleled the region-wide negative trade balances with the West which began in most countries by 1972, and in all by 1974, and which continued to rise throughout the decade (Table 9 of Chapter 5). Nevertheless, not all Eastern countries followed similar policies in this respect. The rapid cumulation of debt, was general. But whereas, for example, in Poland trade deficits on technology items were the predominant factor in its growth, they made a much more modest contribution in the case of, for example, Czechoslovakia (Table 6 of Chapter 5).

The recession in the West severely disrupted the Eastern countries' plans to expand exports to the West in order to finance the debts incurred with respect to imports of Western technology (and grain and meat) after 1970. The effects of the recession on Western markets were compounded by the Eastern countries' inability to utilise Western technology imports effectively[7]. Hence, the slackening demand on Western markets for East European exports was difficult to offset because their manufactures (in contrast to the Soviet Union, their primary export commodity group) were, for the most part, uncompetitive (see Chapter 8). This continues to be the case.

The second external development which adversely affected the East European countries most directly was the successive increase of world prices for raw materials – notably fuel. In addition to the deterioration of trade imbalances with the West noted above, this led, following the 1975 adjustment in the CMEA foreign trade pricing mechanism and the subsequent increase in the price of Soviet oil, to a serious deterioration in the East European countries' terms of trade and balance of trade with the Soviet Union[8]. During this period, rising trade and current account deficits continued until the beginning of the 1980s, and hard currency debt on this account rose to a high point of $94 billion gross ($79 billion net) in 1981 (Table 8 of Chapter 1).

Trade data show that the annual Soviet trade surplus in intra-CMEA trade grew from a modest 112 million rubles in 1970 to a little over one billion rubles (Table 6 of Chapter 1) in 1979. It averaged over 1 billion roubles in 1976-1980. The overall trade deficits of the East European countries with the Soviet Union peaked in 1981 at somewhat over 3 billion rubles. Though they fell in 1982 and 1983, they still totalled 1½-2 billion in 1983 and 1984. These developments are all the more significant if account is taken of the fact that intra-CMEA trade commands the preponderant share of the Eastern countries' total trade and that within this total, the share of the Soviet Union has been increasing since 1980 (Table 3 of Chapter 1).

Given the above-mentioned developments in trade (East-West and intra-CMEA) and especially the very big rise in net debt to the West, import levels and patterns had to be readjusted as early as the final years of the 1976-1980 period. The readjustments continued during the 1981-1985 plan period. This rebalancing, and the need to cut imports and shift resources towards exports (especially on the part of the East European countries) have had serious effects on the domestic availability of industrial and other inputs – one of the basic causes of slower economic performance.

## 2. Principal Developments During Successive Plan Periods: The Emergence of a Trend?

### a) *1971-1975*

The optimistic plan objectives which had been foreseen for this period reflect the high rates of growth which, on the whole, had still been achieved during 1966-1970. Indeed, the expectation that high rates of growth would be maintained was, by and large, justified, some noteworthy results having been registered during the 1971-1975 period. Most main growth targets were over-fulfilled in most countries. Some countries not only over-fulfilled plans, but achieved record or near record growth performance. The primary goal of raising living standards was met.

The Soviet Union is the striking exception to this generally rosy picture. Its economy developed much less favourably than the economies of the East European countries and it was the only country which failed to reach the planned output for industrial production[9]. Its performance in agriculture (where, as was the case for the majority of the Eastern countries, the growth target was not met) was amongst the poorest. As far as NMP was concerned, not only was the rate of growth below planned projections, but it was below the rates of growth that had been achieved during the previous plan period[10]. This was undoubtedly the reason why the intention that "for the first time in the history of Soviet planning, the growth rate of the consumer-goods sector in industry [would] be faster than that of producer goods"[11], was already abandoned after the second year of the plan period.

Indeed, throughout the area there were some general reasons for caution, which were to lead to the more prudent growth strategies ultimately incorporated into the Five-Year Plans for 1976-1980. Among these were the previously mentioned Soviet agricultural failures – notably in 1972 and 1975 – which had given rise to the first big shipments of grain since the mid-1960s. This, together with the big upsurge in technology imports (both of investment and technology-based intermediate goods) brought about very significant and unplanned deficits in the Eastern countries' trade with the West.

The fact that the active population was growing much more slowly was one of the principal reasons for the fact that, as noted earlier, planners began to emphasise improvement in efficiency (intensive growth) as the principal means by which future economic growth was to be achieved. As far as overall labour productivity was concerned, the record varied among the seven countries. In the majority of cases planned growth rates were not achieved and (as indicated in Table 5 of Chapter 2) only in Czechoslovakia, Hungary and the Soviet Union did industrial labour productivity grow faster than during the 1966-1970 period. More disturbing was the fact that labour productivity rose more slowly than fixed assets in several countries.

This reflected the failure of capital productivity to improve. It is of interest to note, that despite the massive increase in technology imports, capital productivity fell in absolute terms in most countries of Eastern Europe in 1971-1975 (the exceptions were Czechoslovakia and Poland) – in contrast to the small though decelerating improvements of the 1960s. In the Soviet Union, the slight decline in industrial capital productivity that had occurred during the previous plan period, turned into a substantial fall during 1971-1975.

Equally disturbing were developments in investment and total production. Contrary to planned objectives, investments did not stabilise. Furthermore, the relationship between investment and NMP growth worsened – that is, NMP growth, measured in relation to investment expenditures, became more costly (see Table 4 of Chapter 2). The pattern was similar for industry[12], where capital productivity growth rates either decelerated or declined compared with the earlier period in every country except Poland.

It was generally recognised in most Eastern countries that successful growth performance in the first half of the 1970s had, as in the past, been achieved predominantly by increases in the supply of inputs – which included a much higher share of Western investment and intermediate goods with a high technology content than in the past. It should be recalled here that the objectives written into the Five-Year Plans for the 1971-1975 period gave no indication that the Eastern countries were about to undertake a very substantial move towards more open policies with regard to trade with the Western countries. Indeed, East-West trade was assigned a secondary role. The plan provisions related to trade continued to stress, as a priority, the development of economic integration and specialisation within the CMEA area. Unless the plans represented a deliberate policy of omission, it seems clear that subsequent increases in the Eastern countries' imports of Western technology did not originally form part of the development strategies adopted for the 1971-1975 period.

A combination of circumstances favoured the import-intensive growth strategy which evolved during this period. As was noted earlier, all of the CMEA countries were faced with an increasing need for investments goods required to restructure and modernise their economies. All were also faced – albeit to varying degrees – with shortages of capital goods of the right quality and technical performance. Détente and the Western recession were to provide a short-term solution, and perhaps were viewed as at least a temporary alternative to economic and management reforms. Western firms were eagerly looking for markets while Western financial institutions were more than willing to recycle the rapid accumulation of OPEC-owned petro-dollars to countries which had borrowed little up to then and were, hence, considered to be good credit risks. Given the green light of détente, previously existing Western restrictions on credit facilities had been lifted. In short, the Eastern countries turned to Western credit in order to meet their import financing needs[13]. In Bulgaria, Poland, Hungary and the Soviet Union, a major element in this strategy was to expand imports of Western produced investment goods.

In addition to the continued availability of domestic inputs, economic performance had benefited from the availability of relatively low-priced fuel and raw materials within the area and from the rest of the world. But by 1975, the OPEC oil price shock was beginning to affect the price of oil traded between the Soviet Union and its East European partners. This opened prospects for a steady deterioration in the terms of Eastern Europe's trade with the Soviet Union. By the mid-1970s, therefore, both domestic and external conditions had changed substantially for the Eastern countries. In effect, a turning point had been reached.

b) *1976-1980*

Falling indicators of economic efficiency, particularly noticeable in the Soviet Union, and the need to moderate the growth of recently incurred debts with Western institutions, provided a dual motive for adopting more modest overall output growth and related targets during 1976-1980. These developments also suggested that a major readjustment of policies would take place.

In evident recognition that economic conditions were deteriorating and that the pace of overall economic expansion could not be maintained, emphasis was placed in the Five-Year Plans of all seven countries on the need to redress economic imbalances (in particular, trade deficits). Furthermore, a more effective utilisation of those factors involved in furthering intensive as opposed to extensive growth, became an explicit requirement.

Growth targets projected for this period indicate that all countries planned a slower growth of NMP than the rates of growth that had actually been achieved during the 1971-1975 period. As far as industrial production is concerned, only Bulgaria anticipated that growth rates would exceed those registered for the preceding period (see Tables 1 and 2 of Chapter 2).

The most remarkable change of emphasis took place with regard to investments. Not only did the policy towards investment mirror the reorientation of an economic policy based on slower growth, but the intent to restrain, or cut back, investment growth indicated that this was intended as the primary means by which external imbalances would be redressed. Thus final versions of the 1976-1980 plans (though not earlier drafts) projected growth rates for gross investment that were significantly below those registered for the 1971-1975 period – the only exceptions being the German Democratic Republic and Romania (see Table 4 of Chapter 2). Investment slowdowns were particularly pronounced in those countries where investment outlays had previously grown faster than NMP.

In fact, as can be calculated from Table 4 of Chapter 2, all countries but Czechoslovakia (which appears to have abandoned its fairly ambitious Five-Year Plan strategy shortly after the 1976-1980 plan period started) and Romania set their investment targets lower than the expected growth rates of aggregate output. This was a significant departure in a group of countries hitherto committed to investment-led growth[14]. The new policy clearly reflected increasing concern with growing trade deficits and external indebtedness and provides a succinct explanation of the fundamental change in policy towards Western countries, as far as trade was concerned[15]. Trade policy was henceforth to focus on export stimulation, as opposed to import substitution or on curtailing the economies' import propensity. Thus, "in a number of countries sharp improvements in the balance of trade or, (notably in Poland) the full elimination of the trade defict by the end of the decade, were made part of the five-year targets"[16].

Finally, mention must be made of the emphasis that was placed, for this plan period, on the need to pursue a strategy of more efficient growth. Concerning investment, it is interesting to note that Eastern planners were not anticipating a reduction in the medium-term investment costs of growth as expressed in ICORs for the years 1976-1980 (Table 4 of Chapter 2). However, in most countries, the deceleration planned in investment was bigger than that planned for NMP growth. The "decoupling" of capital formation from overall economic growth – or to put it differently, reduction of the role assigned to gross investment as a tool for promoting growth – illustrates that planners clearly hoped to move towards more intensive growth in general, and a more effective utilisation of investment resources in particular.

The fact that the 1976-1980 plans were based on a strategy of more efficient output growth is also illustrated by the fact that "the most prominent role, as an element in the planned growth of national product in all the Five-Year Plans, was assigned to labour productivity"[17]. "Targets implying a 90 to 100 percentage share in the increase of NMP attributable to increases in social labour productivity are stipulated in almost all countries"[18]. Nevertheless, such pronouncements notwithstanding, the planned targets indicate that there was little expectation that labour productivity itself could be accelerated. In fact, the figures in Table 5 of Chapter 2 show that the majority of countries anticipated slower industrial labour productivity growth than during the previous plan period. The same thing may be said concerning the forecasts for labour productivity in the economy as a whole.

In any event, the outcomes for 1976-1980 were uniformly below the target figures and indicate a clear and pronounced deceleration of economic growth throughout the region. That the goals of the plans would not be reached had already become apparent by 1979, the year for which lowest growth was recorded in the whole of the post-war period[19]. For the East European countries, the marked decline in the growth of NMP can be largely attributed to growing problems in the energy sector – both the increase in oil prices and the reduction in imports.

However, it is not possible to give any single reason for the growth slowdown. The deterioration in employment growth in the material sphere and in industry was clearly important in most countries – particularly as it was not fully compensated by improvements in labour productivity. This key indicator increased much more slowly than had been the case during the previous two plan periods. Rates of growth were also below growth of capital intensity. This implies that the increases in capital equipment could not be adequately translated into an increase in labour productivity. The decline in capital productivity levels also, therefore, accelerated significantly during the 1976-1980 period. The reasons for this varied between countries, but foremost among them is certainly declining capacity utilisation rates which followed from shortages of material inputs. The increased cost of imported goods, a mode of management and organisation which could not meet the needs of a more complex economy and increasing technical requirements all contributed negatively to attempts to use capital and labour more effectively.

The deceleration in growth was, of course, also partially caused by the strategy pursued towards trade. While aiming to reduce external deficits, cutbacks were imposed on imports – investment goods in particular, but also supplies of industrial inputs. Shortages of such inputs contributed to slower capacity utilisation, and hence to slower growth.

Contrary to the Five-Year Plan strategies, the rates of investment growth actually achieved were higher than output growth rates in all countries of the region during 1976-1980. That is, the investment ratio rose, whereas it was, in most countries, planned to fall. In some countries, rising investment ratios reflected, however, below-plan output growth performance rather than above-plan allocations to investment. In those countries where output growth was higher than planned (USSR and the German Democratic Republic), it reflected above all higher-than-planned investment outlays. Whichever the cause, the investment costs associated with growth were, as during 1971-1975, higher than expected[20].

Bearing in mind the commitment to the initial, more cautious, output growth plans designed to facilitate redressment of trade deficits during 1976-1980, and within this the cutback slated in most countries for investment growth, it seems fairly clear (though this is stated nowhere in the Five-Year Plans of any country) that capital goods were one of the main import commodity groups to be restricted. Even so, the scale of readjustment incorporated into the original Five-Year Plans seems to have been inadequate. This is shown by the second readjustment which took place in 1977-1978. Output growth performance fell back further and restrictions on faster-than-planned investment were tightened up considerably in all countries without exception.

The timing of the two readjustment rounds was somewhat different as between Eastern Europe and the Soviet Union. In current value and constant price terms, Eastern European imports of Western capital goods remained below 1975 levels in both 1976 and 1977 in the region as a whole[21]. In the case of the Soviet Union, capital goods imports (measured in current value terms) only began to fall in 1980 – though the level of imports remained above that of 1975. Measured in constant terms, import levels fell below 1975 already in 1979. It is interesting to note here that Eastern imports of Western intermediate goods with a significant technology content continued to rise in both value and constant terms, and that these imports remained higher, on average, during the 1976-1980 period, than in 1975[22].

Nevertheless, as regards embodied technology as a whole, developments after 1975 signified a reversal of the trend, particularly strong in the early 1970s, towards increasing reliance of most Eastern countries on Western investment goods. Estimates of the volumes of East-West capital goods trade suggest that the import dependence of investment as a whole rose sharply up to 1975 in most countries. The movements described since then resulted in a volume level of imports by 1980 which was lower relative to investment than at the beginning of the decade in all countries but Hungary and also (because of the stagnation of investment between 1975 and 1980) in Poland.

It may also be noted that from 1980 the rate of growth of East-West trade as a whole slowed. The CMEA countries were able to check imports from the OECD countries and their exports to these countries increased more rapidly than their imports from them. The Eastern countries were thus able to diminish trade deficits with the West. However, the external financial situation of the East European countries worsened in 1980. This was primarily due to increased debt servicing and, because export earnings were insufficient, further borrowing could not be avoided.

c) *1981-1985*

The principal characteristic of the plans adopted for the first half of the 1980s is clearly the common reformulation of the growth concept[23]. Thus the strategy of "intensification" is now, even more than for 1976-1980, the central theme of the Five-Year Plans of all the countries in the area. Not only has the need to achieve more growth with lower factor inputs been reiterated, but the very notion of intensive growth has been given a different emphasis. Intensive growth now implies much more than increasing the share of labour productivity in the growth of NMP. Rather, it was seen as a development strategy which, in addition to labour productivity growth, implies: "*a)* the rational use of productive potential, including increases in capital productivity through optimal exploitation of existing capacities; *b)* the rational use of all material inputs with a view to decreasing material input per unit of output and *c)* improvement in the quality of output"[24].

The growth targets themselves all reflect that Eastern planners have indeed accepted slower growth as a fact of economic life. The German Democratic Republic is the only country which projected a higher rate of growth for NMP than was achieved during the previous plan period. The planned targets for industrial production also reflect the anticipation of moderate growth and are likewise (with very few exceptions) set below outcomes registered for the previous plan period. Savings in the use of energy and raw materials are stressed throughout. Furthermore, attention was again drawn to the need for better adjustment of industrial production to the requirements of foreign markets. Those industrial branches which are more competitive and best able to produce import substitutes were to be given priority.

Although comprehensive figures were not published, consumption and investment indicators show evidence of a further shift away from domestic demand in favour of restoring external balances. It was not specified whether trade deficits were to be eliminated by export expansion or import cuts. Existing debt levels clearly ruled out any further recourse to large-scale credit to finance investment. Moreover, official credit (though this is not extended by all Western countries) was becoming more expensive[25]. Within total resources for domestic use, consumption was expected to increase its share. Policy announcements relating to the plans justified the consequent limitation of investment growth by stressing the need to improve the efficiency with which investment funds were used – notably by cutting excessive lead times and concentrating resources on the completion of existing projects and cutting new starts. Low or zero growth rates were in fact projected for gross investment in all countries. The latter aspect of growth strategy would inevitably entail a further downgrading of Western technology imports.

Since the principal trade imbalances in 1976-1980 concerned East-West trade, the Five-Year Plans placed increasing stress on furthering intra-Eastern area trade and economic co-operation. The emphasis also placed on improving the R&D systems, the increasing role to be played by science and technology in economic growth and perhaps also a more critical or realistic view of the role which Western technology can play in the development of their economies under given conditions, led to the expectation that the improvements in technology levels called for by the factor efficiency targets, given or implied in the plans, were to derive largely from indigenous sources.

With regard to actual progress relative to Five-Year Plans, it is evident from developments during the first four years of the current plan period, as well as from the generally modest growth targets foreseen in the 1985 plans, that most of the CMEA countries will not meet the NMP and industrial output goals set for the current Five-Year Plan period (Bulgaria and possibly the Soviet Union are the exceptions).

As far as overall economic growth is concerned, historically low NMP growth rates characterized the area as a whole in 1981-1982. Better performance in 1983 and 1984, reflected also in the plans for 1985, on the other hand, shows that all the CMEA countries are experiencing some improvement, though in some cases it is small. In Poland, growth of NMP was positive in 1983 – indeed it was twice as high as planned – for the first time in several years, and it remained on target in 1984. Even so, its level remained some way below 1975. Further, albeit modest growth, is planned for 1985 – still a considerable improvement over the four years of continuously declining output in 1979-1982.

The very poor growth performance in 1981-1982 in the region as a whole was due to several factors of which the continuing decline in Polish NMP levels was only one. Inadequate agricultural performance in the Soviet Union and in Eastern Europe was a prime direct effect. But industrial performance also stagnated. This probably reflected slow growth of output of energy-intensive branches such as metallurgy, as well as a slowdown in both domestic production and imports of fuel which, apart from the direct effects of the former, held back the growth in output of other branches as well. In addition, imports of raw-materials and other industrial inputs were restricted. As noted later in this study, imports of Western intermediate goods with a high technology content were generally maintained, though they were sharply reduced by the Soviet Union after 1980. While continuing input shortages were experienced in 1983 as well, the generally modest but slightly accelerated NMP growth achieved in that year continued in 1984. Some results in easing shortages and freeing production bottlenecks were, moreover, achieved in 1984. Improvements in domestic production of basic material inputs such as steel, fuel and energy were registered, and the fall in imports was stabilized or reversed.

Trade adjustment polices in the 1980s led, as anticipated in the plans, to slower growth of resources for domestic use. The investment slowdowns foreseen in the Five-Year Plans were implemented with increasing severity in most of Eastern Europe. Although, after initial cutbacks in growth, investment is likely to be higher than planned for the five-year period in Bulgaria and the Soviet Union, it will be lower even than planned in the other countries of the region. The ratio of gross fixed investment to NMP continued to fall uninterruptedly (Table 9 of Chapter 2) in virtually all years in all countries. It is thus not surprising that imports of capital goods from the West continued to stagnate in most of Eastern Europe. The share of capital goods imports in total imports – and particularly in imports from the West – has also fallen sharply.

Clearly, a future upturn in Eastern technology imports from the West, and imports of capital goods in particular, is conditional in the first instance on the reversal of the investment slowdown. The costs of the latter have probably been high in terms of future Eastern economic growth prospects. A recent study gives evidence indicating that the deterioration in the age structure of the stock of fixed assets stemming from slower growing or declining investment in recent years has been intensified by falls in the rates of retirement of fixed assets – perhaps in response to shortages of new investment resources[26]. Moreover, as noted above, the share of imports of Western capital goods, which presumably embody technology types and levels which are already scarce in the Eastern countries, has also fallen. Thus the influence of the Eastern readjustment process on the quality of fixed capital stocks, and hence on medium-term growth projects, in the Eastern countries has been demonstrably negative. The Five-Year Plans for 1986-1990 will have to address this problem with vigour. To offset it, more buoyant investment growth will be required. This would, in itself, indicate a substantial resurgence in Eastern demand for Western technology.

How far higher investment growth can be maintained will depend in the first instance on how far the upturn in the growth of NMP in 1983-1984 represents a sustainable recovery. In fact, the acceleration of output growth in recent years may reflect to a large extent the taking-up of slack capacity which built up in 1981-1982. No Eastern country has indicated that it expects growth rates to return to the rates of the early 1970s. Moreover, even if NMP growth is maintained at the rates of recent years, a substantial part of the extra resources available will continue to be channelled to the repayment of debt in some countries. There is probably not much likelihood of a significant squeeze on consumption for social and political reasons. Thus, only if current output growth rates can be at least maintained, will the Eastern leaders find it possible to raise the growth of resources for investment purposes significantly in 1986-1990.

The improved Eastern current balance of trade situation is a favourable augury for the new plan period. While prospects vary between Eastern countries, it is significant that the Soviet Union and Bulgaria, which have generally been the Eastern countries least affected by the readjustment squeeze, already moved to increase imports of Western capital goods in 1982 and 1983. Some other countries have followed since, albeit at much more modest rates. This development has not, in contrast to the previous Five-Year period, yet been paralleled by a rise in imports of other kinds of technology items – notably intermediate goods with a high technology content (see Part II, Chapter 5 below). For the future, many imponderables remain. The downward movement of world oil prices could constrain the Soviet market for Western technology. Better Soviet harvests could reduce the country import bill for grain, and hence free resources for the purchase of technology items from the West.

For Eastern Europe, the prospects are also difficult to gauge. Falling or stable fuel prices would, in contrast to the Soviet case, free resources for imports of Western technology. Eastern Europe's import possibilities will depend, in the long-term, on whether the six countries can recover their falling share of the OECD market for manufactures – and, in particular, meet the challenge of the newly-industrialized countries on that market. This may be the more difficult, given that investment and imports of Western technology, both of which may be deemed as essential for the modernisation of their export structures, were cut more sharply in Eastern Europe than in the Soviet Union. This would, in itself, provide a strong motive for accelerating investment and capital goods imports, where progress in debt repayment permits, in 1986-1990.

Table 1 (Chapter 1)

TRENDS IN TOTAL FOREIGN TRADE OF THE 7 EASTERN COUNTRIES AND SHARE OF EACH EASTERN COUNTRY IN TOTAL AND IN INTRA-EASTERN AREA TRADE
(values at current prices and per cent)

I. TOTAL TRADE TURNOVER (EXPORTS PLUS IMPORTS)

| | Total Seven | Bulgaria | Czecho-slovakia | German Democratic Republic | Hungary | Poland | Romania | USSR |
|---|---|---|---|---|---|---|---|---|
| Total trade turnover (million T.R.) | | | | | | | | |
| 1975 | 120 090 | 7 521 | 12 171 | 15 930 | 8 717 | 17 075 | 7 972 | 50 700 |
| 1979 | 189 277 | 11 562 | 18 240 | 23 307 | 18 470 | 23 538 | 13 853 | 80 309 |
| 1980 | 215 483 | 13 219 | 20 213 | 25 718 | 20 902 | 24 821 | 16 512 | 94 100 |
| 1981 | 241 965 | 15 245 | 21 746 | 28 464 | 22 866 | 21 744 | 22 158 | 109 739 |
| 1982 | 261 330 | 16 812 | 23 686 | 31 073 | 24 973 | 26 766 | 18 446 | 119 576 |
| 1983 | 282 129 | 18 295 | 25 856 | 34 352 | 28 426 | 28 869 | 17 864 | 127 480 |
| 1984 | 313 306 | 19 815 | 28 496 | 37 238 | 30 941 | 35 689 | 21 864 | 139 711 |
| Share of each country in total trade turnover (%) | | | | | | | | |
| 1975 | 100 | 6.3 | 10.1 | 13.3 | 7.3 | 14.2 | 6.6 | 42.2 |
| 1979 | 100 | 6.1 | 9.6 | 12.3 | 9.8 | 12.4 | 7.3 | 42.4 |
| 1980 | 100 | 6.1 | 9.4 | 11.9 | 9.7 | 11.5 | 7.7 | 43.7 |
| 1981 | 100 | 6.3 | 9.0 | 11.8 | 9.4 | 9.0 | 9.2 | 45.4 |
| 1982 | 100 | 6.4 | 9.1 | 11.9 | 9.6 | 10.2 | 7.1 | 45.8 |
| 1983 | 100 | 6.5 | 9.2 | 12.2 | 10.1 | 10.6 | 6.3 | 45.2 |
| 1984 | 100 | 6.3 | 9.1 | 11.9 | 9.9 | 11.4 | 7.0 | 44.5 |
| Share of each country in intra-Eastern trade turnover (%) | | | | | | | | |
| 1975 | 100 | 8.4 | 12.4 | 16.2 | 8.9 | 13.2 | 4.7 | 36.3 |
| 1979 | 100 | 8.6 | 12.2 | 15.2 | 9.6 | 12.9 | 4.9 | 36.6 |
| 1980 | 100 | 8.8 | 12.1 | 14.8 | 9.5 | 12.2 | 5.2 | 37.5 |
| 1981 | 100 | 8.8 | 12.0 | 14.8 | 9.6 | 10.8 | 5.7 | 38.3 |
| 1982 | 100 | 9.1 | 12.3 | 14.6 | 9.6 | 10.9 | 4.8 | 38.6 |
| 1983 | 100 | 9.4 | 12.4 | 14.3 | 9.5 | 11.1 | 4.6 | 38.7 |
| 1984 | 100 | 9.1 | 12.6 | 14.1 | 9.2 | 11.6 | 4.5 | 38.8 |

Table 1 (Chapter 1) contd

## II. TOTAL EXPORTS

| | Total Seven | Bulgaria | Czecho-slovakia | German Democratic Republic | Hungary | Poland | Romania | USSR |
|---|---|---|---|---|---|---|---|---|
| Total exports (millions T.R.) | | | | | | | | |
| 1975 | 56 551 | 3 493 | 5 831 | 7 517 | 3 999 | 7 694 | 3 986 | 24 030 |
| 1979 | 94 965 | 5 898 | 8 770 | 11 225 | 8 817 | 11 305 | 6 527 | 42 426 |
| 1980 | 108 190 | 6 847 | 10 020 | 12 234 | 10 111 | 11 691 | 7 652 | 49 635 |
| 1981 | 122 136 | 7 585 | 10 961 | 14 117 | 11 156 | 10 029 | 11 180 | 57 108 |
| 1982 | 136 148 | 8 369 | 11 914 | 16 109 | 12 480 | 13 988 | 10 122 | 63 165 |
| 1983 | 148 172 | 9 090 | 12 980 | 18 036 | 14 389 | 15 591 | 10 196 | 67 891 |
| 1984 | 165 024 | 9 973 | 14 279 | 19 358 | 15 921 | 18 731 | 12 378 | 74 384 |
| Share of each country in total exports (%) | | | | | | | | |
| 1975 | 100 | 6.2 | 10.3 | 13.3 | 7.1 | 13.6 | 7.0 | 42.5 |
| 1979 | 100 | 6.2 | 9.2 | 11.8 | 9.3 | 11.9 | 6.9 | 44.7 |
| 1980 | 100 | 6.3 | 9.3 | 11.3 | 9.3 | 10.8 | 7.1 | 45.9 |
| 1981 | 100 | 6.2 | 9.0 | 11.6 | 9.1 | 8.2 | 9.2 | 46.8 |
| 1982 | 100 | 6.1 | 8.8 | 11.8 | 9.2 | 10.3 | 7.4 | 46.4 |
| 1983 | 100 | 6.1 | 8.8 | 12.2 | 9.7 | 10.5 | 6.9 | 45.8 |
| 1984 | 100 | 6.0 | 8.7 | 11.7 | 9.6 | 11.4 | 7.5 | 45.1 |
| Share of each country in intra-Eastern area exports(%) | | | | | | | | |
| 1975 | 100 | 8.2 | 11.9 | 16.0 | 8.5 | 13.6 | 4.8 | 37.1 |
| 1979 | 100 | 8.3 | 11.7 | 15.5 | 9.3 | 13.0 | 4.7 | 37.5 |
| 1980 | 100 | 8.5 | 11.8 | 14.5 | 9.5 | 11.4 | 5.3 | 39.0 |
| 1981 | 100 | 8.2 | 11.8 | 14.5 | 9.9 | 9.3 | 5.6 | 40.6 |
| 1982 | 100 | 8.6 | 12.0 | 14.5 | 9.8 | 10.4 | 4.9 | 39.8 |
| 1983 | 100 | 9.0 | 11.9 | 14.7 | 9.6 | 10.7 | 4.6 | 39.5 |
| 1984 | 100 | 8.8 | 12.0 | 14.4 | 9.4 | 11.1 | 4.5 | 39.7 |

Table 1 (Chapter 1) contd

III. TOTAL IMPORTS

| | Total Seven | Bulgaria | Czecho-slovakia | German Democratic Republic | Hungary | Poland | Romania | USSR |
|---|---|---|---|---|---|---|---|---|
| Total imports (million T.R.) | | | | | | | | |
| 1975 | 63 535 | 4 027 | 6 340 | 8 413 | 4 718 | 9 381 | 3 986 | 26 671 |
| 1979 | 94 312 | 5 664 | 9 470 | 12 082 | 9 653 | 12 234 | 7 326 | 37 883 |
| 1980 | 107 294 | 6 371 | 10 193 | 13 484 | 10 791 | 13 130 | 8 860 | 44 465 |
| 1981 | 119 825 | 7 660 | 10 785 | 14 347 | 11 710 | 11 715 | 10 978 | 52 631 |
| 1982 | 125 184 | 8 443 | 11 772 | 14 963 | 12 493 | 12 778 | 8 323 | 56 411 |
| 1983 | 133 960 | 9 205 | 12 877 | 16 316 | 14 037 | 14 268 | 7 669 | 59 589 |
| 1984 | 148 730 | 9 842 | 14 217 | 17 880 | 15 020 | 16 958 | 9 485 | 64 327 |
| Share of each country in total imports (%) | | | | | | | | |
| 1975 | 100 | 6.3 | 10.0 | 13.2 | 7.4 | 14.8 | 6.3 | 42.0 |
| 1979 | 100 | 6.0 | 10.0 | 12.8 | 10.2 | 13.0 | 7.8 | 40.2 |
| 1980 | 100 | 5.9 | 9.5 | 12.6 | 10.1 | 12.2 | 8.3 | 41.4 |
| 1981 | 100 | 6.4 | 9.0 | 12.0 | 9.8 | 9.8 | 9.2 | 43.9 |
| 1982 | 100 | 6.7 | 9.4 | 12.0 | 10.0 | 10.2 | 6.6 | 45.1 |
| 1983 | 100 | 6.9 | 9.6 | 12.2 | 10.5 | 10.7 | 5.7 | 44.5 |
| 1984 | 100 | 6.6 | 9.6 | 12.0 | 10.1 | 11.4 | 6.4 | 43.9 |
| Share of each country in intra-Eastern area imports (%) | | | | | | | | |
| 1975 | 100 | 8.7 | 12.8 | 16.5 | 9.3 | 12.8 | 4.6 | 35.4 |
| 1979 | 100 | 9.0 | 12.7 | 14.9 | 9.8 | 12.8 | 5.1 | 35.7 |
| 1980 | 100 | 9.0 | 12.4 | 15.0 | 9.5 | 13.0 | 5.1 | 35.9 |
| 1981 | 100 | 9.3 | 12.2 | 15.2 | 9.3 | 12.3 | 5.8 | 35.9 |
| 1982 | 100 | 9.6 | 12.7 | 14.8 | 9.4 | 11.5 | 4.7 | 37.4 |
| 1983 | 100 | 9.7 | 12.9 | 14.0 | 9.3 | 11.6 | 4.5 | 37.9 |
| 1984 | 100 | 9.4 | 13.2 | 13.8 | 9.0 | 12.1 | 4.5 | 37.9 |

Table 1 (Chapter 1) contd

IV. TOTAL TRADE BALANCES

| | Total Seven | Bulgaria | Czecho-slovakia | German Democratic Republic | Hungary | Poland | Romania | USSR |
|---|---|---|---|---|---|---|---|---|
| Total trade balances (million T.R.) | | | | | | | | |
| 1975 | -6 980 | -534 | -509 | -896 | -718 | -1 687 | - | -2 637 |
| 1979 | +657 | +233 | -700 | -857 | -834 | -929 | -800 | +4 545 |
| 1980 | +897 | +475 | -172 | -1 250 | -680 | -1 439 | -1 207 | +5 172 |
| 1981 | +2 309 | -75 | +177 | -230 | -555 | -1 686 | +202 | +4 477 |
| 1982 | +10 964 | -74 | +142 | +1 146 | -13 | +1 210 | +1 799 | +6 754 |
| 1983 | +14 212 | -114 | +103 | +1 720 | +352 | +1 323 | +2 527 | +8 305 |
| 1984 | +16 294 | +132 | +62 | +1 478 | +902 | +1 772 | +2 893 | +9 056 |

Source: CMEA Yearbooks and Economic Commission for Europe.

Table 2 (Chapter 1)

TRENDS IN INTRA-EASTERN AREA TRADE OF THE 7 EUROPEAN CMEA COUNTRIES AND SHARE OF INTRA-EASTERN AREA IN TOTAL TRADE
(values at current prices and per cent)

I. INTRA-AREA TRADE TURNOVER (EXPORTS PLUS IMPORTS) AND BALANCES OF TRADE

| | Total Seven | Bulgaria | Czecho-slovakia | German Democratic Republic | Hungary | Poland | Romania | USSR |
|---|---|---|---|---|---|---|---|---|
| Turnover of intra-Eastern area trade (million T.R.) | | | | | | | | |
| 1975 | 63 883 | 5 375 | 7 899 | 10 387 | 5 622 | 8 433 | 2 989 | 23 178 |
| 1979 | 98 505 | 8 501 | 12 019 | 14 992 | 9 420 | 12 718 | 4 815 | 36 040 |
| 1980 | 106 839 | 9 351 | 12 948 | 15 772 | 10 144 | 13 041 | 5 569 | 40 014 |
| 1981 | 118 825 | 10 431 | 14 259 | 17 626 | 11 437 | 12 827 | 6 794 | 45 451 |
| 1982 | 131 143 | 11 928 | 16 184 | 19 163 | 12 607 | 14 328 | 6 315 | 50 618 |
| 1983 | 146 362 | 13 693 | 18 175 | 20 958 | 13 847 | 16 290 | 6 720 | 56 680 |
| 1984 | 162 000 | 14 750 | 20 480 | 22 867 | 14 929 | 18 781 | 7 320 | 62 872 |
| Share of intra-Eastern area trade turnover in total trade turnover (%) | | | | | | | | |
| 1975 | 53.2 | 71.5 | 64.9 | 65.2 | 64.5 | 49.4 | 37.5 | 45.7 |
| 1979 | 52.0 | 73.3 | 65.9 | 64.3 | 51.0 | 54.0 | 34.8 | 44.9 |
| 1980 | 49.6 | 70.7 | 64.1 | 61.3 | 48.5 | 52.5 | 33.7 | 42.5 |
| 1981 | 49.1 | 68.4 | 65.6 | 61.9 | 50.0 | 59.0 | 30.7 | 41.4 |
| 1982 | 50.2 | 70.9 | 68.3 | 61.7 | 50.5 | 53.5 | 34.2 | 42.3 |
| 1983 | 51.9 | 74.8 | 70.3 | 61.0 | 48.7 | 54.6 | 37.6 | 44.5 |
| 1984 | 51.6 | 74.4 | 71.9 | 61.4 | 48.3 | 52.6 | 33.5 | 45.0 |
| Balances of intra-Eastern area trade (million T.R.) | | | | | | | | |
| 1975 | -21 | -161 | -269 | -187 | -292 | +277 | +57 | +554 |
| 1979 | +503 | -273 | -469 | +362 | -214 | +178 | -139 | +1 058 |
| 1980 | +515 | -253 | -246 | -156 | +30 | -813 | +129 | +1 824 |
| 1981 | +869 | -577 | -137 | -282 | +445 | -1 645 | -84 | +3 149 |
| 1982 | +1 009 | -590 | -306 | -51 | +403 | -576 | +157 | +1 972 |
| 1983 | +1 252 | -402 | -572 | +680 | +332 | -550 | +140 | +1 624 |
| 1984 | +1 322 | -319 | -805 | +674 | +475 | -666 | -67 | +1 916 |

Table 2 (Chapter 1) contd

II. INTRA-AREA EXPORTS AND IMPORTS

| | Total Seven | Bulgaria | Czecho-slovakia | German Democratic Republic | Hungary | Poland | Romania | USSR |
|---|---|---|---|---|---|---|---|---|
| Intra-Eastern area exports (millions T.R.) | | | | | | | | |
| 1975 | 31 931 | 2 607 | 3 815 | 5 100 | 2 665 | 4 355 | 1 523 | 11 866 |
| 1979 | 49 504 | 4 114 | 5 775 | 7 677 | 4 603 | 6 448 | 2 338 | 18 549 |
| 1980 | 53 677 | 4 549 | 6 351 | 7 808 | 5 087 | 6 114 | 2 849 | 20 919 |
| 1981 | 59 847 | 4 927 | 7 061 | 8 672 | 5 941 | 5 591 | 3 355 | 24 300 |
| 1982 | 66 076 | 5 669 | 7 939 | 9 556 | 6 505 | 6 876 | 3 236 | 26 295 |
| 1983 | 73 807 | 6 645 | 8 802 | 10 819 | 7 089 | 7 870 | 3 430 | 29 152 |
| 1984 | 81 661 | 7 215 | 9 838 | 11 771 | 7 702 | 9 048 | 3 694 | 32 394 |
| Intra-Eastern area exports as a % of total exports | | | | | | | | |
| 1975 | 56.5 | 74.6 | 65.4 | 67.8 | 66.6 | 56.6 | 38.2 | 49.4 |
| 1979 | 52.1 | 69.8 | 65.8 | 68.4 | 52.2 | 57.0 | 35.8 | 43.7 |
| 1980 | 49.6 | 66.4 | 62.4 | 63.8 | 50.3 | 52.3 | 37.2 | 42.1 |
| 1981 | 49.0 | 65.0 | 64.4 | 61.4 | 53.2 | 55.7 | 30.0 | 42.6 |
| 1982 | 48.5 | 67.7 | 66.6 | 59.3 | 52.1 | 49.2 | 32.0 | 41.6 |
| 1983 | 49.8 | 73.1 | 67.8 | 60.0 | 49.3 | 50.5 | 33.6 | 42.9 |
| 1984 | 49.5 | 72.3 | 68.9 | 60.8 | 48.4 | 48.3 | 29.8 | 43.5 |
| Intra-Eastern area imports (million T.R.) | | | | | | | | |
| 1975 | 31 952 | 2 768 | 4 084 | 5 287 | 2 957 | 4 078 | 1 466 | 11 312 |
| 1979 | 49 001 | 4 387 | 6 244 | 7 315 | 4 817 | 6 270 | 2 477 | 17 491 |
| 1980 | 53 162 | 4 802 | 6 597 | 7 964 | 5 057 | 6 927 | 2 720 | 19 095 |
| 1981 | 58 978 | 5 504 | 7 198 | 8 954 | 5 496 | 7 236 | 3 439 | 21 151 |
| 1982 | 65 067 | 6 259 | 8 245 | 9 607 | 6 102 | 7 452 | 3 079 | 24 323 |
| 1983 | 72 555 | 7 048 | 9 374 | 10 139 | 6 757 | 8 420 | 3 290 | 27 528 |
| 1984 | 80 339 | 7 535 | 10 642 | 11 097 | 7 227 | 9 733 | 3 627 | 30 478 |

Table 2 (Chapter 1) contd

II. INTRA-AREA EXPORTS AND IMPORTS

| | Total Seven | Bulgaria | Czecho-slovakia | German Democratic Republic | Hungary | Poland | Romania | USSR |
|---|---|---|---|---|---|---|---|---|
| Intra-Eastern area imports as % of total imports | | | | | | | | |
| 1975 | 50.3 | 68.7 | 64.4 | 62.8 | 62.7 | 43.5 | 36.8 | 42.4 |
| 1979 | 52.0 | 77.5 | 65.9 | 60.5 | 49.9 | 51.3 | 33.8 | 46.2 |
| 1980 | 49.5 | 75.4 | 64.7 | 59.1 | 46.9 | 52.8 | 30.7 | 42.9 |
| 1981 | 49.2 | 71.9 | 66.7 | 62.4 | 46.9 | 61.8 | 31.3 | 40.2 |
| 1982 | 52.0 | 74.1 | 70.0 | 64.2 | 48.8 | 58.3 | 37.0 | 43.1 |
| 1983 | 54.2 | 76.6 | 72.8 | 62.1 | 48.1 | 59.0 | 42.9 | 46.2 |
| 1984 | 54.0 | 76.6 | 74.9 | 62.1 | 48.1 | 57.4 | 38.2 | 46.7 |

Source: CMEA Yearbooks and Economic Commission for Europe.

Table 3 (Chapter 1)

SHARE OF THE USSR IN FOREIGN TRADE OF THE EAST EUROPEAN COUNTRIES, 1975-1984
(per cent)

| | Exports | | | | | | | Imports | | | | | | |
|---|---|---|---|---|---|---|---|---|---|---|---|---|---|---|
| | 1975 | 1979 | 1980 | 1981 | 1982 | 1983 | 1984 | 1975 | 1979 | 1980 | 1981 | 1982 | 1983 | 1984 |
| Bulgaria | 54.6 | 52.4 | 49.9 | 48.4 | 51.6 | 56.0 | 55.7 | 50.7 | 58.9 | 57.3 | 54.7 | 55.9 | 58.1 | 59.2 |
| Czechoslovakia | 33.0 | 35.6 | 35.6 | 37.6 | 40.6 | 41.8 | 43.4 | 32.1 | 35.5 | 36.0 | 39.9 | 43.4 | 46.2 | 46.8 |
| German Democratic Rep. | 35.5 | 37.4 | 35.7 | 36.5 | 35.2 | 36.5 | 37.8 | 35.8 | 34.8 | 35.3 | 38.6 | 41.0 | 39.5 | 39.5 |
| Hungary | 38.9 | 28.1 | 29.3 | 33.4 | 33.5 | 31.6 | 30.1 | 34.9 | 29.3 | 27.7 | 28.6 | 29.5 | 28.5 | 29.1 |
| Poland | 31.5 | 35.4 | 31.2 | 32.4 | 30.1 | 31.2 | 29.6 | 25.3 | 31.3 | 33.1 | 41.7 | 37.8 | 36.8 | 36.4 |
| Romania | 19.9 | 16.9 | 19.6 | 18.1 | 17.0 | 17.4 | 16.0 | 17.2 | 14.7 | 15.6 | 18.2 | 18.4 | 22.3 | 19.8 |

Source: Based on data reported by the East European countries shown.

Table 4 (Chapter 1)

INTRA-EASTERN AREA TRADE IN 1984
(million rubles)

| | X to the USSR | X to Eastern Europe | M from USSR | M from Eastern Europe | Balance vis-à-vis the USSR | Eastern Europe |
|---|---|---|---|---|---|---|
| Bulgaria | 5 559 | 1 656 | 5 829 | 1 705 | -270 | -29 |
| Czechoslovakia | 6 201 | 3 637 | 6 654 | 3 989 | -453 | -233 |
| German Dem. Republic | 7 313 | 4 458 | 7 057 | 1 040 | +256 | +372 |
| Hungary | 4 795 | 2 907 | 4 371 | 2 856 | +423 | +143 |
| Poland | 5 543 | 3 505 | 6 164 | 3 569 | -622 | +7 |
| Romania | 1 984 | 1 710 | 1 881 | 1 746 | +103 | -169 |
| Six | 31 394 | 17 873 | 31 956 | 17 904 | -562 | +91 |

Note: X = Exports.
M = Imports.

Source: Based on data reported by the East European countries shown.

Table 5 (Chapter 1)

SHARES OF EASTERN EUROPE AND THE SOVIET UNION
IN TOTAL WORLD EXPORTS
(per cent)

| | 1970 | 1975 | 1980 | 1981 | 1982 | 1983 | 1984 |
|---|---|---|---|---|---|---|---|
| Eastern Europe | 5.7 | 5.1 | 4.0 | 4.0 | 4.2 | 4.5 | 4.5 |
| Soviet Union | 4.1 | 3.8 | 3.8 | 4.0 | 4.7 | 5.1 | 4.8 |
| Total | 9.8 | 8.9 | 7.8 | 8.0 | 8.9 | 9.6 | 9.3 |

Source: Based on data published in the UN *Monthly Statistical Bulletin*, United Nations, New York, June 1983, pp. xxx-xxxii, *Ibid.*, June 1984 pp. xviii-xx and *Ibid.*, June 1985, pp. xix-xx.

Table 6 (Chapter 1)

EASTERN EUROPE'S TRADE WITH THE SOVIET UNION (a)
(million transferable rubles)

| | 1970 | 1975 | 1976 | 1977 | 1978 | 1979 | 1980 | 1981 | 1982 | 1983 | 1984 |
|---|---|---|---|---|---|---|---|---|---|---|---|
| Bulgaria | | | | | | | | | | | |
| Exports | 972.5 | 1 931.2 | 2 188.8 | 2 494.6 | 2 997.4 | 3 173.7 | 3 438.9 | 3 696.9 | 4 288.1 | 5 053.3 | 5 608.0 |
| Imports | 844.0 | 2 059.6 | 2 276.7 | 2 658.7 | 3 144.4 | 3 312.7 | 3 660.2 | 4 374.5 | 4 884.6 | 5 510.8 | 6 124.4 |
| Balance | +128.5 | -128.4 | -87.9 | -164.1 | -147.0 | -139.0 | -221.3 | -677.6 | -596.5 | -475.5 | -516.4 |
| CSSR | | | | | | | | | | | |
| Exports | 1 110.5 | 1 891.7 | 2 222.8 | 2 436.9 | 3 058.6 | 3 183.4 | 3 535.9 | 4 104.8 | 4 731.9 | 5 420.4 | 6 016.5 |
| Imports | 1 082.7 | 2 019.5 | 2 320.5 | 2 680.4 | 3 002.0 | 3 362.9 | 3 648.1 | 4 382.3 | 5 047.5 | 5 871.6 | 6 590.8 |
| Balance | +27.8 | -127.8 | -97.7 | -243.5 | +56.6 | -179.5 | -112.2 | -277.5 | -315.6 | -451.2 | -574.3 |
| German Dem. Rep. | | | | | | | | | | | |
| Exports | 1 556.9 | 2 643.1 | 2 779.3 | 3 066.3 | 3 711.2 | 3 917.0 | 4 326.6 | 5 154.6 | 5 776.2 | 6 595.7 | 7 367.2 |
| Imports | 1 738.1 | 2 980.3 | 3 217.9 | 3 661.2 | 3 982.0 | 4 216.5 | 4 873.4 | 5 526.1 | 6 419.6 | 6 797.8 | 7 481.4 |
| Balance | -181.2 | -337.2 | -438.6 | -594.9 | -270.8 | -299.8 | -546.8 | -371.5 | -643.4 | -201.1 | -114.2 |
| Hungary | | | | | | | | | | | |
| Exports | 721.6 | 1 616.0 | 1 720.8 | 1 960.1 | 2 429.9 | 2 413.8 | 2 756.6 | 3 300.4 | 3 746.4 | 4 007.0 | 4 434.4 |
| Imports | 758.3 | 1 657.7 | 1 771.3 | 2 066.5 | 2 396.4 | 2 741.3 | 2 981.6 | 3 306.7 | 3 707.2 | 4 058.0 | 4 320.8 |
| Balance | -36.7 | -41.7 | -50.5 | -106.4 | +33.5 | -327.5 | -225.0 | -6.3 | +39.2 | -51.0 | 113.6 |
| Poland | | | | | | | | | | | |
| Exports | 1 134.9 | 2 406.1 | 2 484.9 | 2 872.1 | 3 600.0 | 3 735.5 | 3 596.1 | 3 220.8 | 4 097.0 | 4 786.7 | 5 296.8 |
| Imports | 1 214.9 | 2 447.2 | 2 750.1 | 3 195.9 | 3 449.6 | 3 837.5 | 4 405.9 | 4 931.3 | 4 812.9 | 5 274.3 | 6 069.2 |
| Balance | -80.0 | -41.1 | -265.2 | -323.8 | +150.4 | -102.0 | -809.8 | -1 710.5 | -715.9 | -487.6 | -772.4 |
| Romania | | | | | | | | | | | |
| Exports | 474.0 | 823.7 | 829.7 | 1 021.9 | 979.9 | 1 067.8 | 1 441.2 | 1 673.1 | 1 683.4 | 1 665.3 | 1 755.2 |
| Imports | 444.6 | 702.1 | 770.2 | 1 003.5 | 971.3 | 1 077.8 | 1 350.3 | 1 779.1 | 1 423.6 | 1 639.6 | 1 807.2 |
| Balance | +29.4 | +121.6 | +59.5 | +18.4 | +7.7 | -10.0 | +91.0 | -106.0 | +259.8 | +25.7 | -52.0 |
| Total | | | | | | | | | | | |
| Exports | 5 970.4 | 11 311.8 | 12 226.3 | 13 851.9 | 16 776.1 | 17 491.2 | 19 095.3 | 21 150.6 | 24 323.0 | 27 528.4 | 30 478.1 |
| Imports | 6 082.6 | 11 866.4 | 13 106.7 | 15 266.2 | 16 945.7 | 18 548.7 | 20 919.5 | 24 300.0 | 26 295.4 | 29 152.1 | 32 393.8 |
| Balance | -112.2 | -544.6 | -880.4 | -1 414.3 | -169.6 | -1 057.5 | -1 824.2 | -3 149.4 | -1 972.4 | -1 623.7 | -1 915.7 |

Note: a) All trade, including trade settled in hard currency.

Source: Vneshnynya Torgovlya SSSR, various years.

Table 7 (Chapter 1)

SHARE OF THE WEST IN THE FOREIGN TRADE OF SEVEN EASTERN COUNTRIES, 1970-1984
(per cent)

| | 1970 | 1975 | 1976 | 1977 | 1978 | 1979 | 1980 | 1981 | 1982 | 1983 | 1984 |
|---|---|---|---|---|---|---|---|---|---|---|---|
| **Exports** | | | | | | | | | | | |
| Bulgaria | 14.2 | 9.3 | 10.5 | 9.6 | 9.8 | 14.4 | 15.8 | 13.5 | 11.4 | 10.5 | 9.1 |
| Czechoslovakia | 20.3 | 19.8 | 18.2 | 18.4 | 18.0 | 19.9 | 21.8 | 19.6 | 17.8 | 16.4 | 16.1 |
| German Democratic Republic | 21.9 | 22.4 | 24.3 | 20.6 | 19.7 | 20.8 | 24.1 | 27.4 | 29.0 | 29.9 | 30.0 |
| Hungary | 27.2 | 21.5* | 30.9 | 29.1 | 29.8 | 32.3 | 34.0 | 29.7 | 29.7 | 32.6 | 34.8 |
| Poland | 28.4 | 31.5 | 32.0 | 31.3 | 31.3 | 31.1 | 34.4 | 29.5* | 32.7 | 32.4 | 34.5 |
| Romania | 31.9 | 34.8 | 34.7 | 30.1 | 32.8 | 36.3 | 34.8* | 33.6 | 32.2 | 34.6 | 42.1 |
| Total Eastern Europe | 23.7 | 24.1 | 25.8 | 23.9 | 24.0 | 26.0 | 27.9 | 26.3 | 26.4 | 27.1 | 28.9 |
| Soviet Union | 18.7 | 25.6 | 28.0 | 26.5 | 24.4 | 29.5 | 32.0 | 30.2 | 29.8 | 29.0 | 28.7 |
| Total Eastern Europe and the Soviet Union | 21.6 | 24.7 | 26.7 | 25.0 | 24.2 | 27.5 | 29.8 | 28.1 | 28.0 | 27.9 | 28.8 |

| | 1970 | 1975 | 1976 | 1977 | 1978 | 1979 | 1980 | 1981 | 1982 | 1983 | 1984 |
|---|---|---|---|---|---|---|---|---|---|---|---|
| **Imports** | | | | | | | | | | | |
| Bulgaria | 19.1 | 23.6 | 18.4 | 15.6 | 15.0 | 15.5 | 17.2 | 20.0 | 16.6 | 13.9 | 13.7 |
| Czechoslovakia | 24.5 | 24.7 | 24.9 | 23.5 | 23.2 | 24.1 | 24.4 | 22.2 | 19.0 | 16.7 | 15.1 |
| German Democratic Republic | 26.7 | 29.1 | 31.8 | 26.4 | 25.4 | 30.8 | 30.5 | 29.5 | 27.3 | 28.9 | 29.0 |
| Hungary | 27.1 | 27.2* | 36.3 | 37.0 | 39.1 | 38.1 | 39.5 | 40.0 | 36.2 | 34.3 | 34.6 |
| Poland | 25.8 | 49.3 | 48.9 | 43.3 | 40.5 | 37.9 | 35.1 | 28.9* | 30.9 | 28.8 | 29.5 |
| Romania | 39.5 | 42.8 | 36.2 | 36.9 | 38.5 | 35.9 | 31.2* | 32.0 | 22.7 | 17.2 | 20.3 |
| Total Eastern Europe | 26.7 | 34.1 | 34.7 | 31.2 | 31.4 | 31.6 | 30.8 | 29.4 | 26.3 | 24.7 | 25.0 |
| Soviet Union | 24.1 | 36.4 | 37.7 | 33.0 | 31.8 | 35.0 | 35.4 | 34.4 | 33.5 | 31.4 | 30.0 |
| Total Eastern Europe and the Soviet Union | 25.7 | 35.1 | 35.9 | 32.2 | 31.5 | 32.9 | 32.7 | 31.6 | 29.5 | 27.7 | 27.2 |

Note: * National statistics are converted at official transferable ruble exchange rates. These are not consistent with ruble/dollar cross rates in Hungary (after 1975); Poland (after 1981) and Romania (after 1980).

Source: ECE data files based on national statistical publications.

Table 8 (Chapter 1)

ESTIMATES(1) OF GROSS AND NET DEBT OF EASTERN EUROPE AND THE SOVIET UNION
(hard currency debt - $ millions)

| | 1971 | 1975 | 1976 | 1977 | 1978 | 1979 | 1980 | 1981 | 1982 | 1983 | 1984 |
|---|---|---|---|---|---|---|---|---|---|---|---|
| Bulgaria | | | | | | | | | | | |
| Gross | 743 | 2 640 | 3 198 | 3 707 | 4 263 | 4 032 | 3 550 | 3 082 | 2 907 | 2 437 | 2 135 |
| Net | 723 | 2 257 | 2 756 | 3 169 | 3 710 | 3 292 | 2 760 | 2 272 | 1 947 | 1 267 | 705 |
| Czechoslovakia | | | | | | | | | | | |
| Gross | 485 | 1 132 | 1 862 | 2 616 | 3 206 | 4 099 | 4 920 | 4 439 | 3 858 | 3 522 | 3 070 |
| Net | 160 | 827 | 1 434 | 2 121 | 2 513 | 3 050 | 3 670 | 3 369 | 3 128 | 2 582 | 2 070 |
| German Dem. Rep. | | | | | | | | | | | |
| Gross | 1 408 | 5 388 | 5 856 | 7 828 | 9 666 | 12 300 | 14 100 | 15 264 | 12 858 | 12 101 | 11 267 |
| Net | 1 205 | 3 748 | 5 047 | 6 842 | 8 320 | 10 340 | 11 600 | 13 114 | 10 978 | 8 711 | 6 727 |
| Hungary | | | | | | | | | | | |
| Gross | 1 071 | 3 135 | 4 049 | 5 020 | 7 290 | 8 140 | 9 090 | 8 699 | 7 952 | 8 250 | 8 836 |
| Net | 848 | 2 195 | 2 852 | 3 856 | 6 349 | 6 910 | 7 000 | 7 799 | 7 222 | 6 920 | 7 306 |
| Poland | | | | | | | | | | | |
| Gross | 1 138 | 8 014 | 11 483 | 13 967 | 17 844 | 22 669 | 25 000 | 25 468 | 24 700 | 26 440 | 26 800 |
| Net | 764 | 7 381 | 10 680 | 13 532 | 16 972 | 21 500 | 24 350 | 24 708 | 23 730 | 25 170 | 25 250 |
| Romania | | | | | | | | | | | |
| Gross | 1 227 | 2 924 | 2 903 | 3 605 | 5 221 | 6 950 | 9 400 | 10 159 | 9 766 | 8 880 | 7 198 |
| Net | 1 227 | 2 449 | 2 528 | 3 388 | 4 992 | 6 650 | 9 100 | 9 859 | 9 466 | 8 370 | 6 558 |
| USSR | | | | | | | | | | | |
| Gross | 1 800 | 10 500 | 14 700 | 15 600 | 16 400 | 18 100 | 17 800 | 26 548 | 26 361 | 24 031 | 23 185 |
| Net | 600 | 7 400 | 10 000 | 11 200 | 10 400 | 9 300 | 9 200 | 18 098 | 16 331 | 12 991 | 11 845 |
| Total Seven | | | | | | | | | | | |
| Gross | 7 872 | 33 733 | 44 051 | 52 343 | 63 890 | 76 290 | 83 860 | 93 659 | 88 402 | 85 661 | 82 491 |
| Net | 5 527 | 26 257 | 35 297 | 44 108 | 53 256 | 61 042 | 67 680 | 79 219 | 72 802 | 66 011 | 60 461 |

Notes : 1. Note the break in the series after 1980. OECD estimates are available only from 1981 onwards. This break affects, in particular, debt estimates for the USSR. For instance, according to CIA figures, gross debt in 1981 was $20.9 billion (net debt $12.5 billion) and, in 1984, gross debt was estimated at $20.2 billion (net debt $10.4 billion). For the East European countries, the difference in the estimate is not significant.

2. Net debt is gross debt less deposits in BIS-area banks (for all countries, 1981-1984).

Sources : For 1971-1980: CIA, Handbook of Economic Statistics (for the years 1982, 1984 and 1985).

For 1981-1984: Financial Market Trends, No. 33, OECD, Paris, March, 1986, p. 25.

Table 9 (Chapter 1)

DEBT SERVICE RATIO (1)

| | 1981 | 1982 | 1983 | 1984 |
|---|---|---|---|---|
| Bulgaria | 32 | 32 | 29 | 26 |
| Czechoslovakia | 22 | 21 | 24 | 20 |
| German Dem. Rep. | 55 | 49 | 37 | 33 |
| Hungary | 52 | 47 | 48 | 48 |
| Poland | 171 | 202 | 147 | 106 |
| Romania | 32 | 43 | 35 | 27 |
| USSR | 22 | 19 | 15 | 16 |

Note : 1. Interest and amortisation on medium- and long-term debt as per cent of one year's exports. Amortisation on rescheduled debt is excluded.

Source : *Financial Market Trends*, No. 33, *op.cit.*, p. 30.

## NOTES AND REFERENCES

1. Net material product (NMP), as defined in the "material product balance system" of national accounts used in all the centrally planned economies, is the difference between gross output of goods and material services (i.e. those services which contribute to material output such as freight transport, but excluding those contributing to private consumer demand) on the one hand, and intermediate consumption including the consumption of fixed assets on the other. This measure approximates most closely to the net national product (i.e. gross national product *minus* depreciation) of the Standard National Accounting system of the United Nations, though the omission of most service activities from the Eastern NMP system needs to be borne in mind throughout. For the most part, United Nations data are used as they are based on official (Eastern) statistics, plans and plan fulfilment reports. A comparison of data based on various methods of calculation is given by Paul Marer, "Economic Performance and Prospects in Eastern Europe", in *East European Economic Assessment, Part 2 – Regional Assessments, A Compendium of Papers*, Joint Economic Committee, Congress of the United States, Washington, D.C., July 10, 1981. For a detailed comparison of GNP and NMP growth rates, see also Thad. P. Alton, "Production and Resource Allocation in Eastern Europe: Performance, Problems, and Prospects", in *Ibid.*
2. Unless otherwise stated, industrial and other sectoral industrial branch data are based, unlike the global NMP aggregate, on gross output, i.e. including inter-industry transfers. This tends to inflate output *levels*. Also, a rise in cost per unit of output tends to *inflate* gross output relative to net output and vice versa.
3. Wharton Econometric Forecasting Associates, *Centrally Planned Economies Outlook*, Washington, D.C., April 1982, p. 31. See also announcement in *Hospodorske Noviny*, 6th August, 1982 and *Neues Deutschland*, 18th/19th December, 1982.
4. Very substantially below-plan harvests were experienced in only 2 years during the 1960s (1963 and 1965) while in four years (1972, 1975, 1979 and 1980) during the following decade. Performance was also disappointing in all years 1981-1984, growth having averaged about 1.2 per cent annually compared with a Five-Year Plan target of 2.5 per cent annually over the average of 1976-1980. As far as the East European countries are concerned, their total growth of output has, with the exception of Romania, increased more modestly than anticipated since the 1971-1975 plan period. For a detailed discussion of trends and prospects in the agricultural sector see, for instance: *Prospects for Agricultural Production and Trade in Eastern Europe*, Volumes 1 and 2, OECD, Paris, 1981 and 1982 (respectively); *Prospects for Agricultural Production and Trade in Eastern Europe*, OECD, Paris, March 1984 ; *Prospects for Soviet Agricultural Production and Trade*, OECD, Paris, 1983, and *The Soviet Feed-Livestock Sector and the Meat Processing Industry*, OECD, Paris, March 1984.
5. Parallel attempts were also made to strengthen intra-CMEA integration by adopting a series of measures designed to ensure economies of scale, to stimulate technological progress and to further intra-bloc specialisation and trade – notably, the adoption in 1971 of the "Comprehensive Programme for the Further Intensification and Improvement of Co-operation and Development of Socialist Economic Integration". The series of measures taken since 1971 and their effectiveness, as well as likely significance for East-West trade, are taken up in Chapter 4.
6. For a detailed analysis of the failure of the import-led growth strategy, see Zbigniew Fallenbuchl, *East-West Technology Transfer: Study of Poland, 1971-1980*, OECD, Paris, 1983.

7. Eastern planners' expectations that a rapid expansion in the production of sophisticated and efficiently produced commodities and improvements in the quality of the capital stock and its productivity would ensue from the import of Western technology, did not, by and large, materialise. This issue is discussed in that part of the analysis dealing with impact.

8. The East European countries had been initially shielded from the OPEC oil shocks which were plunging the Western economies into recession. This was because prior to the 1975 adjustment of their pricing mechanism, intra-CMEA foreign trade prices were held constant during five-year periods and were based on averages of world market prices during the preceding five years.

9. Bulgaria, while not reaching the high end of the scale, was, nevertheless, within the projected range.

10. NMP growth rates have progressively, and almost continuously, declined since the mid-fifties. Soviet official statistics indicate that average annual growth of national income (in constant terms) was 11.3 per cent in 1951-1955, dropping to 9.1 per cent in 1956-1960. Calculated from *Narodnoye Khozyaystvo SSSR za 60 let*, Moscow, 1977, p. 485. For subsequent plan periods see Table 1 of Chapter 2.

11. *Economic Survey of Europe in 1971, Part II, The European Economy in 1971,* United Nations, Economic Commission for Europe, New York, 1972, p. 124.

12. It is interesting to note that industrial production grew fastest in those countries which still had significant labour reserves (Bulgaria, Poland and Romania). Since industrial labour productivity improvements were relatively small, growth rates have since decelerated markedly and virtually all countries are faced with labour shortages, one can conclude that present industrial structures and management "procedures" are directly responsible for the increasingly poor labour productivity performance.

13. While all the CMEA countries believed that expanded economic relations with the West would bring economic benefits, only Poland adopted a development strategy based entirely on Western technology; the so-called "new development strategy". This strategy entailed accelerating the import of Western technology on the basis of Western credit. For the Soviet Union, an additional impetus for expanding trade with the West was the possibility of selling raw materials and fuels for higher prices than was possible on CMEA markets and earning, thereby, hard currency.

14. All countries foresaw investment growing faster than output in 1971-1975, and all countries, except the German Democratic Republic, achieved this objective.

15. It also took into account such factors as more costly energy and raw materials supplies.

16. *Economic Survey of Europe in 1981, United Nations, Economic Commission for Europe, New York*, 1982, p. 283.

17. *Economic Survey of Europe in 1976, Part II, The Five-Year Plans for 1976-1980 in Eastern Europe and the Soviet Union*, United Nations, Economic Commission for Europe, New York, 1977, p. 17.

18. *Ibid.*

19. National income for the area as a whole grew only 2.4 per cent, of which 2.2. per cent was accounted for by the Soviet Union. Gross industrial production grew by a mere 3.7 per cent. See *Economic Survey of Europe in 1981*, *op.cit.*, Table 3.1.1.

20. During the 1976-1980 period the efficiency of capital input progressively diminished. It is estimated that a given addition to NMP requires a greater input of investments than in the past. At the same time, since economic growth rates are either stagnating or increasing more slowly, investment cannot be significantly increased in order not to further squeeze the consumer sector's share. See G. Fink and F. Levcik, *Die Wirtschaftskrise im RGW-Raum*, Wiener Institut für Internationale Wirtschaftsvergleiche, Sonderdruck No. 55, August 1981, p. 43.

21. The striking exception is Hungary, whose imports of capital goods (in both value and constant terms) increased until 1978. While declining in 1979 and in 1980, in both years the level was higher than that attained in 1975. See Part II below.

22. See Part II.

23. The Five-Year Plans of the individual CMEA countries are discussed in more detail in Chapter 2.

24. *Ekonomicheskaya Gazeta*, No. 39, Moscow, 1981, p. 3 as quoted in *Economic Survey of Europe in 1981*, *op.cit.*, p. 162.

25. According to the "Arrangement on Guidelines for Officially Supported Export Credits" which was amended in 1982 and 1984, certain countries were reclassified according to their "GNP" income. Relative to the Eastern countries, the Soviet Union, Czechoslovakia and the German Democratic Republic were reclassified as "relatively rich" countries. This raises minimum interest rates to 12.15 per cent for credits up to 5 years and, for credits over 5 years, to 12.40 per cent.

26. *Economic Survey of Europe in 1984-1985*, United Nations, Economic Commission for Europe, New York, 1985, pp. 137-138.

*Chapter 2*

# EAST-WEST TECHNOLOGY TRANSFER WITHIN THE CONTEXT OF THE 1981-1985 FIVE-YEAR PLANS:

## Implications for the 1986-1990 Period

### 1. Comment

The set of goals formulated in the Five-Year Plans [and which have the status of law[1]] define the nature of economic and social development and spell out the principal means by which the objectives set for that particular period are to be attained. They also provide the basis for formulating the annual economic plans which are designed to implement the five-year development strategy.

The frequent need to adjust to unanticipated or unfavourable economic developments means that the announced goals of the Five-Year Plans are often modified in favour of *ad hoc* measures, usually designed to re-establish short-term equilibrium. Poland provides an extreme – and exceptional – example of this. Presumably, developments leading up to the declaration of martial law in December 1981 (and the subsequent aftermath of this action) were responsible for the fact that planners have not attempted to formulate a plan for the 1981-1985 period taken as a whole.

Nevertheless, despite frequent adjustments made, the Five-Year Plans are a "statement of intent". That is, they indicate overall policy objectives and the strategies to be pursued to reach them. As such, they also reflect the problems facing the Eastern countries. For the Western analyst, they provide a framework for judging the likely demand for Western technology; as well as the role to be played by Western technology in the pursuit of a given set of objectives.

As has been noted in the previous chapter, the CMEA countries are faced, albeit to varying degrees and with different consequences, with a set of similar problems. These problems (low level of efficiency of the economic systems, structural problems, a difficult foreign trade situation with the accompanying vulnerability to external developments and problems in the supply and cost of energy) underlie the near economic stagnation of recent years and, in some cases, absolute declines in output. To be sure, the gap between plan and actual achievement has varied considerably from country to country.

It is also evident (notwithstanding the upswing in economic performance in 1984 – notably as regards Eastern Europe), that there is now a greater differentiation in the effects of these problems – particularly between the energy and raw materials rich Soviet Union and the countries of Eastern Europe. Consequently, there are some differences in the type of adjustment policies applied and the degree to which they have been imposed in the countries concerned.

In the discussion that follows, the Five-Year Plans of the CMEA countries are examined. Since the primary focus of this study is on the role and impact of technology transfer, only those aspects of the plans have been chosen which are thought to be most relevant for this optique. Poland will not be treated here. It is very difficult to discuss a long-term development strategy since Poland did not publish a Five-Year Plan for 1981-1985. While it did produce three-year targets for 1983-1985, these were essentially short-term objectives retained as guidelines for its post-crisis adjustment programme. However, Poland's recent economic performance was discussed briefly in the previous chapter and relevant data are included in the tables of this chapter.

## 2. The Country Perspectives

### a) *BULGARIA*

#### 1. *Development Strategy*

In comparison with the preceding plan period, the sharpest decline in the growth rate of NMP for 1981-1985 was foreseen by Bulgaria (see Table 1 of the chapter). Despite the modest growth projection, Bulgaria's past economic performance has not been significantly worse than any of its CMEA partners – on the contrary (see Tables 1 and 2 of Chapter 2). Rather, it seems to reflect a realistic assessment of Bulgaria's current economic situation and has been geared to a level which that country is likely to attain. Moreover, if the 1985 target is reached (see Table 6 of Chapter 2), Bulgaria will achieve a faster rate of NMP growth than had been anticipated by the Five-Year Plan.

On the whole, the 1981-1985 Five-Year Plan does not signal a significant departure from previous plans. Emphasis was again placed on checking investment and the Plan foresaw no investment growth. A difficulty, which is implicit in the tight investment policy (and which is likely to have an adverse affect on the policy of modernisation), is the evident strain placed on the ability to continue to import machinery and equipment from the West at the same volume levels as in the past. In the event that Bulgaria must increasingly depend on domestic production or that requirements for advanced equipment and machinery cannot be met by its CMEA partners, its modernisation programme will be acutely challenged.

It is interesting to note that during 1981-1984 investment growth in fact exceeded the five-year planned target – a development also forecast in the 1985 annual plan (see Tables 4 and 9 of Chapter 2). It can be assumed that Bulgaria's inability to shift to the intensive growth path at the pace desired (as witness, for instance, the continued decline in capital productivity – see Tables 5 and 8 of the chapter), the consequent concern with safeguarding NMP growth and the intent (or ability) to implement those structural changes designed to modernise the economy, stand behind the deviation from the original investment growth target.

As is the case for its CMEA partners, greatest emphasis has been placed on increasing the efficiency and quality of production. Consequently, the lion's share of investment (some 73 per cent) was to be allocated to the material sphere, and within this share, some 70 per cent was to be devoted to reconstruction and modernisation.

Relative to the structural adjustments which are indicated in the Plan, it was foreseen that heavy industry would remain as the centre of gravity for overall economic development, although the gap in growth rates between heavy and light industry was to be narrowed. Priority (in terms of growth rates to be achieved) was given to the machine-building and chemicals industries, the energy sector and to the ferrous metallurgy branches. The allocation of investment resources foreseen by the 1985 annual plan, indicate that these priorities have been maintained.

Within engineering, the leading sub-branches were (and will continue) to be heavy and metal-working machines, computers and associated equipment, industrial robots, automation equipment and electronic elements. In the 1985 annual plan, even greater emphasis was placed on developing, first and foremost, the engineering and electronics branches, particularly mini- and micro-computer systems and industrial robots. Discussion in the Bulgarian literature suggests that during the 1986-1990 period a greater share of total investments will be devoted to the production of advanced machinery and electronics since output of the engineering and electronics branches themselves is to be doubled – primarily to increase their export share[2].

All of the Eastern countries intend, of course, to increase agricultural output. In the case of Bulgaria, the attention given to the agricultural sector reflects its importance as a determinant of overall economic performance. For instance, the sharp drop in the NMP growth rate registered in 1983, was essentially due to the decrease (-7.2 per cent) in agricultural output (see Table 7 of this chapter). However, in view of the fact that foodstuffs are also an important source of hard currency earnings, improving this sector's performance is increasingly linked to enhancing export potential – a consideration which will be reinforced given the agreements reached at the CMEA Summit conference held in June, 1984 (see Chapter 4).

Apart from modernising the sector, a series of organisational and management reforms have been progressively introduced. For instance, the New Economic Mechanism adopted in 1979, as well as the formation of a National Agro-Industrial Union (which now oversees all aspects of agricultural production) were designed to increase efficiency and output. The number of indicators imposed on agricultural enterprises has been reduced, greater autonomy in decisions related to production has been given to agricultural organisations and the importance of private plots in increasing output of food products is officially acknowledged. Purchase prices of many agricultural products have been raised in order to more closely reflect actual production costs and developments in world markets.

Scarcity of domestic natural resources, other than land and coal (the latter is of poor quality – the worst in the CMEA area), has made Bulgaria highly reliant on intra-CMEA trade and explains essentially the high concentration of its foreign trade on the CMEA region, particularly on the Soviet Union. Concern with meeting its energy needs (Bulgaria is also the most dependent of the Six on imported energy), would account for the fact that of the 74 per cent of total investments (18.9 billion leva) allocated to the material sphere in 1984, the preponderant share is said to have gone for developing the raw materials and energy branches[3].

Since energy expenditures per unit of production are very high, energy conservation continues to take priority and regulations to reduce energy (notably electricity) consumption have been imposed – though to date with little effect. Priority was also given to developing indigenous and alternate sources of energy. As coal production is to be augmented, restructuring and modernisation of the industry is foreseen – the long-term goal being to substitute coal for oil. While hydro-thermal power was also to be increased, the overriding priority is clearly the development of nuclear energy (in close co-operation with the Soviet nuclear energy programme) and it was to account for some 30 per cent of total electricity generated by the end of the 1981-1985 plan period (60 per cent by the end of the century).

Finally, mention must be made of efforts undertaken since January 1982, to strengthen the New Economic Mechanism and to introduce it throughout the economy[4]. Essentially, these are measures intended to make planning more flexible and, in the interest of augmenting economic efficiency, to increase the responsibility of enterprises in attaining production targets. For instance, greater economic independence has been granted to economic units in elaborating their production and investment plans. Subsidies to enterprises are to be gradually withdrawn, thereby also shifting economic accountability to them.

In order to encourage enterprise self-sufficiency and production of higher quality goods, wages are to be linked to output. Domestic prices are to be adjusted to changes in world market prices. The role of central planning is to be reduced. This primarily entails decreasing the number of mandatory indicators. Industrial production remains essentially subject to Ministerial control, even though individual enterprises have the right to make their own views known by submitting "counterplans" as to how general targets are to be met.

In the area of foreign trade, export-oriented enterprises will have more independence in decision-making relative to production targets, direct "co-operation links" between domestic and foreign enterprises may be established and larger production units have been given the right to directly buy from and sell to abroad. However, trade transactions will continue to be co-ordinated by the Ministry of Foreign Trade and the branch ministries.

While the measures taken or envisioned appear to have been based on the Hungarian model, they are in fact much more limited. For one thing (and contrary to the Hungarian reforms), the scope of the Bulgarian private sector is unlikely to be significantly increased. The basic inconsistency in the Bulgarian reforms is that planners are retaining under central control those elements essential for meaningful decentralisation. Among these are the allocation of investments, foreign exchange, manpower, raw materials and energy, choice of output assortment – foreign and domestic – and price determination.

Furthermore, in view of the fact that branch ministries will retain control over the functioning of the enterprises under their responsibility (including the right to appoint and dismiss enterprise managers), it is questionable how much independence will be exercised. Since price setting remains under central control, it is evident that there is no real intention to create market conditions for the exchange of commodities. In short, the new measures which have been undertaken seem primarily designed to refine central planning rather than to permit significant new departures.

### 2. *Science and Technology*

In keeping with the central theme of the 1981-1985 Five-Year Plan (achieving "intensive" economic growth), much has been made of the fact that this Plan is to usher in an era of scientific and technological progress. However, apart from indicating principal orientations designed to counteract the slow application of new technologies – one of the major weaknesses of the economy – few concrete measures were actually proposed. According to the Plan, a host of national-level projects (some 950) were to be initiated. These are designed to create new and innovative products and new technologies essentially for reducing energy wastage, increasing automation and speeding up the pace of innovation. It is also stated that scientific and technological progress is to be achieved primarily by means of close intra-CMEA co-operation as well as by "fully exploiting" the experience of other countries. While such guidelines are not very informative, they do indicate, at least implicitly, the general line of Bulgarian planners – achieving qualitative improvements on the basis of applying domestic and foreign technology to production.

Since 1984, furthering S&T progress has become the central theme of Bulgarian planning. As is frequently noted by Bulgarian commentators, development of advanced technologies and their application in production has become indispensable in order to raise capital productivity, improve product quality and to strengthen export potential. In March 1984, a long-term programme designed to raise the technological level of producion was adopted. While in 1984 the State budget allocated 0.6 billion were for scientific research and application, some 2.6 billion leva were earmarked for 1985. As is evident from sub-section 1 above, S&T development and application will be focused primarily on the engineering branches. The sum to be allocated in 1985 also includes foreign currencies. One can assume, therefore, that purchases from hard currency (notably Western) trading partners is foreseen.

Thus, despite the stated objective of developing Bulgaria's scientific and technological activities primarily within the CMEA framework, it has also been explicitly acknowledged that Western technology will continue to be needed for Bulgaria's development effort. It is noted, for instance, that "in carrying out economic relations with non-socialist countries, accent will be put on scientific-technical collaborations"[5]. Furthermore, the decree on "Establishing Conditions for the Accelerated Development of Small- and Medium-Sized Enterprises" (adopted on July 2nd, 1984) explicitly states that one of the objectives of these enterprises is to further S&T progress and to accelerate S&T application[6]. Subsequent discussions in the Bulgarian literature suggest that in order to achieve these objectives such enterprises are also to seek co-operation with Western firms in order to gain access to "high level technology and modern equipment".

### 3. *Trade*

For the 1981-1985 plan period, the most important element of Bulgaria's trade strategy remained co-operation with CMEA partners. It was foreseen that overall trade turnover would rise by some 40 per cent (an average annual increase of 7.0 per cent which is higher than that planned for NMP). Trade turnover with the Eastern countries was expected to increase by up to 80 per cent. If Bulgaria reaches this goal, it would reverse a falling trend during the 1970s. By 1983-1984, this share had risen to 74 per cent in ruble terms. The Soviet Union will remain Bulgaria's most important partner and, according to the 1985 annual plan, is to account for 63 per cent of Bulgaria's intra-Eastern area trade (a sizeable increase over the 57 per cent recorded in 1984).

Bulgaria's development strategy suggests that a greater effort will be made to increase the volume of exports to the Soviet Union and to restrain imports of energy and raw materials. The sizeable trade deficits in its trade with the Soviet Union (0.5 billion rubles in 1984, more than a two-fold increase over 1980) would provide the major motivation for this – see Table 6 of Chapter 1.

In terms of the commodity structure of foreign trade, discussion relative to the 1985 annual plan indicates that the export share of mechanical engineering products in total trade is to increase to some 60 per cent and that engineering and electronic goods will constitute more than 82 per cent of Bulgaria's trade with its Eastern partners[7].

As far as trade with the West is concerned, balance of payments considerations have led Bulgaria, like its East European partners, to reduce imports from the West. Since (together with Czechoslovakia) Bulgaria is least oriented towards trade with the West, it has been relatively protected from unfavourable economic developments external to the CMEA area. Net debt to the West is not significant and fell to an estimated $ 0.7 billion by 1984 (see Table 8 of Chapter 1). In 1984 debt servicing represented 26 per cent of Bulgaria's exports in hard currency (see Table 9 of Chapter 1). This ratio is considered to be manageable.

Nevertheless, Bulgaria's situation is far from secure. OECD trade data indicate that Bulgaria's trade balance with the Western countries remains negative (some $-0.9 billion in 1983 and $-0.8 billion in 1984)[8]. This is all the more serious given the commodity structure of Bulgaria's trade. Bulgaria is one of the East European countries which has been most active in selling refined oil products based on low-priced oil imports from the Soviet Union – a policy presumably adopted to counteract its inability to increase exports of manufactures. However, the share of refined oil products in its exports have been declining since 1982 – apparently due to cut-backs in Soviet oil supplies as well as increases in the price of Soviet oil. This would imply that Bulgaria's ability to finance imports of Western technology has been limited.

The intent to constrain imports from, while increasing exports to, the West has already been noted. However, the continuing need for Western technology limits the extent to which such imports can be reduced. Capital goods imports from the OECD, which were cut by half in volume terms between 1975-1980, recovered sharply thereafter, but were still smaller in 1983 than their 1975 peak. Constraints on the ability to purchase Western technology has led Bulgaria, like the majority of its CMEA partners, to stress the importance of pursuing and strengthening economic, industrial, and scientific and technological co-operation with the Western countries. It has also been specifically noted that increased participation in licence and know-how trade will be sought in the interest of expanding exports. The above-mentioned financial constraints are likely to ensure that this policy is pursued vigorously.

## b) *CZECHOSLOVAKIA*

### 1. *Development Strategy*

The growth rate projected for NMP (1.6-2.2 per cent) in the 1981-1985 Five-Year Plan (only adopted towards the end of 1981) is the lowest ever targeted in that country's five-year plans and is close to the poor growth performance recorded during the 1961-1965 period (see Table 1 of Chapter 2). It reflects the continuation of problems which have led to virtual economic stagnation. Foremost among them are labour supply problems, declining capital productivity (see Table 8 of Chapter 2) and ineffective (according to Czechoslovak commentators) use of investments to bring about goals of changing the industrial production structure in order to ease the limited supply and selection of raw materials resources, especially of oil and gas[9]. The latter is particularly serious since those industrial branches which have preponderant weight in terms of resource allocation (for instance, heavy and electrical engineering, iron and steel, non-ferrous metals and chemicals – not to mention the energy industry itself), are both highly capital- as well as energy-intensive.

Furthermore, Czechoslovakia's capital stock is, by Western standards, largely outdated. The resultant inefficiency in production has exacerbated already declining rates of productivity. It has also made Czech machinery and equipment (the predominant export item to the West and in intra-CMEA trade) increasingly uncompetitive on Western markets; thereby depriving Czechoslovakia of needed hard currency earnings. It is noted, for instance, that since "machinery production is not supported by a sufficiently strong R&D base, it is lagging in the most progressive areas of engineering – automation, control and electronic equipment. The product mix is frequently not of the highest level and, since Czech engineering products are generally below top world standards, they can either not be sold at world market prices, or not at all on developed markets"[10].

Thus, Czechoslovakia finds itself in a vicious circle. Czechoslovakia's trade deficits with the Soviet Union have risen since 1980 (see Table 6 of Chapter 1). In order to pay for increasingly expensive raw materials and energy imports, exports of machinery and equipment had to be greatly augmented. However, inability to increase production sufficiently not only led to shortfalls in exports, but also created bottlenecks in the domestic economy. Subsequent cutbacks in imports of raw materials (as well as imports of machinery and equipment from the West in order to check trade deficits) has further aggravated these problems.

In order to meet the development objectives foreseen by the Plan, closer co-operation with its CMEA partners (particularly with the Soviet Union) will be sought. According to Secretary-General Husak, this policy will also be a central focus of the Plan drawn up for the 1986-1980 period. The country's high level of dependence on its Eastern partners for supplies of energy and raw materials (particularly on the Soviet Union from which Czechoslovakia imports almost all of its oil), is clearly a major factor in shaping this policy.

The Five-Year Plan foresaw an investment target well below that retained for the previous plan period. Even this objective has not been reached. Indeed, investment in 1984 has been running below its 1980 level, but the annual target for 1985 suggests that the modest growth of 1983 and 1984 will continue (see Tables 4 and 9 of this chapter).

Investment priority was given in 1981-1985 to the energy sector, to those industrial sectors which account for Czechoslovakia's largest share of exports (machine-building and wood-processing) and also to electrical engineering[11]. It is noteworthy that the share of fuel and energy combined in total industrial investment was to rise to 36 per cent by 1985 (it was 28 per cent only in 1980 and rose to 32 per cent in 1983). It is evident from the 1985 targets and from discussion of the Plan for 1986-1990 that these priorities will be maintained, though an even greater share of investments will be devoted to modernising present industrial structures.

In this context, it is interesting to note that the 1985 annual plan provided for a higher allocation of foreign exchange to finance imports of Western licences, technology, machinery and equipment in order to speed up industrial modernisation and improvement of product quality. The stringent credit regulations imposed by the State Bank in 1982 and which were designed to restrict hard currency purchases are, therefore, being eased. Nevertheless, hard currency credits granted to enterprises will have to be repaid within a period of three years.

The industrial sector was to continue to set the pace for overall economic development in 1981-1985. Consequently, the share of total investments allocated to industry was to be increased compared with the previous plan period (by roughly 1 to 3 per cent). This has been achieved. Within industry, engineering and electronics were to be the leading output and export branches. Available data suggests that these priorities will be retained during 1986-1990. Furthermore, to boost output of the engineering and electronics branches (as well as to encourage wider application of electronics in production and management) a "Long-term Complex Programme for the Electronisation of the National Economy up to 1995" was adopted in October 1984, and is expected to play an important role in boosting labour productivity and reducing energy and raw materials consumption[12].

As can be seen from the big rise in investment allocations to fuel and energy noted above, these sectors were accorded highest priority in the 1981-1985 Five-Year Plan[13]. The central objective was to restructure and increase output in order that, with the help of economies in use, fuel imports could be reduced.

In terms of energy production, coal and nuclear power are to become the primary source for power and heat production. Coal output is to be increased. Given the problems in this industry (low-grade coal and increasing extraction costs), it is doubtful, however, whether Czechoslovakia can reduce its import dependence to a substantial extent. The main element of restructuring the energy sector was the increased role of atomic energy production. This is continued in the available indicators for 1986-1990, which foresee that some 40 per cent of investments will be devoted to the energy sector with the predominant share slated for the development of atomic energy. Nuclear power is to be developed in close co-operation with the Soviet Union. At present, Czechoslovakia has installed two atomic reactors, six are in the process of construction and by 1990, 13 reactor blocs with a planned capacity of 6 280 MW are to be in operation. By the end of the century atomic power is to provide some 70 per cent of Czechoslovakia's energy requirements.

However, the cost of nuclear power plant construction has increased much more than was anticipated. This is primarily due to failures in meeting construction deadlines and, given insufficient technological levels, increased cost in producing high quality materials for construction. Co-operation with the Soviet Union is also likely to become more costly, given that country's intention to raise the price of nuclear fuel. Further, despite increases in the wholesale prices of fuels, energy and certain raw materials, energy consumption has not been significantly reduced. Consequently, the energy issue is likely to remain a major factor in the formulation of Czechoslovakia's development policy.

Inability to effectively counteract those problems which have constrained economic growth led Czechoslovakia to adopt "Measures for Improving the System of Planned Management of the National Economy after the Year 1980". In effect since 1981, these measures continue to be amended and supplemented. Broadly speaking, they are designed to make planning more flexible, improve management of the economic system, motivate enterprises to produce better quality and less costly goods and to bring about a more realistic pricing structure. For instance, the various plans (long, medium and short-term) are to be linked within the framework of a "complex programme" in order to fulfill the priorities foreseen in the Five-Year Plan. The role of quality indicators is to be increased (while decreasing that of volume targets) in order to assure better output. Wages are to be linked to output and enterprises put on a "cost accounting" basis.

It is evident from numerous official comments that the reform measures are not being well received by enterprise managers. This leads to the conclusion that planners are facing difficulties in implementing them. Furthermore, attention continues to be drawn to the fact that the measures which have been instituted are not meeting goals for structural output changes or achieving energy savings. Central planning has seemingly not been diminished, despite the fact that the role of Ministries in production-related decision-making was to be gradually reduced and that of enterprises in proposing "counter-plans" to be increased.

2. *Science and Technology*

On the whole, relatively little is said in the Plan itself concerning the development of science and technology. That which is said is extremely general and gives little insight as to how the general goals which are given are to be practically realised.

As may be expected, achievement of the objectives indicated in the Five-Year Plan has been linked to the "accelerated exploitation of the results of research and development." To this end, it is urged that research and development work be more closely linked to production and that new products and technologies be developed in order to improve substantially the techniques and costs of production. It is also underlined that "the share of new products with high technical-economic parameters is to be drastically increased in order to help increase profitable exports to world markets and better satisfy the demands of the domestic market"[14].

It was acknowledged that these goals would be difficult to achieve. It is said, for instance that R&D has not been well concentrated with the result that the technical level of products was still too low. Lead-times between research, development and application remain too long, planning is not well co-ordinated and, due to the negative attitudes of enterprise managers, innovation targets are not being met.[15]

As of 1983, a new State Commission for Scientific-Technological and Investment Development has (in consultation with the State Planning Board) been given responsibility for matters concerning S&T planning and the R&D work of relevant institutions. The stated objective is to overcome the above-mentioned problems. Consequently, specific scientific-technological development objectives were incorporated into the general plans of the individual production entities. Furthermore, in order to accelerate S&T development new regulations governing the financing of R&D were brought into effect as of 1985. Among others, these regulations set up technological development funds within enterprises, ease credit rules for the purchase of machinery and equipment for technological development and seek to increase the responsibility of the enterprises in meeting S&T development targets and the application of R&D results[16]. Intra-CMEA scientific and technological co-operation (particularly with the Soviet Union), is also to be strengthened.

For the 1981-1985 period, the plan for scientific and technical development was to be allocated a total of Kcs 22 billion in non-investment funds. Of this, 23 per cent was earmarked for the electrical engineering industry and 29 per cent for developing mechanical engineering and metallurgy. In order to meet the priorities of the Five-Year Plan, 30 per cent of the "assignments" were designed to improve labour productivity, 25 per cent dealt with increasing suitable or decreasing unjustifiable exports and almost 20 per cent were earmarked for improving the fuel and energy balance[17].

Particular emphasis was placed on developing those new products and technologies which would economise fuel, raw materials and energy resources, investments and foreign currency. In order to increase the efficiency of overall production, stress was placed on automation and on the wider application of mechanised and automated production methods. Some 3 000 industrial robots were, for instance, to be in use by 1985. In the electronics branch, R&D was to be directed towards the production of computers and computer parts, numerically controlled machine tools, microprocessors and laser techology.

According to the annual 1985 plan, some Kcs 20 billion (only in the non-investment sphere) were to be spent on financing R&D – a 7 per cent increase over 1984 – and Kcs 8 billion were to be allocated from the State budget. Some 22 per cent (of non-investment funds) were to be allocated to the electro-technical industry and some 30 per cent for developing the engineering industry and metallurgy[18].

Presumably, these priorities reflect the intent to improve the output structure of industry, particularly of the engineering sector. Czechoslovakia's domestic requirements for electronic products have had to be met primarily by imports. Thus the objectives stipulated seem to aim not only at reducing imports, but at recapturing a share of the international market.

### 3. *Trade*

The overall policy which was to govern trade relationships during the 1981-1985 plan period was restoring external equilibrium by means of a faster growth of exports than imports. The expansion of foreign trade turnover that was foreseen, (6.2-7.0 per cent, annual average) was to be primarily directed toward the CMEA area. According to the Five-Year Plan, by 1985 trade with the CMEA countries was expected to increase by 25 per cent (in constant terms) and, compared to the previous plan period, trade with the Soviet Union was to increase by 36 per cent. This theme is repeated in the 1986-1990 Plan. Based on comments made by Secretary-General Husak, the expansion of intra-CMEA trade remains a priority.

By 1984, some 72 per cent of Czechoslovakia's total trade (compared with 64 per cent in 1980) was accounted for by its Eastern partners, with the Soviet Union's share alone representing some 45 per cent. This shift toward intra-CMEA trade is the natural outcome of Czechoslovakia's inability to expand its exports to the West – particularly equipment and machinery – and dependence (notably on the Soviet Union) for needed supplies of raw materials and energy.

Persistent trade deficits with the Soviet Union also explain Czechoslovakia's drive to expand trade with that country. Despite significant increases in exports of machinery and equipment (machinery and equipment account for about two-thirds of Czechoslovakia's exports to the Soviet Union), the trade deficit again grew between 1983 and 1984 – from 0.5 to 0.6 billion rubles (see Table 6 of Chapter 1). The primary reason for this was the price increase in Soviet energy and raw materials. These could not be compensated for (despite increased exports) since Czechoslovakia's export prices rose less than the prices of imports from the Soviet Union.

Trade with the West is based on the policy that foreign borrowing will not be resorted to as a means of redressing negative trade balances. As has been noted, "...for fundamental reasons we cannot finance by credits that difference between our exports and imports which arises from the fact that our economy is not able to export with adequate profit. To offset this difference by credits is a very dangerous policy"[19]. Despite the fact that Czechoslovakia has decreased its net hard currency debt considerably (to some $2.0 billion – see Table 8 of Chapter 1) and debt service obligations (20 per cent – see Table 9 of Chapter 1) are considered to be low, discussion relative to the 1985 plan indicates that this policy will be adhered to. It is said to be dictated by the fact that Czechoslovakia continues to encounter "considerable problems" with its exports to the West.

However, one significant change in policy should be noted. Since 1980, Czechoslovakia has achieved increasing surpluses in its trade with the OECD countries. This surplus amounted to some $0.6 billion in 1984. However, until 1983, these surpluses were brought about by a simultaneous reduction of imports and exports – although in 1984, exports did increase somewhat[20].

According to the 1981-1985 Plan, Czechoslovakia's restrictive import policy affected primarily machinery and equipment. These had been Czechoslovakia's largest import commodities and, until 1981, accounted for some one-third of that country's imports from the West. According to official estimates, imports of machinery and equipment were to decline up to 41 per cent by 1985[21]. That this policy, which is officially acknowledged to have had a detrimental effect on economic development, would not be pursued was substantiated by the 1985 plan which specifically specified that imports of machinery and equipment from the West – in order to modernise production – would increase by 14 per cent.

Hard currency considerations and inability to penetrate Western markets explain the increasing interest that is displayed in pursuing economic and scientific and technological co-operation with the West. Relative to the objectives which were to be met during the 1981-1985 plan period, it was noted that "advantage must be taken of higher forms of co-operation, production and research development co-operation and to strengthen these relations for the long range effect"[22]. This signals a pronounced change since Czechoslovakia's attitude towards East-West co-operation had been rather restrictive.

A number of measures have been taken to facilitate co-operation with Western countries and to encourage enterprises to conclude co-operation agreements. For instance, hard currency will be made available to finance business trips. Furthermore, foreign trade organisations which fulfil their contracts and have a positive payments balance, will have freer access to foreign currency. The same applies to production organisations which meet their targets. To facilitate execution of co-operation agreements with Western countries, foreign exchange accounts have been set up.

Co-operation will primarily be sought for the production of machinery and to develop the engineering sector – notably for the production of electronic semi-conductors, automatic measuring and control instruments and equipment for the nuclear engineering sector. However, it is also noted that any project which would expand export capability would receive preferential consideration as far as the allocation of hard currency is concerned.

## c) *GERMAN DEMOCRATIC REPUBLIC*

### 1. *Development Strategy*

Though not the highest projection given for NMP growth for the 1981-1985 plan period in the region, the target figure in the Five-Year Plan (5.1 per cent, annual average – see Table 1 of Chapter 2) was the only one which largely exceeds recorded outcomes for the 1976-1980 period. As such, it was perhaps the most optimistic target within the CMEA area. In fact, it is unlikely to be reached – despite the marked acceleration in NMP growth registered in 1984. Nevertheless, if the planned target for 1985 is attained, the German Democratic Republic's economy will have been the most dynamic in the area during this plan period.

The economic development strategy pursued during 1981-1985 reflected major pressures on the economy. Among these were unfulfilled consumer expectations coupled with labour supply problems. Over the past 20 years population growth has stagnated. Since the participation rate of the labour force is already high, it is unlikely that any significant increase in employment has been achieved.

Trade is potentially the most serious pressure on the economy. In many respects, the provisions of the 1981-1985 Plan can be explained in terms of trade constraints and their resolution could be decisive as far as future economic development is concerned. The overriding importance of trade for economic development is, in part, a function of trade dependence: exports and imports each equal roughly one third of the German Democratic Republic's national income. Equally important, its dependence on foreign sources for most of its energy and raw materials supplies (other than brown coal and potash) is virtually total. For a number of branches in the manufacturing industry (notably, machine tools, foodstuffs, instruments, electrical engineering and electronics) the ability to expand production is closely linked to increasing foreign trade – partly because they are import intensive and also because a large share of their output is traditionally export oriented. The second, and most serious, factor contributing to the German Democratic Republic's foreign trade problems, is its need to reduce trade deficits with the East and maintain trade surpluses with the West. This places a great burden on the economy.

It is evident from the provisions contained in the 1981-1985 Plan and from the discussions preceding its formulation (in particular, the Directives of the 10th SED Party Congress, addressed specifically to the problems of economic development during the 1981-1985 period), that the NMP growth target was formulated on the assumption that input savings could and must be increased[23]. This was the central condition for sustaining economic growth. Attainment of this goal has been closely linked to harnessing the "achievements of the scientific-technological revolution"[24]. As was noted by Secretary General Honecker, "science has grown to assume a new role and answerability.... Each significant step forward in the development of modern productive capacity has and will have its roots in scientific knowledge.... The key to basic improvement in the effectiveness of the economy in the 1980s rests on higher achievements in science and technology"[25].

The planned investment policy was a significant departure from the past and demonstrates the very tight conditions under which planners were operating. According to the Five-Year Plan targets (see Table 4 of Chapter 2), investments were to be reduced from the level which had been expended during the 1976-1980 period. This policy has been adhered to. While the State Budget for 1985 indicates that M56 billion in current prices will be earmarked for investments (as opposed to the M49 billion allocated in 1984), the anticipated cost of investment growth at constant prices in 1981-1984 (see Table 9 of Chapter 2) would imply a drop of investments in real terms, compared with 1980.

According to the Five-Year Plan, absolute priority in the allocation of investment resources was to be given to renovation, modernisation and reorganisation of fixed capital assets. The share of investment in construction work was to be substantially reduced – in industry to no more than 25 per cent. Furthermore, R&D and investments were to be so linked and utilised as to assure that products of the highest possible technological level were brought on stream: not only to improve the productive base of the domestic economy, but to strengthen export potential.

Discussion relative to the 1985 plan and the Directives for the 1986-1990 Plan indicate that these priorities have been and continue to be largely adhered to. Even greater emphasis is now being placed on modernising industry and on increasing output of technologically advanced products. As well as encouraging exports, it has been made clear that this strategy is designed to reduce still excessive energy and raw materials consumption[26].

The structural changes foreseen for industry in 1981-1985 – in terms of branch development and changes in the output structure – illustrate, above all, the central place assigned to science and technology in achieving economic growth on the basis of reduced material inputs and increased efficiency.

Accordingly, within industry, engineering was accorded absolute priority and an annual average growth rate of 7 per cent was set. Within engineering, particular emphasis was placed on the manufacture of robots and on developing electrical engineering and electronics. The output of this branch was to increase by 9.6 per cent. As is noted in the Plan, electrical engineering – especially micro-electronics – was expected to raise the technical level of the entire economy (hence its productive capacity) and to guarantee a significant growth of exports. According to the 1985 annual plan, output of the electronics and engineering branches was to rise twice as fast as that for industry as a whole. Above average growth rates were also foreseen for the production of machinery, equipment and machine-tools. Again, the primary consideration was export expansion.

As far as energy is concerned (a sector also accorded priority in terms of ensuring overall economic growth), the overriding objectives were, as already mentioned, to reduce energy and raw materials consumption and, in the interest of reducing import dependence, to enhance the development of indigenous sources of energy. These objectives will remain as the focal point for this sector during the 1986-1990 period.

Relative to energy conservation, the 1981-1985 Plan stipulated that energy consumption was to decrease by 6.1 per cent (annual average) – double the rate of decline compared to the 1976-1980 period. Concerning energy development, emphasis was placed on increasing coal production and output was to be raised to 285-290 million tons. Furthermore, brown coal was to be used increasingly as a substitute for oil, as a raw material source for the chemical industry (refinement of chemical products) and was to become the basis for the production of electricity and for thermal energy generation. Further development of atomic energy was also foreseen (in close co-operation with the Soviet Union) and nuclear energy was to become an increasingly important element in the energy substitution strategy that is being pursued. At present four power blocs are said to be in operation, generating some 12 per cent of the country's total electric energy[27].

As far as management of the economy is concerned, there is no evidence in the Plan that reforms – in the sense of modifying central planning – were being contemplated. On the contrary, the series of measures which have been taken since 1982, indicate that even greater control has been exercised over the economy, particularly in the distribution and use of factor inputs.

Even prior to the adoption of these new measures, the changes which had already been instituted were not intended to alter the orthodox centrally planned nature of the system. For instance, the creation of *Kombinate* (Combines or trusts) was designed to concentrate scientific-technological resources, as well as the industrial labour force under unified leadership. In order to accelerate technical progress in industry, almost the totality (90 per cent) of industry's research and development capability has now been incorporated into the *Kombinate*. In essence, these organisations presently have jurisdiction over all operations involved in production: supply, R&D and manufacturing, although of course, under ministerial supervision.

The basic purpose of giving increasingly greater responsibility to the *Kombinate* is to guarantee that investment and R&D decisions with greater practical utility will be taken than would be the case if planning for such activities were centered in independent R&D institutions.

Since 1981, modifications in the mechanisms used for regulating foreign trade have also been introduced. Previously, all foreign trade activity was conducted by organisations under the Ministry of Foreign Trade. The new structural arrangement provides for a dual subordination system. Namely, foreign trade organisations are now to be answerable to both the Ministry of Foreign Trade and to the *Kombinate* to which they are attached. Given the expected closer co-ordination between industrial producers and exporters, it is evident that one purpose of this measure is to bring about better export performance. Even if that objective is achieved (and deadlocks and delays associated with dual decision-making minimised), it is difficult to see how such perennial difficulties as improving (or providing adequate) after-sales service, devising more effective marketing mechanisms or responding more quickly to changing international prices and shifts in demand, will be resolved.

The series of administrative measures which have been instituted since 1982 are designed to bring about even stricter control over the use of investments, decrease energy and raw materials consumption, assure that the scientific and technological base is effectively developed with reduced investment inputs (and economic development closely tied to it) and, export oriented output efficiently increased.

Consequently, investment activity is being more closely scrutinised, as are R&D plans. Criteria on which scientific and technological development are judged have been tightened. Tighter norms have also been set for production and efficiency indicators have been more closely aligned to net cost of production. A stricter policy towards credits has been adopted and credits and interest payment linked to output. Greater accountability for meeting production goals is demanded of the enterprises and, in particular, of the *Kombinate*.

In essence, stricter central control is being exercised over all aspects of production – be this related to cost determination and control, financing of production or the determination of output. In short, planners seem to have opted for the policy of steering the economy by means of "finetooling" central management – that is, attempting to find a better combination between administrative and economic instruments[28]. In the short run, such a policy may be instrumental in forcing more efficient output, as has been the case in the early 1980s. Whether success in the long run is possible by such means remains to be seen.

## 2. *Science and Technology*

Given the constraints (external and internal) on maintaining economic growth, the central role accorded to science and technology to modernise and restructure the economy is not surprising. This strategy has become an absolute imperative – as was again underlined at the 11th SED Congress held in November, 1984. While the 1981-1985 Plan did not indicate the volume of financial resources to be allocated to S&T development, it is said that some

5 per cent of national income will be earmarked for R&D in 1985 (compared with 4.4 per cent in 1981 shown in the UNESCO yearbook).

The benefits expected from, and the tasks assigned to the S&T system, are enormous. In the 1981-1985 Plan greatest emphasis was placed on the need to shorten lead times between research, development and practical application. R&D activity was to be primarily directed toward those areas which assure a high and fast payoff in terms of increasing export potential. R&D efforts were also expected to be of particular benefit to those branches of industry which have traditionally shown above average growth rates so that they will continue to expand even faster. Energy and raw materials needs were to be met by developing appropriate technologies. Such technologies were also, of course, to lead to big energy savings and, in turn, allow import substitution.

The 1981-1985 plans for labour productivity growth were based on increased automation. Accordingly, emphasis was placed on developing electronic control and electronic computer technologies as well as industrial robots: some 45 000 robots were expected to be in use by 1985.

In order to realise these goals, emphasis was placed on the need to expand scientific and technological co-operation within the CMEA area, particularly with the Soviet Union. Co-operation with the latter now takes place primarily within the framework of the Long-Term Programme for Scientific, Technical and Economic Co-operation concluded between the Soviet Union and the German Democratic Republic in October, 1984 and the specialisation and production agreements (see Chapter 4). S&T co-operation will be centred on developing techniques for processing raw materials, on the electronics branch and on the development of automation systems – notably robotics. The reason for seeking co-operation with the Soviet Union has been clearly stated. Close co-operation with the Soviet Union is designed to speed up scientific and technological progress and, most importantly, to ensure "stable and long-term supplies of raw materials, fuels and equipment needed for the development of the economy"[29].

## 3. *Trade*

The 1981-1985 Five-Year Plan did not give any quantified growth targets for total trade turnover. The Directives of the 10th SED Congress merely noted that trade with the Soviet Union and the other CMEA countries would continue to be the base of the country's total trade and that trade turnover with the CMEA area would increase by an annual average of 7.5 per cent during the 1981-1985 period.

It is interesting to note in this regard that while in 1984 some 61 per cent of the German Democratic Republic's total trade was directed towards the CMEA area, the evolution of intra-Eastern area trade during 1981-1984 suggests that this share has remained more or less constant (see Table 2 of Chapter 1). This would imply that the German Democratic Republic has not shifted its trade more towards Eastern partners. In fact, in 1983 the trade share with the Western countries increased faster than that with its CMEA partners.

As has been noted, foreign trade is probably the most serious problem area of the economy. Relative to the East, the primary concern is the Soviet Union and restoring trade balances and reducing its trade deficits with the USSR (which had reached some 0.6 billion rubles in 1982 and were brought down to 0.1 billion rubles in 1984 – see Table 6 of Chapter 1) has been – and will continue to be – the major preoccupation. The need to bring its trade with the Soviet Union in balance is two-fold. For one thing, the Soviet Union is the German Democratic Republic's principal trading partner, accounting for nearly 40 per cent of its total trade turnover. For another, it is the principal supplier of raw materials and fuels. Raw materials and fuels supplies from the Soviet Union currently cover some 70 per cent of the German Democratic Republic's import requirements – including 90 per cent of its oil and all of its natural gas[30]. In view of the fact that Soviet oil deliveries to the German Democratic Republic were reduced by some 10 per cent in 1982 (accompanied by annual price increases), the need to bring its trade in balance assumes increasing urgency particularly since unfavourable price developments necessitate a marked increase in the volume of exports in order to secure needed supplies of energy and raw materials. That is, price increases of machinery and equipment, the German Democratic Republic's predominant export items to the Soviet Union, are smaller than the price increases of Soviet oil and raw materials.

It is evident that within the CMEA area, the German Democratic Republic's trade with the Soviet Union will expand at a higher rate than trade with its East European partners. The intent to increase trade with the Soviet Union was re-emphasised at the 11th SED Congress. In view of the fact that re-exports of petroleum products were primarily responsible for the German Democratic Republic's ability to increase exports to the West (particularly up to 1983), it can be assumed that securing supplies of oil is a primary consideration.

As far as trade with the West is concerned, restoring trade balances and improving its financial situation is also the thrust of the German Democratic Republic's trade policy. As is noted in the 1981-1985 Five-Year Plan trade with the West must, in the interest of economic development, be expanded. Particular emphasis was placed on the need to increase exports. While export expansion had already been given a high priority during the 1976-1980 plan period, its importance was upgraded for 1981-1985.

The policy of rigorously reducing imports (particularly of capital goods) which became noticeable after 1979, combined with the enormous effort made to expand exports, led to a marked reduction in trade deficits with the Western countries in 1981. In 1982, a trade surplus of $0.5 billion with the OECD countries was recorded for the first time in over a decade. This surplus, though smaller, (some $0.3 billion) was maintained in 1983 and 1984. Furthermore, these surpluses were achieved by means of faster export expansion as opposed to import reduction. Nevertheless, while this trend continued the data also indicate that exports fell in 1984[31]. In future, the German Democratic Republic is also likely to be faced with the problem of modifying the commodity structure of its exports since, given its dependence on fuel imports, a policy of export expansion based primarily on re-exports of petroleum products may become increasingly vulnerable.

Redressing external balances by reducing imports or expanding exports or by some combination of the two, was also dictated by the high debt level which had been incurred. Net debt reached $13.1 billion in 1981, and debt servicing represented some 55 per cent of hard currency exports[32]. While the net debt is estimated to have fallen to $6.7 billion by 1984 and debt servicing stood at 33 per cent (see Tables 8 and 9 of Chapter 1), the debt structure is such that credits must, in principle, be repaid within a short period. Supplier credits (with maturities under two years) still represent approximately half of the German Democratic Republic's Western debt. However, it should be noted in this regard, that the new credits extended to the German Democratic Republic during 1984 (a total of some $0.9 billion) are long-term credits and most have maturities between four to six years.

So far, import restrictions have affected capital goods more than other items. Whether they will be as rigorously restricted as they have been in the recent past, seems doubtful. Data for 1983 indicate that an upswing of such imports occurred (see Table 6 of Chapter 5). As noted previously, import reduction of capital goods hindered industrial modernisation and contributed considerably to the decline in the growth rate of NMP. Nevertheless, imports are likely to remain highly selective and to be geared predominantly to those items best suited for modernising domestic industrial capacity.

The manner in which intra-German trade develops will have a pronounced influence on trade with OECD countries as a whole, since the former alone accounts for about half of the German Democratic Republic's trade with the OECD area. The expansion of intra-German trade is likely to remain particularly attractive to the German Democratic Republic.

This is mainly due to the very substantial hard currency earnings it obtains each year for services to West Berlin and the Federal Republic, as well as to the "swing credit", an interest-free credit facility amounting to DM850 million[33]. This clearing system enables the German Democratic Republic to import vitally needed goods without using convertible currency. More flexible payment arrangements for imports and trade advantages (for instance, customs duties are not applied), also makes the Federal Republic of Germany a most attractive market.

Based on the official literature, the German Democratic Republic is also interested in pursuing industrial co-operation with the West. However, this interest seems largely confined to pursuing forms of compensation arrangements which allow for substantial export expansion.

## d) *HUNGARY*

### 1. *Development Strategy*

The Five-Year Plan for 1981-1985 reflected the new economic policy which was initiated in 1979 – a policy designed essentially to restore equilibrium in Hungary's external payments situation while protecting the standard of living. It entailed a planned deceleration of economic growth and the re-introduction of systemic reforms. As was noted, the cut-back in overall economic growth that had been foreseen was designed to allow the economy to "gain strength, reorganise and to establish a production structure that can better adapt to international requirements"[34].

Accordingly, and in contrast to the previous Five-Year Plan, the growth rate set for NMP was very modest (an annual average of 2.7-3.2 per cent – see Table 1 of Chapter 2). Furthermore, the NMP growth target contained in the annual plans was set on the basis of estimates of potential hard currency export capability, import elasticities and foreign debt targets. The reasons for adopting this approach were clearly stated: "We have no free hand in choosing the growth rate... It is influenced by such factors as rational import opportunities, the rate of increase of economical and competitive exports, domestic purchasing power..., the directions of economic development and the resources available for them. The forcing of growth divorced from these circumstances leads to grave economic problems, influencing detrimentally the fulfilment of other economic goals"[35].

Outcomes for the first four years of the plan period show that even this modest growth target will not be met even if the NMP growth rate envisaged by the 1985 annual plan were to be achieved (see Table 6 of Chapter 2). Preliminary figures for the 1986-1990 Plan, which project an annual NMP growth rate of 2.7-3.2 per cent, do not indicate greater optimism – despite the economic upswing which occurred in 1984.

In keeping with the policy of austerity and in order to improve the balance of trade and payments (which became a major objective of economic policy for the 1981-1985 period) investments were to be reduced. As can be seen from Table 4 of Chapter 2, no investment growth was foreseen relative to the 1976-1980 period. In fact, investments were actually cut during 1981-1984 – although by less than indicated in successive plans. According to the 1985 annual plan, a modest investment increase is forecast (see Table 9 of Chapter 2). However, it is underlined that this expansion will only occur if export targets are met. As a matter of policy, priority in the allocation of investments was given to those activities which serve to increase exports and reduce imports, result in energy and raw materials savings and lead to more efficient production. Large-scale development projects were to be cut to a minimum and no new programmes requiring significant investment outlays were foreseen. Instead, effort was to be concentrated on finishing existing projects.

As already mentioned, the Five-Year Plan had as its principal objectives the restoration of external equilibrium and the maintenance, to the extent possible, of living standards. However, due to the continuing though less marked deterioration of Hungary's terms of trade, emphasis since 1982 has shifted clearly to the former. Thus, while within the framework of the development strategy adopted for the 1981-1985 quinquennium, private consumption was to be the second claimant on output, its share was being progressively reduced. The Five-Year Plan had envisaged that consumption was to increase by 7-9 per cent, a projected growth target which was already less than half the rate of growth that had been foreseen for the 1976-1980 period. The growth target for personal consumption indicated in the 1985 annual plan (1-1.5 per cent) suggests that containing the fall in living standards within acceptable limits will be a major pre-occupation.

As far as industry is concerned, gross output was to increase by 3.5-4.0 per cent during 1981-1985, a target that, judging by the 1985 annual plan, will not now be reached (see Tables 2 and 6 of Chapter 2). The preliminary data for the 1986-1990 Five-Year Plan forecast an annual average output growth of 2.5-3.0 per cent, suggesting that stabilisation as opposed to accelerated industrial output growth will become the major policy objective for the industrial sector.

Given the overriding importance attached to restoring external equilibrium, the role which industry was expected to play in Hungary's economic development was decisive. Industry accounts for some 45 per cent of national income produced, contributes 34 per cent to GDP and commands some 37 per cent of total investments, while about 22 per cent of total industrial output is exported. It is thus not surprising that the principal tasks assigned to industry for the 1981-1985 period were to improve trade balances and to increase import substitution (in particular, by decreasing the hard currency component of imports). Higher efficiency of production was stressed as was the need to reduce material inputs. Expanded industrial output was to be achieved on the basis of increased productivity, and modernisation of the production structure was given priority. Current discussion in the Hungarian literature suggests that industrial policy for 1986-1990 will adhere to these guidelines. Industrial development is also to be more selective and more closely aligned to anticipated demand on domestic and world markets. Thus, priority will be given to those branches and products which already have, or could obtain a relative edge in terms of competitivity and market shares. Disinvestment in those branches and products which have no market potential (or have lost it) will also be undertaken[36].

Priority was, and continues to be, given to developing the engineering and chemical industries, as these are expected to lead the hard currency export drive. Relative to the former, particular emphasis in the 1981-1985 Plan was placed on increasing the output of semi-finished products, electronics, tools, spare parts, and machines and production lines for the food processing industry and for agriculture. More than half of the gross output of the machinery branch was slated for export and exports to the convertible currency area were to increase by 14-16 per cent (annually) during the 1981-1985 period. In 1985, a programme designed to promote the production and application of electronics, computers and robots during 1986-1990 was adopted – a programme clearly designed to speed up technological progress and to increase the competitivity of industrial products. For the chemical industry, emphasis was placed on increasing the production of pharmaceuticals and plant protectives which are also to be the primary export items.

A further major objective of Hungary's industrial development strategy was to develop domestic sources of energy and to reduce energy consumption. To this end an Energy Management Programme was adopted in 1981. One aspect stressed was the introduction of more energy-efficient techniques and equipment in industry, agriculture and in transport. Another was to further the development of alternative energy sources in order to reduce the consumption of oil and oil products. Thus targets to increase coal extraction and electric energy generation may also be complemented by further extending the Paks atomic power station beyond the planned four power blocs – of which each has a capacity of 440 MW[37].

Nevertheless, the thrust of the Energy Management Programme was primarily to control energy demand through a pricing and financing policy designed to encourage a more rational use of energy resources. Thus, domestic prices of raw materials, fuels and basic intermediate products are now set on the basis of their convertible currency prices and are adjusted periodically to reflect changes in world market prices and exchange rates. On the financing side, new credits are given for the installation of more energy-efficient technology. Investments in energy saving products will be partially covered by loans and partially by budgetary grants.

Relative to Hungary's development strategy, mention must also be made of the agricultural sector. Its importance lies in the fact that agricultural production is basically export oriented – to both Western and Eastern markets. Relative to the latter, some 40 per cent of food products exported to the CMEA area are paid for in convertible currency – an important aid in improving Hungary's overall convertible currency balance. In order to further agricultural production, reforms have been widely and consistently applied, primarily to increase the efficiency of large-scale enterprises. However, agricultural performance has also benefited from private initiatives which have long been encouraged by means of a variety of economic incentives. Small-scale farms are now increasingly recognised and supported (financially) by the Government. According to Hungarian data, small farms produce roughly one-third of gross agricultural output and account for some 25 per cent of all agricultural exports to hard currency markets. However, it is important to note that the agricultural sector (notably grain production) has also benefited considerably from Western technology – in the form of co-production of agricultural machinery, licence purchases and the acquisition and dissemination of Western production systems[38]. Indeed, Hungary is the only Eastern country where capital goods for the agricultural sector have occupied a significant share in capital goods imports from the OECD (about 6 per cent of the total).

The economic management reforms undertaken by Hungary since 1979 are the most comprehensive of all the Eastern countries. They involve industrial re-organisation, pricing formation and the financial system[39].

Relative to industry, a single Ministry of Industry was created in 1981 which took over the functions of the Ministries of Light Industry, of Heavy Industry and of the Metallurgy and Machine Industry, but with a much reduced staff. The principal objective was to reduce hitherto frequent ministerial interventions in enterprise operations and to strengthen central industrial development policy. This seeming concentration of authority represents, in fact, a further decentralisation of decision-making, since the new Ministry will not concern itself with detailed production plans. Furthermore, several large trusts which had enjoyed monopoly positions were broken up in such a way that their subordinate enterprises were transformed into independent economic units, with a view to increasing competition and flexibility to adjust to market requirements. Unprofitable firms are either to be liquidated or merged with profitable enterprises. Further steps were taken in 1985 to discontinue direct regulation of enterprises by central organisations. Enterprise autonomy in decision-making will be increased and supplemented by a degree of self-management. Also significant is that more producing enterprises are receiving direct foreign trading rights and that, gradually, the monopoly over foreign trade of Foreign Trade Organisations specialising in certain products is being abolished. Equally significant is the move toward establishing and expanding small- and medium-sized enterprises and co-operatives. Small-scale private initiatives are now welcome in many areas.

Concerning pricing, as of 1980, a "competitive pricing system" was introduced. That is, domestic prices of most manufactured goods were tied to price movements on hard currency markets. From 1984, administrative restrictions on price setting will be eased to allow more enterprises (particularly those which are competitive) to be guided by market considerations.

Among the reforms concerning the financial system, are the right given to enterprises and financial institutions to buy and issue bonds and enterprises are allowed to extend commercial loans to one another. Venture capital is now given to small enterprises and private firms. Financial institutions with foreign capital shares may be founded with government permission. Perhaps most important is the intent to continue the progressive decentralisation of the banking system – encouraging thereby the creation of commercial banks. In short, the thrust of the measures which are being undertaken is designed to strengthen the role of money and credit in economic relations and to create a unitary monetary and credit policy.

The very magnitude of the reforms poses an increasingly difficult problem of maintaining public support while the measures have their effect. Ability to sustain economic growth by means of increased productivity and a favourable foreign trade situation will have a considerable influence on the ability to implement successfully – and sustain –the development strategy which has been chosen.

2. *Science and Technology*

The 1981-1985 Five-Year Plan for scientific research and development was primarily concerned with raising the quality of output and with reducing the material intensiveness of production. This was the first national plan for research and development worked out in parallel with a Five-Year Plan.

Three per cent of national income was to be earmarked for R&D. Some was set aside for those small- and medium-sized enterprises which do not have their own research facilities. Research priorities were to be worked out in co-operation with enterprises to strengthen links with producers. The activities of research institutes and enterprises were to be more closely co-ordinated. Expenditures for technological development were to be decided on the basis of market conditions and export possibilities.

General priority was given to those R&D activities which lead to energy savings, the creation and production of new and up-to-date materials, streamlining of major new technologies and, which increase agricultural output. In terms of research application, particular emphasis was placed on improving automation processes and on the mass production of components and spare parts.

In order to give greater flexibility to the enterprises and to encourage them to become more closely involved in R&D, some new initiatives were subsequently taken. Contributions to the technical development funds (which the enterprises were compelled to form and out of which some 75 per cent of industrial R&D was funded) are now being progressively reduced. For instance, in 1984, 65 per cent of the enterprises were exempted from this compulsory contribution and from 1985, with the exception of the machine-building and chemical industries, no compulsory contributions will remain in effect. Howerever, enterprises will continue to pay taxes to be used by the central authorities to support specific R&D tasks. The purpose of eliminating compulsory contributions was to encourage individual enterprises to develop competitive R&D outputs since, those enterprises whose products are commercially successful, will receive preferential tax treatment and other economic concessions[40].

In order to speed up the application of R&D, a number of industrial research institutes were transformed into R&D companies and given sufficient incentives to become self-financing. Innovation is also receiving greater attention. For instance, one aspect of the banking reform involved changing three of the National Bank's institutions into independent banks. One of these is the "Innovation-Monetary Institution" (formerly called the Innovation Fund) which is to finance innovation.

In the interest of accelerating technological progress, it is also urged that the "possibilities inherent in international co-operation, including technical and commercial contacts with foreign enterprises" be more fully exploited. Joint research and joint technological development is being encouraged[41]. Furthermore, sizeable funds have been allocated for the purchase of foreign licences and selected enterprises given the *ad hoc* right to participate in licence trade. Available data indicate that licence purchases have increased since 1981 and that some 90 per cent of licences bought in 1983 came from the West.

3. *Trade*

The overriding importance attached to restoring external equilibrium in 1981-1985 was reflected in the growth targets for exports and imports. Hence, exports were to increase by an annual average of 6.5-7.0 per cent while the upper limit for import growth was set at 3.5 per cent. It was estimated that achieving the planned net surplus in exports would take up 8-9 per cent of Hungary's national income (calculated in 1980 prices)[42]. The enormous burden this represented for the economy explains the austerity policy adopted vis-à-vis both consumers and investment.

Hungary's trade policy is essentially aimed at achieving a balance in its trade with the West and the East. While in 1984, 48 per cent of Hungary's total trade turnover was with the Eastern countries (the Soviet Union alone accounting for about 30 per cent), trade turnover with the West was 35 per cent – a share which, as pointed out in Chapter 4, makes Hungary the most Western trade oriented country in East Europe. Nevertheless, its trading relationships with the CMEA area are of crucial importance and will remain so. According to the draft 1986-1990 Five-Year Plan, the expansion of trade with the CMEA countries (in particular with the Soviet Union) will be a priority and intra-CMEA trade is to increase by an annual average of 3 per cent.

Part of the significance of intra-CMEA trade to Hungary is that it is an important source of hard currency earnings. Items such as food products, pharmaceuticals and specialised machinery which Hungary sells to its Eastern partners are considered as "hard goods". That is, they are paid for in hard currency and can be traded above the quantities agreed to in bi- and multilateral trading agreements. Since 1977, some 21 to 25 per cent of Hungary's exports to the CMEA were settled in convertible currency, whereas only some 9 to 13 per cent of total imports were paid for in this way. This surplus has been a crucial input in helping to finance the trade deficits incurred with the Western countries.

Increasing exports to the Soviet Union was, and will continue to be, a major preoccupation. Export expansion is, of course, dictated by the Soviet Union's general unwillingness to continue to carry sizeable ruble deficits and, as Table 6 of Chapter 1 indicates, Hungary achieved a surplus in its ruble trade with the Soviet Union in 1984. Furthermore, since 1976, Hungary has sold grain and meat to the Soviet Union on a dollar basis in exchange for oil – a deal which the Soviet Union apparently does not intend to maintain after 1985[43]. This will affect export expansion plans since price increases for

Hungary's machinery and industrial goods exports (which account for a high share of Hungary's exports) have not kept pace with the price increases of the energy and raw materials imported from the Soviet Union.

Ability to meet its energy and raw materials requirements (particularly since roughly half of Hungary's raw materials imports are paid for in convertible currency) also explains the importance of CMEA trade for Hungary. However, the primary concern here will also be the Soviet Union, since it is not only Hungary's largest supplier, but accounts for almost the totality of Hungary's gas and most of its crude oil imports.

That a major pre-occupation for the 1980s will be securing adequate supplies of energy and raw materials has been made very clear. As was noted, "in developing bilateral and multilateral co-operation with the CMEA countries, it is an especially important task for the 1980s to arrive at a long-term solution for a secure supply of fuels, energy and raw materials"[44]. It is this aspect which will most likely (despite expressed criticism of the present system of intra-CMEA co-operation) push Hungary to more actively support current efforts being made to strengthen intra-CMEA economic co-operation (see Chapter 4).

As far as trade with the West is concerned, Hungary's policy continues to be focused on reducing its trade deficits and improving its financial situation. Developments since 1981 indicate that outcomes have been mixed. As regards trade, OECD data indicate that deficits have been reduced and fell to $0.15 billion in 1984[45]. However, until 1983, this reduction was achieved primarily by squeezing imports – particularly machinery and equipment where imports declined by some 22 per cent. In 1984, Hungary, like most of the East European countries succeeded in increasing its exports. Nevertheless, net debt has not been significantly reduced (estimated at $7.3 billion in 1984 – see Table 8 of Chapter 1) and debt servicing commitments remain high (equivalent to 48 per cent of hard currency exports – see Table 9 of Chapter 1).

Consequently, a significant improvement of the East-West balance of trade continues to be a major objective of the 1985 annual plan. Apart from the measures taken to achieve a more efficient and competitive commodity structure, a number of steps were also taken to directly increase exports. For instance, investment credits will be given primarily to those enterprises which undertake to increase their output of convertible export goods (or cut imports). Export promotion credits will, however, have to be repaid more quickly. The forint has been progressively devalued with the objective of increasing export competitivity in Western markets. A single exchange rate has been introduced in order to facilitate meaningful economic calculation, providing thereby a single measure for linking domestic and foreign prices.

The second element in achieving balanced foreign trade is import reduction. While this was a major goal of the 1981-1985 plan period, it is also evident that reductions were to be selective. For instance, in the interest of implementing the economic development strategy (particularly industrial development), the 1981-1985 Five-Year Plan made specific provision for importing "basic and raw materials, semi-products and up-to-date machinery and equipment which are needed for the development of production". It should also be noted in this regard, that imports of intermediate goods with a high technology content have remained relatively stable since 1980.

Furthermore, since 1984 the stringent import restrictions which were imposed in 1982, have been relaxed. Following the discussions at the Party Congress held in March, 1985, it appears that Hungary intends to increase imports of Western technology (notably, machinery and equipment) – provided that the hard currency surplus achieved in 1984 can be maintained. The draft 1986-1990 Five-Year Plan indicates that this is to be a major objective throughout the period.

Due to hard currency considerations, Hungary has become increasingly active in pursuing the "expanded economic relations – including commercial ties" – foreseen by the 1981-1985 Five-Year Plan. In addition to trade agreements (where Hungary is particularly involved), export growth is being pursued through a variety of industrial co-operation agreements. Among the areas where future co-operation is to be most actively sought, mention is made of specialised machinery construction, energy technology and raw materials processing.

## e) *ROMANIA*

### 1. *Development Strategy*

Rapid industrialisation has been the traditional economic goal pursued by Romanian planners. The strategy was to create a broad industrial base, which, in turn, was to provide the foundation for modernisation. It was an economic development strategy based essentially on "extensive growth".

The growth target for NMP given in the 1981-1985 Five-Year Plan (7.5 per cent, annual average – see Table 1 of Chapter 2) was the highest in the area. However, it was well below the target set for 1976-1980. The cuts reflected the general downturn of the economy after 1978. More significantly, the growth strategy adopted, notably the fact that the 1981-1985 Plan foresaw an even sharper deceleration of investment growth, indicates that Romania is now also attempting to shift away from the still extensive growth strategy laid down for 1976-1980.

It is evident from outcomes registered to date that the 1981-1985 NMP target will not be attained, even if the growth rate foreseen by the 1985 annual plan were to be reached. In any event, the 1985 target of 10.0 per cent (see Table 6 of Chapter 2) seems highly unrealistic despite the marked recovery recorded in 1984. That the 1985 growth target is over-optimistic seems to be implicitly acknowledged in that the annual NMP growth rate for 1986-1990 has been forecast at 7.6-8.3 per cent[46].

The intent to restrict investment growth in 1981-1985 was particularly significant in that Romania, more than any other country in the Eastern bloc, had maintained consistently high rates throughout previous five-year plan periods. The projected annual growth rate of 4.4 per cent for the 1981-1985 period was less than half that for the two previous plan periods (see Table 4 of Chapter 2). Restrictions on investment growth were sustained until 1984 when expenditures increased sharply. A further acceleration is expected in 1985 (see Table 9 of Chapter 2). This contrasts sharply with the 1.8-2.5 per cent annual rate foreseen in 1986-1990. The latter may reflect, in part, the intent to increase capacity utilisation, but it also reflects the big shift of resources which will be necessary to service external debt. Slower investment growth seems incompatible with the 10 per cent annual labour productivity increase planned for industry in 1986-1990. The investment slowdown up to the end of this decade also indicates that imports from the West, and in particular capital goods, which have fallen more than in any other Eastern country but Poland since 1980, will not recover rapidly to previous levels.

As far as the allocation of investment resources is concerned, the 1981-1985 Plan gave highest priority to industry – some 53 per cent of the total. Within industry, the chemicals, machine-building, metallurgical and the energy and raw materials branches were given highest priority. The share of agriculture continued to be the second biggest after industry. Resources were to be used primarily for the completion of existing projects and for modernising existing production capacities. Furthermore, priority was to be given to those branches which require less energy and raw materials inputs, which raise the technical level of industry as a whole and which contribute most to export potential. These priorities are to be adhered to in the 1986-1990 Five-Year Plan which indicates that the lion's share of available investment resources will be allocated for the development of energy and raw materials resources and for modernising industry. However, the agricultural sector will be given higher priority than before.

Industrial growth during 1981-1985 was planned at a reduced rate (7.6 per cent annual average – see Table 2 of Chapter 2) compared with 1976-1980. The outcomes recorded to date however (see Table 6 of Chapter 2), indicate that even these more modest goals will not be attained. The industrial growth target for 1986-1990 (6.0-6.5 per cent – marketable production) appears to be more realistic given past shortfalls in meeting production targets in key industrial sectors and the continuing difficulties which Romania is encountering in meeting its energy and raw materials needs.

For 1981-1985, the chemical industry was designated as a key branch, since it was to be a leading hard currency earner. Particular emphasis was placed on increasing production of synthetic rubber, various synthetic plastics and resins, pharmaceuticals and chemical fertilizers. These priorities are to be maintained in 1986-1990. However, greater emphasis will be placed on improving the quality of chemical products (for instance, production of ultra-pure materials and speciality chemicals for use in the electronics industry is to be increased) in order to expand exports. In the interest of increasing exports (as well as decreasing imports of raw materials) Romania will also shift production towards small tonnage chemical products. These not only command higher prices per ton output but are also less energy-intensive.

Machine-building also continues to be treated as a key branch. It is, first of all, the principal domestic source of technical progress. For the 1981-1985 period, more than 85 per cent of the machinery and equipment for the investment programme were to be provided by this branch. Its products were also to provide more than 32 per cent of Romania's total exports by 1985[47]. In the interest of promoting intensive economic growth, particularly high growth rates were foreseen for electronically controlled machine-tools, electronics and micro-electronics, optical hydraulics, pneumatic equipment, precision equipment and industrial robots. For the latter a programme has been drawn up by Romania's National Council for Science and Technology to develop and produce industrial robots. Both the annual plan for 1985 and the 1986-1990 Five-Year Plan indicate that emphasis in engineering will continue to be placed on accelerating the development of electronics for automation and other industrial uses. The 1981-1985 Five-Year Plan had stipulated that output of sophisticated electronic products would increase by 18-20 per cent. During 1986-1990, output of the electronics industry is to grow by 62-67 per cent (including computers) and that of precision engineering by 34-36 per cent. According to the 13th Party Congress, this should enable Romania to base 90 per cent of its machine-building production on sophisticated technologies, ensuring thereby increased competitiveness on world markets.

Regarding metallurgy, Romania has a high per capita steel output and is one of the world leaders in steel production. Nevertheless, Romania continues to be highly dependent on imports of high quality steel. During the 1986-1990 period, Romania not only intends to pursue its traditional policy of increasing steel production, but also aims to increase investment for the production of high quality steel. However, plans to increase exports to hard currency markets may be jeopardised by the fact that demand for steel products in these markets is weak.

As mentioned earlier, development of agriculture, which accounts for some 38 per cent of NMP, 10 per cent of exports and employs about 30 per cent of the labour force, will receive greater attention during 1986-1990 than in 1981-1985. The reasons for this are clear. Romania is a net exporter of food products to the West, earning thereby much needed hard currency to finance its imports[48]. Domestic concerns are also becoming increasingly pressing since exports have been achieved by holding down domestic demand and shortages in food supplies are chronic.

For 1981-1985, a separate plan for agriculture and for the food industry, was drawn up for the first time. During 1986-1990, Romania intends to modernise the agricultural sector. Investments are to be substantially increased. Whether agricultural production can reach the annual growth target of 5.4-5.8 per cent (compared to the 4.5-5.0 per cent increase targeted for 1981-1985 – see Table 3 of Chapter 2) remains to be seen. As Table 7 of Chapter 2 indicates, output has varied greatly during the 1980s. The 1985 annual plan target was set well below the output in 1984. Furthermore, since Romania intends to reduce State subsidies and small producers are to be compelled to deliver specified quotas at low prices, the private sector – which cultivates some 15 per cent of farm land – could reduce its output.

In terms of its overall development strategy, the performance of the energy sector was, and is likely to remain, a primary pre-occupation. A major problem is that Romania has tended to develop industrial branches which are highly energy and raw materials intensive. While it has met its crude oil needs by increased imports, hard currency constraints have severely curtailed this option. Within the energy sector itself, Romania is now saddled with excess oil refining capacity which is likely to remain since crude oil output has stagnated since 1981. Coal production is hampered by obsolescent machinery and equipment.

The energy programme drawn up in connection with the 1981-1985 Five-Year Plan was designed to make Romania largely energy independent by 1990 – a goal which seems to have been discarded. Nevertheless, highest priority continues to be placed on the more economical use of fuel and on developing domestic and alternate sources of energy in order to minimise hard curency imports of fuels and raw materials.

According to the 1981-1985 Five-Year Plan, output of domestic crude oil and gas was to be increased. Consequently, financial resources for oil and gas geological research were to be augmented by some 32 per cent over the resources allocated in the previous plan period. The emphasis placed on increasing coal production reflected the intent to reduce petroleum consumption and to increase the share of electric and thermal power generation based on coal. According to the 1981-1985 Five-Year Plan, production of bituminous coal was to grow by at least 55 per cent at the end of the plan period (compared to 1980). Electricity generation was to reach some 82 billion kWh by 1985 and 110 billion kWh by the end of the decade – primarily on the basis of increased coal extraction. As part of the effort made to extend and improve the national electric power system and to introduce new sources of energy, nuclear power sources were to be developed. However, the power bloc currently under construction is not expected to become fully operational until 1990.

The 1986-1990 Plan indicates some departures from the previous Five-Year Plan strategy. Greatest emphasis will still be placed on increasing oil and coal production. More investment resources will be made available for geological exploration, mining and to increase the recovery rate from existing oil wells. However, natural gas production is to be reduced to safeguard reserves. The electricity output target (95-97 billion kWh) has also been scaled down compared to the previous plan period. As was the case during the previous plan period, increased attention is to be paid to reducing the consumption of energy and raw materials and tighter consumption norms have been brought into effect. In order to control energy and raw materials consumption, progressive price increases have been introduced. These apply both to domestic and industrial consumers.

As far as management of the economy is concerned, various attempts have been made since 1978 to introduce a measure of flexibility into the system of economic management[49]. In 1978, a New Economic Mechanism was brought into effect with the major purpose of increasing enterprise efficiency. Accordingly, enterprises were given a greater role in the formulation of their production plans. The main enterprise performance criterion (hence the basis for bonus payments) was changed from gross to net output. In effect, these measures have been ineffective since rigid central control was maintained over all aspects of output and the allocation of resources. While additional measures have been taken since 1978 to strengthen the New Economic Mechanism, these are basically designed to promote greater hard currency earnings rather than relax central planning.

From 1983, economic activity was to be geared to increasing profit. Self-financing and self-management of the economic units was to be strengthened and greater discretion in the use of financial resources was to be extended to them. Workers were to be given the right to buy shares amounting to 30 per cent of an enterprise's assets. The purpose was to generate more investment funds, provide an incentive to increase labour productivity and to absorb excess purchasing power. These measures did not alter the basic character of the system, but attempted to introduce the profit motive within the framework of rigid centralised planning.

Apart from the adjustments made to the New Economic Mechanism, a series of additional measures have been taken to improve Romania's balance of payments situation. For instance, domestic raw materials prices are now closely related to world market prices. The net effect has been sizeable increases in energy, raw materials and industrial prices. Special incentives are given to enterprises producing for export. They are entitled to higher revenues and profits than those producing for the domestic market. A new system of wage determination has also been introduced. In essence, wages are no longer guaranteed, but tied to performance – notably, to output plan fulfilment and export targets.

The Economic Memorandum drawn up by Romania at the request of its creditors, released in July 1982 as part of the attempt to assure further rescheduling of its debt is, in essence, an attempt to stabilise the economy. However, since systemic reforms are not touched upon (nor in the Memorandum submitted in February, 1983), the measures listed in it are essentially short-term arrangements whose net effect may, in the absence of far-reaching reforms, do no more than reduce growth for a period.

### 2. *Science and Technology*

According to the 1981-1985 Five-Year Plan, scientific research was to be intensified and effort focused on assimilating new technologies. Priority was to be given to: the development of new products requiring low fuel and raw materials consumption; enlarging the domestic primary materials base and developing new sources of energy; designing and expanding the utilisation of up-to-date technologies, and to increasing the competiveness of goods produced. Similar objectives are assigned to the S&T system by the 1986-1990 Plan although even greater emphasis is placed on developing, above all, electronics and micro-electronics. It is interesting to note that the above-mentioned goals closely approximate the tasks assigned to science and technology in a decree published by the Central Committee and the Council of Ministers in July 1960[50]. It may be presumed, therefore, that the various attempts made since 1960 to reorganise the S&T system and to upgrade the role to be played by R&D in economic development still leave much to be done.

Despite the importance attached to science and technology in economic development, neither of the Five-Year Plans give concrete guidance on how the stated goals are to be achieved. Nor do they indicate how modern technologies are to be developed or how the overall technological level of the country is to be raised.

According to the 1981-1985 Five-Year Plan, 34 billion lei were to be allocated to scientific research and development – a 50 per cent increase over the resources devoted to these activities during the preceding plan period. However, in view of the fact that investment levels were to increase at a much lower rate than before, it is difficult to see how the overall goal of raising the share of highly advanced products (however defined) in industrial output to over 90 per cent by 1985, was to be achieved. Discussion in the Romanian literature now targets their share at 95 per cent by 1990.

Furthermore, the absence of concrete guidelines (or yardsticks by which to measure past performance) renders virtually meaningless the official pronouncement that "the programme for scientific research, technological development and introduction of technical progress for the 1981-1985 period, comprises more than 3 000 objectives of national interest, 2 600 of which are to be finalised and applied in the economy by the end of the present Five-Year Plan. It is estimated that in 1985, new and modernised products will account for at least 45 per cent of production of the processing industry"[51]. It may be noted in this regard that the "Programme for Scientific Research, Technological Development and the Introduction of Technical Progress for the 1986-1990 Period" has roughly the same objectives.

According to the 1981-1985 Five-Year Plan, co-operation is intended to play a significant role in achieving the goals which have been set for scientific and technological development. As is noted, "economic and scientific-technical co-operation will take a primary place in our country's economic relations". The importance of furthering S&T co-operation was reiterated at the 13th Party Congress. Based on Secretary-General Ceausescu's address to the Congress, Romania will seek to expand S&T co-operation primarily within the CMEA framework.

### 3. *Trade*

Foreign trade unquestionably dominates Romania's develoment strategy. According to the 1981-1985 Five-Year Plan, total trade turnover was expected to grow by 11.9 per cent (annual average) – two-thirds higher than that foreseen for NMP. As noted in the Plan, this would represent an increase of 75.5 per cent (in volume terms) over the 1976-1980 period. The 1981-1985 Plan had set the reduction of hard currency debt and achievement of a positive trade balance as major objectives to be achieved. Consequently, exports were to increase faster than imports with an annual average growth rate of 15.7 per cent foreseen for the former.

A similar policy is reiterated for 1986-1990 though more modest goals for trade growth are envisaged. Foreign trade turnover is to increase by an annual average of 7.1-7.7 per cent and exports by 8.7-9.3 per cent. Nevertheless, a positive and rising hard currency trade balance is expected for the entire period. All hard currency debt is to be repaid prior to the end of the plan period.

As far as the commodity structure of foreign trade is concerned, mechanical engineering is to play a central role in export promotion. Machine-building is to account for roughly one-third of total exports and the share of machinery, equipment and transport vehicles for 60 per cent by 1990.

Trade policy foresees that a greater share of Romania's trade will be carried out with its CMEA partners. Of all the East European countries, Romania's trade has been least oriented toward the East and it has most rigorously resisted the type of supra-national planning favoured by the Soviet Union. While, as mentioned in Chapter 4, this attitude toward intra-CMEA intergration has not changed, closer intra-CMEA trading and co-operative relationships are now being supported. The intent to increase trade with the CMEA area, notably with the Soviet Union, was already underlined at the 37th session of the CMEA held in 1983, and was reiterated at the 13th Party Congress. This signals a pronounced change in attitude on the part of Romanian officials, since the share of Romania's total trade with the CMEA countries progressively diminished since 1965, averaging about 35 per cent in 1981-1984. In 1984, some 34 per cent of Romania's total trade was with the CMEA countries, the Soviet Union accounting for about 18 per cent (see Table 3 of Chapter 1).

The intent to expand trade with its Eastern partners and the greater readiness to participate more actively in the various forms of intra-CMEA co-operation, particularly in specialisation and co-operation in production, is directly (and officially) linked to Romania's desire to obtain Soviet fuel, energy, and raw materials on a long-term basis. This would replace imports from the Western and Third World countries.

Within the CMEA area, the Soviet Union is Romania's main trading partner. Nevertheless, trade with the Soviet Union has not expanded at the rate foreseen by the 1981-1985 Five-Year Plan (an increase of some 70 per cent compared to the previous plan period). Furthermore, Soviet data indicate that Romania's terms of trade with the Soviet Union have deteriorated since 1982. Moreover, Romania again registered a small deficit with the Soviet Union in 1984 (see Table 6 of Chapter 2). Consequently, and in the interest of securing greater amounts of Soviet energy (notably oil), it can be expected that increased intra-CMEA co-operation will be centred above all on the Soviet Union.

As already mentioned, trade with the West is based on the policy of export expansion in order to eliminate all hard currency debt prior to the end of the 1986-1980 Plan period. Romania has reduced its net debt from an all time high of $9.9 billion in 1981 to an estimated $6.6 billion in 1984 (see Table 8 of Chapter 1) and the debt service ratio has fallen to 27 per cent (see Table 9 of Chapter 1). However, it is doubtful whether the debt repayment goal can be attained. Furthermore, the structure of Romania's debt (short-term liabilities are high and roughly one-third of its outstanding debt will mature by 1985) indicate that Romania's vulnerability to international monetary and financial movements, as well as to the attitudes of Western lending institutions, remains high.

As far as trade developments are concerned, OECD data indicate that Romania has increased its East-West trade surpluses since 1981 – to $2.1 billion in 1984[52]. However, the strategy which Romania intends to (or must) pursue, is likely to lead to severe long-term economic problems. The manner in which Romania was able to achieve its trade surplus with the OECD countries during the past four years illustrates the problem. Despite the fact that the 1984 trade surplus was achieved by means of increased exports, imports from the West continue to remain at a very low level. This implies a squeeze on production potential, hence harming the country's long-term export potential.

As early as 1980, Romania concluded a trade agreement with the EEC – the first and only East European country to do so. Official statements continue to voice the desirability of expanding trade relations with the Western countries and pursuing economic and technical co-operation with them. In view of the ambitious growth target which has been set for exports, export promotion is being rigorously pursued and is the primary purpose of the various reform measures that have been undertaken. As has been noted, "foreign trade, and chiefly export trade, is essential to the development of Romanian economic activity [and] export development [is becoming] a priority consideration in Romania's trade policy"[53].

Countertrade, in fact, plays a major role in Romania's East-West economic relations. Its use is compulsory in the event of hard currency shortages, in specific transactions such as those involving imports of machinery and equipment and for any items which fall outside the import plan. That imports would primarily be transacted on the basis of compensation trade, was already made explicit by Mr. Ceausescu in the draft plan for 1983 submitted for consideration by the Party Plenum in 1982. It has been incorporated into Romania's foreign trade legislation. Such rigid requirements underlie the pressures felt (and denounced) by Western businessmen in their transactions with Romania, particularly in recent years.

## f) *SOVIET UNION*

### 1. *Development Strategy*

The Soviet Union entered the 1980s with a series of underlying constraints which are reflected in the modest growth target for NMP used (no target was published for NMP produced) for 1981-1985. As noted in Chapter 1, these include a slower growth rate of the labour force, acute labour shortages in the European areas of the country – the main centres of industry, depletion of easily accessible oil and gas reserves with a resultant sharp increase in the costs of energy extraction, persisent declines in capital productivity and continuing shortfalls in agricultural output. Since grain shortages are compensated for by large-scale imports, they are a major drain on hard currency holdings as well.

It should be noted that planners have attempted to counteract the above constraints on economic growth within the framework of the existing system of economic planning and management. It remains to be seen whether the approach – or the development strategy adopted for the 1981-1985 Five-Year Plan – will be maintained under the present leadership. Furthermore, at the time of writing only sketchy information about the content and thrust of the 1986-1990 Five-Year Plan was available. The provisions of the 1985 annual plan are not necessarily a good indication of the likely directions of policy up to the end of the 1980s.

Turning first to the 1981-1985 Five-Year Plan, it appears that Soviet planners accepted the view that the decline in economic growth could be a long-term phenomenon, since the planned growth rate for NMP used (3.8 per cent, annual average) was set below what was actually achieved during the preceding plan period. This was also the case for the 1985 annual plan (see Tables 1 and 6 of Chapter 2). The draft 12th Five-Year Plan calls for an annual NMP growth rate of 3.5-4.1 per cent for 1986-1990. This would indicate that Soviet planners are now less pessimistic despite the downturn in overall economic growth which occurred in 1984.

For 1981-1985, priority was again given to shifting the economy on to an intensive growth path – an objective which the two preceding plans failed to achieve. Consequently, the average annual growth rate foreseen for investments (1.5 per cent, a rate which is about one-third below that for 1976-1980, see Table 4 of Chapter 2) was exceptionally low by past standards. For the first time investment growth targets were set below those planned for NMP. This, it was stressed, required an essentially new approach to investment allocation. The essence of this new approach lay in the intent to raise the efficiency with which investments are used and to reduce both the number of new projects started and the amount of unfinished construction. Furthermore, greater emphasis was placed on investing to change the output structure of production, to augment the manufacture of new products and to increase labour productivity. Thus capital investments were to be used primarily for renovation (modernisation) and for the technical re-equipment of existing enterprises. The 1985 annual plan maintained these priorities and preliminary indications for the 12th Five-Year Plan imply that they will characterise the investment policy for 1986-1990. Even greater emphasis is placed in the 12th Five-Year Plan on the need to use investment resources for modernising existing plants and upgrading technological equipment in order to improve the quality of output and to reduce inputs of energy, metals and labour.

As can be seen from Table 9 of Chapter 1, investment in 1981-1984 did, in fact, rise well above the Five-Year Plan targets for those years and a further increase was foreseen for 1985

(3.4 per cent in volume terms). According to the draft 12th Five-Year Plan, which sets investment growth at 3.4-4.1 per cent, annual average, this will also be the case during 1986-1990.

The increase in the volume of investments in 1981-1984 – despite a sharp deceleration in the latter year – already indicates that Soviet planners were unable, or unwilling, or simply did not need to cut back investments to the extent originally planned. Investment fell only marginally as a share of NMP produced, probably by much less than planned. Since NMP used by 1984 was slightly above the Five-Year Plan target, there remained some free resources for investment, over and above what was expected. This clearly shows the persistence of high investment demand and the leadership's desire to accommodate it when possible. It also reflects the fact that attempts to secure the intensification of the investment process have had little effect – the stock of unfinished construction continues to increase.

As to the sectoral allocation of investments, industry was given the first priority. According to the 1981-1985 Five-Year Plan, capital investment in industry was to grow by an average annual rate of 4.2 per cent. Altogether industry was to receive more than 40 per cent of the total volume of investments. The share set aside for the development of the agro-industrial complex was targeted at 27 per cent, while for the fuel-energy complex some 19 per cent of total investments was foreseen. The 1985 annual plan suggests that this general ordering of priorities will be maintained. The 1985 State Budget again sets aside a high share of resources for agriculture – which includes the agricultural sector proper, the "agro-industrial" complex and also expenditures under the Long-Term Programme of Land Reclamation which was adopted in October, 1984.

Within industry, priority was given to the machinery branch, notably to engineering. The growth rate foreseen by the 1981-1985 Five-Year Plan (7.0 per cent, annually) was greater than that planned for the industrial sector as a whole. (According to the draft 12th Five-Year Plan, which targets gross industrial output at 3.9-4.4 per cent annually and that of the machine-building sector at 6.9-7.7 per cent annually, this will again be the case.) This branch was supposed to furnish new and highly efficient machinery and equipment to all sectors of the economy. Presumably to counteract the structural problems which plague Soviet industrial production (obsolete production equipment, metal-intensiveness of machinery and equipment, etc.), particular attention was to be paid to automation and to design improvement. As to the commodity structure of output, priority was to be given to steam turbines, drilling equipment, electrical machines and motors, computers, trucks and specialised equipment for the chemical and petro-chemicals industry as well as for the agricultural sector. Current discussion in the Soviet literature suggests that the faster development and application of electronics, automation equipment and industrial robots is becoming a major pre-occupation – an aspect taken up in subsection 2 below.

As far as agriculture is concerned, its development is crucial since shortfalls in meeting output targets have a negative effect on the economy as a whole. The slow-down in NMP growth in 1984 was related to the poor harvest (which diverted a sizeable share of hard currency imports to grain) and to reduced supplies of raw materials of agricultural origin for industry. While gross agricultural output did not increase in 1984, a still higher target (6.7 per cent – see Table 7 of Chapter 2) was forecast in the 1985 annual plan (for the 1986-1990 period, output targets of – 2.7-3.0 per cent annually are given). Judging by the provisions contained in the 1985 annual plan, the importance attached by Mr. Brezhnev to developing the "Food Programme" (adopted in 1982), designed to provide the country with its domestic food needs and to reduce excessive reliance on grain imports, will be maintained. Ability to meet grain requirements for food and fodder had been singled out as being, both politically and economically, the central problem facing the Soviet Union for the coming decade[54]. Thus investment outlays for the agro-industrial complex will be increased by 7.0 per cent in order to boost the "Food Programme". The agricultural sector is to benefit from increased deliveries of technically superior machinery and equipment, fertilisers and improved supplies of plant protection agents. The "Land Reclamation Programme", in many respects an extension of the "Food Programme", will begin to be implemented in 1985. Its purpose is to increase the area of irrigated land by 50 per cent by the end of the next decade and to double current grain output.

It is interesting to note that the Food Programme had also singled out land improvement as a key component for intensifying production. The area of irrigated land was to increase between 23-25 million hectares by 1990 and drained land by 18-19 million hectares. According to the Land Reclamation Programme (at least the version adopted by former Secretary-General Chernenko), the extension of irrigated land will reach 30-32 million hectares by the year 2000 and drained land 19-21 million hectares. This would imply a considerable downscaling.

Particular stress was placed in the 1981-1985 Five-Year Plan on the need to develop the fuel and power sectors of the economy[55]. This included not only energy production, but energy conservation as well. This orientation also reflects the fact that oil and gas are the Soviet Union's primary export items – with the former alone representing some 60 per cent of total Soviet exports to the West. Stagnating oil production (an absolute decline in output occurred in 1984) and the greater costs associated with oil extraction, appear to have influenced the energy output targets given for the 1981-1985 period. Furthermore, the growth targets which have been set also reflect the intention to shift to alternate and potentially less costly (in terms of raw material inputs) sources of energy – notably, nuclear power.

The highest growth rate was foreseen for gas (6-8 per cent, annual averages) with output targeted to reach 600-640 billion cubic meters by 1985. Lower growth rates were planned for coal (roughly 2 per cent) and for crude oil (approximately 1 per cent) with a planned output by 1985 of 770-800 and 620-640 million tons respectively. The increased output foreseen for electrical power (1 555 to 1 600 billion kilowatt-hours), was to come primarily from the accelerated development of atomic and hydro-electric power stations in order to reduce the demand for organic fuels. Expanded exploitation of Siberian fields was to yield 60 per cent of total petroleum production and approximately 50 per cent of total gas output. The proportion of petroleum in fuel production was to drop.

Despite sizeable investment increases (compared to the 1976-1980 period) energy production rates are falling. It is also evident from the 1985 annual plan, that the above-mentioned growth targets will not be reached since the corresponding output targets were scaled down. It has been officially stated that oil extraction levels are not expected to rise beyond the present level. For 1985, gas output is expected to reach 632 billion cubic meters, coal 726 million tons, oil 628 million tons and electrical power 1 540 billion kWh. The draft 1986-1990 Five-Year Plan indicates that oil output is expected to remain at the level foreseen by the previous Five-Year Plan, while coal output is to reach 780-800 million tons by 1990, gas 835-850 billion cubic meters and electrical power 1 840-1 880 billion kWh.

In 1984, a Long-Term Energy Programme was adopted (to be fully implemented by the turn of the century) designed to alter radically the structure of energy output. Its core is the development of electric power engineering and stringent energy conservation. According to the programme, crude oil production will, where possible, be replaced by gas. Oil and organic fuels (and eventually gas) are to be entirely eliminated from electricity and heat production and replaced by nuclear energy, which is to become the country's main source of electric and thermal power. Consequently, nuclear power capacity is expected to reach between 65 000 to 70 000 MW by 1990 (current nuclear power capacity is said to be 18 000 MW). Coal extraction is to be expanded by developing open-cast mining and, in order to ensure high quality coal, its processing will be increased.

These are ambitious targets and the switch from oil to gas as well as the expansion of atomic energy will be costly. The development of coal basins will require large-scale investment in new and sophisticated equipment, implying not only a strain on financial resources but long lead times. The same is true for gas. In addition, the ability to increase production of gas is greatly dependent on Soviet capability to lay pipelines. The 1981-1985 Plan had already foreseen extremely high growth rates for synthetic resins and plastics, both of which are designed to insure increased production of thermal plastic pipe as well as to decrease dependence on imports of steel pipe. Given the ambitious plans for gas development, it can be assumed that the production of synthetic resins and plastics will also receive priority in the subsequent Five-Year Plan. As far as electric power generation is concerned, nuclear capacity is not progressing as planned and the development of hydro-electric capacities is also encountering technical and infra-structural difficulties.

As far as management of the economy is concerned, the Decree adopted in July, 1979 remains the most comprehensive attempt to improve the system of economic planning[56]. The measure was designed to increase the efficiency of management and planning only within the traditional and highly centralised system. It foresaw no change in the long-standing principle of subordinating enterprises to central planning and administrative authorities – limiting thereby possibilities of responding flexibly to market requirements. Rather, it sought to boost economic performance by changing and diversifying the structure of output, by reducing the cost of production and by a better allocation of financial, labour and capital resources, within the framework of centralised, but more efficient, planning.

In 1984, new measures were adopted (on an experimental basis and applicable to five industrial Ministries) designed to stimulate the quality and volume of production by broadening the involvement of enterprises in planning and management decisions. The basic goals of the experiment are to link the balance sheet of the enterprises directly with production responsibility, notably by reducing the scope of ministerial directives, granting enterprises the possibility of allocating their own resources and correlating wages directly with production results/performance. The measures also provide for payment of special bonuses for the successful development and application of advanced technologies and enterprises will have more independence in using the Development Fund in order to undertake technical re-equipment of production[57]. In January, 1985, a further 26 All-Union, Union and Republican Ministries were transferred to the system and all told some 2 500 enterprises are participating in the experiment. Some 11 per cent of industrial production – and more than 62 per cent of machine-building output – is being produced in enterprises participating in the economic experiment. Of the additional changes which were also introduced perhaps the most interesting, in terms of its potential impact on technology transfer, is that engineering Ministries may now retain a portion of their hard currency earnings.

Discussion in the Soviet literature suggests that to date the results of the experiment have been mixed. It is underlined that labour productivity has increased, quality of output improved and that contract deliveries are better fulfilled. Nevertheless, it is also pointed out that Ministries have not, in practice, relinquished control over the enterprises and have, therefore, limited the impact of the experiment. On-going ministerial interference in setting (and changing) plan indicators has reduced enterprise initiatives to undertake more demanding production targets. Continuing supply problems limit the use of the enterprise Development Fund with the result that S&T progress is insufficiently promoted.

It should be pointed out that in launching this experiment it was made very clear by the preceding leadership that the experiment was designed to find an optimal combination between central management and greater enterprise rights, but that the principle of central management of the economy was to remain intact. Criticisms continue to be voiced by Secretary-General Gorbachev concerning the efficiency of the present economic system, and there is an apparent concensus that further change is needed. Put simply, the question is whether to continue the attempt to render centralised planning more efficient by means of incremental changes – likely to result in a limited jump in factor productivity and enterprise performance – or to undertake more far-reaching systemic reforms which could lay the foundation for a longer-term acceleration in the pace of such changes. To the extent that incremental change can, and has improved economic performance (for instance in the German Democratic Republic in the early 1980s and in certain sectors of the Soviet economy – the oil industry being a case in point), it is likely that this option will prevail. Given the position of the Soviet Union within the Eastern bloc, successful implementation of this option could influence the future evolution of the Eastern systems, particularly if the reform path undertaken by Hungary fails to achieve its long-term objectives.

2. *Science and Technology*

A prime objective of current Soviet economic policy is to offset reduced labour and other factor inputs by improving capital productivity. This substitution is clearly linked to the accelerated development and application of science and technology. As noted in the 1981-1985 Five-Year Plan, "the development of science and technology must be subordinated, to an even greater extent, to the accomplishment of the economic and social tasks of Soviet society, the acceleration of the economy's transition to a path of intensive development and increasing the efficiency of social production".

The 1985 annual plan echoes this theme and links directly the achievement of the growth targets stipulated for NMP and industrial output to scientific and technical progress. The 1985 State Budget allocates 27.5 billion rubles to scientific research, a 3.6 per cent increase over 1984 (whether this sum includes military R&D expenditures is not known). In his presentation of the budget, Finance Minister Garbuzov also indicated that a common fund for the development of science and technology was being put at the disposal of ministries and departments, as well as an incentive fund of 5.3 billion rubles to encourage the creation and introduction of new equipment.

Thus S&T are directed, above all, towards achieving resource savings. Consequently, emphasis was, and continues to be placed on the need to develop more efficient machinery and equipment as well as less costly and material-intensive products – for instance, by accelerating the substitution of plastics for steel or nuclear energy and synthetic fuels for organic fuels. Technologies conducive to developing alternate sources of energy were to be brought on stream, with priority given to the development of fast-neutron reactors and magneto-hydro-dynamic methods of transforming energy.

Productivity was also to be increased, by the widespread use of automation for managing technological processes: computers, micro-processors and industrial robots. According to the official literature, "some 30 000 industrial robots – eight times more than during 1976-1980" were to be produced over the 1981-1985 period[58]. It was anticipated that by 1985, some 40 000 robots would be available.

Investment costs were to be reduced by introducing technologically superior machinery and equipment, re-equipping industrial branches and by introducing advanced techniques – particularly in the metal-processing industries. During 1981-1985, some "19 000 new types of machines, equipment, instruments and materials were to be introduced, improving thereby the parameters of production output and making it possible to introduce essentially new manufacturing processes"[59]. According to the 1985 annual plan, an output of 4 200 new types of machinery, equipment, instruments and materials was foreseen by the end of 1985.

In 1984, a new division for informatics, computing and automation was added to the USSR Academy of Sciences, underscoring the current emphasis placed on automating production processes. However, even if the above-mentioned targets are reached, improvements in industrial structures and efficiency could be jeopardised by conservatism on the part of enterprise management. The incorporation of new technologies, sophisticated machinery and industrial robots often involve organisational changes and requirements for more sophisticated inputs which have adverse effects on the attainment of enterprise plan targets. As a result, available supplies of new technology are not fully used.

In order to improve co-ordination of S&T with economic development, "comprehensive programmes" and "specific-purpose programmes" were incorporated into the 1981-1985 Five-Year Plan. The former are basically designed to improve management of the economy, in the broadest sense of that term, and are slanted toward basic research. The USSR Academy of Sciences has been given foremost responsibility in the development and execution of these programmes. This decision reflects the stress that is being placed on "big science" for the solution of key economic problems and on the perceived need for a more active participation of the Academy in improving the entire system of scientific research. The specific purpose programmes are goal-oriented research-development-innovation programmes. They are designed to help forge closer links between R&D and application. Some 170 scientific-technical programmes, including 41 complex target programmes – all with the aim of introducing the most advanced S&T achievements into the economy – became an integral part of the 1981-1985 Five-Year Plan. According to the Decree "On Measures to Accelerate Scientific-Technical Progress in the National Economy" issued in 1983, goal oriented S&T planning will also be permanently incorporated into subsequent Five-Year Plans.

The 1981-1985 Plan had also focused on the need to improve the system of evaluating the technical and economic level of products and to shorten substantially the lead time between research and assimilation. Thus it prescribed a series of other organisational measures designed to further closer and more productive co-operation among all entities involved in research and development – inclusive of enterprise institutes.

None of the goals assigned to science and technology is new. Nor have the various reforms undertaken since 1968 brought about significant improvements in the research-development-innovation process – whether by breaking down institutional barriers, creating an effective incentive system to further innovation, or in speeding up lead times between research and application. These problems have led, among others, to technological lags in crucial sectors of the economy and to large-scale import of Western technology. It has also led to participation in a variety of agreements with Western firms designed to provide access to know-how and to accelerate scientific and technological progress. The intent to pursue scientific and technical co-operation with Western countries can be clearly inferred from the 1981-1985 Five-Year Plan. Furthermore, Soviet commentators make frequent reference to the desirability of active participation in foreign trade in order to gain access to "scientific and technological break-throughs and inventions". There is, to date, no evidence which would indicate a change from this perspective – despite the current stress placed on furthering intra-CMEA co-operation in order to curtail imports of Western technology.

### 3. *Trade*

In the 1981-1985 Five-Year Plan, the section devoted to trade is brief and imprecise. No quantitative targets for exports and imports were given in the Plan itself, nor was precise information provided as to the commodity structure of foreign trade for the 1981-1985 period. It was emphasised that foreign trade was to contribute to economic development by raising the quality and efficiency of domestic production. The structure of exports was to be adjusted by expanding exports of machinery and manufactures. Imports were to accelerate scientific and technological progress and to meet domestic needs for advanced equipment, latest production processes and raw materials more fully.

As far as trade with the other Eastern countries is concerned, the major thrust of the Soviet Union's policy will be to improve the distribution and balance of trade. By 1984, trade with the Soviet Union accounted for a higher share of the East European countries' intra-area trade turnover (imports plus exports) than in 1975 or in 1980 – some 64 per cent compared with 57 per cent and 60 per cent respectively. This was some 36 per cent of their total trade turnover with the world in 1984. Since trade with other East European countries accounted for another 20 per cent, a total of 57 per cent of Eastern European trade turnover, was conducted within the Eastern area. Conversely, Eastern Europe accounted for some 45 per cent of total Soviet trade turnover (Table 1 of Chapter 1). Thus the Soviet Union was (and continues to be) less dependent on Eastern Europe than was (and is) Eastern Europe on intra-area trade.

From the East European side, there has been increasing concentration of both exports to and imports from the Soviet Union since 1975, relative to trade within Eastern Europe. This was particularly marked between 1978 and 1981, though it slowed down thereafter. The movement is well illustrated by the fact that the Soviet Union accounted for about the same percentage share of Eastern Europe's total exports to the world in 1975 and in 1984 (about 34 per cent), whereas the Soviet share in Eastern Europe's total imports from the world rose from 32 to 38 per cent between the same two years.

The results of these movements are evident from the Soviet Union's balance of trade position vis-à-vis Eastern Europe (Table 2 of Chapter 1). This asymetrical development of imports and exports within the most important intra-area flow resulted in a sizeable Soviet trade surplus, which, as already noted, reached a high point of some 3 billion rubles in 1981. Since that year, the East European countries' exports to the Soviet Union have grown faster than the corresponding import flow and the deficits have been reduced. Their reduction has, however, been hampered by a continuing deterioration in the East European terms of trade with the Soviet Union. In the case of Bulgaria, Czechoslovakia, Poland and Romania, deficits with the Soviet Union were much higher in 1981-1984 than in 1976-1980 (Table 6 of Chapter 1).

The effort made by the Six to reduce their trade deficits with the Soviet Union in recent years from the highpoint reached in 1981, is, however, notable. As is described in Chapter 4, a continuation of this process will entail primarily increasing the volume and quality of East European machinery and equipment exports, as well as provision of food products, to the Soviet Union.

The commodity structure of intra-CMEA trade is also a factor in this. It is well known that the Soviet Union imports primarily machinery and equipment from its Eastern partners. Its primary exports are fuels, the price of which has risen continuously since the mid-1970s. Although oil prices on world markets have fallen in recent years, the pricing formula in effect in the CMEA area means that these declines have not yet benefited the Six. This contributed to the post-1980 deterioration of these countries' terms of trade with the Soviet Union. It has thus increased pressures on Eastern Europe to expand the volume of exports to the Soviet Union.

Given the economic problems facing the East European countries, it is evident that the USSR would not, in the interest of preserving political and economic stability, squeeze its partners too hard. However, balance of payments considerations, due to the inability of the Six to maintain their exports at a level sufficient to pay for their fuel and energy imports from the Soviet Union, explain the Soviet Union's emphasis on the need to strengthen intra-CMEA economic co-operation and the particular stress placed on increasing co-production and sectoral specialisation. The attempt to increase intra-regional co-operation is long-standing – and to date, not very effective. However, the current emphasis on making co-operation more selective and focusing it increasingly on key areas is new. This effort could, if successful, have significant long-term implications for East-West trade.

As far as Soviet trade with the West is concerned, it is evident from the formulation of the 1981-1985 Five-Year Plan and from the development of Soviet trade with the West, that such trade is seen as an important input of Soviet economic development: to relieve bottlenecks, improve factor productivity and to develop industry. The share of the Western countries in Soviet trade increased steadily from 1970 to 1982: rising from 21 per cent to 32 per cent of the USSR'S total trade turnover. While this share has declined since, it was still 29 per cent in 1984 (see Table 7 of Chapter 1). Apart from grain, capital goods (machinery and equipment) are the Soviet Union's major import items. While such Soviet imports slowed since 1976, they remain predominant, especially in imports from the West. Furthermore, imports of capital goods again increased rapidly in 1982 and 1983 – notably, but not exclusively, equipment for oil and gas transportation (see Table 6 of Chapter 5).

As far as the evolution of the Soviet Union's trade with the Western countries is concerned, OECD trade data indicate that the Soviet Union has achieved surpluses since 1981 – although these declined until 1983. In 1984, this surplus again increased (to $2.1 billion). The data also indicate that imports have continued to decline and, while exports increased in 1984, they grew at a lower rate than was the case for the majority of the Six[60].

The Soviet Union's net debt decreased between 1983 and 1984, falling to $11.8 billion in the latter year (though still the second highest level in the region), and debt servicing is low – equivalent to 16 per cent of hard currency earnings (see Tables 8 and 9 of Chapter 1). While the Soviet Union's credits are large in absolute terms, its debt servicing capacity is good and the Soviet Union is still considered as a "good" credit risk. Furthermore, the USSR is under less financial pressure than the East European countries since it can supplement its hard currency earnings by gold sales, if need be.

However, Soviet ability to generate hard currency will become more difficult in the near future because of the commodity structure of its exports. Eighty per cent of its exports to the OECD are fuel. As mentioned previously, oil accounts for some 60 per cent of hard currency earnings. While Soviet oil deliveries reached a peak of 65 million tons in 1982, Soviet officials have, as noted above, clearly stated that the upper limit of oil output has been reached. Global oil exports are likely to, at best stagnate, and supplies to Eastern Europe have even been reduced. Thus, although the Soviet Union has increased the share of gas in its energy exports, this does not, necessarily, assure increased hard currency earnings from fuel as a whole – in the long term.

The ability of the Soviet Union to sustain its trade involvement with the West is, therefore, dependent on its long-run capability to diversify its export pattern. This is an important aspect of the emphasis placed on changing and improving its industrial output structure. In view of the evident difficulties which the Soviet Union has encountered in attaining this goal, it can be assumed that a vigorous attempt will be made to pursue forms of co-operation which ensure increased trade volumes. As noted in the 1981-1985 Five-Year Plan, the "Soviet Union will continue to implement existing agreements and will seek new agreements [in order] to carry out large-scale projects in the fuel, metallurgical and chemical industries as well as other branches of the economy. New paths for developing co-operation with the developed countries will be sought".

It is clear from the Soviet literature, that compensation agreements are seen as the most advantageous form of co-operation. To date, compensation trade has been predominantly resorted to for fuels and chemicals. For methonal alone, a doubling of exports, on the basis of past compensation, is foreseen for the 1980s[61]. In future, compensation trade is also to be pursued in a wider variety of sectors, particularly engineering. Industrial co-operation is also to be expanded. Emphasis is placed on turnkey-type arrangements and on co-production and specialisation agreements.

Table 1 (Chapter 2)

CHANGES IN GROWTH NET MATERIAL PRODUCT
(average annual percentage change) (d)

| | 1961-1965(c) | 1966-1970 | 1971-1975 (Plan) | 1971-1975 | 1976-1980 (Plan) | 1976-1980 | 1981-1985 (Plan) |
|---|---|---|---|---|---|---|---|
| Bulgaria | 6.7 | 8.7 | 7.7-8.5 | 7.8 | 7.7 | 6.1 | 3.7 |
| Czechoslovakia | 1.9 | 6.9 | 5.1 | 5.7 | 5.0 | 3.7 | 1.6-2.2 |
| German Democratic Republic | 3.5 | 5.2 | 4.9 | 5.4 | 5.0 | 4.2 | 5.1 |
| Hungary | 4.1 | 6.8 | 5.5-6.0 | 6.2 | 5.4-5.7 | 3.2 | 2.7-3.2 |
| Poland | 6.2 | 6.0 | 6.7-6.8 | 9.8 | 7.0-7.3 | 1.2 | 3.5-5.6 (b) |
| Romania | 9.1 | 7.7 | 11-12 | 11.3 | 11.0 | 7.3 | 7.5 |
| Soviet Union | 6.5 | 7.8 | 6.5-7.0 (a) | 5.7 | 4.7 | 4.3 | 3.8* |
| Eastern Europe | 4.9 | 6.5 | 6.7 | 7.8 | 7.0 | 3.9 | 3.3 |
| Soviet Union and Eastern Europe | 6.0 | 7.4 | 6.7 | 6.2 | 5.3 | 4.2 | 3.7* |

Notes:

* Estimate (in all the tables of this chapter).
a) Distributed NMP.
b) 1983-1985.
c) Figures taken from the Economic Survey of Europe in 1971 are only original data and may, therefore, differ from data in other ECE publications. For a detailed description of the composite index see Economic Survey of Europe in 1969, Part I, United Nations, Economic Commission for Europe, New York, 1970.
d) Annual percentage changes: unless otherwise stated, average annual percentage changes over five-year periods are average annual percentage changes between final year of period shown and final year of the previous five-year period. This applies to all the tables in the chapter.

Sources:
1961-1965 -- Economic Survey of Europe in 1971, Part II, The European Economy in 1971, United Nations, Economic Commission for Europe, New York, 1972, p. 108.
1966-1970 -- Economic Survey of Europe in 1976, Part II, The Five-Year Plans for 1976-1980 in Eastern Europe and the Soviet Union, United Nations, Economic Commission for Europe, New York, 1977, p. 8.
1971-1975 (Plan & Actual) -- Economic Survey of Europe in 1981, United Nations, Economic Commission for Europe, New York, 1982, p. 133.
1976-1980 (Plan & Actual) -- Economic Survey of Europe in 1983, United Nations, Economic Commission for Europe, New York, 1984, Table 3.1.1.
1981-1985 (Plan) -- Economic Survey of Europe in 1984-1985, United Nations, Economic Commission for Europe, New York, 1985, Table 4.1.1.

Table 2 (Chapter 2)

GROSS INDUSTRIAL PRODUCTION
(average annual percentage change)

| | 1961-1965 | 1966-1970 | 1971-1975 (Plan) | 1971-1975 | 1976-1980 (Plan) | 1976-1980 | 1981-1985 (Plan) |
|---|---|---|---|---|---|---|---|
| Bulgaria | 11.7 | 10.9 | 9.2-9.9 | 9.0 | 9.2 | 6.0 | 5.1 |
| Czechoslovakia | 5.2 | 6.8 | 6.0 | 6.6 | 5.7-6.0 | 4.7 | 2.7-3.4 |
| German Democratic Republic | 5.9 | 6.5 | 6.0 | 6.5 | 6.0 | 5.0 | 5.1 |
| Hungary | 8.0 | 6.2 | 5.7-6.0 | 6.4 | 6.0 | 3.4 | 3.5-4.0 |
| Poland | 8.5 | 8.3 | 8.2-8.5 | 10.4 | 8.2-8.5 | 4.7 | 3.8-5.4(a) |
| Romania | 13.7 | 11.9 | 11.0-12.0 | 12.9 | 10.2-11.2 | 9.5 | 7.6 |
| Soviet Union | 8.6 | 8.5 | 8.0 | 7.4 | 6.3 | 4.5 | 4.7 |
| Eastern Europe | 7.6 | 8.1 | 7.7 | | 6.4 | 5.6 | 3.7-3.8 |
| Soviet Union and Eastern Europe | 8.3 | 8.4 | 7.9 | 7.8 | 6.3 | 4.8 | 4.4 |

Note: a) 1983-1985.

Sources:

| | |
|---|---|
| 1961-1965 | Economic Survey of Europe in 1971, Part II, op. cit., p. 121. |
| 1966-1970 | Economic Survey of Europe in 1976, Part II, op. cit., p. 14. |
| 1971-1975 (Plan) | Economic Survey of Europe in 1976, Part II, op. cit., p. 71. |
| 1971-1975 | Economic Survey of Europe in 1981, op. cit., p. 215. |
| 1976-1980 (Plan) | Economic Survey of Europe in 1983, op. cit., Table 3.1.1. |
| 1976-1980 (Actual) 1981-1985 (Plan) | Economic Survey of Europe in 1984-1985, op.cit., Table 4.3.1. |

Table 3 (Chapter 2)

GROSS OUTPUT IN AGRICULTURE AND CONSTRUCTION
(average annual percentage change)

| | Agriculture | Construction |
|---|---|---|
| **Bulgaria** | | |
| 1961-1965 | 4.3 | 7.4 |
| 1966-1970 | 4.8 | 12.0 |
| 1971-1975 Plan | 3.2-3.7 | .. |
| 1971-1975 | 2.2 | 5.8 |
| 1976-1980 Plan | 3.7 | .. |
| 1976-1980 | 2.1 | 5.9 |
| 1981-1985 Plan | 3.4 | .. |
| **Czechoslovakia** | | |
| 1961-1965 | 0.4 | 2.7 |
| 1966-1970 | 3.5 | 7.9 |
| 1971-1975 Plan | 2.5* | 6.7 |
| 1971-1975 | 2.9 | 8.2 |
| 1976-1980 Plan | 2.7-2.8 | 6.2-6.5 |
| 1976-1980 | 1.5 | 3.0 |
| 1981-1985 Plan | 1.4-1.9 | .. |
| **German Democratic Republic** | | |
| 1961-1965 | 1.8* | 5.3 |
| 1966-1970 | 1.9 | 8.3 |
| 1971-1975 Plan | 2.4 | 4.9 |
| 1971-1975 | 2.1 | 5.1 |
| 1976-1980 Plan | 2.6 | 5.0 |
| 1976-1980 | 1.2 | 3.3 |
| 1981-1985 Plan | 1.1 | .. |
| **Hungary** | | |
| 1961-1965 | 1.6 | 4.6 |
| 1966-1970 | 2.8 | 10.7 |
| 1971-1975 Plan | 2.8-3.0 | 7.3 |
| 1971-1975 | 3.4 | 5.4 |
| 1976-1980 Plan | 3.2-3.4 | 5.5-6.0 |
| 1976-1980 | 2.9 | 2.4 |
| 1981-1985 Plan | 2.3-2.8 | .. |
| **Poland** | | |
| 1961-1965 | 2.6 | 6.8 |
| 1966-1970 | 2.9 | 8.6 |
| 1971-1975 Plan | 3.4-3.9 | .. |
| 1971-1975 | 3.1 | 13.9 |
| 1976-1980 Plan | 3.0-3.5 | 7.4-7.7 |
| 1976-1980 | 0.6 | -0.5 |
| 1983-1985 Plan | 2.2* (a) | .. |

Table 3 (Chapter 2) contd

| | Agriculture | Construction |
|---|---|---|
| Romania | | |
| 1961-1965 | 3.2 | 12.5 |
| 1966-1970 | 4.2 | 11.5 |
| 1971-1975 Plan | 6.3-8.3 | .. |
| 1971-1975 | 4.6 | 6.9 |
| 1976-1980 Plan | 5.1-7.6 | .. |
| 1976-1980 | 4.8 | 5.8 |
| 1981-1985 Plan | 4.5-5.0 | .. |
| Soviet Union | | |
| 1961-1965 | 2.3 | 5.5 |
| 1966-1970 | 3.9 | 7.4 |
| 1971-1975 Plan | 3.9 | ca.7 |
| 1971-1975 | 2.5 | 6.0 |
| 1976-1980 Plan | 3.0 | .. |
| 1976-1980 | 1.7 | 2.6 |
| 1981-1985 Plan | 2.5 | .. |

Notes:

.. Not available.

a) Presumably 1983-1985.

1961-1965 Agriculture: Annual percentage change within five years over the previous five years.
1966-1970 Agriculture: For five-year periods average annual percentage changes calculated on the basis of previous five-year period total.

Sources:

| | |
|---|---|
| 1961-1965 | Economic Survey of Europe in 1971, Part II, op. cit., p. 108. |
| 1966-1970 | Economic Survey of Europe in 1976, Part II, op. cit., p. 14. |
| 1971-1975 (Plan) | Economic Survey of Europe in 1971, Part II, op. cit., p. 108. |
| 1971-1975 (Actual) | Economic Survey of Europe in 1976, Part II, op. cit., p. 14. |
| 1976-1980 (Plan) | Economic Survey of Europe in 1983, op. cit., Table 3.1.1. |
| 1976-1980 (Actual) 1981-1985 (Plan) | Ibid., and Economic Survey of Europe in 1984-1985, op. cit., Table 4.2.1 (agriculture only). |

Table 4 (Chapter 2)

GROSS INVESTMENT RATIO (IR)
(percentage of NMP)

| | 1961-1965 | 1966-1970 | 1971-1975 (Plan) | 1971-1975 | 1976-1980 (Plan) | 1976-1980 (e) | 1981-1985 (Plan) (e) |
|---|---|---|---|---|---|---|---|
| Bulgaria | 29.7 | 35.1 | 33.7 | 36.6 (a) | 35.8 (a) | 35.4 | 31.2 |
| Czechoslovakia | 29.8 | 31.1 | 32.0 | 31.7 (b) | 34.0 (b) | 33.7 | 29.4 |
| German Democratic Republic | 22.8 | 27.6 | 28.1 | 30.1 (c) | 29.3 (c) | 30.7 | 25.8 |
| Hungary | 26.4 (1) | 32.9 | 29.3 | 35.3 (d) | 33.6 (d) | 37.2 | 33.2 |
| Poland | 27.3 | 25.3 | 31.3 | 34.3 (b) | 31.5 (b) | 30.7 | .. |
| Romania | 29.5 | 28.8 | 34.1 | 37.9 (b) | 40.9 (b) | 41.1 | 39.9 |
| Soviet Union | 29.1 | 27.0 | 28.9 | 29.1 (d) | 28.8 (d) | 30.3 | 28.4 |

GROSS INCREMENTAL CAPITAL-OUTPUT RATIO (ICOR)
(co-efficient)

| | 1961-1965 | 1966-1970 | 1971-1975 (Plan) | 1971-1975 | 1976-1980 (Plan) | 1976-1980 | 1981-1985 (Plan) |
|---|---|---|---|---|---|---|---|
| Bulgaria | 4.43 | 4.0 | 4.16 | 4.8 | 4.6 | 5.7 | 8.4 |
| Czechoslovakia | 15.68 | 4.5 | 6.27 | 5.7 | 6.9 | 8.6 | 12.9 |
| German Democratic Republic | 6.51 | 5.3 | 5.73 | 5.7 | 5.9 | 7.5 | 5.2 |
| Hungary | 6.44 | 4.8 | 5.05 | 5.7 | 6.0 | 9.9 | 11.4 |
| Poland | 4.40 | 4.2 | 4.67 | 2.9 | 4.4 | 10.1 | .. |
| Romania | 3.24 | 3.7 | 2.97 | 3.3 | 3.7 | 4.9 | 5.6 |
| Soviet Union | 4.48 | 3.5 | 5.2 | 5.7 | 6.0 | 6.4 | 8.8 |

Table 4 (Chapter 2) contd

GROSS FIXED CAPITAL FORMATION
(annual percentage change)

| | 1961-1965(3) | 1966-1970 | 1971-1975(3) (Plan) | 1971-1975 | 1976-1980 (Plan) | 1976-1980 | 1981-1985 (Plan) |
|---|---|---|---|---|---|---|---|
| Bulgaria | 14.2 | 12.5 | 6.4 | 6.7 | 4.7 | 4.1 | 0.9 |
| Czechoslovakia | 5.7 | 7.2 | 6.2 | 7.8 | 5.4 | 4.1 | -1.7 |
| German Democratic Republic | 7.0 | 9.9 | 5.2 | 3.8 | 4.2 | 4.7 | 0.7 |
| Hungary | 9.2 | 11.6 (2) | 5.6 | 5.6 | 2.1 | 3.8 | - |
| Poland | 8.2 | 8.1 | 7.3 | 17.1 | 0.2 | -0.4 | .. |
| Romania | 14.6 | 11.3 | 10.4 | 10.7 | 12.9 | 9.7 | 4.4 |
| Soviet Union | 7.7 | 7.5 | 7.2 | 6.8 | 3.2 | 3.9 | 1.6 |

Notes:

- \- No growth.
- .. Not available.
- 1. Excluding private investment.
- 2. At constant prices for the whole economy.
- 3. Annual percentage change within five years over the previous five years.

Based on constant prices:

a) 1980 prices. Series for 1970-1980 are calculated from index numbers rebased on current 1980 prices.

b) 1977 prices.

c) 1975 prices. Including investments in other socialist countries amounting, in million marks, to 1 374 annually during 1976-1980, 1 080 in 1979 and 1 325 in 1980.

d) 1976 prices. Socialist sector only. Figures for Plan 1981-1985 are based on published figures in current prices deflated by an estimated 4 per cent annual price rise.

e) For the 1976-1980 period (Actual) and 1981-1985 Plan the price base used is:

Bulgaria: 1980 prices;
Czechoslovakia: 1977 prices;
German Democratic Republic: 1980 prices;
Hungary: 1981 prices;
Poland: 1982 prices;
Romania: 1981 prices;
Soviet Union: 1976 assumed prices.

Sources :

| | |
|---|---|
| 1961-1965 | Economic Survey of Europe in 1971, Part II, op. cit., pp. 108 and 116. It is noted in the Source cited that the price base for fixed capital formation (and NMP) is:<br>Bulgaria - current prices; Czechoslovakia - 1967 prices; German Democratic Republic - 1967 prices; Hungary - 1959 prices; Poland - 1961 prices; Romania - 1963 prices; Soviet Union - 1969 prices for fixed investment, 1965 prices for NMP. |
| 1966-1970 | Economic Survey of Europe in 1976, Part II, op. cit., pp. 10 and 94. It is noted in the Source cited that all value data are at current prices for Bulgaria and Hungary and at constant prices for all other countries; all data for Hungary are for the socialist sector; all figures in value terms relating to five-year periods are annual averages; all planned annual rates of growth are calculated on the basis of the previous five-year period total. |
| 1971-1975 (Plan) | Economic Survey of Europe in 1971, Part II, op. cit., pp. 108 and 116. (See Note for 1961-1965 above). |
| 1971-1975 | Economic Survey of Europe in 1981, op. cit., pp. 138 and 239-240. It is noted in the Source cited that where ranges were given for annual or five-year plan targets, investment figures (and NMP growth rates used to calculate ICORs) are mid-points of the range. Geometric averages were used to compute average levels and percentage changes over five-year periods. The latter are based on the final year of the preceding five-year period. Five-year plan objectives are based on secretariat [ECE] estimates of the year-by-year evolution of investment and NMP objectives where these were not shown in the plan, on the assumption of constant annual percentage growth rates. |
| 1976-1980 (Plan) | Economic Survey of Europe in 1981, op. cit., Table 3.4.1. |
| 1976-1980) (Actual) 1981-1985) (Plan) | Economic Survey of Europe in 1984-1985, op. cit., Tables 4.4.1 and 4.4.2. |

Table 5 (Chapter 2)

CHANGES IN GROSS OUTPUT AND FACTOR INPUTS IN INDUSTRY
(average annual percentage change)

| | Gross Output | Employment | Labour Productivity | Capital Productivity | Fixed Assets |
|---|---|---|---|---|---|
| <u>Bulgaria</u> | | | | | |
| 1961-1965 | 11.7 | 4.4 | 7.0 | .. | .. |
| 1966-1970 | 10.9 | 4.0 | 6.6 | -2.5 | 13.6 |
| 1971-1975 Plan | 9.2-9.9 | 1.5-2.1 | 7.6 | -1.4 | 11.2* |
| 1971-1975 | 8.9 | 2.7 | 6.1 | -0.6 | 9.4 |
| 1976-1980 Plan | 9.2 | .. | 8.5 | 0.7 | .. |
| 1976-1980 | 6.0 | 1.5 | 4.6 | -1.8 | 7.9 |
| 1981-1985 Plan | 5.1 | .. | .. | .. | .. |
| <u>Czechoslovakia</u> | | | | | |
| 1961-1965 | 5.2 | 1.9 | 3.2 | .. | .. |
| 1966-1970 | 6.8 | 1.3 | 5.4 | 2.3 | 4.4 |
| 1971-1975 Plan | 6.0 | 0.7 | 5.3 | 0.8 | 5.2* |
| 1971-1975 | 6.7 | 0.5 | 6.1 | 1.1 | 5.5 |
| 1976-1980 Plan | 5.7-6.0 | .. | 5.3 | -1.0 | .. |
| 1976-1980 | 4.7 | 0.7 | 4.2 | -1.3 | 6.1 |
| 1981-1985 Plan | 2.7-3.4 | 0.2. | .. | .. | .. |
| <u>German Democratic Republic</u> | | | | | |
| 1961-1965 | 5.9 | 0.2 | 5.7 | .. | .. |
| 1966-1970 | 6.5 | 0.4 | 6.0 | 1.3 | 5.2 |
| 1971-1975 Plan | 6.0 | -0.2 | 6.2 | 1.2 | 4.9* |
| 1971-1975 | 6.3 | 0.3 | 6.0 | -0.2 | 6.5 |
| 1976-1980 Plan | 6.0 | .. | 5.5 | 0.5 | .. |
| 1976-1980 | 5.0 | 0.5 | 4.3 | -0.5 | 5.5 |
| 1981-1985 Plan | 5.1 | .. | .. | .. | .. |

Table 5 (Chapter 2) contd

| | Gross Output | Employment | Labour Productivity | Capital Productivity | Fixed Assets |
|---|---|---|---|---|---|
| Hungary | | | | | |
| 1961-1965 | 8.0 | 2.8 | 5.1 | .. | .. |
| 1966-1970 | 6.2 | 2.7 | 3.4 | -0.7 | 7.0 |
| 1971-1975 Plan | 5.7-6.0 | 1.5 | 4.1-4.4 | -0.2 | 6.1* |
| 1971-1975 | 6.5 | 0.2 | 6.3 | -1.6 | 8.1 |
| 1976-1980 Plan | 6.0 | .. | 6.0 | -2.5 | .. |
| 1976-1980 | 3.4 | -1.6 | 5.1 | -3.8 | 7.5 |
| 1981-1985 Plan | 3.5-4.0 | .. | .. | .. | .. |
| Poland | | | | | |
| 1961-1965 | 8.5 | 3.2 | 5.1 | .. | .. |
| 1966-1970 | 8.3 | 3.3 | 4.9 | 0.2 | 8.0 |
| 1971-1975 Plan | 8.2-8.5 | 2.5 | 5.6-5.9 | 0.2 | 8.2 |
| 1971-1975 | 10.6 | 2.8 | 7.5 | 0.6 | 9.7 |
| 1976-1980 Plan | 8.2-8.5 | .. | 7.4-7.8 | -3.0 | .. |
| 1976-1980 | 4.7 | 0.4 | 4.5 | -3.5 | 8.5 |
| 1981-1985 Plan | 3.8-5.4 (a) | .. | .. | .. | .. |
| Romania | | | | | |
| 1961-1965 | 13.7 | 6.0 | 7.3 | .. | .. |
| 1966-1970 | 11.9 | 4.3 | 7.4 | -1.2 | 13.3 |
| 1971-1975 Plan | 11.0-12.0 | 3.5-4.4 | 7.3 | .. | .. |
| 1971-1975 | 13.2 | 6.5 | 6.4 | 0.7 | 12.3 |
| 1976-1980 Plan | 10.2-11.2 | 1.6-2.1 | 8.5-9.0 | 3.0 | .. |
| 1976-1980 | 9.5 | 3.5 | 5.8 | -0.6 | 10.2 |
| 1981-1985 Plan | 7.6 | 0.6 | .. | .. | .. |

Table 5 (Chapter 2) contd

| | Gross Output | Employment | Labour Productivity | Capital Productivity | Fixed Assets |
|---|---|---|---|---|---|
| Soviet Union | | | | | |
| 1961-1965 | 8.6 | 3.9 | 4.5 | | |
| 1966-1970 | 8.5 | 2.9 | 5.5 | -0 | |
| 1971-1975 Plan | 8.0 | 1.1 | 6.8 | -1.0 | 8.3 |
| 1971-1975 | 7.4 | 1.5 | 5.9 | -1.1 | 8.6 |
| 1976-1980 Plan | 6.3 | 0.8 | 5.5 | -0.6 | .. |
| 1976-1980 | 4.5 | 1.6 | 2.8 | -2.8 | 7.5 |
| 1981-1985 Plan | 4.7 | 0.5 | .. | .. | .. |

Notes:

.. Not available.
a) 1983-1985.

Sources:

1961-1965 Economic Survey of Europe in 1971, Part II, op. cit., p. 21.

1966-1970 Economic Survey of Europe in 1976, Part II, op. cit., p. 71.

1971-1975 (Plan) Ibid.

1971-1975 (Actual) Economic Survey of Europe in 1980, op. cit., p. 119.

1976-1980 (Plan) Economic Survey of Europe in 1980, op. cit., p. 119, Economic Survey of Europe in 1983, op. cit., Table 3.3.1 and Economic Survey of Europe in 1984-1985, op. cit., Tables 4.3.2 and 4.3.6. (N.B. 1976-1980 Plan for capital productivity: various Eastern sources).

1976-1980) (Actual) Economic Survey of Europe in 1983, op. cit., Tables 3.3.1 and 3.3.2.

1981-1985) (Plan) Economic Survey of Europe in 1983, op. cit., Table 3.3.1 and Economic Survey of Europe in 1984-1985, op. cit., Table 4.3.1.

Table 6 (Chapter 2)

CHANGES IN GROWTH IN NET MATERIAL PRODUCT AND GROSS INDUSTRIAL PRODUCTION
(average annual percentage change)

NMP

| | 1980 | 1981 (Plan) | 1981 | 1982 (Plan) | 1982 | 1983 (Plan) | 1983 | 1984 (Plan) | 1984 | 1985 (Plan) |
|---|---|---|---|---|---|---|---|---|---|---|
| Bulgaria | 5.7 | 5.1 | 5.0 | 3.6 | 4.2 | 3.8 | 3.0 | 3.8 | 4.6* | 4.1 |
| Czechoslovakia | 2.9 | 2.5 | 0.1 | -0.2 | 0.2 | 1.6 | 2.3 | 2.8 | 3.0 | 3.0 |
| German Dem. Rep. | 4.4 | 5.0 | 4.8 | 4.8 | 2.6 | 4.2 | 4.4 | 4.4 | 5.5 | 4.4 |
| Hungary | -0.8 | 2.0-2.5 | 2.5 | 1.0-1.5 | 2.6 | 0.5-1.0 | 0.3 | 1.5-2.0 | 2.8-3.0 | 2.3-2.8 |
| Poland | -6.0 | -3.7 | -12.0 | -1.6* | -5.5 | 2.0-2.5 | 6.0 | 2.6 | 5.1 | 3.0-3.5 |
| Romania | 2.9 | 7.0 | 2.2 | 5.5 | 2.7 | 5.0 | 3.7 | 7.3 | 7.7 | 10.0 |
| Soviet Union | 3.9 | 3.4 | 3.3 | 3.0 | 3.9 | 3.3 | 4.2 | 3.1 | 3.0* | 4.0* |
| Eastern Europe | 0.1 | 2.8 | -1.9 | 2.4 | 0.1 | 3.2 | 3.9 | 3.9 | 5.1 | 4.8 |
| Soviet Union and Eastern Europe | 2.7 | 3.2 | 1.7 | 2.8 | 2.8 | 3.3 | 4.1 | 3.7 | 3.6 | 4.2 |

Industrial Production

| | 1980 | 1981 (Plan) | 1981 | 1982 (Plan) | 1982 | 1983 (Plan) | 1983 | 1984 (Plan) | 1984 | 1985 (Plan) |
|---|---|---|---|---|---|---|---|---|---|---|
| Bulgaria | 4.2 | 5.6 | 4.9 | 4.5 | 4.6 | 4.8 | 3.9 | 5.0 | 4.6 | 5.2 |
| Czechoslovakia | 3.5 | 2.7 | 2.1 | 0.6* | 1.1 | 2.4 | 2.8 | 2.9 | 3.9 | 3.0 |
| German Dem. Rep. | 4.7 | 5.0 | 4.7 | 4.6 | 3.1 | 3.8 | 4.1 | 3.6 | 4.2 | 3.8 |
| Hungary | -2.2 | 3.0-3.5 | 2.4 | 2.0-2.5 | 2.5 | 1.0-2.0 | 1.4 | 1.5-2.0 | 3.0 | 3.0 |
| Poland | 0.0 | .. | -10.8 | 0.6* | -2.1 | 3.7-4.0 | 6.4 | 4.5 | 5.3 | 4.0-4.5 |
| Romania | 6.5 | 8.1 | 2.6 | 4.7 | 1.1 | 6.6 | 4.7 | 6.7 | 6.7 | 7.5 |
| Soviet Union | 3.6 | 4.1 | 3.4 | 4.7 | 2.9 | 3.2 | 4.2 | 3.8 | 4.2 | 3.9 |
| Eastern Europe | 3.0 | 3.9 | -0.6 | 2.8 | 1.2 | 4.1 | 4.4 | 4.3 | 4.9 | 4.6 |
| Soviet Union and Eastern Europe | 3.4 | 4.0 | 2.2 | 4.2 | 2.4 | 3.4 | 4.3 | 4.0 | 4.4 | 4.1 |

Note: .. not available

Sources: Economic Survey of Europe in 1983, op. cit., Table 3.1.1 and Economic Survey of Europe in 1984-1985, op. cit., Tables 4.1.1 and 4.3.1.

Table 7 (Chapter 2)

CHANGES IN GROSS OUTPUT IN AGRICULTURE AND CONSTRUCTION
(average annual percentage change)

Agriculture

| | 1980 | 1981 (Plan) | 1981 | 1982 (Plan) | 1982 | 1983 (Plan) | 1983 | 1984 (Plan) | 1984 | 1985 (Plan) |
|---|---|---|---|---|---|---|---|---|---|---|
| Bulgaria | -4.5 | 4.7 | 5.9 | 2.2 | 5.2 | 2.7 | -7.2 | 3.1 | 6.8 | 3.2 |
| Czechoslovakia | 4.8 | 2.6 | -2.5 | 3.2 | 4.4 | 2.7 | 4.2 | - | 3.6 | -1.1 |
| German Dem. Rep. | 0.0 | 0.8 | 1.6 | 0.2 | -4.0 | 1.0 | 4.1 | 6.4 | 8.5 | -1.9* |
| Hungary | 4.3 | 3.0 | 2.0 | 4.0-4.5 | 7.3 | 1.0-2.0 | -2.7 | 3.4-4.1 | 2.5-3.0 | 1.0 |
| Poland | -10.7 | 5.0 (a)* | 3.8 | 5.0* | -2.8 | 1.5-2.4 | 3.3 | 1.5-2.0 | 5.7 | -1.4 |
| Romania | -4.4 | 9.0 | -0.9 | 6.0-7.9 | 7.6 | 5.1-5.6 | -1.6 | 5.4-6.0 | 13.3 | 6.0-6.8 |
| Soviet Union | -1.9 | 11.7 | -1.0 | 10.2 | 5.5 | 10.5 | 6.1 | 6.4 | - | 6.7 |
| Eastern Europe | -3.9 | 4.5 | 1.8 | 4.1 | 1.6 | 2.5 | 0.8 | 2.2 | 6.9 | 1.4 |
| Soviet Union and Eastern Europe | -2.6 | 9.2 | -0.1 | 8.1 | 4.2 | 7.5 | 4.3 | 5.0 | 2.2 | 5.0 |

Construction

| | 1980 | 1981 (Plan) | 1981 | 1982 (Plan) | 1982 | 1983 (Plan) | 1983 | 1984 (Plan) | 1984 |
|---|---|---|---|---|---|---|---|---|---|
| Bulgaria | 3.6 | .. | 5.9 | 3.8* | 3.0 | .. | 3.1 | 3.9 | 3.6 |
| Czechoslovakia | 1.9 | 2.8 | -1.8 | .. | -3.8 | 2.0 | 2.8 | .. | 1.7 |
| German Dem. Rep. | 1.4 | 2.8 | 4.1 | 2.5 | 3.0 | 1.4 | 2.9 | 2.1 | 2.5 |
| Hungary | -3.0 | -1.0 | -0.2 | -(1-2) | 0.5 | -(2-3) | - 3-4* | -(2-3) | -(4.5)* |
| Poland | -10.0 | .. | -21.0 | .. | -6.1 | .. | 6.4 | .. | 6.9 |
| Romania | -0.8 | .. | -4.8 | .. | 1.0.. | .. | .. | .. | .. |
| Soviet Union | 1.2 | .. | 1.8 | .. | 1.9 | .. | 3.9.. | .. | 3.0 |

Notes: .. Not available.
- No growth.
a) Data refer to targets of the draft plan, at current prices, as reported in Trybunu Ludu, 20th December, 1981, p. 2 and 15th January, 1982, p. 3.

Sources: Economic Survey of Europe in 1983, op. cit., Table 3.1.1 and Economic Survey of Europe in 1984-1985, Tables 4.2.1 and 4.4.5.

Table 8 (Chapter 2)

CHANGES IN LABOUR AND CAPITAL PRODUCTIVITY
(average annual percentage change)

MATERIAL SPHERE

| | Labour Productivity | | | | Capital Productivity | | | |
|---|---|---|---|---|---|---|---|---|
| | 1980 | 1981 | 1982 | 1983 | 1980 | 1981 | 1982 | 1983 |
| Bulgaria | 5.5 | 3.9 | 3.5 | 2.7* | -2.6 | -2.3 | -2.5 | -3* |
| Czechoslovakia | 2.5 | -0.4 | -0.1 | 1.7 | -0.3 | -6.0 | -5.0 | -2.4* |
| German Dem. Rep. | 4.0 | 4.2 | 2.0 | 3.7* | -0.8 | -0.7 | -2.4 | -0.8 |
| Hungary | 0.9 | 3.1 | 1.9 | 1.8 | -6.9 | -3.2 | -1.8 | -3* |
| Poland | -5.2 | -11.9 | -1.6 | 5* | -7.0 | -14.8 | -8.7 | 2-3* |
| Romania | 2.0 | 2.4 | 2.4 | 2.8* | -5.7 | -5.8 | -6.8 | -4* |
| Soviet Union | 3.2 | 2.6 | 3.2 | 3.5 | -3.4 | -3.2 | -2.6 | -2.6* |

INDUSTRY

| | Labour Productivity (a) | | | | | Capital Productivity (b) | | | |
|---|---|---|---|---|---|---|---|---|---|
| | 1980 | 1981 | 1982 | 1983 | 1984 | 1980 | 1981 | 1982 | 1983 |
| Bulgaria | 3.1 | 2.0 | 2.1 | 2.4 | 4.3* | -5.4 | -3.0 | -4.7 | -3.3 |
| Czechoslovakia | 2.8 | 1.5 | 0.7 | 2.3 | 3.3 | -2.5 | -4.3 | -3.9 | -2.0 |
| German Dem. Rep. | 4.5 | 4.2 | 2.9 | 3.9 | 3.8* | -0.8 | -1.2 | -2.8 | -1.4 |
| Hungary | 0.6 | 5.0 | 4.9 | 3.6 | 3.0* | -7.9 | -1.4 | -2.3 | .. |
| Poland | -0.2 | -10.6 | 2.7 | 6.5 | 5.8 | -4.4 | -13.7 | -4.4 | 3.5 |
| Romania | 3.2 | 0.5 | -1.0 | 3.0 | 6.7* | -3.2 | -6.0 | -7.7 | -4.2 |
| Soviet Union | 2.5 | 2.5 | 1.9 | 3.6 | 3.8 | -3.3 | -3.4 | -3.7 | -2.5 |

Notes: a) Gross output per employee.
b) Gross output -- fixed assets ratio.

Sources: Labour and capital productivity Material Sphere: Economic Survey of Europe in 1983 op. cit., Tables 3.1.3 and 3.1.5. Labour and capital productivity in Industry: Economic Survey of Europe in 1983, op. cit., Table 3.3.2 and Economic Survey of Europe in 1984-1985, op. cit., Tables 4.3.2 and 4.3.6.

Table 9 (Chapter 2)

GROSS INVESTMENT RATIO (IR)
(percentage of NMP)

| | 1980 | 1981 (Plan) | 1982 | 1982 (Plan) | 1983 | 1983 (Plan) | 1984 | 1984 (Plan) | 1985 (Plan) |
|---|---|---|---|---|---|---|---|---|---|
| Bulgaria | 33.9 | 35.7 | 36.9* | 35.4 | 31.8 | 34.6 | 33.4 | 33.1 | 33.8 |
| Czechoslovakia | 33.1 | 31.6 | 30.5 | 30.8 | 30.0* | 30.3 | 29.5* | 29.8 | 29.6 |
| German Dem. Rep. | 29.1 | 28.6 | 26.9* | 26.4 | 24.7* | 25.3 | 24.6* | 24.0 | 23.0 |
| Hungary | 36.0 | 33.7 | 32.0* | 32.3 | 27.8 | 31.0 | 26.1 | 28.2 | 27.7 |
| Poland | 28.2 | 24.8 | 29.2* | 23.1 | 23.0 | 23.8 | 23.0 | 24.5 | 23.7 |
| Romania | 41.5 | 39.2 | 37.6* | 36.9 | 34.4 | 36.4 | 34.0 | 35.9 | 35.3 |
| Soviet Union | 28.7 | 29.9 | 28.1 | 29.7 | 29.7* | 30.2 | 29.7* | 29.9 | 29.6 |

GROSS INCREMENTAL CAPITAL-OUTPUT RATIO (ICOR)
(co-efficient) (a)

| | 1980 | 1981 (Plan) | 1982 | 1982 (Plan) | 1983 | 1983 (Plan) | 1984 | 1984 (Plan) | 1985 (Plan) |
|---|---|---|---|---|---|---|---|---|---|
| Bulgaria | 5.9 | 7.1 | 10.3* | 8.4 | 8.4 | 11.5 | 8.8 | 7.2 | 8.2 |
| Czechoslovakia | 11.3 | -284 | 60.9 | 131 | 18.7* | 12.5 | 10.6* | 9.9 | 9.9 |
| German Dem. Rep. | 6.6 | 5.9 | 5.6* | 10.2 | 5.9* | 5.7 | 5.6* | 4.4 | 5.2 |
| Hungary | -42.6 | 13.4 | 24.6* | 12.2 | 37.1 | 103 | 14.9 | 9.7 | 10.9 |
| Poland | -4.7 | -2.1 | 19.5* | -4.2 | 10.2 | 4.0 | 8.4 | 4.9 | 7.3 |
| Romania | 13.9 | 18.8 | 6.8* | 13.3 | 6.9 | 9.4 | 4.7 | 4.7 | 3.5 |
| Soviet Union | 7.6 | 9.1 | 7.2 | 7.4 | 9.0* | 7.2 | 9.8* | 9.9 | 6.6 |

GROSS FIXED CAPITAL FORMATION
(annual percentage change)

| | 1980 | 1981 (Plan) | 1982 | 1982 (Plan) | 1983 | 1983 (Plan) | 1984 | 1984 (Plan) | 1985 (Plan) |
|---|---|---|---|---|---|---|---|---|---|
| Bulgaria | 7.5 | 10.5 | 3.6* | 3.6 | -6.9* | 0.7 | 8.1 | -* | 6.1(a) |
| Czechoslovakia | 1.4 | -4.6 | -3.3 | -2.3 | -1.7* | 0.6 | -* | 1.5* | 2 |
| German Dem. Rep. | 0.3 | 2.8 | -* | -5.1 | -1.3* | - | -* | -* | -* |
| Hungary | -6.7 | -4.3 | -1.7 | -1.6 | -10.0 | -3.7 | -10.0 | -(6-7) | 1.1 |
| Poland | -12.3 | -22.4 | -* | -12.1 | 1.6 | 9.4 | -2.8 | 8 | - |
| Romania | 3.0 | -7.1 | 5.0* | -3.1 | 0.7 | 2.5 | 4 | 6.1 | 8.3 |
| Soviet Union | 2.4 | 3.8 | 0.9 | 3.5 | 3.2 | 5.8 | 3.9* | 2 | 3.4 |

Notes: - No growth.
a) At current prices.

N.B. The price base used is not uniform for 1980-1985 (actual and plan). See sources cited. For the price base used for 1981-1984 (actual) and the 1985 plan, see notes to Table 4.

Sources: Economic Survey of Europe in 1982 and 1983, op. cit., Table 3.4.1 and Economic Survey of Europe in 1984-1985, op. cit., Tables 4.4.1 and 4.4.2.

## NOTES AND REFERENCES

1. It may be noted that for the German Democratic Republic, the Five-Year Plan is not legally binding on the enterprises.
2. Increasing exports applies, of course, to all the traditional export branches. As far as the chemical industry is concerned, it may be noted that Bulgaria has undertaken a major reorganisation of this industry in order to improve its foreign trade performance. This reorganisation essentially involves eliminating organisational duplication. Consequently, a National Corporation responsible for the entire chemical industry has been set up and all matters relating to trade have been concentrated in one organisation – Khimcomplekt.
3. The importance of the energy sector for Bulgarian economic development is evident from the creation, in 1980, of a Council on Power and Raw Materials' Problems. It has been placed under the surveyance of the Council of Ministers and is to co-ordinate investment strategy in the energy field.
4. For a detailed discussion of the New Economic Mechanism see Ilse Grosser, *Der Neue Ökonomische Mechanismus in Bulgarian*, Forschungsbericht Nr. 98, Wiener Institut für Internationale Wirtschaftsvergleiche, September, 1984, and K. Khristov, "Minister Examines New Economic Mechanism-Foreign Trade Relationship", *Vunshna Torgoviya*, Nos. 5 and 6, 1984, as given in *Joint Publications Research Service* (EEI-84-100), *East Europe Report*, 5th September, 1984, pp. 3-20.
5. "For New Successes in the 1984 Jubilee Year", *Politicheska Provesta*, No. 11, 1983, as given in *Joint Publications Research Service* (EEI-84-009), *East Europe Report*, 17th January, 1984, p. 6.
6. *Durzhaven Vestnik*, 24th August, 1984, as given in *Joint Publications Research Service* (EEI-84-117), *East Europe Report*, 25th October, 1984, p. 6.
7. *Presseschau Ostwirtschaft*, No. 2, Österreichiches Ost-und Südosteuropa-Institut, 1985, p. 41.
8. OECD, Foreign Trade Statistics.
9. The decline in capital productivity is said to be particularly serious in those branches which were accorded priority in the Five-Year Plan. Namely, fuels, energy and construction materials, as well as in the wood-processing industry. See K. Ujhazy, Deputy Chairman, State Planning Commission, "Plan Implementation Steps for the Years 1983-1985", *Investicni Vystavba*, No. 7-8, 1982, as given in *Joint Publications Research Service* (L/11106), *East Europe Report*, 28th January, 1983, p. 16.
10. "Czechoslovak-Soviet Economic Co-operation in Engineering and Electrical Engineering on a Higher Level", *Jemna A Optika*, No. 11, 1982 (unsigned lead article), as given in *Joint Publications Research Service* (82875), *East Europe Report*, 16th February, 1983, p. 22.
11. K. Ujhazy, *art. cit.*, p. 11.
12. *Radio Free Europe Research*,. Vol. 10, No. 15, 12th April, 1985, pp. 17-18.
13. See also, K. Ujhazy, *art. cit.*
14. "Law No. 122/1981 of the Collection of Laws on the State Plan for the Development of the National Economy of the Czechoslovak Socialist Republic for the 1981-1985 Period", *Hospodarske Noviny*, 29th December, 1981, as given in *Joint Publications Research Service* (80171), *East Europe Report*, 24th February, 1982, p. 17.

15. See Z. Bilak, "Research and Development in the Seventh Five-Year Plan for Development of the National Economy", *Automatizace*, No. 8-9, 1982, as given in *Joint Publications Research Service* (L/11103), *East Europe Report*, 28th January, 1983, p. 2 and E. Klacova, "Two Reflections of Progress", *Hospodarske Noviny*, No. 49, 1985, as given in *Joint Publications Research Service* (EEI-85-026), *East Europe Report*, 5th March, 1985, pp. 34-44.

16. S. Sourek (Deputy Minister of Finance), "New Methods of Financing Technological Development", *Technicke Noviny*, No. 46, 1984, as given in *Joint Publications Research Service* (EEI-85-006), *East Europe Report*, 16th January, 1985, pp. 19-22.

17. Z. Bilak, *art. cit.*, p. 3. As is noted by the author, the plan for scientific and technological development contains some 392 assignments, of which 75 are obligatory by law.

18. L. Ler (Minister of Finance), *Pravda* (Bratislave), 12th December, 1984, as given in *Joint Publications Research Service* (EEI-85-001), *East Europe Report*, 2nd January, 1985, p. 27 and Z. Bilak, "More Tasks, More Funds: Science and Technology the Key to an Effective Economy", *Hospodarske Noviny*, No. 4, 1985, as given in *Joint Publications Research Service* (EEI-85-029), *East Europe Report*, 12th March, 1985, p. 63.

19. J. Strba (Deputy Chairman of the State Planning Commission), "Plan 1981: We shall Pay for Imports with Exports", as given in *Joint Publications Research Service* (80419), *East Europe Report*, 26th March, 1982, p. 4.

20. *OECD, Foreign Trade Statistics.*

21. K. Ujhazy, *art. cit.*, p. 20.

22. "Main Trends of the CSSR's Economic and Social Development for the 1981-1985 Period", *Hospodarske Noviny*, No. 16, 1981, as given in *Joint Publications Research Service* (78261), *East Europe Report*, 9th June, 1981, p. 32.

23. See "Directives of the 10th SED Congress on the Five-Year Plan for the Development of the GDR Economy During 1981-1985", *Neues Deutschland*, 18th-19th April, 1981, as given in *Joint Publications Research Service* (78163), *East Europe Report*, 27th May, 1981, p. 28.

24. *Ibid.*, p. 4.

25. *Bericht des Zentralkomitees der Sozialistischen Einheitspartei Deutschlands an den X. Parteitag der SED*, Dietz Verlag, Berlin (East), 1981, p. 51.

26. See "XI. Parteitag der SED Einberufen", *Neues Deutschland*, Nr. 279, 24th/25th November, 1984.

27. See G. Flach, H. Ullmann and R. Rockstroh, "Evaluation of Various Types of Nuclear Energy Sources under the Conditions of the GDR", *Kernenergie*, No. 10, 1984, as given in *Joint Publications Research Service* (EEI-85-004), *East Europe Report*, 11th January, 1985, pp. 34-41 and A. Kachanov, "Energy for Peace and Creation", *Foreign Trade* (English edition), No. 11, Moscow, 1984, p. 13.

28. For a discussion of the various reform measures undertaken see, *inter alia*: D. Cornelsen, M. Melzer, A. Scherzinger, "DDR-Wirtschaftssystem: Reform in kleinen Schritten", *Vierteljahrshefte zur Wirtschaftsforschung*, No. 2, Deutsches Institut für Wirtschaftsforschung, Berlin, 1984; *Sozialistische Wirtschaftssyteme vor neuen Reformen*?, Bundesinstitut für Ostwissenschaftliche und Internationale Studien, Köln, August, 1984, and, "Wirtschaftsreformen in Osteuropa: Effiziensteigerung order Sackgasse?", *Wochenbericht*, No. 8, Deutsches Institut für Wirtschaftsforschung, Berlin, 21st February, 1985.

29. "Directives of the 10th SED Congress on the Five-Year Plan for the Development of the GDR Economy During 1981-1985", *art. cit.*, p. 47.

30. H. Sölle, "35 Years of the GDR's Successful Foreign Trade Policy", *Foreign Trade* (English edition), No. 10, Moscow, 1984, p. 3.

31. *OECD, Foreign Trade Statistics*. It should be noted, however, that OECD data do not include trade with the Federal Republic of Germany which accounts for almost 50 per cent of the German Democratic Republic's trade with the West.

32. For a detailed discussion of the German Democratic Republic's trade, economic and financial situation, see also *OECD working document on the German Democratic Republic* (Trade Directorate).

33. This credit is to be progressively reduced to DM 600 million by the end of 1985. Use of this facility has varied over the years, but has, in general, been declining since 1975. In 1984, for instance, only about a third (of the allowed DM690 million ceiling) was used.

34. Interview with Bela Csikos-Nagy, (then State Secretary and Chairman of the Hungarian Economics Society), *Heti Vilaggazdasag*, 8th May, 1982, as given in *Joint Publications Research Service* (81096), *East Europe Report*, 21st June, 1982, p. 100.

35. B. Szikszay, "The Hungarian National Economy-Changing to a New Source of Economic Growth", *Marketing in Hungary*, No. 4, 1982, p. 11.

36. See Paul Marer, *East-West Technology Transfer: Study of Hungary, 1968-1984*, Chapter III, OECD, Paris, 1986. See also J.C. Brada and J.M. Montias, "Industrial Policy in Eastern Europe: A Three-Country Comparison", *Journal of Comparative Economics*, Vol. 8, No. 4, December, 1984.

37. *Heti Vilaggazdasag*, No. 39, 29th September, 1984, as given in *Presseschau Ostwirtschaft*, No. 10, Österreichiches Ost-und Südosteuropa Institut, 1984, p. 26.

38. Paul Marer, *op. cit.*, Chapter V.

39. For a detailed discussion of Hungary's New Economic Mechanism and evolution of the reforms undertaken, see *Ibid.*, Chapters II and III.

40. E. Jovorka, "Paid R&D", *Heti Vilaggazdasag*, No. 17, November, 1984, as given in *Joint Publications Research Service* (ESA-85-005), *East Europe Report*, 4th February, 1985, pp. 50-53.

41. L. Mehes, Minister of Industry, "To Remain in Competition by Improving Performance", *Figyelo*, 13th January, 1983, as given in *Joint Publications Research Service, (82971), East Europe Report*, 28th February, 1983, p. 28.

42. Akos Belassa, "The Rate of Economic Development and the Chief Ratios Thereof in the Sixth Five-Year Plan", *Kozgazdasagi Szemle*, No. 4, 1981, as given in *Joint Publications Research Service* (78163), *East Europe Report*, 27th May, 1981, p. 61.

43. G. Kerekgyarto, "Our Possibilities for Economic Development", *Kozgazdasagi Szemle*, No. 10, 1985, as given in *Joint Publications Research Service* (EI-840133), *East Europe Report*, 10th December, 1984, p. 53.

44. Bela Szikszay, "Development of External Relations, Development Tasks", *Tarsadalmi Szemle*, No. 7, 1982, as given in *Joint Publications Research Service* (81633), *East Europe Report*, 26th August, 1982, p. 43. See also G. Kerekgyarto, *op. cit.*, p. 52.

45. *OECD, Foreign Trade Statistics.*

46. The 13th Party Congress which was held in November, 1984 debated – and ratified – the draft directives for Romania's economic development in the 1986-1990 Five-Year Plan period and up to the year 2000 which were issued in September, 1984. All growth targets as well as the country's development strategy for the post 1981-1985 period are taken from these draft directives.

47. Stelian Teodorescu, "Deputy Minister Details Machine Building Industry Programs", *Romania Libera*, 24th December, 1982, as given in *Joint Publications Research Service (82987), East Europe Report* 2nd March, 1983, p. 100.

48. If, however, Romania's imports of fertiliser and farm equipment are included, it is a net agricultural importer.

49. See, for instance, Michael Kaser and Iancu Spigler, "Economic Reform in Romania in the 1970s" in, Alec Nove, Hans-Hermann Höhman, Gertraud Seidenstecher (editors), *The East European Economies in the 1970s*, Butterworth & Co., 1982 and Gerhard Fink, Gabriele Tuitz, *Wirtschafts-und Systempolik in Rumänian*, Reprint Serie No. 78, Wiener Institut für Internationale Wirtschaftsvergleiche, January, 1985.

50. Michael Kaser, *The Flow of Technology from the OECD Countries to Romania*, working document, OECD, Paris, 9th November, 1981.

51. *Romania During the 1981-1985 Five-Year Plan Period*, Bucharest (n.d.), p. 35.

52. *OECD, Foreign Trade Statistics.*

53. A. Constantinescu, V. Babiuc and N. Murgu, "Ways of Stimulating the Export Trade under the New Economic-Financial Mechanism", *Revista Economica*, No. 22, 29th May, 1981 and No. 23, 5th June, 1981, as given in *Joint Publications Research Service (78620), East Europe Report*, 28th July, 1981, p. 39. The authors are citing Secretary-General Ceausescu.

54. Statement by Mr. Brezhnev in his report to the CPSU Central Committee, May 24th, 1982, relative to the launching of the Food Programme, designed to run through 1990.

55. For a detailed discussion of the Soviet Union's energy problems and prospects, see Ed. A. Hewett, *Energy Economics and Foreign Policy in the Soviet Union*, The Brookings Institution, Washington, D.C., 1984.

56. Decree "On Improving Planning and Strengthening the Effect of the Economic Mechanism on Raising Efficiency of Production and Quality of Work". See *Ekonomicheskaya Gazeta*, No. 32, August, 1979. For a discussion and evaluation of the measures entailed in improving the economic mechanism, see M. Bornstein, "Improving the Economic Mechanism", *Soviet Studies*, Vol. XXXVII, No. 1, January, 1985.

57. Decree "On Additional Measures to Expand the Rights of Industrial Production Associations (Enterprises) in Industry Planning and Economic Activity and to Increase Their Responsibility for the Results of Their Work". See, for instance, A. Milyukov, "New Elements in the Cost-Accounting Independence and Responsibility of Enterprises", *Voprosy Ekonomiki*, No. 12, 1983.

58. N. Baybakov, then Deputy Chairman of the USSR Council of Ministers, Chairman of Gosplan, "Eleventh Five-Year Plan, Its Features and Tasks", *Planovoye Khosyaystvo*, No. 1, 1982, p. 7.

59. *Ibid.*

60. OECD, Foreign Trade Statistics.

61. See, for instance, Tatyana Kuznetsova, "The Capitalist Methanol Market: Peculiarities of its Development", *Foreign Trade* (English edition), No. 11, 1981, p. 25.

*Chapter 3*

# DOMESTIC R&D AND THE TECHNOLOGY GAP

## I. THE SOVIET MODEL

The Soviet organisation of research, development and innovation (RDI) has served as a model for the East European countries, providing them with needed total political control over this aspect of their economies. The following discussion of the Soviet model describes the general characteristics of the Soviet RDI system and summarizes the evaluations of this system made by several Western specialists.

### A. Location of the Soviet R&D Effort

Most Soviet research and development work is conducted at facilities subordinate to the Academy of Sciences, the Ministry for Higher and Secondary Education (VUZy), the Ministry of Defense or one of the industrial ministries. Very little research and development work is conducted outside of these established organisations. Some social organisations, such as the All Union Society of Inventors and Rationalisers (VOIR), the Scientific-Technical Societies (NTO), and the All-Union Voluntary Society for Assistance to the Army, Air Force, and Navy (DOSAAF) support extra-curricular research. However, this support does not constitute a major programme for these organisations.

Garage-inventors, i.e. individuals working independently of any major organisation, do exist in the Soviet Union; yet, their research is likely to be far below the technical levels of organised research and is conducted without access to any moderately sophisticated laboratory equipment. Invention statistics seem to bear out the minor role played by independent individuals in Soviet innovation. According to published Soviet information, only 7 per cent of granted inventors' certificates were not from organisations; whereas, in the United States, almost one-fourth of the patents were granted to individuals[1].

Soviet writings about the management of science consistently stress the superiority of research done as a collective or group in an established organisation and presumably reflect an official policy of discouraging research by private individuals[2]. Private research is also discouraged by the Soviet economic system itself. Important state research institutes often have difficulties in obtaining proper laboratory equipment and frequently have to make their own[3]. The difficulties that private individuals or groups would have in purchasing such equipment are likely even greater.

Traditionally the institutes of the USSR Academy of Sciences and some facilities of the Ministry of Higher and Secondary Education have conducted the bulk of Soviet fundamental research. Central research facilities directly subordinate to ministries or facilities subordinate to industrial enterprises have provided the bulk of Soviet applied research. The present emphasis on bringing science to industry has likely blurred this general distinction. Most notably, research at the Republic Academies of Sciences has become more applied in nature, as has the research at many of the Ministry for Secondary and Higher Education's facilities, especially the polytechnical institutes.

An examination of the origins of registered Soviet discoveries and inventors' certificates illustrates this organisational separation of basic and applied research. The USSR Academy of Sciences, while accounting for 51 per cent of all Soviet discoveries registered up to 1975, accounted for only slightly more than 1 per cent of the inventions used in the economy[4]. The Republic Academies and the Ministry for Secondary and Higher Education, reflecting their dual nature, accounted for about 6 and 18 per cent respectively of registered discoveries and about 7 per cent of the inventions used in the economy. The bulk of Soviet applied research is conducted by research and design institutes that are either directly subordinate to an industrial ministry (branch institutes) or an enterprise. While branch institutes still accounted for about 30 per cent of registered discoveries, they and the enterprise institutes accounted for the overwhelming majority of inventions used in the economy.

The distribution of scientific workers among the various types of R&D facility has remained relatively constant over the past two decades. The Academies of Sciences and the Ministry of Secondary and Higher Education employ about 44 per cent of all scientific workers, and branch institutes and enterprise facilities employ about 56 per cent (see Table 1).

Table 1
PROPORTION OF SCIENTIFIC WORKERS,
BY PLACE OF WORK, 1962-1976
(in per cent at end of year)

| | Academies and VUZy | Branch and enterprises facilities |
|---|---|---|
| 1962 | 44.6 | 55.4 |
| 1965 | 42.6 | 57.4 |
| 1970 | 46.8 | 53.2 |
| 1976 | 43.8 | 56.2 |

*Source*: Louvan E. Nolting and Murray Feshbach, "R&D Employment in the USSR-Definitions, Statistics, and Comparisons", in *Soviet Economy in a Time of Change*, Vol. 1, *op. cit.*, p. 729.

## B. State Management of Research, Development and Innovation (RDI)[5]

The first Soviet Five-Year Plan extended central economic planning to the laboratory bench and gave the state the responsibility of overseeing the entire innovative process, from directing the creative work of individual scientists and engineers to utilising new technologies in the economy. Central planning subordinated most R&D work to the state's economic goals. "Don't invent what you want, but what socialist construction needs" became the slogan.

Prior to World War II the successful fulfilment of production plans reigned paramount in the eyes of Soviet central planners, and the state's effort in directly managing RDI remained largely undeveloped. Prewar innovation programmes remained at the level of economic campaigns, with only modest central administrative support.

After World War II Soviet leaders recognised the strategic importance of efficient RDI and the Party became more actively involved in the promotion of science and technology. Yet, a centrally dictated RDI programme contained many pitfalls. While enterprise managers often shirked necessary technological innovation by choosing trivial changes that would not disrupt production, a centrally dictated RDI programme did not always bring better results. Central officials were often too far removed from the factory floor, and their innovational schemes often disrupted production without bringing the expected improvements.

The postwar period has witnessed an increased emphasis on RDI programmes that attempt to steer a course between the excesses caused by meddling central officials and the technological stagnation caused by overly timid industrial officials. Special central organisations for the management of RDI were established. These special organisations were not to dictate specific innovations, but they were to pressure ministries and enterprises to increase significant RDI work.

### a) *Principal State Organisations*

#### i) *The State Planning Commission (Gosplan)*

Gosplan formulates economic plans according to the political goals of the central administration. These plans are translated by a hierarchy of state organisations into production requirements for industrial enterprises. Thus, Gosplan's directives ultimately stipulate the kinds and quantities of products produced. Gosplan's primary task – balancing the supply and demand of goods – has probably led it to emphasise the successful fulfilment of the quantitative elements of the enterprise plans over their qualitative elements. In the same way, Gosplan has tended to shun long-term innovation if it carried a risk of disrupting short-term supplies. An enterprise's fulfilment of its production plan probably wards off any questions about innovation by Gosplan.

The strategic pressures created by the atomic age forced Soviet central administrators to remedy Gosplan's inherent neglect of technical innovation. In addition to the obvious military importance of supporting industrial innovation, central administrators recognised that economic growth would increasingly depend on strong RDI performance. Consequently, Soviet economic policy makers increased the importance of plan indicators for science and technology in Gosplan's industrial plans and created a number of specific all-union organisations dedicated to promoting RDI.

#### ii) *The State Committee for Science and Technology (GKNT)*[6]

The primary central organisation for the management of science and technology is the State Committee for Science and Technology. This organisation functions as a bridge over many of the bureaucratic boundaries that often inhibit a rational deployment of scientific and engineering resources. It co-ordinates work among the industrial ministries, between industrial ministries and academic facilities, and between itself and counterpart organisations in other CMEA countries. Co-ordination consists largely of planning research priorities, forecasting technological developments, providing opinions on planned technological acquisitions from the West, and establishing organisational standards. The GKNT may issue recommendations, as distinguished from binding directives. Hence, it is not allowed to dictate policies on science and technology. It only has the power to force other organisations to justify their decisions.

The GKNT also provides a valuable service to the RDI community by managing all-union programmes for the dissemination of scientific-technical information. At first, Soviet officials had expected scientific-technical information to flow freely among socialist industrial enterprises and research institutes, since, theoretically, socialism had no competitive pressures to cause industrial secrecy. Yet, early Soviet press reports show that technical information did not flow automatically among enterprises and was neglected for many of the same reasons as was technical innovation itself[7]. The GKNT's dissemination of technical information occupies a prominent position in Soviet management programmes for innovation and has required considerable co-operation on the part of all Soviet industrial establishments[8].

iii) *The State Committee for Inventions and Discoveries*

Western patent offices, in granting patents, provide incentives for individuals and companies to conduct costly R&D projects; because, if something new is developed and patented, the patentee is given an exclusive right to make, sell or use the patented innovation. This exclusive right can be very rewarding to the patentee, since it acts as a barrier to his competitors, who often are forced to seek licenses from the patentee or compete with him on disadvantageous terms. In managing their patent systems, the primary functions of most Western patent offices center on the processing of patent applications and the publication of the relevant technical documentation.

Soviet socialism demands more of its patent office, the State Committee for Inventions and Discoveries[9]. (The Soviet Union has a dual system of protection for inventions. Traditional patents, which give the exclusive rights to the invention to the inventor, are granted primarily to foreigners. Inventors' certificates, which give the rights to the invention to the state in return for a bonus if the invention is used by industry, are primarily granted to Soviet citizens.) Since there is no free market prodding Soviet industrial enterprises to innovate, the Soviet patent office participates actively in the state's RDI programmes. The State Committee's duties require that it:

- promote the use of new inventions in industrial enterprises, including recommendations that specific inventions be included in ministerial plans;
- establish a general social awareness of the economic importance of RDI programmes;
- organise, translate, and distribute foreign patent literature to industrial enterprises;
- organise and manage numerous local depositories of patent literature;
- provide training for industrial research personnel in the legal technical and information science aspects of patenting;
- manage a centralised bonus system for inventors;
- aid industrial enterprises in verifying that proposed exports will not infringe on foreign patents and,
- recommend that specific inventions be approved or rejected for patenting abroad.

The State Committee, as an organisation outside the influence of most industrial managers, was enlisted by central economic officials to play an important role in evaluating the technical levels of R&D institutes and of enterprise innovation programmes. These officials established the number of inventors' certificates received, the proportion of inventors' certificates accepted, and the number of inventions implemented as important success indicators for innovation.

iv) *The State Committee for Standardisation (Gosstandart)*

Most Western countries sponsor engineering standardisation to promote product specialisation and ensure an acceptable level of manufacturing quality[10]. The Soviet standardisation programme resembles its patent programme in being more ambitious and tied directly to centrally planned industrial innovation.

Soviet economic officials pressure enterprises to innovate by raising the level of state standards. Production failing to meet the levels set by state standards either does not count towards the fulfilment of enterprise plans or subjects the enterprise to a penalty. However, the enterprises have considerable opportunity during obligatory consultations to thwart unrealistic and overly demanding standardisation requirements. In this way, the powers of Gosstandart resemble the modest powers of the other two national science and technology organisations. In 1966 the Soviet government made state standards a part of the Five-Year Plan for National Economic Development, and in 1968, state standards became a chapter of the annual national economic plan. Presently, Gosstandart's "Mark of Quality" plays a prominent role in state innovation programmes.

b) *State Direction of Industrial R&D Efforts*

Under Soviet socialism market pressures do not guide a facility in its RDI choices. Instead, the state must articulate the country's industrial RDI priorities. The most important overall guides are contained in the "Basic Scientific-Technological Problems", a list of slightly more than 200 research areas announced with each Five-Year Plan. These research areas reflect the economic policy makers' priorities and are compiled by special problem councils that are staffed by members of Gosplan, the GKNT and the Academy of Sciences. An industrial ministry usually co-ordinates the actual work on one of the list's problems by assigning tasks to other R&D organisations and by overseeing the completion of the assigned work.

The industrial ministries determine the direction of most of their internal R&D work, usually through a Scientific-Technical Council. Such internal research reflects the general production concerns of the ministry's management. Enterprises also conduct some R&D work independent from either the national research areas or the ministry's particular research topics. Enterprise R&D is largely dedicated to solving operational problems on the factory floor[11].

One Western specialist estimates that centrally planned R&D accounts for about 40 to 50 per cent of total official R&D, half of which is accounted for by work on the approximately 200 basic science and technology problems. Ministerially planned R&D comprises about 30 per cent of the total, with the remainder being done at lower levels[12].

c) *State Financing of Research, Development and Innovation*

Expenditures on science have grown strongly over the past decades, outpacing both the growth recorded for industrial production and national income. (In 1958 the Soviet statistical handbook reported expenditures on science to be 2.4 billion rubles, and in 1982, 24.6 billion rubles)[13]. The structure of Soviet science expenditures differs markedly from the structure of US expenditures. Whereas approximately 60 per cent of Soviet expenditures were used for applied research and 27 per cent for development, the US structure was almost the reverse, with 23 per cent devoted to applied research and 62 per cent for development[14].

Soviet RDI work is funded either directly from state budget allocations – estimated to cover slightly less than half of total RDI – or from the "resources of enterprises and economic organisations". State budget resources are generally used to support fundamental research or research relating to one of the state's announced scientific-technological problems. Branch resources for science are generally used for ministerial or enterprise RDI expenses, but they only cover part of total expenses. Many of a ministry's or enterprise's RDI costs – especially prototype construction and other more development related costs – are handled as production costs.

The conclusion of R&D contracts between organisations has recently been a relatively dynamic form of financing RDI. Contracting has helped bridge the bureaucratic gaps existing between many research institutes and industry. Furthermore, contracting has tended to involve a number of the basic science oriented organisations in more applied and developmental problems.

## C. Problems and Reform Efforts

The Soviet economy has yet to perform well in its creation and use of new technologies. Innovational performance is especially important now that the Soviet Union faces a decline in the growth of its industrial workforce.

While Soviet ideology champions the all-encompassing nature of the state's management of RDI as a source of strength, the resulting bureaucratic jungle is, perhaps, a major source of weakness. In Western countries, individual entrepreneurs or small, newly-formed companies, are often responsible for considerable technological innovation. In fact, the head of corporate R&D for a major US pharmaceutical firm once stated: "perhaps the miracle is that large organisations ever invent anything."[15] Yet, as was noted above, the Soviet economy discourages innovational efforts outside of state organisations.

In fact, the very nature of scientific and technological development, its unpredictability or creative roots, seems incompatible with the Soviet need to fit all R&D programmes into neatly defined, existing institutional arrangements. Consequently, it is not surprising to read discussions in the Soviet literature about the unfortunate consequences resulting from the separation of R&D facilities from the production process or, about how state support seems to be handed out on the basis of institutional reputation instead of on the basis of the technical merits of the specific projects in question.

Another related problem is the rather inflexible Soviet system of central planning. RDI is often an exploration of the unknown. Many creative discoveries or technological solutions are surprises. Thus, the choice of important research themes at an all-union or ministry level can easily work to the disadvantage of creative, pathbreaking research. Furthermore, even ongoing RDI work takes unforeseen turns. Yet, a planning system that requires R&D officials to order all of their equipment a year in advance and adjusts with painful slowness to most programme changes can even act as an impediment to long-standing industrial innovation programmes.

Distinguishing economically unsound innovations from profitable ones poses a major problem for all industrial managers. The Soviet economic system, with sets of inconsistent success indicators and an arbitrary price stucture, has made such sorting extremely difficult. The Soviet press has carried many accounts of success indicators guiding technical innovation in directions that make little economic sense[16]. Other press accounts describe Soviet managers who had to reject extremely valuable innovations because of the negative effect that these innovations would have had on the enterprise's success indicators.

The Soviet leadership's reform efforts have remained consistently cautious. While reforming its technical innovation programmes, the Party has carefully guarded its political power and rejected any erosion of central economic planning. As mentioned above, new central science and technology organisations were created. The GKNT was to balance the tendency of Gosplan to avoid significant innovation, and the State Committees for Standardisation and Inventions and Discoveries were to wield standards and patents as new carrots and sticks for innovation. Scientific Production Associations were created[17]. New financing arrangements were established[18]. R&D contracting grew in importance. Finally, increased purchases of technology from the West allowed entire branches to modernise. All are reforms that incurred no significant domestic changes in the economic system.

### D. Western Evaluations of Soviet Innovational Performance and the Technology Gap

A number of recent Western studies have evaluated Soviet technical innovation. Joseph Berliner analysed how the Soviet economy's structure – organisation, prices, incentives – affects the innovational decisions of industrial enterprises. The overall tone of Berliner's analysis is negative, and he concluded that "there are potent forces for the promotion of technological progress, but they are blunted by the pervasive obstacles to innovation that inhere in the economic structure"[19].

Other Western studies shared Berliner's rather negative conclusion. One study examined Soviet lead-times – defined as the period from the filing for an inventor's certificate to the state's certification of the invention's first use – and found them significantly greater than Western industrial lead-times. More surprisingly, the study found that after adjusting for the technical areas of the inventions, the average lead-time for inventions implemented in the Soviet defense-industrial sector was not statistically different from the average lead-time for inventions implemented in the civilian machine-building sector[20].

Another study found that the performance of Soviet industry in assimilating Western technology imports largely paralleled its poor performance in using domestic technologies. A series of interviews with Western suppliers of technology gave a rather unimpressive picture – the assimilation period and lead-times for follow-on innovations remained long, and planned production levels were often not achieved[21].

The Centre for Russian and East European Studies at the University of Birmingham recently completed a detailed two-volume evaluation of Soviet technological innovation. The first volume compares the levels of Soviet technology to levels found in the West in the following key industrial sectors: iron and steel, machine tools, high voltage electric power transmission, chemicals, industrial process control, computers, military equipment, and automobiles. The authors concluded: "in most of the technologies we have studied there is no evidence of a substantial diminution of the technological gap between the USSR and the West in the past 15-20 years, either at the prototype/commercial application stages or in the diffusion of advanced technology"[22].

## II. EASTERN EUROPE

The Soviet model has served as a pattern for the organisation of RDI in Eastern Europe. The postwar government changes in Eastern Europe, led to the imposition of Soviet-type economic structures. Specifically, central economic planning was introduced and the industrial means of production were transferred to state ownership. These changes led directly to greater state responsibilities for research, development and innovation. Namely, the East European states now had to manage the entire RDI process, from the creation of new technology to its industrial implementation.

### A. Location of East European R&D

Most East European R&D is conducted in state organisations, as property restrictions and economic supply difficulties, similar to those described for the Soviet Union, discourage R&D by private individuals. The Academies of Sciences, educational establishments, and industrial sectors constitute the major support for R&D[23].

R&D workforces seem more directly subordinated to industry in Eastern Europe than is the case in the USSR. In Czechoslovakia, for example, the Academy of Science and university facilities employed only 12 per cent of scientific workers during 1975[24]. In the German Democratic Republic, the proportion was somewhat greater: 18 per cent, but still far below the proportion of over 40 per cent reported for the Soviet Union (see Table 1 above)[25].

A very detailed analysis of technical innovation in the German Democratic Republic noted the following significant difference between the location of German Democratic Republic and Soviet R&D personnel: "Soviet R&D was mainly centralised in facilities geographically distant from the place of production. By contrast, about 60 per cent of the German Democratic Republic's industrial R&D manpower was based in the enterprise and the remaining 40 per cent, that employed by the institutes, was also based for the most part near the place of production"[26]. It is not clear if the other East European countries have also organised their R&D efforts in a similar manner, but such an organisational structure in Eastern Europe might be expected to lessen their innovation problems.

## B. State Management of Research, Development and Innovation (RDI)

### a) *Principal State Organisations*

Each of the East European countries has a central state organisation, resembling closely the Soviet GKNT, for establishing a unified science policy and for co-ordinating RDI programmes among the industrial ministries, Academies of Sciences and educational institutes[27]. The scope of responsibilities varies among these organisations, but all work closely with their respective state planning agency in subordinating RDI work to the goals of the political leadership. In fact, central planning is constantly touted as being the locus of strength for a socialist management of RDI in Eastern Europe[28].

The Bulgarian, Czechoslovakian and Romanian central science and technology organisations have, perhaps, the broadest responsibilities, since each organisation has direct control over patenting, standardisation and the dissemination of scientific-technical information. The other state organisations, while having some role in the dissemination of scientific-technical information, are organisationally divorced from their countries' efforts at patenting and standardisation[29]. None of the East European central science and technology organisations has supra-ministerial powers, and their roles are limited to championing better innovational performance by administrative pressures rather than by fiat[30].

### b) *State Financing of Research, Development and Innovation*

Each of the East European countries reported strongly growing expenditures for science, with Czechoslovakia, Hungary and Poland reporting growth of over 10 per cent per annum. Although the exact composition of the reported figures is not clear, thus making comparisons with other countries difficult, the reported growth rates outpaced the growth for industrial production and national income, and at the very least, indicate similar state policies of stressing scientific-technical progress.

The pattern of expenditure for science and technology in both the German Democratic Republic and Czechoslovakia differs from that in the USSR. The majority of funds in the German Democratic Republic and Czechoslovakia were dedicated to development work (52 and 60 per cent respectively), while Soviet expenditures were concentrated in applied research. One Western specialist on the German Democratic Republic warns, however, that many of the funds reportedly spent on development work were actually siphoned off to support current production[31]. Consequently, it is difficult to judge if some of the East European countries differ from the Soviet Union by emphasizing the importance of development work over applied work.

## C. Problems and Reform Efforts

### a) *Barriers to Innovation*

East European economists have written openly about the many difficulties their economies have encountered in the creation and use of new technologies. One Hungarian report to the UN stated tersely: "it is thought that the adoption, in practice, of scientific and technological achievements has not been fast enough. The management system is not yet sufficiently effective, and there are still deficiencies in using all of the possibilities of the international division of labour"[32].

The major barriers to technical innovation in Eastern Europe resemble closely those barriers that exist in the Soviet economy. The most detailed look at technical innovation in an East European country reveals that a number of familiar Soviet impediments to innovation have taken root in the German Democratic Republic's economy. Other studies have uncovered some of these same impediments in Czechoslovakia, Poland and Hungary, suggesting that they are systemic barriers to innovations in the East European economies[33].

One of the strongest barriers to innovation in the German Democratic Republic is the stress placed on fulfilling production plans, a stress that often overwhelms an enterprise's plans for the introduction of new technologies[34]. Other barriers also undermine the East European drive for greater technological progress. For example, serious pricing difficulties occur when German Democratic Republic or Polish enterprises produce new products[35]. (Entirely new products require the creation of new centrally set prices, and these new prices do not necessarily work to the benefit of the enterprise. For example, with the high start-up costs often associated with the production of a new product, an enterprise might easily discover that a new product's price was too low to offset these costs and that, consequently, the new product had lowered its profitability. The German Democratic Republic had made some clever attempts at avoiding these problems, but none solved all the problems inherent in setting new prices.) Success indicators for German Democratic Republic or Polish enterprises are often inconsistent and penalise innovation[36]. Planning periods can be too short for a German Democratic Republic enterprise to implement significant technologies[37]. The rigid supply system frequently retards a German Democratic Republic's facility's R&D project if the project happens to take an unexpected change in direction[38]. Finally, considerable problems arise in German Democratic Republican or Hungarian innovation programmes from poor communication between the customers and the supplier of new technologies[39].

### b) *Reform Efforts*

Although the German Democratic Republic's economic reforms for technical innovation have been given different names and are tailored for a smaller country, they resemble closely the reform programmes of the Soviet Union. The German Democratic Republic Fund for Science and Technology and its allied financial reforms parallel closely the Soviet Unified Fund for the Development of Science and Technology[40]. The German Democratic Republic's Board of Standards "Mark Q" for quality is a direct copy of the Gosstandart's "znak kachestva"[41]. Finally, the German Democratic Republic has implemented a serious effort at greater R&D contracting; much as has the Soviet Union[42].

Both Hungary and Poland have instituted reforms that affect the innovational performance of their industries. These reforms center on the attraction of Western technical know-how or the support of small projects rather than on changing the existing domestic organisation of R&D. The Hungarian government recently initiated support for venture capital investments. There are at present 6 or 7 venture groups associated with banks and other financial organisations. The Hungarian National Exchange Bank, for example, has underwritten about 200 venture projects. Most of these projects relate to the development of new technologies for production that can either reduce Hungary's hard currency imports or increase hard currency exports. One Western concern has invested about $100 000 in various Hungarian venture capital projects and will receive a share of the hard currency earnings if these projects succeed[43]. Polish legislation on direct investment established in 1976 a form of investment unique to Poland, the "Polonia" firms. While originally established for individuals with personal ties to Poland, many owners of these firms do not have such ties. While much of the production of the Polonia firms is aimed at the domestic consumer market, a recent Polish press report gave these firms credit for the recent growth in Poland's minicomputer industry[44]. Quite possibly the Polish government will be turning increasingly to the Polonia firms to provide flexible responses to its industrial high-technology needs.

### D. East European Innovational Performance and the Technology Gap

Since the East European economies experience the same type of innovational problems as the Soviet economy, it is not surprising that Western references to East European innovational performance resemble the largely negative Western evaluations of Soviet innovational performance. Some Western specialists have attributed the East European need for Western technology to the inadequacies in domestic East European RDI performance[45].

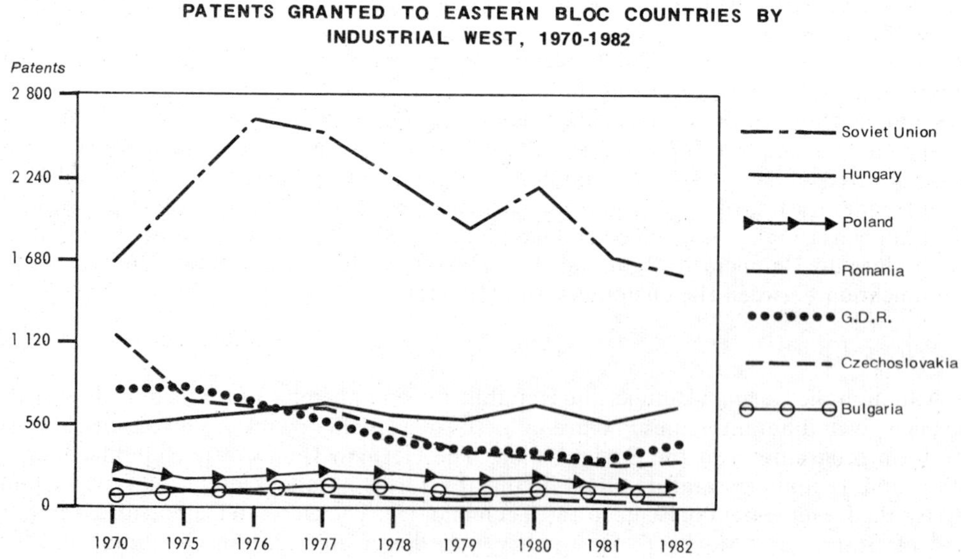

The European CMEA countries have campaigned vigorously for several decades to improve their economies' efficiency in the creation and use of new technologies. Yet, an examination of the patents granted to the East European CMEA countries by the major industrial Western countries shows that the number of grants have stagnated[46]. In fact, the European CMEA countries have never accounted for slightly more than 3 per cent of total industrial Western patent grants. This low level of patenting suggests that the past decades of campaigns have failed to raise significantly the technical level of East European and Soviet commercial R&D output[47].

There have been no studies of the technological gap in Eastern Europe similar to the studies published on the Soviet Union. References to the existence of a gap in Eastern Europe correspond closely to those conclusions made in the studies on the Soviet Union, i.e., the gap is likely widening in most areas of technology.

In this light, two experts on the Czechoslovakian economy wrote: "Czechoslovak data show a deterioration in the technical level of Czechoslovak production during the 1970s...the number of (new products) classified as being at 'world technical and economic level' almost halved between 1970 and 1977"[48].

The German Democratic Republic appears to have achieved, in general, the highest technological level of the East European countries, although Hungary is a technological leader in computers and pharmaceuticals and the Bulgarians make quite competitive fork-lift trucks. A thorough study of the German Democratic Republic concluded: "the German Democratic Republic appears to have had an especially pronounced (technological) lag behind the Federal Republic of Germany in the more modern branches of technology such as electronics, data processing, instrument building, synthetic fibers and plastics. The lag is probably due more to endogenous than exogenous factors..." Yet, the author of the study carefully reminds us that common types of economic systems do not necessarily engender similar economic performance and noted: "with the exception of electrotechnology and electronics, civilian industry in the Soviet Union seems to have lagged behind that in the German Democratic Republic by a substantial technological margin"[49].

## NOTES AND REFERENCES

1. "From organisations" means that the invention was filed through the patent department of an R&D or industrial facility. Many smaller Soviet institutes and factories do not have separate patent departments, and inventions developed at these organisations would not be included in the "from organisations" category. Consequently, the 7 per cent figure is an overstatement of the proportion of inventions coming from private individuals.

   Data from Ye. I. Artem'yev and L. G. Kravets, *Izobreteniya–Uroven' Tekhniki–Upravleniye*, 2nd revised edition, Izdatel'stvo "Ekonomika", Moscow, 1977, p. 47. Soviet data is from 1974, and US data is from 1983.

2. Both Soviet law and the Soviet economy discourage private R&D work. First, to obtain the exclusive rights to the results of any R&D work, a Soviet citizen may not have obtained any material assistance from the state. Such assistance likely includes the use of any state property. Second, the costs of obtaining exclusive rights are beyond the means of most Soviet citizens. Finally, the Soviet economy does not provide private individuals with the opportunity to purchase personal laboratory equipment. See The Statute on Discoveries, Inventions and Innovation Proposals of August 21st, 1973.

3. A prominent, former director of a leading Soviet military design bureau recounted in his memoires the difficulties he experienced in equipping his laboratories. See N. P. Fedoseyev, *Zapadnya*, Posev, Frankfurt, 1979. Clearly, if priority military institutes have difficulties in obtaining equipment, then one can reasonably assume that the problems for individuals are insurmountable.

4. Discoveries are carefully defined by Soviet law as "the determination of hitherto unknown objective laws, properties and phenomena of the material world bringing about fundamental changes in the standard of knowledge". The certification of discoveries by the State Committee for Inventions and Discoveries is meant to provide recognition for the fundamental sciences. As of 1975 the State Committee had certified relatively few (slightly more than 200) discoveries. The Soviet institutes obtaining the greatest number of discoveries were the: Joint Institute of Atomic Research at Dubna(19), Institute of Chemical Physics(15), Moscow State University(14), Institute of Radiotechnology and Electronics(7), Yoffe Physical-Technical Institute(7), and the Institute of Applied Mathematics(7). See V. A. Lisichkin and F. D. Novruzov, "Analiz tendentsii razvitiya nauki na osnove analiza reestra nauchnykh otkrytii", in *Organizatsionnye i metodologicheskiye printsipy prognozirovaniya nauchno-teknicheskogo progressa (Tezisy republikanskogo seminara 17 iyunya 1982)*, Kishnev, 1982.

   Inventors' certificates are also carefully defined and certified by the State Committee. The criteria for inventors' certificates resemble closely most Western definitions of patentability and include the stipulation that the invention be economically useful. As of 1983 the Soviet Union had certified over one million inventors' certificates. Data on inventors' certificates reflects the results of a random sample of inventions published in the Soviet journal *Vnedrennye izobreteniya*.

5. For exhaustive treatments of Soviet science policy, see E. Zaleski *et al.*, *Science Policy in the USSR*, OECD, Paris, 1969 and *Science Policy: USA/USSR, Volume II: Science Policy in the Soviet Union*, National Science Foundation, Directorate for Scientific, Technological and International Affairs, Washington, D.C., 1980.

6. See Louvan E. Nolting, *The Structure and Functions of the USSR State Committee for Science and Technology*, Foreign Economic Report No. 16, US Department of Commerce, Bureau of the Census, 1979.

7. These problems were continually discussed during the 1930s in a journal of the People's Commissariat of Heavy Industry, *Tekhnicheskaya informatsiya*.

8. The GKNT has sought to speed the flow of scientific-technical literature through large investments in VINITI (All-Union Institute for Scientific and Technical Information) and the State Public Scientific and Technical Library. It has also supported the technical information programmes being simultaneously carried out for patents and standards.

9. For a general description of the Soviet patent system see John A. Martens, "Soviet Patents and Inventors' Certificates", in *Soviet Economy in the 1980's: Problems and Prospects, Part 1*, Joint Economic Committee, Congress of the United States, Washington, D.C., December 31, 1982.

10. A good description of the State Committee for Standardisation is contained in M. R. Hill, "The Administration of Engineering Standardisation in the USSR," *The Quality Engineer*, Vol. 37, No.2, February 1973, pp.39-43.

11. The distinction between enterprise level and ministry level R&D is not always clear. Ministry level R&D usually refers to work done at central research institutes directly subordinate to a ministry, and enterprise level R&D, to work directly subordinate to an enterprise. Yet, the structure and duties of R&D facilities at very large enterprises and production associations often resemble more the central ministry facilities than the smaller enterprise facilities.

12. See Louvan E. Nolting, *The Planning of Research, Development, and Innovation in the USSR*, Foreign Economic Report No. 14, US Department of Commerce, Bureau of the Census, 1978, p. 1.

13. For a detailed discussion of the Soviet science budget see Robert W. Campbell, *Soviet R&D Statistics 1977-1980*, Report prepared for the National Science Foundation, Order No. 80-SP-0727, 1983.

14. See Louvan E. Nolting, *Sources of Financing the Stages of the Research, Development and Innovation Cycle in the USSR*, Foreign Economic Report No. 3, US Department of Commerce, 1973, p. 15.

15. L. H. Sarett, "Research and Invention", talk delivered at the National Academy of Sciences Symposium, April 26th, 1982, p. 4.

16. In the early 1930s tractors from the Krasnyi putilovets factory incorporated a suggestion for saving scarce materials that resulted in repair costs that equalled the original cost of the tractor. D. Nikitin, "Vrednye izobreteniya," *Izobretatel*, No. 1, 1932, p. 28. Some forty years later the Yefremov "Red Proletariat" Plant in Moscow produced a new machine tool that decreased somewhat the frequency of repairs but tripled the tool's production costs. Joseph S. Berliner, *The Innovation Decision in Soviet Industry*, The MIT Press, Cambridge, 1976, p. 354.

17. A fuller account of Scientific-Production Associations (NPO) is contained in Ronald Amann and Julian Coopen (ed.), *Industrial Innovation in the Soviet Union*, Yale University Press, New Haven, 1982, pp. 456-470.

18. *Ibid.*, pp. 496-497.

19. Berliner, *op. cit.*, p. 503.

20. John A. Martens and John P. Young, "Soviet Implementation of Domestic Inventions: First Results," in *Soviet Economy in a Time of Change*, Vol. 1, *op. cit.*, pp. 472-509.

21. Phillip Hanson, *Trade and Technology in Soviet-Western Relations*, Columbia University Press, New York, 1981, pp. 144ff.

22. Amann and Cooper, *op. cit.*, p. 66.

23. Little is known about the roles of the East European defense ministries' support for R&D.

24. "Tchecoslovaquie", in *Politiques scientifiques et technologiques nationales en Europe et Amerique du Nord: Etat actuel et perspectives*, UNESCO, Paris, 1978, p. 352.

25. Raymond Bentley, *Technological Change in the German Democratic Republic*, Westview Press, Boulder, 1984, pp. 174 and 226.

26. *Ibid.*, pp. 30-31.

27. The central science organisations are as follows: Czechoslovakia, the Federal Ministry of Technical Development and Investments; Poland, the Ministry of Science, Higher Education and Technology; German Democratic Republic, Ministry of Science and Technology; Bulgaria, Committee for Science, Technological Innovation and Higher Education; Hungary, the National Committee for Technical Development and Romania, the National Council for Science and Technology.

28. "Hungary", in *Politiques scientifiques et technologiques nationales en Europe et Amerique du Nord: Etat actuel et perspectives*, *op. cit.*, p. 216 and Bentley, *op. cit.*, p. 65.

29. On the whole, the use of standards and patents as administrative measures for domestic innovative success does not seem as widely used in the East European countries as in the Soviet Union.

30. For a discussion of decision-making and its effect on S&T see Paul Marer, *East-West Technology Transfer, Study of Hungary: 1968-1984*, *op. cit.*

31. Bentley, *op. cit.*, p.37.

32. "Hungary", *op. cit.*, p. 217.

33. Bentley, *op. cit.* See also Friedrich Levcik and Jiri Skolka, *East-West Technology Transfer: Study of Czechoslovakia*, OECD, Paris, 1984, Zbigniew Fallenbuchl, *East-West Technology Transfer: Study of Poland, 1971-1980*, *op. cit.*, and Paul Marer, *East-West Technology Transfer, Study of Hungary: 1968-1984*, *op. cit.*

34. Bentley, *op. cit.*, pp. 69, 75, 104 and 122.

35. *Ibid.*, pp. 79 and 110-115; and Z.Fallenbuchl, *East-West Technology Transfer: Study of Poland, 1971-1980*, *op. cit.*, p. 79.

36. Bentley, *op. cit.*, p. 73.

37. *Ibid.*, p. 73.

38. *Ibid.*, p. 68.

39. *Ibid.*, pp. 66-67 and P.Marer, *East-West Technology Transfer: Study of Hungary, 1968-1984*, *op. cit.*

40. Bentley, *op. cit.*, pp. 104-107.

41. *Ibid.*, p. 108.

42. *Ibid.*, pp. 168ff.

43. Unpublished interview with John W. Kiser of Kiser Research, Inc., March 1985, Washington,D.C.

44. "Will We Make It Without Sinclair?", *Politikya*, January 26th, 1985.

45. Agota Dezsenyi-Gueullette, "The Utilisation and Assimilation in Hungary of Advanced Technology Imported from the West," *Soviet Studies*, No. 2, Vol. XXXV, April 1983, p. 204 and P.Marer, *East-West Technology Transfer: Study of Hungary, 1968-1984*, *op. cit.*

46. The major industrial Western countries are as follows: Austria, Benelux, Canada, Denmark, France, Federal Republic of Germany, Japan, Norway, Sweden, Switzerland, United Kingdom and the United States.

47. The East European countries do produce some commercially competitive technologies. See John W. Kiser, *Commercial Technology Transfer From Eastern Europe to the United States and Western Europe*, prepared for the United States Department of State, January, 1980.

48. Levcik and Skolka, *op. cit.*, pp. 68-71.

49. Bentley, *op. cit.*, pp. 199 and 202.

*Chapter 4*

# INTRA-CMEA CO-OPERATION: IMPLICATIONS FOR TECHNOLOGY TRANSFER

## INTRODUCTION

The adoption in 1971 of the "Comprehensive Programme for the Further Intensification and Improvement of Co-operation and Development of Socialist Economic Integration" was the culmination of a renewed effort initiated by the Soviet Union in 1969 to achieve, over a 15-20 year period, the long-standing goal of intra-CMEA economic integration. This Programme laid out a set of broad measures which, though their emphasis has been nuanced or changed, remain essentially as the focal point for current efforts to improve and strengthen intra-CMEA co-operation[1].

Western analysts generally agree that the interplay between domestic problems and external pressures has stimulated closer bloc relations. External problems, particularly those related to trade with the West, are also believed presently to support the position of those in the East who have always argued for closer relations with the Soviet Union as a means of ensuring a more secure and stable market for exports.

Opinions differ, however, as to the likely effect of block integration on East-West relations and, in particular, on technology transfer. Differences stem partly from differing assessments of whether or not closer co-operation, or a higher degree of economic integration, will be achieved. They thus generally incorporate judgements made about the likely evolution of intra-CMEA trade.

In this chapter, no attempt will be made to *evaluate* regional integration *per se*. This is an issue which has been extensively discussed[2]. However, as the literature makes eminently clear, measuring the degree of, or progress towards integration, is extremely difficult and fraught with methodological difficulties. Furthermore, opinions differ as to what constitutes economic integration. There is even little agreement, either among Western or between Eastern and Western analysts, on how to *define* integration[3]. In any event, whether a certain measure taken to further co-operation does or does not lead to increased economic integration sheds little light on the issue of technology transfer as such.

However, by examining the nature and state of intra-CMEA co-operation from the Eastern perspective – general priorities laid down, specific sectoral or industrial branches singled out for co-operation and so forth – the likely significance of intra-CMEA co-operation for East-West technology transfer can be highlighted.

This is the principal purpose of this chapter. The chapter is also designed to provide background and perspective to the detailed statistical review and analysis of East-West and intra-CMEA trade and technology transfer undertaken in Part II.

This chapter is divided into two sections. The first seeks to set intra-CMEA co-operation into perspective. It discusses, in a rather general fashion, the motives for co-operation:

protection from international economic disruptions ; the desire to promote more efficient economic growth, the intent to become competitive on Western markets, and so forth. It also discusses those factors which constrain co-operation and then seeks to draw some general conclusions about the implications of co-operation for East-West technology transfer.

Co-operation, a much less ambitious form of economic interaction, is presently the most active. Thus, Section II examines those co-operative mechanisms currently being emphasised as being most germane for meeting the objectives of the Eastern countries. In the discussion, attention is also drawn to some major problems inherent to these mechanisms. Finally, some general conclusions are drawn about the likely significance for the Eastern countries of current efforts made to strengthen intra-CMEA economic co-operation.

## *Section I.* PERSPECTIVES ON INTRA-CMEA CO-OPERATION

### 1. Developments Promoting Co-operation

#### a) *Economic Growth Problems*

The "Economic Conference of the CMEA Member Countries at the Highest Level" which took place in Moscow in June, 1984, was the first Summit meeting of the CMEA since April, 1969. The declarations published subsequent to this Summit meeting as well as to the CMEA Council session held in Havana at the end of October 1984, indicate that the CMEA countries regard closer (and more effective) intra-CMEA co-operation as vital in order to solve some of their most pressing economic problems. Indeed, this theme was consistently evoked at the 37th session of the CMEA (October, 1983). Since the wording of the Summit declaration "On the Guidelines for Further Developing and Deepening the Economic, Scientific and Technological Co-operation of the CMEA Member Countries" and that of the communiqué published following the 37th session are similar, it appears that the content and thrust of the Summit declaration was basically approved at the 37th session[4].

The problems to be solved relate essentially to the achievement of lower costs per unit of output and greater efficiency in the use of resources – particularly fuel and energy. Given the tasks to be achieved (increased economic efficiency, more rational and thrifty use of existing material and manpower resources, better use of fixed assets, raising of technical standards and so forth), it is evident that greater concentration of effort is seen as the remedy to the long-term trend of slower or stagnating economic growth which has plagued the region.

Thus, the push for increasing co-operation reflects those problems and developments which have made the pursuit of "intensive growth" the central theme of the development strategies adopted by all the countries of the region. As noted in Chapter 1, these include labour supply problems and scarcer investment resources which have placed a premium on more efficient resource allocation.

Ability to sustain improved economic performance, as witnessed by the growth rates achieved in NMP and industrial production in 1983 and 1984, is closely related to keeping up recent improvements in labour productivity and to reversing the actual decline in capital productivity. Consequently, closer co-operation in developing science and technology, and particularly, accelerated application of R&D results, also continue to be singled out as major policy objectives. To this end, the Summit conference endorsed that a "Comprehensive

Programme of Scientific and Technological Progress for a Period of 15-20 years", be drafted. This programme is to be "formulated on the basis of national programmes and is to serve as a basis for developing a co-ordinated, and in some cases a single, scientific and technological policy". In sum, co-operation is seen as a primary vehicle for increasing economic efficiency as well as, of course, "broadening the possibilities of achieving socialist economic integration"[5].

b) *External Economic Disruptions*

i) *Energy and Raw Materials Issues*

Much of the present debate about co-operation relates, in fact, to the CMEA countries' preoccupation with energy and raw materials supply problems; an issue of common concern but with very different implications for the individual members of the CMEA.

For the Soviet Union, the problem is one of how best to adjust its fuel exports in such a way as to meet its own hard currency requirements and also the energy needs of its Eastern partners. The observation made at the 37th CMEA session to the effect that "co-operation in achieving energy savings, in the more rational use of energy and raw materials resources and in developing new sources of energy is being dictated by the need to adapt to changed conditions on all foreign markets", highlights the issue[6].

For the East European countries the problems are to secure adequate and stable supplies of energy and raw materials (especially oil) and to find the resources with which to pay for these. However, the general deceleration of the growth rate of Soviet fuel production will force some difficult adjustments within the CMEA area. Additional reductions in Soviet oil deliveries and further increases in the price of Soviet oil and gas cannot be excluded. As far as deliveries of oil are concerned, the Soviet Union had already made it clear in 1980 that these would be held at the 1980 level for the duration of the 1981-1985 Five-Year Plan period. The 10 per cent cut in oil deliveries which took place in 1982 can, therefore, be considered as representing an even greater squeeze on its partners than had evidently been foreseen. Concerning prices, in 1983 the price which most of the East European countries paid for Soviet oil, corresponded to world market prices. Gas prices also increased in 1983.

However, the tenuous economic situation of the East European countries, as well as their continuing negative trade balances with the Soviet Union (with the exception of Hungary), limits the extent to which the Soviet Union can resolve its energy problem by "exporting" it to its CMEA partners – either by raising prices or by cutting deliveries[7].

While the "energy problem", particularly if placed within the context of trade relations (East-West and intra-CMEA) provides a very strong motive for strengthening co-operation, it also seems that it provides an equally strong impetus for restructuring that co-operation.

Such a development is implicit in the position adopted by the Soviet Union already at the 1983 CMEA session; a position which reflects the commodity structure of intra-CMEA trade and, in particular, the dependence of the East European countries on Soviet energy and raw materials. These dependences will not be lessened due to the hard currency shortages which continue to plague the Eastern countries. On the contrary, it leaves the Soviet Union, regardless of price fluctuations, in the position of principal supplier – at least for the near future.

The Summit declaration concerning the main directions of the CMEA countries' economic development again called on the Member countries to jointly solve their energy and raw materials problems. To this effect energy and raw materials are to be used more rationally, energy consumption is to be reduced and energy conservation to be increased. At the same time, co-operation to develop alternate sources of energy is to be enhanced. As noted at the Summit, the structure of energy production is to be changed and co-operation to be expanded to include the development of atomic power generation.

The Havana meeting evidently endorsed this suggestion since it was agreed to draw up a long-term programme (up to 1990 and beyond) for co-operation in energy and raw materials development.

However, based on Soviet statements, it is evident that the improvement of joint efforts (more effective co-operation) is to be a *quid pro quo* arrangement. Put simply, the East European countries are henceforth expected to deliver more and better quality goods (and more quickly) to cover their energy and raw materials imports from the Soviet Union.

For instance, while the Soviet Union took due note of the CMEA countries' interest in receiving Soviet energy and raw materials and "would continue to make efforts" to make such supplies available in the future, these shipments would be "within the framework of her possibilities and depend greatly on the ability of the other CMEA countries to supply goods which are important to the Soviet Union"[8].

ii) *Machinery and Equipment*

The above-mentioned viewpoint, expressed relative to the 37th session of the CMEA, was reiterated at the Summit meeting. It is also clear from the Summit declaration that "meeting Soviet needs" is virtually an all-inclusive concept in that the Eastern countries are expected to increase deliveries of machinery and equipment, finished products as well as parts and components, spare parts for delivered machinery, food and consumer goods.

Furthermore, given the emphasis that is placed on speeding up the development of science and technology and on concentrating co-operation "more on applied research and on problems whose solution need a multilateral framework and which have a decisive effect on [raising] economic efficiency", it is likely that the East European countries will also be expected to orient their own R&D effort to meet Soviet needs[9]. This assertion is substantiated by the insistence on the part of the Soviet Union that "the production of internationally competitive machinery and equipment be increased" and that "a complete end be put to the practice of supplying obsolete products."

This standpoint was mirrored in the Summit declaration, in that, it was specified that the Eastern countries were to make the necessary capital investments in their industries in order to supply the Soviet Union with high quality machinery and equipment meeting world technical standards. It is, of course, the Soviet Union which is the principal client for capital goods traded within the CMEA area (nearly 60 per cent of the total) – though it is also the biggest supplier (40 per cent) of Eastern Europe's capital goods imports (see Table 7 of Chapter 6).

iii) *General External Factors*

Although energy developments in the 1970s have been a major disruptive factor on the CMEA economies, other developments, also external to their systems, have likewise posed a number of economic problems making co-operation more attractive – at least in principle.

One problem is, of course, the need to service the Western debt – particularly onerous for some of the East European countries. Debt servicing has already preempted a large part of the hard currency trade surpluses achieved during 1982-1984: surpluses which, until 1983, were mostly made possible by import reduction.

Furthermore, a recent study notes that East European exports to the West are losing many of their market shares to newly industrialised developing countries. If this trend continues, the East European countries will be hard pressed to sustain, let alone increase, their hard currency earnings[10].

Closely linked to financial considerations, are developments which have resulted in even less receptivity on Western markets for Eastern manufactures – commodities which, as previously noted, have always encountered marketing difficulties due to their general lack of competitivity.

Among these are political disturbances which had adverse effects on the East-West political climate and, in fact, led to the adoption on the part of some Western countries, of measures designed to regulate more closely technology transfers to the East[11]. The cumulative effect of these developments has been the main factor in constraining the import capacity of the East European countries.

It is evident that the Soviet Union's position differs from that of its East European partners. First, its gold and hard currency exports and reserves enable it to meet debt obligations relatively easily. Also, the very size of its economy allowed the Soviet Union to insulate itself from the worst of adverse external economic developments – particularly since the high share of fuels in exports ensured favourable terms of trade development in the 1970s and into the early 1980s.

Nevertheless, the USSR is far from invulnerable to them. The commodity structure of Soviet exports is heavily biased toward fuel, fuel products and raw materials: some 85 per cent of total exports, with oil and oil products alone accounting for some 66 per cent of hard currency earnings in 1983. Export earnings are thus very sensitive to conditions on world primary commodity markets – a factor from which the Soviet Union benefited in the 1970s, but which could prove a weakness under conditions of falling commodity prices.

iv) *Concluding Observations*

Whatever the ultimate effect of the various pressures on the future of bloc integration, it seems that the Eastern countries intend to achieve a greater measure of independence from the West.

The desirability of distancing the bloc countries economically from the West, notably by decreasing Eastern dependence on Western technology, is an issue that has been frequently debated in the past. At present, the debate is less ideologically based and more associated with justifying the merits/needs of bloc integration. In a *Pravda* editorial which broadly outlined the Soviet concept of future intra-CMEA developments, the point was made that "in developing and improving socialist integration, the countries of our community are strengthening their technical and economic independence from the capitalist West"[12]. This theme was reiterated at the 37th CMEA session with emphasis placed on the fact that the CMEA countries had at their disposal "all those instruments which would, by joint effort, ensure inviolability in the field of technology and [for] the economy" enabling them to "counteract the discriminatory measures taken against the CMEA countries in terms of trade, credit and economic blockades".

Some degree of disillusionment with Western technology may also have played a role in shaping these attitudes. Inability to utilise Western technology effectively (which may be due to one or more of several causes such as systemic barriers to innovation, the separation of R&D from production or inadequate level of personnel training and management ability), obviated many of the expectations which had led some of the Eastern countries to adopt an import-led growth strategy. The most extreme example of this is Poland; as expressed both in terms in which previous reliance on Western technology is now evaluated and continued reliance now unequivocally rejected: "our bitter experiences of the past years have clearly shown how dangerous over-extensive economic relations with capitalist countries are... By means of a closer inter-action with the CMEA countries we shall force a solution of our economy's key problems"[13].

It is significant that the need to become more independent from capitalist suppliers is most vigorously advocated by Soviet authorities and that the means to achieve independence is expressly linked to strengthening intra-CMEA co-operation and to changing the division of labour currently practiced within the CMEA. Particular emphasis is placed on increasing the "mutual exchange of machinery and equipment in order to counteract the tendency of relying on the Western countries for such commodities: a tendency which became marked during the 1970s"[14].

However, it is important to underline that the Summit declarations relative to further economic development and "on Safeguarding Peace and International Economic Co-operation", as well as the communiqué of the Havana meeting, emphasise the importance of maintaining mutually beneficial economic relations with the West. The Havana communiqué noted, in fact, that the CMEA countries were ready to strengthen commercial co-operation with Western firms. This clearly reflects the viewpoint (incorporated in the Summit declarations) that continued economic relations with the West remain vital for domestic economic growth.

## 2. General Trade Developments: Some Implications

Indications of the likely effects of increased intra-CMEA co-operation on East-West economic relations need to be set in the context of recent trade flows – that is, changes in the regional pattern and in the commodity structure of trade.

Consequently, the primary purpose of the following sub-sections is to highlight certain developments which set the stage for the discussion of intra-CMEA co-operation undertaken in Section II of this chapter. Given the detailed analysis of trade flows undertaken in Part II of this study, the discussion here is limited to general issues.

### a) *Foreign Trade Involvement*

It is well known that assessing the significance of foreign trade for the Eastern countries is difficult due to statistical problems and that differences in methodological approaches used may lead to different answers.

For instance, the measurement of a basic indicator – the share of imports or exports in national income – cannot be directly undertaken, since outcomes vary according to whether foreign trade in "valuta" national currencies is revalued in domestic or in dollar prices[15].

However approximate such data may be, they do lead to the conclusion that, *in general*, foreign trade represents a relatively small share in the economic activity of most of the Eastern countries. Moreover, the share of the European CMEA countries and the Soviet Union in total world trade (value terms) fell rather sharply during the 1970s – from 9.8 per cent to 7.8 per cent between 1970 and 1980 – and they had still not recovered their earlier share by 1984 (see Table 5 of Chapter 1). For imports, an initial rise up to the mid-1970s was followed by a sharp decline by 1980, though their 1970 share was also more or less re-established by 1984.

It is, therefore, important to recall that the Eastern countries' involvement in foreign trade *has* been maintained since the early 1970s. The decline in the Eastern countries' growth rates of trade with different areas of the world in the late 1970s-early 1980s was not uniform and in sum, not as great as those for world trade as a whole. This was due to the more buoyant performance of both intra-area trade and exchanges with the developing countries relative to East-West trade.

Nevertheless, the participation of the Eastern countries as a whole in foreign trade remains limited in comparison with their economic potential. This is important for the purposes of this discussion, since it nuances all observations or conclusions which may be drawn concerning the significance of the socialist international division of labour and the relationship of that development to the international division of labour in general.

After all, the underlying issue relative to assessing the future of bloc integration is the notion of dependence. Thus, it may be asked: given the adjustments which the Eastern countries must make in face of the problems currently encountered in East-West economic relations, will they turn increasingly "inward"? Notably, will they return to some version of the old Stalinist idea of "two world markets"?

Given the presently published Eastern attitudes towards the relative merits of turning "inward" or "outward", some of which are quoted in subsection 1*b) iv)* above, it is important to note that it remains an issue that is still debated in the East itself. While this issue is unlikely to be resolved in the near future (nor is it likely that it will be resolved in favour of either of the extreme alternatives), it is evident that the very differences in viewpoints among the countries concerned will themselves affect the evolution of intra-CMEA co-operation and the nature of bloc integration.

b) *Intra-CMEA Trade*

i) *General Developments*

Many Western analysts expect that East-West economic relations will be increasingly affected by the evolution of trade among the seven CMEA countries under review (hereafter abbreviated to intra-CMEA trade) particularly if the principal changes in the foreign trade strategies foreseen in the 1981-1985 Five-Year Plans are realised. As noted in Chapter 2, among these is the intent to increase considerably the share of intra-CMEA trade in total trade; a resolution reiterated at the Summit meeting of the CMEA.

For the purpose of providing some perspective on this objective (which implies a policy of turning "inwards") as well as assessing current developments, it is useful to recall some basic features of intra-CMEA trade.

Trade of the seven CMEA countries has always been bloc oriented, although the degree of bloc concentration has also varied greatly among the members of the CMEA. During the 1970s, such intra-CMEA trade accounted for the major share of the Eastern countries' total trade. It did, however, grow more slowly than trade flows with countries outside the area after 1975 (see Table 2 of Chapter 1)[16], and also over the decade as a whole.

Within intra-CMEA trade, flows between the Soviet Union and its six East European partners have been more important than trade among the Six. Also, while the East European countries are the Soviet Union's major trading partners, their share in total Soviet trade turnover (imports plus exports) has been declining since the 1970s while the Soviet Union's share in the East European countries' total trade has been increasing. As was noted in Chapter 2, this asymmetrical development in intra-CMEA trade led (until 1981) to increasing deficits in the trade balances of the East European countries vis-à-vis the Soviet Union.

The changes which occurred in terms of overall growth in intra-CMEA trade are not easy to evaluate since the time-span is too short to reach anything more than an interim conclusion. However, it does seem that these changes are less the outcome of a deliberate strategy (as forecast in the Five-Year Plans) than a response to external factors. Namely, the East

European countries' need to generate hard currency surpluses in order to reduce and service hard currency debt, and to reduce soft currency trade deficits with the Soviet Union.

Turning to the general development of Eastern area trade in the 1980s, the data indicate that between 1980 and 1984 intra-CMEA trade turnover increased by just under 11 per cent in value terms, which is slightly faster than in 1975 to 1980 (see Table 2 of Chapter 1). The growth rate of intra-CMEA trade was smaller than the growth of total trade in 1975 to 1980, but it was higher in 1980 to 1984.

The consequent but small rise in the share of the seven countries' mutual deliveries in their total trade turnover since 1980 is, in fact, only apparent on the import side. The share of intra-CMEA exports in the seven countries' total exports has stayed more or less constant over the last four years. This contrasts with a corresponding 5 percentage point rise on the import side. The increased stress on intra-CMEA trade has, moreover, affected some countries only – on the export side, Bulgaria, Czechoslovakia and the Soviet Union, but in the case of imports, all countries except Hungary were affected (see Table 2 of Chapter 1).

The shift toward CMEA partners in the total trade of the seven countries has occurred largely as a result of an increase in Eastern Europe's exports to the Soviet Union rather than from mutual trade with other Eastern European partners. That is, the radial pattern of exchanges centred upon the Soviet Union, which is a traditional feature of mutual trade, is being strengthened. By 1984, it can be estimated that some 64 per cent of the East European countries' intra-bloc trade was accounted for by the Soviet Union (Table 4 of Chapter 1).

ii) *Probable Future Developments*

It is likely that a continuing effort will be made to increase trade among the Seven since the share of intra-CMEA flows in total trade of the Seven in 1984 was still well below its share in 1970 and 1975. Nevertheless, as the pattern of intra-CMEA trade indicates, the burden of this lies clearly with the East European countries. In part, this is the natural outcome of the need to curtail imports from the West. In part, it is due to the fact that the Soviet Union is becoming increasingly unwilling to accumulate vast surpluses in its trade with the Six: surpluses which, given the inconvertibility of the transferable ruble and the nature of the trading system operating within the CMEA, constitute, in effect, credits[17]. Clearly the Soviet Union intends henceforth to treat surpluses/deficits as loans to be repaid by the delivery of more (and better) goods.

Moreover, the East European need for oil and raw materials also indicates that the Soviet role in intra-CMEA trade will continue to strengthen. Increased dependence on the Soviet Union is also likely for those countries encountering greatest financial difficulties with the West; notably Romania and Poland.

Finally, closer trading relations within the CMEA area will unquestionably be favoured by those regimes intrinsically fearful of the political and social consequences which closer economic relations with the West may bring; among these being the systemic reforms required for the effective assimilation of Western technologies[18].

Nevertheless, the relative buoyancy which intra-CMEA trade growth has exhibited during the past four years, must be placed within a wider perspective. First of all, the growth achieved for total intra-CMEA trade between 1980 to 1984 is no more than marginally higher than between 1975 and 1980 (see Table 2 of Chapter 1). This is not a particularly dynamic performance given the intention (as noted in the discussions surrounding the 1981-1985 Five-Year plans) that intra-CMEA trade was to become "the basis on which overall foreign trade strategies are to be formulated".

For the East European countries, the retreat from Western markets was clearly a defensive measure. For the Soviet Union, it is interesting to note that while growth of Soviet trade with the West slowed down or fell in value, both imports and exports have expanded (in volume terms) since 1980, albeit at modest rates. OECD data indicate that in 1984, 63 per cent of Western trade with the CMEA countries was accounted for by the Soviet Union; a considerable increase over 1975, when that share was in the neighbourhood of 49 per cent. While due in large part to successive declines in OECD exports to Eastern Europe in the four years 1981-1984, this hardly indicates, nonetheless, a retreat from East-West trade by the country currently most vociferously advocating greater independence from the West.

Finally, it must be borne in mind that one of the primary objectives of the Eastern countries for trading with the West have been to ensure, in the short-term, the maintenance of living standards by importing grain, and, for longer-term purposes (though at much reduced rates in recent years), to gain access to Western technology. The relative weight of these priorities over time will be an important determinant of the manner in which intra-CMEA trade develops.

c) *East-West Trade*

Taking the OECD area as a whole, trade with the East has only small economic significance for the Western countries, although there is variation among them as to their reliance on the East as an export market. Looked at from the viewpoint of trade shares, these also indicate that the West's involvement with the Eastern countries is modest. OECD data indicate that in 1984 only some 3 per cent of the OECD countries' total trade was accounted for by the Eastern countries.

The predominant position of the Soviet Union (and the diminishing participation of the Six) in the CMEA countries' trade with the West is particularly striking relative to Western exports. Western reported trade data show that while in 1980, the share of the Six in the Western countries' exports to destinations outside the OECD area was still 5.6 per cent, it had dropped to 3.6 per cent in 1984. In contrast, the Soviet Union's share which had been 5.8 per cent in 1980, increased to 6.5 per cent in 1984. Indeed, the USSR has become the second largest non-OECD partner of the Western countries taken as a whole[19].

That trade with the West is much more important for the Eastern countries than it is for the West can be illustrated by developments relative to imports. Thus, despite the policy of import reduction, purchases from the West still accounted for 27 per cent in the region's total imports in 1984 (about a quarter for Eastern Europe alone – see Table 7 of Chapter 1).

The policy of the CMEA countries toward trade with the West, as well as the degree of reliance on imports of Western technology, vary considerably. Turning first to some general features of East-West trade by Eastern country, Table 7 of Chapter 1 indicates that the percentage share of the West in the foreign trade of Bulgaria and Czechoslovakia is smallest, estimated within a range of 9 to 16 per cent for both imports and exports in 1984. For the German Democratic Republic and the Soviet Union it was moderate (25-30 per cent). In contrast, the degree of openness to the West is relatively high for such countries as Poland and Hungary. In 1984, these countries were most oriented towards the West with between 30 and 35 per cent of their imports and exports accounted for by the Western countries. The Western countries also purchased 42 per cent of Romanian exports in the same year, but their contribution to Romanian imports have halved since the late 1970s.

At the same time a growing divergence in the evolution of East-West trade between exports and imports, both of the Soviet Union and Eastern Europe, is becoming much more noticeable. The Western share of total Eastern imports has fallen markedly (from 35 per cent in 1975 to 27 per cent in 1984). This clearly reflects the efforts made by all the Eastern countries to eliminate balance of trade deficits and to increase net hard currency export earnings to service debt.

Bearing in mind the issue of turning "inwards", as well as the significance which Western technology holds for the Eastern economies, it is relevant here to briefly sum up some of the major findings (discussed in detail in Part II of the study) regarding developments relative to imports of Western embodied technology.

Thus, while the level of Eastern imports of Western capital goods and intermediate goods with a significant technology content fell in value and in volume terms since 1980, they still accounted for slightly over 60 per cent of the seven countries' total imports from OECD countries in 1983. The share stabilised in 1981-1983, reversing steady declines since 1975 (Table 6 of Chapter 5). The share of capital goods in total imports from the OECD fell sharply during the 1975-1980 period. In contrast, the corresponding share of imports of intermediate goods varied very little for Eastern Europe though it feel rather sharply in the Soviet Union after 1980. The data indicate that in 1983, capital goods still represented some 55 per cent of total Soviet embodied technology imports from the West, while for the Six, the capital goods share was 40 per cent (see Table 6 of Chapter 5). Thus, the priority awarded, or the resources available, to import Western capital goods were much more evident in the Soviet Union than in Eastern Europe.

Some first indications of the significance of Western technology for the Eastern countries is also given, on the one hand, by the distribution of the OECD countries' exports of technology among the Eastern countries and, on the other, by their relative shares in each country. With regard to the former, it is not, of course, surprising that in the total of the OECD countries' technology exports to the East, more than half is accounted for by the Soviet Union (value terms). However, if one looks at the shares of technology in the total OECD exports by country, a very different picture emerges. For instance, Bulgaria and Czechoslovakia, which have always been least involved in East-West trade as a whole (see Table 7 of Chapter 1), were among those where the share of technology in total OECD exports was among the highest – over 70 and 60 per cent respectively in 1983 (see Table 6 of Chapter 5).

It is clear that much less importance has been given to the *purchase* of Western technology within a reduced volume of imports as a whole. This caution has been dictated by financial constraints, particularly as far as the Six are concerned. The evolution of East-West technology transfer since 1975, confirms that the boom period of the first half of the 1970s, led by very buoyant investment plans and performance, is unlikely to be repeated.

Nevertheless, the commitment made by all Eastern countries to pursue intensive growth strategies will increase the need for productivity-enhancing technologies. This indicates that they will continue to rely on Western technology, to the extent that neither the RDI nor the Eastern management systems, as presently constituted, can produce the broad range of such technologies. Failure to produce sufficient domestic technology and, in particular the continued reliance on Western intermediate technology goods, will likely push the Eastern countries back to the West as customers. However, the rate at which technology imports pick up in the future will be determined by the growth of output as a whole and also by domestic allocation policies – notably, the availability of investment resources following a period of severe restraint. This will be a central issue in drafting the Five-Year Plans for the 1986-1990 period.

It is maintained here that, despite the concerted effort that is to be undertaken by the CMEA countries to increase and improve their production of those commodities vital for safeguarding economic growth and sustaining the production process, a continuing (perhaps even increasing) effort will also be made to have access to Western technology: whether the objective is to meet consumer demands or to increase exports. The latter is a long-term necessity if hard currency earnings are to be ensured.

## 3. Factors Inhibiting Co-operation

Since the adoption of the Comprehensive Programme in 1971, the CMEA official policy line has related the economic welfare of the Eastern countries directly to the degree of economic integration achieved. Integration is, in other words, viewed as an important component of the transition to intensive economic development.

However, the Summit declaration of June 1984 (which noted that "there are still considerable reserves that could be used to expand mutual co-operation") and discussion in the Eastern literature suggest that the policy of expanding intra-CMEA co-operation continues to be circumscribed. Indeed, it has even been maintained that "a long-term common strategy of economic development does not exist and this prevents concentration of forces and means on the decisive lines of scientific, technical and industrial development"[20].

The Hungarians have, in general, been most critical of intra-CMEA co-operation. One such criticism is unusual in its candour and illustrates the range of problems impeding effective economic interaction:

> "Although the Comprehensive Programme called for it, we have not really succeeded, to the extent desired, in bringing into operation the economic means of co-operation, price incentives, the foreign exchange system, the credit mechanism, the export stimulating and structure-oriented role of the market or – to put it very broadly – the material incentive in co-operation....The economic means exist, but only formally and not in their true function and, therefore, they live in a shadow life"[21].

This section summarizes some of the major problems which are holding back intra-CMEA co-operation. The aim here is essentially to place into perspective those measures (discussed subsequently) now being singled out in the official literature as the "key" instruments for bringing about effective intra-CMEA co-operation. It should be noted that these problems are widely discussed in the Eastern literature. However, it is also evident that there is little agreement on ways of resolving them. This in itself illustrates that economic integration will not only be difficult to achieve, but that the very concept continues to be a contentious issue.

### a) *Institutional Shortcomings*

One basic obstacle is that the Council for Mutual Economic Assistance (the CMEA) is not a supra-national economic power of the kind advocated by the Soviet Union in 1962 – a conception which continues to be vigorously opposed by the majority of its CMEA partners. Consequently, the Council is not vested with decision-making powers and can only make recommendations. Only when recommendations are accepted by all members do they become, at least in principle, binding economic agreements.

It is presumably for this reason that the Summit declaration included the observation that the CMEA Member countries find it necessary to enhance the role of the Council. However, it is questionable whether this is more than a palliative designed to accommodate the leadership in those countries who most rigorously advocate the need to forge closer and more effective links within the community.

Different management systems within the CMEA countries also hinder the formulation and execution of a joint economic development policy. At a basic level, the problem hinges on the ability of the enterprises themselves to conclude co-operative agreements.

With the exception of Hungary, the CMEA countries have directive economic management systems. The right of making economic (production) and commercial decisions thus belongs to the organs of the State administration, while the role of enterprises is limited to executing those decisions. It is pointed out that, "this presents a very strong restraint on co-operation because there is a possibility for settling only problems of outstanding importance at the State level – although this is the smaller part of co-operation"[22]. The practical result is that enterprises are effectively wedded to fulfilling *national* plans and, any agreements which might be concluded among them, must be grafted on to those plans. It is evident that this effective lack of enterprise decision-making capability and their consequent inability to conclude agreements, is a severe brake on industrial integration.

The Summit conference did draw attention to the desirability of establishing direct links between enterprises (as well as granting them decision-making powers) and to creating "favourable conditions" for setting up joint firms. This idea, which harks back to the Comprehensive Programme, was proposed again by Mr. Brezhnev at the 26th Soviet Party Congress in 1981 but has not been implemented to date.

### b) *The Pricing Mechanism*

Differences in national price systems and in the method of translating foreign prices into domestic prices are a serious practical obstacle to integrating joint activities at enterprise level on a clear-cut economically rational basis. Moreover, centralised price formulation affecting enterprise output costs and production reserves obviously limit the degree of freedom of enterprises engaged in joint activities.

The heterogeneity of pricing systems operating within the CMEA stems essentially from the different price relatives between different products and between sectors and branches of the economy in individual countries. This continues to be a fundamental obstacle to implementing multilateral projects and, as has been pointed out by a number of Eastern officials, "does not help cement integrative bonds".

Intra-CMEA foreign trade prices are set according to a different mechanism than other foreign trade prices. Current world prices determine trade transactions outside the CMEA area. However, for intra-CMEA trade, prices are only linked to world market prices via retrospective adjustments based on "moving averages" of past world prices – the so-called "Moscow formula" first adopted in 1975. Since trade prices follow – either directly or with a lag – those on world markets, they do not adequately take into account domestic costs nor do they reflect relative scarcities of CMEA goods within the region.

Thus trade transactions become a function of the overall bargaining strength of the countries involved – and not necessarily the financial outcome. The pricing system is one explanation for the fact that intra-CMEA trade relations continue to rest essentially on bilateral quota agreements. This leads to transactions not far removed from barter, an aspect taken up below[23].

One consequence of such pricing systems is that they adversely affect the ability to articulate a production strategy (notably, specialisation in production) which is economically "sound" and mutually advantageous; nor can projects be ranked according to rational criteria based on financial costs and benefits.

The need to reform the existing foreign exchange and internal pricing systems is the object of much discussion in the Eastern literature. In the event that world energy prices (notably oil) continue to fall, it can be anticipated that the pressure to do so will increase. However, the rather lame reference in the Summit declaration to the effect that the "existing system of price-setting in mutual trade and the monetary and financial instruments of co-operation will be improved further" gives no insight as to what decisions (if any) have been taken in this regard.

c) *Currency Problems: Implications for trade*

The "transferable ruble" (created in 1964) is the official intra-bloc clearing currency. It was intended to become a fully convertible currency, having real purchasing power, stimulating thereby the growth of multilateral trade among the CMEA countries.

In fact, although gold parity is claimed for it, the transferable ruble has neither currency nor commodity convertibility (see note 15). It cannot, therefore, satisfactorily represent the market value of goods traded[24].

In short, in contrast to the role foreseen for it, the transferable ruble is essentially only an accounting unit – an artificial currency. It is not used for payment either inside or outside the bloc. Its basic function is simply to express *bilateral* trade transactions in terms of a common unit of value.

The lack of an effective foreign exchange system is the major reason why within the CMEA area bilateral commodity exchanges, based on annual trade protocols or long-term trade agreements, continue to predominate. This severely constrains the possibility of achieving multilateral settlements – an important precondition for achieving a higher level of economic integration. It also places obstacles in the way of co-ordinating economic plans, an aspect again emphasized at the Summit session since such co-ordination is considered vital for achieving effective production specialisation.

It is evident that the lack of multilaterality will accentuate bilateral links both among the Six and between the Soviet Union and individual East European countries. However, since the volume and price of intra-CMEA trade is currently determined by annually agreed quotas, the trade position of the Soviet Union will unquestionably be strengthened. It is, after all, the strongest bargainer since it is the principal possessor of fuels and raw materials which also have a hard currency (real market) value. The East European countries can, in general, not offer goods with the same sales possibilities (as Soviet oil and raw materials) on Western markets and, therefore, cannot easily reduce their dependence on Soviet supplies.

d) *Divergence in National Interests*

An intrinsic constraint in achieving effective co-operation is the disparity in motivations among the members of the CMEA.

The Comprehensive Programme set out only the overall tasks for furthering economic co-operation. It did not define the mechanics or instruments of integration. As noted by one Eastern commentator, "in all certainty, this vagueness stems from a great and justified caution....[Namely], the political decision [to adopt the Programme] was not preceded by a basic and objectively scientific investigation and debate of the essence and methods of integration"[25].

The commentator is, clearly, referring to the fact that fundamental incompatibilities in viewpoints and national interests relative to certain kinds of co-operative approaches were not reconciled prior to issuing the Comprehensive Programme. Given the very different and often conflicting problems and policy options currently confronting the East European countries and the Soviet Union in meeting East-West and intra-CMEA trade requirements, such a reconciliation is likely to be increasingly difficult to achieve.

The divergence of national interests is well illustrated by those aspects of co-operation considered to be indispensable for developing a more efficient and more export oriented production structure mutually beneficial to CMEA countries: planning, the Long-Term Target Programmes and specialisation and co-operation in production.

i) *Planning*

Differences in national interests are, of course, the main reason why some East European countries have rigorously opposed supra-national planning. This also explains why the CMEA, as an organisation, is not vested with decision-making powers.

Romanian commentators have been most explicit on this point: "bilateral as well as multilateral plan co-ordination starts from the assumption that the make-up of the national plan as well as the direction and the choice of the application modalities are significant prerogatives of the sovereignty of each CMEA member country"[26]. The principle of respecting national interests and maintaining national economic independence in all co-operative efforts was reiterated in the Summit declaration.

Co-ordination of economic plans continues to be put forward as the basic tool for aligning economic policies in order to strengthen co-operation. At the same time (and perhaps because of the East European countries' insistence on maintaining a margin of manœuvre), the principle of unanimity in decision-making, as articulated in the Comprehensive Programme, will be upheld. Consequently, and as noted in the Summit declaration, plan co-ordination is to be limited to those areas already the subject of co-operation and to "other areas where it meets the interests of the individual countries".

The stress placed on preserving national sovereignty in planning reflects the reluctance on the part of the Six to strengthen the already predominant position of the Soviet Union within the bloc. Should supra-national planning be institutionalised, it would certainly further weaken the bargaining position of those countries which are most dependent on Soviet energy and raw materials supplies or which are most dependent on trade with the Soviet Union.

However, bilateral plan co-ordination and the conclusion of bilateral trade agreements with a country of the size, natural resource endowments and overall economic strength of the Soviet Union, inevitably tips the scales in favour of the latter country even under present arrangements. Nevertheless, the bargaining position of some countries – for instance those which have a high technological level (the German Democratic Republic) or which possess goods commanding hard currency payment (Hungary's grain and meat) – is somewhat stronger under the present system than it would be under supra-national planning.

ii) *Long-Term Target Programmes of Co-operation*

As will be discussed in the second section of this chapter, these programmes are designed to develop on a joint basis the "leading" sectors of the seven countries' economies. Their ultimate aim is to further production specialisation.

Whether the programmes will achieve this goal is problematical. For one thing, the scope of activities to be addressed by them was reduced primarily to resolving energy related problems. Hence, the goal of furthering production specialisation throughout the economies was clearly compromised[27]. Furthermore, they have been critized in the East as being economically unsound.

Some East European countries (in particular Hungary) have argued that joint projects foreseen under the Target Programmes are not economically effective and that economic criteria play a negligible role in their selection. This may be one explanation as to why participation in joint investment projects has been low relative to total national investment allocations – less than one per cent of the participating countries' total investments during the 1976-1980 plan period[28].

The hard currency "element" involved in these projects is also thought to be far too high, particularly since they require substantial imports of Western technology. In the case of Hungary, it is said that the "direct dollar content of joint investment credits granted by

Hungary during the 1970s to other CMEA countries amounted to 54 per cent of all such credits"[29]. Under current conditions, borrowing on Western financial markets to finance joint investment projects (or to purchase the technology needed to construct them) is obviously an expensive option.

Given the commitment made at the Havana session to undertake new joint investments in the sphere of energy, fuel and raw materials, it can be anticipated that the issue of financing will continue to be a bone of contention. As was noted by Czechoslovak Prime Minister Strougal, the estimated cost (some 45 to 55 billion transferable rubles) is very demanding and "represents a new situation".

Financial considerations apart, the conflicting national interests underlying the Target Programmes and their implementation call for radical negotiating and decision-making procedures to reconcile legitimate differences in approach of the countries concerned.

The Long-Term Target Programmes are not supra-national planning instruments, and participation in them is hence voluntary[30]. In the absence of supra-national planning, they are implemented by means of multilateral agreements, each of which is underpinned by a series of bilateral agreements. In essence, therefore, they differ little from ordinary trade co-operation agreements.

Furthermore, since they are co-ordinated, in principle at least, with Five-Year Plans, the resources required for their implementation should be provided within the framework of national plans co-ordinated at the CMEA level.

However, since national planning remains the prerogative of each individual country and the evolution of national plans is "dependent on the actual shape and structure of the planning system",[31] implementation may be far from perfect. In fact, it is this situation which leads to conflict concerning the size of the programmes, the objectives to be achieved by them, the resources to be made available and the measures to be used for their implementation. At issue are conflicting resource allocation requirements in general and, in particular, as they affect domestic needs vis-à-vis requirements for developing trade.

iii) *Specialisation and Co-operation in Production*

The basic purpose of specialisation and co-operation in production is to improve and render more efficient the CMEA countries' production structures.

Since production specialisation will involve policy adjustments at all points in the production chain – from S&T policies to the planning of final output – it is co-operation at this level which will determine the success or failure of the integration process in the long-term. However, the official literature suggests that there is, as yet, little agreement among the CMEA countries even on an operational definition of specialisation and co-operation.

In commenting on the Summit meeting, for instance, Prime Minister Strougal pointed out that "it will be necessary, above all, to agree upon the main directions of the individual states' specialisation within the international socialist division of labour... and to specify the details of production sharing"[32]. As noted also in the Soviet literature, while "agreements specify the general parameters of co-operation, they do not include the basic economic conditions of the products which are the object of specialisation and co-operation – prices and delivery dates"[33].

Difficulties in so doing stem basically from problems inherent to the pricing, financial and management systems, as previously discussed. These problems are said to account for the fact that "deliveries resulting from specialisation agreements are modest, with co-production constituting barely two to three per cent of intra-CMEA trade turnover"[34].

Implementation of specialisation and co-operation in production, even if operational plans at a detailed level are agreed upon, will have immediate effects on the structure of output and demand which national interests might find difficult to accommodate. These difficulties will be even harder to resolve within the limits of factor supply constraints which are now, as noted in Chapter 1, affecting all the countries concerned. Under these conditions, policy makers are less willing to divert resources needed for the production of goods for immediate use in the context of national plans, in favour of international policies whose effects might not be attainable except in a longer-term time frame.

Also, given the difference in the level of economic development among the Eastern countries, specialisation (particularly if institutionalised) could exclude a lesser developed country from participating in newly arising industries. It could also have adverse effects on the use made of production output. For instance, those countries whose production of machinery and equipment is primarily geared for hard currency markets risk to find such commodities diverted from these markets to the CMEA.

Thus, divergent interests and, in particular, the possible diminution of current degrees of self-reliance, account basically for the fact that discussion continues to focus on the *need* to enhance production specialisation: an indication of the fact that its actual implementation is encountering difficulties.

## 4. Complementarity between Intra-CMEA Co-operation and East-West Technology Transfer?

It is the belief of a number of Western analysts – one that is shared by the authors of this study – that a certain complementarity will continue to exist between intra-CMEA co-operation and East-West technology transfer.

The need to decrease their dependence on Western technology (or, for that matter, on items such as spare parts and components for Western investment goods) is not questioned by the Eastern countries. Nevertheless, as was made quite clear at the Summit meeting, the CMEA member countries "have always opposed economic isolationism" and continue to be "prepared to develop mutually advantageous trade, economic and scientific and technological ties with developed capitalist countries".

According to the East European and even, at times, the Soviet literature, continued access to Western technology will remain a long-term requirement for economic development. As has been noted, co-operation with the industrialised Western countries is a "valuable supplementary source for increasing national income and for realising plan targets"[35].

Integration into the world economy also ensures "that the most developed techniques and technologies flow into all of CMEA through the importing countries"[36] since the "turning inward alternative would lead to a situation in which the CMEA countries would lag increasingly behind the world economy"[37].

Even in the Soviet Union, the country currently the strongest advocate of the "turning inward" alternative, the viewpoint is voiced that "increasing the competitive ability of goods being produced within CMEA on world markets presumes precisely stimulating the participation of these countries in worldwide commercial and economic relations"[38].

Such considerations apart, discussion in the Eastern literature suggests that access to Western technology is also seen as an instrument for achieving more effective intra-CMEA co-operation. It is argued, for instance, that "in many respects co-operation with the capitalist countries is an integral part of the CMEA countries' economic stragegy for the 1980s.... Co-operation with Western industrial countries does not run counter to deepening the scope of CMEA integration. Indeed, we assume that there is every chance for intensifying integration"[39].

As far as the role of Western technology is concerned, past experience indicates that its use was decisive in implementing those resource development projects which were the central part of the "Co-ordinated Plan for Multilateral Integration Measures" approved by the CMEA in 1975, for the 1976-1980 Plan period.

A detailed Western analysis of these joint (CMEA) investment projects indicates "that they were designed to incorporate Western inputs. Western equipment, technology and credits played a key role in the success of these projects, serving to alleviate some of the traditional obstacles to intra-CMEA co-operation"[40]. While, as the study points out, Western help was used for a number of them, it was most important for the Orenburg gas complex in terms of equipment supplied, participation in design and construction of facilities, the technology made available and also the funding.

There is little reason to believe that Western inputs will not be needed to implement the joint investment projects foreseen for the post-1985 period. Indeed, even prior to the Summit meeting it was pointed out that current developments in the energy sector dictate that priority be given to projects designed to rationalise the use of energy. That is, efforts must be directed towards bringing about energy conservation. "To accomplish this, however, the CMEA would need equipment and technology that can be acquired mostly only from the West"[41].

As far as intra-CMEA specialisation is concerned, other case studies indicate that economically the most important co-operation contracts concluded, for instance, between Hungarian and Western enterprises, were linked to actual CMEA specialisation agreements[42].

Available evidence indicates that there is also a certain complementarity between intra-CMEA scientific-technological co-operation and West-to-East technology transfer. While the former aims at developing primarily high-intensive technologies (though, hitherto, with mixed results), the latter appears to have been mainly concentrated on the purchase from the West of relatively middle- and low-intensive technologies.

Thus, the commodity structure of technology trade, the types of commodities to be produced on the basis of specialisation and intra-CMEA co-operation programmes in the field of research and development (as well as Eastern interest in pursuing industrial co-operation with the West) all point to the existence of a potential interaction between technology transfer from the West to the East and the modernisation policies adopted by the Eastern countries. While the Eastern modernisation strategy now increasingly emphasises better and more effective intra-CMEA co-operation, continued access to Western technology remains as a key input in achieving it.

## *Section II.* INTRA-CMEA CO-OPERATION: CURRENT EMPHASES AND IMPLICATIONS

This section examines measures currently singled out by Eastern planners as best suited for achieving the long-term economic objective of restructuring and raising the effectiveness of the CMEA countries' production structures: plan co-ordination; the Long-Term Target Programmes of Co-operation; specialisation and co-operation in production and, co-operation in the area of science and technology. The purpose of the discussion is to clarify Eastern policy objectives and, in so doing, illuminate their relevance to East-West technology transfer.

Given the paucity of concrete information, certain of these measures can only be described or analysed in a limited fashion. Concrete "measurable" outcomes of the co-operative mechanisms are precisely those where data limitations are most acute.

## 1. Plan Co-ordination: Overall Objectives

Efforts made to improve plan co-ordination ante-date the adoption of the Comprehensive Programme[43]. Work on co-ordinating national economic plans was already initiated in 1954,[44] and was actually carried out for the first time in connection with the preparation of the Five-Year Plans for 1956-1960[45].

According to official CMEA literature, "co-operation of the CMEA-countries in planning and in co-ordinating plans for the development of the national economies is the main method of organising economic, scientific and technical co-operation and for extending the international division of labour." In the absence of a market mechanism, plan co-ordination becomes the primary administrative method by which the CMEA countries can achieve a measure of co-ordination in economic development. This is, of course, the reason why so much emphasis is placed on it – particularly by the Soviet Union.

The importance of co-ordinating economic plans has always been acknowledged. However, since the mid-seventies, a shift in concept and method seems to have taken place.

First, the range of co-ordinated planning activities has been expanded to include investment, joint construction of production capacities, specialisation and partnership in production and joint research in specific scientific fields. At the same time, and as specified at the Summit conference, co-ordination of national plans will, in future, be focused on the solution of "priority tasks". It is also evident from the Summit meeting that these priority tasks pertain, first and foremost, to resolving the energy and raw materials supply problem.

Secondly, the time horizon for co-ordinating planning has been extended. This reflects the agreement reached in 1976 to formulate Long-Term Target Programmes of Co-operation spanning a 10-15-year period. Current discussion in the Soviet literature even underlines the desirability of extending plan co-ordination well beyond the 1986-1990 plan period.

Finally, persistent problems in achieving the desired degree of long-term co-ordination undoubtedly accounts for the fact that the method of organising plan co-ordination which has already been modified, is still under review. As noted in the Summit declaration, measures will be taken to complete plan co-ordination (essentially assignment of tasks in order to implement the economic development strategy decided upon) before the beginning of a new plan period. The results of this co-ordination will henceforth be formalised in corresponding agreements and included in the national plans. Given some of the pricing policies, conflicting perceptions of national interests and other problems which still impede intra-CMEA integration (described in Section I above), achievement of this objective can be expected to be particularly difficult.

### a) *The Co-ordinated Plan of Multilateral Integration Measures for the 1976-1980 Period*

Adopted in 1975, this Plan was designed to supplement existing forms and methods of co-ordinating the national economic plans of the CMEA Member countries, and was to be integrated into each of the national plans for the 1976-1980 period.

Officially described as providing a new and more effective form of economic co-operation, the Plan's main objectives were: to solve major production problems; provide for specialisation and co-operation on a multilateral basis – primarily for products of the machine-building industry, electro-technology/electronics and the chemical industry; solve specific scientific-technical problems and, provide for further joint research. It was also designed to raise the level of economic development of the less-developed members of the CMEA (notably, Mongolia)[46].

Although the details of this Plan have never been published, it is evident that Section I, which lays down measures for the joint construction of industrial projects designed to resolve the "pressing task of meeting the growing fuel and raw materials requirements of the Member countries", was considered to be the most important part[47].

In particular, joint investment in projects designed to secure raw materials, energy and fuels were clearly seen as constituting the essential element of multilateral integration. The official literature mentions numerous joint construction projects. Figures vary between 28-30 (a sample is given in Table 1 of Chapter 4). However, the financial resources allocated for the realisation of all these multilateral joint investment projects for the 1976-1980 period were, in fact, made available for only ten of them[48]. Of these, the majority were concerned with meeting energy and raw materials needs and most were located in the Soviet Union.

It should be mentioned that few of the multilateral projects retained in the Plan were new as "most had already been agreed to, and were merely incorporated ex-post into the co-ordinated Plan"[49]. As such, these projects can hardly be regarded as representing a new departure in co-operation – the format frequently used in the Eastern literature. Relative to financing, no new approach is evident either. As has been noted by a number of Western analysts, the financing mechanism essentially took the form of traditional bilateral compensation[50].

b) *The Co-ordinated Plan of Multilateral Integration Measures for the 1981-1985 Period*

Finally adopted in 1981, this Plan was also essentially concerned with resolving energy and raw materials supply problems. More modest than its predecessor, only some 2.5 billion transferable rubles were apparently allocated for the implementation of joint investment projects.

Presumably reflecting the fact that the East European countries were increasingly reluctant to commit scarce investment resources to this type of co-operative measure, only three projects were to be undertaken. These are: the Khmel'nitskiy nuclear power plant with a capacity of 4 000MW; the 750 kV transmission line from the Khmel'nitskiy nuclear power plant (USSR) to Rzeszow (Poland), and the Mozyr plant for the production of feed yeasts from paraffin oils of high purity (on the territory of the USSR)[51].

Perhaps more significant for the economic development strategy to be pursued, particularly for the 1986-1990 period, was the intent to achieve closer co-ordination in formulating economic development (notably industrial) policies.

Thus implementation of the Co-ordinated Plan for 1981-1985, was closely linked to execution of the Long-Term Target Programmes and the bilateral long-term co-operation agreements. This was, in fact, singled out as being the new feature of this Plan[52].

c) *Current Developments Relative to Plan Co-ordination*

The "Programme for the Co-ordination of National Economic Plans for 1986-1990" (approved at the 36th CMEA session in 1982) serves as the basic input for preparing the "Co-ordinated Plan for Multilateral Integration Measures for the 1986-1990" period.

Furthering multilateral co-operation, in the interest of solving mutual economic problems, has thus clearly now become a major pre-occupation. Consequently, the draft programme for co-ordinating the 1986-1990 economic plans foresaw (a provision referred to as being fundamentally new) not only the co-ordination of national plans, but the formulation of an "economic policy as a whole"[53]. Furthermore (and also said to be a new approach), plan co-ordination will, henceforth, interlink more closely production and S&T co-operation – an aspect on which particular emphasis is placed in articulating the multilateral plan. As official discussion of the Summit meeting indicates, plan co-ordination for the 1986-1980 period and beyond is designed to improve the structure of production and to further "intensive" economic growth.

Consequently, priority areas for plan co-ordination, already singled out at the 37th session of the CMEA, include:

- across the board reconstruction and technical modernisation of the CMEA countries' production apparatus;
- priority development for such "technical progress embodying" branches as electronics, new materials and biotechnology;
- creation and development of new technologies and machinery for the food-stuff and light industries, for processing of raw materials and agricultural products and for the automation of production processes;
- better use of scientific, technical and production potential in order to eliminate duplication in production and unwarranted imports of raw materials, equipment and technology from capitalist countries,
- creation of increased export potential by the CMEA countries to compete and strengthen their positions in the world economy[54].

Current discussions relative to the co-ordination of national economic plans reflect that the energy issue will again determine the formulation of an integrated economic development strategy for the rest of the decade and possibly beyond.

Even prior to the Summit conference, Soviet authorities had been explicit that the factors which have made energy, notably oil and gas production, more costly to the Soviet Union, "influence the possibilities of increasing the production of petroleum and gas, the main supplier of which to the European CMEA Member countries is the Soviet Union. It is impossible not to take this into account when determining the long-range direction of co-operation". Consequently, in order to bring about a more economic and efficient use of fuel and energy resources "countries must concentrate efforts on the production of technically advanced equipment....[notably] on specialisation and co-operation in the production of machines and equipment"[55].

As mentioned previously in Section I of this chapter, this viewpoint was reiterated both at the Summit conference and at the Havana session of the CMEA and resulted in the commitment made by the East European countries to secure energy and raw materials supplies by means of producing and supplying the Soviet Union with goods "equally significant" for meeting that country's economic development needs.

Despite their reluctance (primarily because of the considerable financial burden entailed), the East European countries have committed themselves to continue their participation in joint investment projects. This commitment reflects the Soviet position that under current conditions plan co-ordination also "includes co-ordinating structural policy in order to provide for effective compensation – both in money and kind – entailed in the production of fuel, energy and raw materials. Thus plan co-ordination must reflect the willingness of interested countries to use jointly natural, manpower and financial resources to meet their common interests"[56].

As agreed to at the Havana meeting, the joint investment projects will include:

- construction of the Krivoy Rog Iron Ore Processing Complex (USSR) with the participation of Czechoslovakia, the German Democratic Republic, Hungary, Romania and the USSR;

- construction of the Yamburg gas pipeline (USSR) to the Western Soviet border and production facilities for the Yamburg gas fields in order to increase gas shipments to Eastern Europe;
- joint use of hard coal in Poland, development of Poland's hard coal fields and modernisation of its coke industry;
- common exploitation and processing of magnesite beds in Czechoslovakia;
- development of cobalt and nickel production in Cuba;
- mining of non-ferrous metals in Vietnam, Mongolia and Cuba;
- development of bauxite mining in Vietnam and of phosphorites in Mongolia,
- increased joint development of atomic energy plants, heat supply stations, installations for the rational and economic use of materials and further integration of power generating systems.

However, despite the emphasis placed on furthering multilateral co-operation, discussions in the Eastern literature suggest that bilateralism in the achievement of stated objectives will prevail.

For instance, plan co-ordination will, as was past practice, be completed by bilateral protocols. A list of mutual goods to be delivered will be a component part of such protocols and they will also include the results of co-ordination in the field of capital construction[57].

The continued reliance on bilateral mechanisms in plan co-ordination is also suggested by the importance attached to developing specialisation agreements on the grounds that "inter-branch specialisation will increasingly determine the growth of mutual trade". Bilateral agreements for specialisation and co-operation in production have, for instance, been drawn up between the Soviet Union and each of its partners and will be operational until 1990. As noted, "these programmes have acquired the nature of political-economic documents which reflect the specific nature of the bilateral integration processes between the countries of the socialist community"[58].

d) *Implications*

The implications of the developments outlined above are essentially three-fold:

Firstly, efforts made to strengthen the planning mechanism and the evident attempt that will, once more, be made to co-ordinate national economic policy at the central level, indicate that the traditional path of steering the Eastern economies by primarily administrative means will be pursued. However, it can also be anticipated that such a policy will continue to be strenuously opposed by those Eastern planners who advocate not only the need for fundamental reforms of the domestic economic systems, but of the CMEA as an institution.

Secondly, it is also evident, that the Eastern countries will pursue a production strategy which is both more selective and increasingly export-oriented. The 37th CMEA session had already confirmed this shift and it is also reflected in the sectors and commodities which have been singled out as priorities in the Target Programmes and in the specialisation agreements.

Finally, to the extent that emphasis is still being placed on bilateral agreements, the Soviet Union's role in formulating a common CMEA development strategy will be greatly strengthened – particularly, since Soviet planners have made it very clear that resolution of the energy problem is the *sine qua non* for the manner in which intra-CMEA co-operation will, in future, evolve.

## 2. The Long-Term Target (Special Purpose) Programmes of Co-operation (LTPCs)

### a) *Types of Programmes Adopted and Overall Objectives*[59]

It should be noted at the outset that these Programmes have never been published. Consequently, the discussion here reflects what is said about them in the Eastern literature.

In essence, the LTPCs extend economic plan co-ordination up to 1990. Their essential goals are to develop key sectors of the economy, advance the articulation and implementation of a long-term (10-15 years) economic develoment strategy, solve the most important economic development (notably production) problems and to provide the CMEA member countries with needed industrial and consumer goods.

Five Programmes have been adopted – the first three in 1978, at the 32nd CMEA session[60]. These cover:

*Energy, Raw Material and Fuels*

This Programme provides for co-operation in developing the most important types of fuels, ferrous and non-ferrous metals, chemical raw materials and electrical energy. It stipulates the development of indispensable raw material deposits and new energy sources; mainly atomic energy.

*Agriculture and Food Industries*

The Programme is directed towards meeting essential requirements for high-quality food items. It includes provisions for developing the technical and material base of agriculture and the application of chemicals in farming and land reclamation. It also envisages: furthering reciprocal deliveries of mineral fertilisers, machinery and equipment for the agriculture and food industry, means of transportation and equipment for refrigeration services; strengthening the fodder base and the production of protein fodders and, the possible establishment of special funds to finance and stimulate the production of farm and food output.

*Mechanical Engineering*

This Programme will be carried out primarily on the basis of specialisation and co-production. It is to result in the introduction of modern engineering and manufacturing technologies into the economies of the CMEA countries. The ultimate objective is to bring about an overall increase in factor productivity. Co-production will also begin for some types of machines and production flow lines not hitherto manufactured in CMEA countries, but which are vital for solving problems related to technical re-equipment and for modernizing the economies of the CMEA countries.

The final two Target Programmes were adopted in 1979 at the 33rd session of the CMEA. They concern:

*Industrial Consumer Goods*

This Programme will be implemented primarily on the basis of specialisation and co-production and production-investment joint ventures – in CMEA as well as in third countries. The principal purpose is to increase the supply of raw materials for production of high-quality industrial consumer goods (as well as a wider assortment of them) and to achieve systematic improvements in living standards. Increased reciprocal deliveries of consumer goods will play a growing role in achieving these goals.

*Transport Systems*

Co-operation within the framework of this Programme consists mainly of establishing a unified transport system in CMEA countries which would integrate different modes of transport (rail, sea, river, air, vehicle and pipe-line transmission). Specialisation and co-production ventures in the field of railroad rolling stock and other transport equipment will play an important role. Operational procedures are also to be standardised.

b) *Strategy and Provisions of LTPCs*

In the general development strategy foreseen for the 1981-1985 Plan period and beyond, the LTPC to develop the energy sector was clearly given priority[61]. The Summit conference and the subsequent agreements reached at the Havana session relative to energy and raw materials development indicate that this priority will be maintained. Nevertheless, in terms of the measures envisaged, it is the programme to develop mechanical engineering which is the most extensive (see Table 2 of Chapter 4).

As early as the 32nd session of the CMEA, attention was drawn to the fact that fulfilment of the measures contained in the Long-Term Target Programmes would depend primarily on the ability of the CMEA countries to "provide new lines of production with the necessary equipment: that is, in the final count, it would depend on the strength of the machine-building industry of the community. The priority task now is... to concentrate the efforts of our machine builders on the major projects outlined in the special purpose programmes"[62].

Enhanced development of the mechanical engineering sector is closely linked to the goal of ensuring long-term economic growth and output of competitive manufactures. Hence the objectives laid down for this LTPC, illustrate that it is regarded as being the keystone for realising the development targets mapped out for *all* the LTPCs. This has been made quite explicit: "the machine-building programme is intended to ensure high growth rates in the production of modern machinery, equipment and instruments for the fuel, power and raw-materials industries, agriculture and the food industry, light industry, transport and to meet the needs of the machine-building sector itself"[63].

The agreement reached at the Summit conference to the effect that co-operation in re-enforcing the machine-building industry would be "comprehensive" in order to meet supply requirements of key production branches, as well as the decision taken at the Havana session to create a new CMEA committee for co-operation in the area of machine-building, suggest that this LTPC will move up the priority ladder.

A similar conclusion can be reached relative to the Agriculture and Food Industries LTPC. As noted in the Soviet literature, the CMEA countries are expected to participate more actively in putting into effect the USSR "Food Programme" and the "Long-Term Programme of Land Reclamation".

It is this expectation which presumably stands behind the adoption by the CMEA (in October, 1983) of the "Comprehensive Measures for Improving the Food Supplies of the Population of the CMEA Member Countries": a set of some 47 measures designed to supplement the original LTPC. Building up the agro-industrial complex (and increasing co-operation in this area) was assigned a "top priority" at the Summit conference. This priority will, presumably, receive a good deal of attention from those East European countries most reliant on Soviet energy and raw materials.

c) *Implementation*

Implementation of the measures foreseen by the LTPCs is primarily by means of multilateral agreements. A total of 232 were foreseen, of which some 120 had actually been signed by the end of 1980 (see Tables 2 and 3 of Chapter 4). Information as to how many are currently in operation is contradictory (and not up to date). Some sources give a figure of 200 by the beginning of 1982, while others maintain the number given for 1980 (i.e. 120 signed agreements)[64].

Multilateral agreements are backed up by bilateral agreements. Of these, increasing importance is being placed on those concerned with specialisation and co-operation in production. By the end of 1980, more than 1 000 such agreements are said to have been concluded[65].

d) *Financing the LTPCs*

Funds under the LTPCs, adopted in 1978-1979, were forecast at 70-90 billion transferable rubles. Presumably they were to cover the period up to 1990, athough this is not clear. For 1976-1980, the budgeted figure was some 20 billion TRs and, as mentioned, for the joint investment projects agreed to at the Havana session of the CMEA, some 40-50 billion transferable rubles will be needed. Again, the time-span of the expenditures is not clear, but presumably it is for the 1986-1990 period and beyond.

To date, financing has, in essence, been by means of joint investments. These investments were, as noted, primarily devoted to securing needed supplies of energy, raw materials and fuels. This emphasis is also illustrated by the allocation of loans made by the International Investment Bank (the IIB, founded in 1970). It is said that by 1983, the Bank had "credited 83 projects, estimated at a cost of 10 billion transferable rubles. About 70 per cent of the credits went to the development of the fuel-power industry, 19 per cent for the machine-building industry and electronics, 9 per cent for metallurgy and chemistry and 2 per cent for transport and communication development"[66]. No information as to the future distribution of IBB credit facilities was found in the literature consulted.

Financial considerations related to the allocation of investment resources are likely to become major policy issues. The manner in which they are resolved will affect not only implementation of the objectives contained in the LTPCs, but also the development strategy mapped out for the 1986-1990 period.

It should be recalled that joint investments are, in fact, little more than barter arrangements. Credits are extended against delivery of products/commodities resulting from realisation of a given project. Uncertainties in such areas as calculating satisfactorily the future costs and benefits of joint investment projects escalates, therefore, the element of risk.

Guidelines for drawing up the 1986-1990 economic development plans indicate that, in future, greater emphasis will be placed on investment co-ordination.

However, this will not solve the problem of financing the ambitious targets foreseen by the LTPCs since what is meant by investment co-ordination is an issue which continues to be widely discussed in the Eastern literature. As pointed out by one commentator, "although co-ordination of capital investments is of great significance from the viewpoint of realising the most important national economic tasks, certain theoretical and methodological problems of such co-ordination, including its basic directions...are among the least investigated"[67].

Given the importance attached to implementing the LTPCs primarily by means of bilateral specialisation agreements, the presumption is that the East European countries will use every opportunity to direct scarce resources towards development of those projects which meet their individual internal and export needs.

Emphasis is indeed placed on the fact that co-ordination of capital investments is to be essentially a "voluntary activity, limited to that part (volume) directly connected to participation in the construction of new projects or in the expansion, reconstruction, modernisation and technical re-equipment of existing enterprises, the outputs of which are used to solve common economic problems, on the basis of developing the international socialist division of labour"[68].

Nevertheless, the Soviet position relative to the future evolution of intra-CMEA co-operation, suggests that the USSR will have a major say in defining "voluntarism" in the allocation of investment resources.

## 3. Specialisation and Co-operation in Production

### a) *Overall Objectives*

The agreements governing specialisation and co-operation in production are designed to implement the tasks contained in the LTPCs and the objectives foreseen by plan co-ordination. The purpose is "to strengthen synchronised development of the national economic complexes of the CMEA member countries"[69], furthering thereby the articulation of a long-term, co-ordinated economic development strategy

As is noted, this type of co-operation is "no longer confined to redistributing production programmes in order to reduce the range of output and to agree upon mutual shipments arising out of this reduction". Rather, the basic objectives are to "reorganise current national production structures and to raise production efficiency by redistributing the range of goods produced and by defining the volume of goods exchanged: creating thereby the most important condition of efficiency – concentration of production"[70].

The idea is, thus, to "eliminate contradictions between the growing production of general-purpose goods and the limited possibilities of the individual countries to produce the entire range of items in optimal series and of suitable quality"[71].

It is argued that this selective approach will, by enabling the individual countries to concentrate their co-operative efforts, ensure maximum economic effect and make the greatest contribution to accelerating technical progress – increasing, thereby, the economic and technical potential of the community. Thus, the agreements which are being put into effect (particularly those signed between the Soviet Union and each of its East European partners) are said to extend the advantages of industrial co-operation (and of large-scale industrial production) by raising the level of mutual complementarity of the CMEA countries' economies[72].

Increasing intra-CMEA trade and assuring stable markets is a parallel goal of production specialisation agreements[73].

Finally, and closely linked to the target of changing the CMEA countries' production structures, are the dual objectives of increasing the quantity and quality of industrial output. Both are also intended to increase the export potential of the Eastern countries (as well as allowing import substitution from the West) and to improve their competitive position on world markets.

Consequently, the agreements for specialisation and co-operation in production are "organically co-ordinated with joint programmes for scientific research, planning and design work"[74]. Particular importance is attached to the fact that the bilateral long-term programmes for specialisation and co-operation in production drawn up between the Soviet Union and the East European countries, include a series of measures designed to accelerate scientific-technical progress[75].

b) *Agreements Signed and Output*

As is the case for the LTPCs, the texts of the various agreements have never been published. While they are widely discussed in the Eastern literature, information on their content is scattered and often contradictory; especially information on the output said to ensue from production specialisation. Available data do not indicate "how much overall production is covered by such agreements, how much specialisation is actually carried out or, to what extent the agreements may apply to already established production patterns"[76].

Nevertheless, discussion in the Eastern literature does give a general idea of the division of "production tasks" foreseen for the CMEA countries, as well as the priority industrial branches (see Annex Tables 2 and 3 of Chapter 4).

The CMEA countries are said to have concluded more than 120 multilateral and 1 000 bilateral agreements dealing with specialisation and co-operation in production[77].

Official data claim that in 1983, some 18 per cent of intra-CMEA exports were the outcome of production specialisation[78]. For 1982, a figure of 17.7 per cent is given[79] and for 1981, 18.3 per cent[80]. The relative share of "specialised output" in the total industrial export output of the individual countries is given as follows:

In percentage

| | 1981[80] | 1982[79] | 1983[78] |
|---|---|---|---|
| Bulgaria | 38.7 | 33.3 | .. |
| Czechoslovakia | 24.1 | 24.4 | 29.0 |
| German Democratic Republic | 30.6 | 34.7 | 43.0 |
| Hungary | 24.5 | 29.0 | 34.5 |
| Poland | 20.9 | 18.0 | .. |
| Romania | 18.3 | 28.0 | 33.5 |
| Soviet Union | 7.2 | 6.2 | .. |

In 1982, machine-building is said to have accounted for 81.6 per cent of the entire trade in products produced on the basis of specialisation (82.5 per cent in 1981) and more than 38 per cent of exports of machines and equipment were products of machine-building based on specialisation[81]. According to the communiqué published subsequent to the Havana meeting, the share of specialised machine-building output in the overall volume of machinery exports reached 41 per cent in 1983.

It is also said that the rate of growth of machine-building exports produced through specialisation (17.7 per cent) considerably exceeded the growth rates of that branch's output (11.6 per cent) in 1982[82]. This is clearly viewed as the beginning of a long-term trend since "trade in machinery, especially that in specialised products, is to develop faster than total intra-CMEA trade"[83].

The primary recipient will obviously be the Soviet Union. The figures quoted above show that the relative share of Soviet products produced on the basis of specialisation in the volume of total exports in intra-CMEA trade is the smallest – and evidently declining. Also, as Table 4 of Chapter 4 indicates, Soviet imports of specialised products from the CMEA Member countries exceed its exports.

c) *Priority Areas*

It would appear from the agreements which have been concluded, particularly those bilateral long-term specialisation and co-operation agreements (extending to 1990) signed between the Soviet Union and each of its East European partners, that their essential objective is to accelerate development of the mechanical engineering sector; notably machine-building (see Annex Table 4 of Chapter 4).

Table 4 of Chapter 4 clearly indicates that the vast majority of the agreements for specialisation and co-operation in production concern machine-building. Of the agreements signed by 1980, "about 75 per cent embraced the production of such branches in mechanical engineering as the automotive and bearing industry, construction and road-building, tractor and agricultural machinery and the machine tool and radio-electronics industry"[84].

The priority attached to developing the mechanical engineering branch is, as mentioned previously, also evident from the number of measures assigned to the corresponding LTPC.

Furthermore, of the approximately 120 multilateral agreements dealing with specialisation and co-operation in production, 74 (in 1984) were accounted for by machine-building, equipment and instruments[85]. By 1983, some 694 bilateral agreements for production specialisation were concerned with machine construction. All told, the multi- and bilateral agreements embrace some 11 200 commodity items[86].

In the interest of furthering inter-branch co-operation, seven priority "directions" for developing specialisation and co-operation in machine-building have been delineated. These "directions" concern the development and production of:

- new "progressive" systems for controlling production processes (on the basis of mini- and micro-processors);
- specialised technological equipment for electronics;
- machine complexes and equipment to reduce manual labour, including standardised industrial robots;
- machine complexes for open-cast mining and pipeline construction;
- equipment for ensuring a more rational consumption of fuels and energy, especially energy equipment, chemical and petro-chemical installations and diesel motors;
- high precision machinery and foundry equipment, and
- machinery systems for mechanising agricultural production[87].

Development of the electronics industry was again singled out at the Summit conference and at the Havana session as constituting priority areas for production co-operation. A general agreement to further co-operation in this area was already signed at the 36th session of the CMEA held in 1982. This agreement is said to cover the gamut of co-operation – "from R&D to sales"[88]. In accordance with the decisions of the 36th session, the following distribution of tasks were agreed to:

- Soviet Union and Bulgaria: machine planning systems;
- Hungary: measuring and testing equipment;
- German Democratic Republic: optical-mechanical equipment;
- Poland: equipment for assembling integrated micro-circuits;
- Romania: equipment for processing semi-conductor materials;
- Soviet Union: equipment for manufacturing large and medium integrated micro-circuits,
- Czechoslovakia and the German Democratic Republic: manufacture of analytic, control and measuring equipment[89].

Relative to the development of industrial robots, no details about the agreement (signed in 1982) to organise specialised co-production of industrial robots could be found. Judging from the discussions at the 36th session of the CMEA, the general objectives of this agreement are to improve the production organisation of the machine-building sector, to utilise manpower resources more effectively and to automate the industrial production process.

Production of components and parts is also to be furthered on the basis of specialisation. This type of co-operation has gone furthest in motor vehicle manufacture (cars and trucks), tractors, ships and boats, agricultural machinery, road building equipment and metal-working machinery. Relative to the latter, it is mentioned that "the agreement on multilateral specialisation and co-operation in the production of metal-cutting machine tools, specialised and heavy machine tools and automated lines also includes mutual deliveries of more than one million components and essential parts"[90].

Finally, it is evident from discussions in the Eastern literature that specialisation and co-operation in production will play a complementary role with joint investments in implementing the Long-Term Programme to develop energy and raw materials resources.

d) *Concluding Observations*

Whether specialisation and co-operation in production can meet Eastern expectations is questionable. Indeed, discussion in the Eastern literature suggests that the given objectives are, on the whole, not being met.

For instance, differences in the level of scientific-technical and economic development among the CMEA countries are said to be responsible for the fact that production specialisation has not favoured economic development *per se* or the development of new production lines. Instead, such agreements seem to have led to "monopoly-type" practices. That is, they have increased dependence on socialist partners while not assuring mutual deliveries. This has, in turn, reduced incentives for further product development by the supplier concerned and has also resulted in a loss of international competitiveness[91].

Output of machinery and equipment is likewise considered to be unsatisfactory. It should be recalled that machinery and equipment occupy a predominant position in intra-CMEA trade. However, since 1979, the share of machinery and equipment in intra-CMEA exports fell. While this is now rising, it is still below the share attained in 1979. Thus, "the tempo of development of reciprocal machinery deliveries still lags behind the rate of growth of mutual trade"[92]. This movement, which mainly reflects a leveling-off or even contraction of investment in the Eastern countries, nonetheless does indicate a lack of dynamism in the progress of specialisation.

It is also important to note that specialisation and co-operation in production is least developed for the manufacture of components and parts. In many instances needed parts and components cannot be replaced either by domestic production or by imports from other CMEA countries. This situation is often alluded to in the Eastern literature and most often (and most severely) criticised by Soviet officials. It is noted, for example, that shortages are particularly severe in casting, forging, drives, clutches and hydraulic equipment, as well as pumps, compressors and measuring instruments.

It is also pointed out that the "proportion of technically new articles in the total deliveries of specialised products is still relatively low. Specialisation and co-operation has been extended thus far to groups of articles already in production for a long time. Steps taken to improve their quality, to quickly replace them with new models or to renovate production procedures are still inadequate"[93].

To rectify this situation, some Eastern officials suggest that production co-operation be increasingly tied in with R&D co-operation. This would, it is thought, focus production specialisation more on new product development[94].

However, presently production and R&D co-operation remain, at best, loosely joined, and most specialisation agreements simply "do not raise the technical level of existing productive capacities, but preserve the predominantly extensive path of development"[95].

Furthermore, in the Eastern literature attention is drawn to the fact that within the mutual deliveries of specialised products, the share of "advanced technological products" remains low – some 10-15 per cent of the total value of trade arising from specialisation and co-operation agreements.

This observation can be substantiated by analysing actual trends in intra-CMEA S&T co-operation. For instance, research carried out by the EEC indicates that the CMEA countries are focusing their effort on the development of high and moderately high R&D-intensive products.

Nevertheless, the EEC analysis also indicates that intra-CMEA co-operation in S&T usually stops at the research level. This conclusion also emerges from discussions in the Eastern – notably Soviet – literature. R&D development, particularly of technologically advanced products, is apparently carried out primarily on a national basis. This seems to be largely due to prevailing national interests which "dictate" strict domestic utilisation of research that has been carried out jointly. The most important consequence of this is that intra-CMEA scientific and technological co-operation does not appear to foster the mutual exchange of high technology embodying products among the CMEA countries[96].

The above-mentioned shortcomings and developments suggest three, rather obvious conclusions. The first is that access to Western technology will remain an important consideration for those CMEA countries which are highly developed technologically and lack compatible partners within the CMEA in order to further the production of technically advanced products.

Secondly, they therefore reduce prospects for achieving the policy objective of greater independence from the West. For instance, developments in East-West technology trade indicate that imports of intermediate (technology based) goods have remained "stable" over the past decade or so. Furthermore, OECD exports of low to medium R&D-intensive goods over the 1972-1983 period have predominated – accounting for three-fourths to four-fifths of the total (see Table 9 of Chapter 6). Case studies also suggest that specialisation and co-operation in production, particularly in the machine-building sector, has been most successful when linked to imports of Western components, licences and technical know-how[97].

Thirdly, and despite their shortcomings, it can be expected that Soviet interest in furthering bilateral specialisation and production agreements will not diminish since it is these agreements which determine "who" is to specialise in "what", stipulate the volume of exports that is to ensue from them and obliges participants in the agreements to adhere to fixed import quotas. Given the position of the Soviet Union within the CMEA (particularly viewed from the vantage point of the energy issue), the voice of the USSR in articulating these agreements can also be expected to remain predominant.

## 4. Scientific and Technological Co-operation

*Comment*

The organisational structure for promoting the co-ordinated development of science and technology is very complex. It includes the highest decision-making bodies of the CMEA itself, CMEA committees for planning and directing RDI co-operation, CMEA permanent commissions and conferences, inter-governmental commissions and sub-commissions, international economic organisations co-ordinating R&D, co-ordination centres for scientific programmes and so forth.

Methods employed to integrate science and technology are equally intricate. Among them are forecasting and planning, funding mechanisms, exchanges of specialists and co-operation in training personnel, joint information processing and exchange of information, co-operation in supplying and equipping joint and co-ordinated RDI projects, co-operation in establishing standards, co-operation in the use of inventions and in developing a patent policy and co-operation in fundamental research.

It is obvious that even a brief discussion of each of these topics is not feasible within the confines of this analysis. All these aspects, as well as the historical development of S&T integration are treated in a detailed study by L. Nolting[98].

We have drawn on this information for the following section. However, our discussion is limited to the recent period and only those aspects are singled out which indicate current Eastern policy concerns – namely, expectations as to what co-operation in S&T is to accomplish and the main priority areas.

a) *Evolution of S&T Co-operation*

Since the early 1970s, the basic pre-occupation underlying efforts made to co-ordinate the S&T systems of the CMEA countries is to link RDI more closely with economic development. The various measures taken (or planned) to achieve this goal are briefly summarised below.

i) *The Comprehensive Programme*

The adoption of this Programme marked the first attempt made to link S&T development organically and systematically with economic development and production. Consequently, it foresaw collaboration in: drawing up plans for developing scientific, technical and economic co-operation; research, planning and design work; scientific forecasts and analyses and, the development and use of joint technological processes. The Programme also provided for co-ordination and joint performance of specific research, planning and design operations, the exchange of technical specifications and S&T information, as well as technical assistance[99].

S&T co-operation was to encompass both fundamental research (for the future development of the national economies) and applied research. Since one aim of the Programme was to bring about "equalisation" of economic development within the CMEA area, provision was made for specialisation and co-operation in production in order to enhance the mutual exchange of products arising from such co-operation.

The Programme also formalised the principle that intra-CMEA scientific and technical co-operation would proceed both against payment as well as on the basis of free exchanges. The question of payment was to be settled by mutual agreement and according to the level and value of research results transferred. Since, technological transfers were shifted to a "paying basis", the idea was clearly to "provide an additional incentive for the development of scientific and technical integration among the fraternal countries"[100].

Some 18 fundamental and applied research areas were listed in the Programme. These dealt with various means to increase agricultural output, (for instance, development of pesticides, further mechanisation within the agricultural sector), use of atomic energy for industrial production as well as the development of such items as plastics, anti-corrosive materials, advanced and automated machinery and equipment (including computers), new semi-conductor materials, high purity metals and so forth[101].

Table 5 of Chapter 4 indicates where application of advanced technologies was apparently envisaged. However, it is important to note that despite the provision made for specialisation and production co-operation, no clear-cut distinction was made in the Comprehensive Programme between those areas of production co-operation where advanced technologies should have been applied and those where their joint application was not foreseen[102].

Furthermore, current developments suggest that specialisation and production co-operation is being confined to an increasingly narrow set of priority areas – notably, those concerned with developing (and conserving) energy and raw materials resources and the machine-building/energy branches. This, together with the fact that intra-CMEA S&T co-operation is primarily confined to fundamental research, would seem to explain why the mutual exchange of high-technology products does not, as mentioned previously, occur to the extent desired.

ii) *Plan Co-ordination*

– Consolidated Plans

Even prior to the adoption of the Comprehensive Programme, the CMEA countries had drawn up so-called "consolidated five-year RDI plans" designed to integrate RDI planning on the CMEA level[103]. The intent to integrate RDI more closely with the CMEA countries' economic plans in order to further the formulation of a long-term economic development strategy, was made evident by the fact that during "the 1970s, these plans were then drafted as integral stages of the Comprehensive Programme"[104].

It is interesting to note in this regard, that the consolidated plan formulated for the 1976-1980 period, was "renamed the plan for co-operation (rather than co-ordination), indicating that it placed greater emphasis on joint research efforts and consultations than previously and was not confined to the meshing of separate country plans and the elimination of duplication"[105]. The number of problem areas and projects (covering for the most part complex inter-branch RDI co-operation) included in this plan and projections for the 1981-1985 consolidated plan, are given in Table 6 of Chapter 4.

– Multilateral Plans

The policy of integrating RDI with economic planning was reinforced by the adoption of the first Co-ordinated Multilateral Plan for 1976-1980. The move towards co-operation on complex problems is said to be its distinctive feature.

The section devoted to scientific and technical co-operation included 17 broad areas, encompassing 339 problems and 1 336 themes. S&T co-operation seems to have been principally designed to meet the growing energy needs of the CMEA countries since "70 per cent of the 565 million transferable rubles allocated for solving S&T problems (actual expenditures came to about 600 million T.R.s.) relates to work connected with the fuel and energy problem"[106].

The Multilateral Plan for the 1981-1985 period extends and develops the concept of focusing S&T co-operation on inter-branch problems. It contains 13 major topics, incorporating some 35 projects. As Table 7 of Chapter 4 indicates, these are essentially centered on developing power engineering and sophisticated machinery and equipment.

Since implementation of this Multilateral Plan is closely tied to the LTPCs, emphasis is placed on the fact that scientific and technological co-operation is now chiefly aimed "at co-ordinating the Member countries' scientific-technical policy as a component part of economic policy"[107]. In order to ensure that S&T co-operation leads to accelerated development of advanced technologies, as well as industrial application, the programmes which are being drafted for it are said to cover "the entire science-technology-production-marketing cycle".

iii) *The Long-Term Target Programmes*

The adoption of the LTPCs also reinforced the policy trend of formulating S&T and economic development plans in tandem. Thus, the RDI goals to be attained by 1990 (contained in the section concerned with S&T co-operation) were integrated into the consolidated plan and the Multilateral Plan for the 1981-1985 period[108]. In order to ensure that S&T co-operation would provide the desired support for achieving the objectives mapped out for the LTPCs, the document "Basic Directions of the CMEA Countries' Scientific-Technical Co-operation Through 1990 and the Technical-Economic Consequences Expected on this Basis" was adopted in 1979. It incorporates the 117 measures contained in the LTPCs related to S&T problems (see Table 2 of Chapter 4).

These "Basic Directions" and the integration of the LTPCs into long-term economic planning are said to have enabled the CMEA countries to "co-ordinate the cardinal problems of science and technology with production problems and to prepare detailed plans for co-operation encompassing scientific research, technical developments, introduction [of research results] and marketing"[109].

iv) *Bilateral Co-operation*

Bilateral co-operation is considered to be the mutual complement of multilateral S&T co-operation. Both forms are said to ensure the necessary flexibility needed to accommodate most fully individual national interests.

In order to further S&T development, RDI co-operation is, as mentioned previously, a component part of the agreements governing specialisation and production co-operation. The bilateral agreements drawn up between the Soviet Union and the East European countries are also included in the above-mentioned "Basic Directions". Furthermore, bilateral S&T and economic co-operation agreements (designed to supplement the specialisation and production co-operation agreements) have also recently been signed between the Soviet Union and some of its East European partners (for instance, Poland, Czechoslovakia and the German Democratic Republic). Consequently (and, particularly, since the application of research results ensuing from joint efforts remains the prerogative of the individual countries), it can be assumed that the evolution of RDI planning and co-operation within the CMEA will not only be influenced essentially by bilateral mechanisms, but will adhere closely to the priorities established by the Soviet Union.

v) *Recent Measures*

Following the 35th session of the CMEA (held in 1981), "Basic Directions for Economic and Scientific-Technical Co-operation for the Period 1981-1985 and until 1990" were adopted. These directions form a component part of the "Programme for the Co-ordination of National Economic Plans for 1986-1990" and the Multilateral Plan for the 1986-1990 period.

As is evident from the discussion on plan co-ordination [see 1*c)* above], priority in this Programme is given to the wider application of science and technology in production.

Consequently, and given the intent to make S&T co-operation part of economic development, the Programme foresees supplementing plan co-ordination with formal agreements governing development of the economic and S&T policies of the CMEA countries[110]. The CMEA Committee on Scientific and Technological Co-operation has also been charged with developing a long-term concept for co-operation, taking into consideration the tasks for economic and social development defined for the Programme for Co-ordinating National Economic plans for the 1986-1990 period. In developing such a concept, particular attention is to be given to improving the division of labour within the CMEA in order to accelerate S&T development.

Accelerating S&T development is also the purpose of the Comprehensive Programme for S&T Progress to be elaborated following the agreement reached at the Summit conference. As mentioned previously, the intention now is to develop a co-ordinated and, in some cases, a unified S&T policy. Even if this objective is realised, it remains to be seen whether this will result in more effective RDI co-operation, more widespread application of advanced technologies or, for that matter, their more rapid development.

b) *The Context of S&T Co-operation*

The Eastern literature emphasises that since the adoption of the Comprehensive Programme, the scope of intra-CMEA co-operation in science and technology has progressively broadened. That is, and as mentioned previously, co-operation is said to have become more comprehensive and now encompasses all sectors of the economy and industrial branches. In other words, S&T co-operation has moved beyond work on individual problems to inter-branch problems.

The number of problems and topics included in the plans for S&T co-operation and subject to multilateral co-operation has greatly expanded (see Table 6 of Chapter 4). Also, by 1982, more than 200 agreements and contracts on research, development, design and experimental projects were said to have been concluded and, by July 1983, 198 multilateral agreements dealing with economic and scientific-technical co-operation had been signed[111].

The number of institutes involved in intra-CMEA S&T co-operation has also greatly increased. "The number of scientific research, planning and design organisations and higher educational institutions of the CMEA countries participating in co-operation, on a bi- or multilateral basis was about 700 in the late 1960s, but is nearly 3 000 at present"[112].

Eastern officials likewise claim that S&T co-operation is becoming increasingly economic (output) oriented. "A gradual transition from planning of individual problems and subjects to planning of economic results of research co-operation" is taking place"[113]. Relative to this, it is said that each year 200 to 300 machines, instruments and units of equipment and 100-150 technological processes are designed or perfected and 100-200 new types of materials and substances developed"[114].

Implementation of scientific and technical co-operation agreements is being improved. That is, "along with co-ordination, joint development of scientific and technical problems on the basis of agreements and contracts is finding increasing application"[115].

A more differentiated approach relative to S&T co-operation is also being worked out. Thus, topics subject to S&T co-operation adopted for the 1981-1985 period are to be divided into three groups. The first includes problems to be completed during the current Five-Year Plan period. The second includes problems to be resolved during the plan period and to be applied during the subsequent plan period. Co-operation for both groups is to proceed on the basis of agreements and contracts. Basic research problems, to be worked out by *individual countries*, are included in the third group. For this group of problems, work is to be "co-ordinated". This classification scheme, plus the gradual transference of S&T co-operation onto a contractual basis will, in future, determine which mode of co-operation for carrying out research is chosen[116].

This manner of proceeding is interesting for a number of reasons. First of all, one underlying objective is clearly to provide greater financial incentives for co-operation. The need to increase "economic levers", in order to promote S&T co-operation is, in any event, one aspect underlined by the author cited above. Secondly, problems included in the first two groups concern primarily the development of sophisticated machinery, equipment and materials as well as the development of new production processes. Since co-operation and joint development work is to be the basic mode of co-operation, it is evident that the primary aim is to ensure that S&T co-operation will meet the production needs of the individual countries. Finally, and given the disparities in the level of S&T development of the CMEA countries, there seems to be a greater recognition that, if S&T is to be effectively developed, more freedom in the choice of basic research problems to be addressed is needed.

### c) *Motives for Increasing S&T Co-operation*

Achievement of the long-term economic development strategy mapped out for the 1981-1985 period and beyond is intimately linked to progress made in science and technology and to the application of S&T results. Consequently, increasing economic efficiency by modernising industrial production structures and, counteracting thereby those factors (discussed in Chapter 1) which have led to slower economic growth, provide the basic motives for increasing and improving intra-CMEA S&T co-operation. As has been succinctly stated, "there is no other means of solving the problems of the 1980s than by accelerating scientific and technical progress"[117].

Financial considerations, be they related to savings (particularly of energy and raw material resources) or achieving economies of scale – since, as is noted, the cost of RDI and the application of S&T results is now so great as to exceed the capacity even of those countries possessing considerable scientific and production potentials – are becoming an increasingly strong motivation for furthering intra-CMEA S&T co-operation.

Closely related to the above, is the intent to improve the competitiveness of Eastern manufactures on world markets. As noted, "maintaining the quality and technical characteristics of industrial articles at the level of world requirements is becoming more expensive and difficult.... Catching up to world [quality] levels in areas such as metal-working, machine tool and instrument manufacture and increasing our export potential depends on the extent to which we are able to stimulate joint efforts in S&T co-operation[118]."

Perhaps most interesting from the viewpoint of this analysis, is that enhanced S&T co-operation within the CMEA area is meant to improve the use of Western technology and reduce costs in the procurement of such technology. Relative to patent and licence trade for instance, it is pointed out that "the lack of co-ordinated S&T policy results in CMEA countries buying the very same licences from Western companies"[119].

It is interesting to note, therefore, that the CMEA countries have recommended co-ordinating their purchases of licences, know-how, instruments, samples and materials from third countries in support of the projects contained in the production-specialisation agreements. This recommendation was reiterated at the 37th session of the CMEA[120].

However, the CMEA countries have not been successful in this co-ordination effort. While some evidence suggests that Western technology is transferred within the bloc, the policy of actually co-ordinating technology purchases has not been realised to date.

d) *Priorities*

The CMEA Summit conference and the Havana session once more confirmed that the major emphasis in the area of scientific and technological co-operation will be placed on the accelerated development of new technologies, machinery, equipment and materials. The CMEA countries' R&D capacity will also be increasingly focused on developing (and modernising) the chemical, energy and raw materials and agricultural sectors (see also Table 8 of this chapter). Such fields as micro-biology and biotechnology will also be strengthened. However, the overriding pre-occupation remains the engineering sector – notably, the enhanced development and application of micro-processors, electronics and robotics.

Tables 9 and 10 (of this chapter) indicate products subject to intra-CMEA R&D co-operation and contained in the co-operative programmes for the 1981-1985 period and beyond. As pointed out by one author[121], it is impossible to assess whether the products (or groups of products) said to be involved in the various R&D programmes represent the full list or are only random samples. It is also impossible to rank the different programmes (or products within each programme) according to their actual importance, since no quantitative or financial data are quoted. Furthermore, the official approval of a CMEA undertaking does not necessarily mean that it is put into operation. These data shortcomings must be borne in mind.

Nevertheless, when these products are compared to trade in technology goods classified according to R&D intensity, it appears that intra-CMEA co-operation in S&T concerns mainly high and moderately-intensive products or groups of products[122]. Also, the share of high R&D-intensive commodities exported by the OECD countries during the 1972-1983 period has been very low, oscillating between 2-3 per cent of total embodied technology exports. On the other hand, moderate R&D-intensive exports have swung between 15 to 20 per cent while, as noted previously, low R&D-intensive exports have varied between three-fourths and four-fifths[123].

In sum, it appears that the Eastern countries are attempting to develop, within the framework of the CMEA, key (or highly sophisticated) technologies. Since it is these technologies which Western countries are least eager to sell or which are most subject to export controls or embargoes, one can also conclude that the primary goal is to reduce import dependence on the West or, as advocated by the Soviet Union, achieve independence from the West. Minimising imports of high technologies from the West could also be related to the fact that it is these technologies which, within the context of present economic and R&D management systems, have proven to be the most difficult to introduce into the Eastern countries' economies.

e) *Problems in CMEA S&T Co-operation: Concluding Observations*

As has been pointed out by L. Nolting in his study, CMEA S&T co-operation has benefited the East European countries. Co-operation has also proven to be beneficial to the Soviet Union, particularly for the development of its computer programme and such advanced technologies as robotics, numerical programming in machine tools and technologies designed to conserve energy and raw materials or to develop alternate sources of energy. Intra-CMEA co-operation has added substantially to the Soviet Union's RDI potential in terms of resources (including investments), manpower, equipment and so forth made available for the development of its S&T system[124].

Nevertheless, the development of intra-CMEA co-operation in the sphere of science and technology continues to be hampered by numerous difficulties. One aspect of the problem is the very nature of the CMEA itself. As noted earlier, the mere fact that the CMEA is not a supra-national institution implies that decisions which are taken at the CMEA level can only be enforced by "voluntary" agreement. Institutional shortcomings – particularly the price and trading systems in effect within the CMEA area – are also hindering effective intra-CMEA S&T co-operation (and the development of science and technology) since market considerations for the individual countries are not taken adequately into account in the formulation of co-operative agreements.

Repeated attention is drawn in the Eastern literature to the fact that economic incentives for furthering CMEA co-operation are insufficient or lacking. Reference is made to the inequitable manner in which costs are determined and assigned and the financing of programmes distributed among participating countries. It is pointed out that the absence of contract commitments and the impossibility of re-distributing expenditures for the performance of expensive research, are particular shortcomings in co-ordination[125].

Effective co-operation is also said to be hindered by the fact that linkage among the various CMEA bodies charged with S&T development is not firmly established, nor is there sufficient co-ordination among their various tasks. Co-ordination of bi- and multilateral co-operation is also inadequate. This has led to duplication in research work and has adversely affected co-operation in the application of research results.

These shortcomings, combined with faulty selection of "priority subjects" (as well as the great diversity in subjects covered by co-operation) and selection of problems without serious economic evaluation, are held responsible for the fact that S&T co-operation has not resulted in adequate output of advanced products and processes.

In many respects, CMEA S&T co-operation is plagued by the very problems which hinder the development and application of S&T within the individual countries. The most frequently cited in the Eastern literature are that:

- "economic structures are characterised by a lack of demand for innovations, which is the result of the existing model of managing and directing the economy and the functioning of the so-called seller's market (a monopoly);
- the system for planning and directing scientific-technical policy is not conducive to efficiency in research, to reduction in the time for applying technical innovations and,
- the existing economic, social and political structures have shown themselves to be insufficiently effective for adapting imported, new technology"[126].

These shortcomings have led to a spate of reform proposals. Foremost among these is the creation of economic incentives (levers). Mention here is made of ending the free exchange of information and other S&T results. Instead, payment would be made for expenditures connected with the performance of scientific and technical research. Other suggestions are that the contract forms of S&T co-operation should be expanded and direct contractual relationships between research institutes developed. Inter-country bonuses should be given to reward outstanding S&T achievements as well as to staff members of institutions who have "stimulated" work of international importance. Another suggestion is "to allocate to scientific research institutes and enterprises an assigned share of foreign trade profits from exporting the results of scientific research and experimental design work"[127].

Perhaps most interesting, given the nature of the planned-administrative economic systems, are those recommendations which call for widespread management reforms (of the S&T system) in order to "adapt management and planning to higher requirments". At the heart of this adaptation are proposals which would render these systems more flexible. Among them is the suggestion (reiterated at the Summit conference) to establish more direct links between enterprises and research institutions.

In essence, it is explicitly recognised that far-reaching systemic reforms are necessary if S&T co-operation is to bring hoped-for results. Whether such reforms will be brought into effect remains, to date, an unanswered question, though there is evidence of movement in that direction.

Table 1 (Chapter 4)

MULTILATERAL CMEA JOINT INVESTMENT PROJECTS LAUNCHED, CONTINUED OR COMPLETED IN THE 1976-1980 FIVE-YEAR PERIOD (a)

| Project | Year of Initial Agreement | Participants (b) | Planned Commis-sionning Date | Planned Date for End of Deliveries |
|---|---|---|---|---|
| 1. Non-energy Projects | | | | |
| a) Projects located in the Soviet Union | | | | |
| i) Ust-Illimsk Wood Pulp Cellulose Combine | 1972 | B,G,H,P,R,S | 1980 | 1982 |
| ii) Kiembai Asbestos Combine | 1973 | B,Cs,G,H,P,R,S | 1980 | 1991 |
| iii) Combines (in Dneprovsk, Gubkinsk, Mikhailovsk & Stoilensk) for production of iron ore pellets and concentrates | 1974 | B,Cs,G,H,P,S | 1979 | 1990 |
| iv) Kingisepp Phosphorous Mining Combine | 1974 | B,Cs,G,H,S | 1978 | 1985 |
| v) Nikopol & Yermak Ferro Alloy Plants (7 plants) | 1974 | B,Cs,G,H,P,R,S | ? | 1990 |
| vi) Mosyr Fodder Yeast Plant | 1979 | Cs,Cu,G,P,S | ? | 1998 |
| b) Projects located elsewhere in the CMEA: | | | | |
| i) Devnya Soda Plant, Bulgaria | ? | B,H,S | 1974 | ? |
| ii) Las Camaricas Nickel/Cobalt Ore Dressing Plant, Cuba | 1975 | B,Cs,Cu,G,H,P,R,S | 1982 | ? |
| iii) Project to expand Cuban Citrus Fruit Production and Processing | 1978 | B,Cs,Cu,G,P,S | 1981 | 1990 |
| iv) Project to develop Cuban Sugar Industry | ? | B,Cu,G,S | 1981 | 1980 |
| 2. Energy Projects | | | | |
| i) Orenburg Gas Concentrate Plant & Orenburg-Uzhgorod Pipeline | 1974 | B,Cs,G,H,P,R,S | 1979 | 2000 |
| ii) Adria (Krk, Y-Szazhalombatta, H-Tupa, Cs) Oil Pipeline | 1974 | Cs,H,Y | 1979 | ? |
| iii) Vinnitsa, S-Albertirsa, H 750kV Power Line | 1974 | B,Cs,G,H,P,S | 1979 | ? |

Notes: a) This table excludes all bilateral projects. Thus, a number of "joint investment" projects have been excluded. For example, the following are not mentioned in this table: the Nikopol-Turnu Magurele Hydro-electricity project (between B&R), the Gabcikovo-Nagymaros Hydro-electricity project (between Cs & H), the Iron Gate II-Derdap II Hydro-electricity Project (between R & Y), and various branch electricity and gas lines.

b) B=Bulgaria; Cs=Czechoslovakia; Cu=Cuba, G=GDR; H=Hungary; P=Poland; R=Romania; S=Soviet Union; and Y=Yugoslavia.

Source: Cam Hudson, "Are CMEA Joint Investment Schemes Being Downgraded", Radio Free Europe Research, RAD Background Report (Eastern Europe), Munich, 10th December, 1982, p. 4.

Table 2 (Chapter 4)

STRUCTURE OF LTPCs BY NUMBER OF MEASURES AND MULTILATERAL AGREEMENTS

| | Number of measures | Measures for which no multilateral agreements need to be concluded (2) | Multilateral agreements to be concluded for the realisation of: | Joint investments (3) | Specialisation and co-operation of production (4) | Co-operation in science and technology | Other | Multilateral agreements signed by 30.4.1980 |
|---|---|---|---|---|---|---|---|---|
| LTPC for Energy, Fuels and Raw Materials | 87 | 35 | 51 | 25 | 9 | 15 | 2 | 12 |
| LTPC for Agriculture and Food Industry | 34 | 7 | 27 | 7 | 9 | 8 | 3 | 10 |
| LTPC for Mechanical Engineering | 123 | 40 | 104 | 9 | 61 | 34 | - | 24 |
| LTPC for Industrial Consumer Goods | 40 | 14 | 26 | 3 | 8 | 7 | 8 | 5 |
| LTPC for Transport | 56 | 32 | 24 | 11 | - | - | 13 | 16 |
| Total | 340 (1) | 128 | 232 | 55 | 87 | 64 | 26 | 67 (5) |

1. Including 117 scientific and technical problems.

2. Columns two and three do not necessarily add up to column one as the conclusion of two or more agreements is required for some measures whereas, on the other hand, one agreement may cover several measures.

3. Concluded, as a rule, by government representatives.

4. Concluded, as a rule, by economic organisations of participating CMEA member countries. Agreements of particular importance such as that on nuclear power station equipment, may also be of inter-governmental character.

5. Eighty-two by 1980, more than 120 by the end of 1980 (Ekonomicheskoe Sotrudnichestvo Stran-Chlenov SEV, No. 4, 1980, p. 68 and Robotnichesko Delo, 17th January, 1981, p. 5.).

Source: Ekonomicheskoe Sotrudnichestvo Stran-Chlenov SEV, No. 3, 1980, pp. 30-35, as given in Raimund Dietz, Ilse Grosser, Comecon Energy Perspectives and the Long-Term Target Programmes, Reprint Serie No. 58, Wiener Institut für Internationale Wirtschaftsvergleiche, November 1981, p. 35.

Table 3 (Chapter 4)

EXAMPLES OF MULTILATERAL AGREEMENTS
CONCLUDED FOR IMPLEMENTATION
OF THE LONG-TERM TARGET PROGRAMMES

- Agreement on the joint construction of nuclear power stations and the production of equipment for them;
- Agreement on multilateral specialisation and co-operation in the production and delivery of nuclear equipment for the period 1981-1990, signed in 1979;
- Agreement on the construction of machinery and equipment for the oil industry, the extraction and transport of natural gas and oil, coal and iron ore and the steel industry;
- Tripartite Soviet/Polish/Czech agreement on the construction of new machinery for underground coal mining (mechanical pit-props, cutter-loaders for thin seams and other mining equipment);
- Tripartite USSR/German Democratic Republic/Czech agreement on the modernisation of open-cast mining (rotary excavators and other mining equipment);
- Agreement on the construction of machinery and equipment for the enrichment of solid fuels and ferrous and non-ferrous ores, and for the manufacture of chemical feedstocks;
- Agreement on the construction of equipment required for refining crude oil into light products for the CMEA countries. All European CMEA countries participate in this agreement;
- Agreement on increasing the production of basic chemicals using energy-intensive processes in the USSR - ammonia, methanol, caprolactam, polyethylene, etc. In exchange, the other CMEA countries will develop less energy-intensive products such as parachemicals, pharmaceuticals, dyes, etc. All European CMEA countries participate;
- Agreement on the construction of equipment for processing plastics. All European CMEA countries plus Yugoslavia participate;
- Agreement on the construction of plant and equipment for the steel and non-ferrous metal industries – sintering, coking, continuous casting and rolling mill equipment. All European CMEA countries plus Yugoslavia participate;
- Agreement on the mechanisation of handling and storage activities and the modernisation of transport and lifting equipment;
- Agreement on the construction of plant and equipment for the biological, mechanical and physico-chemical treatment of water;
- Agreement on co-operation and specialisation in the production of electric and diesel locomotives, goods-wagons and passenger coaches;
- Agreement on co-operation and specialisation in the production of equipment, machinery and apparatus for the textile, knitwear and clothing industries;
- Agreement to design and produce standardized foundry equipment with the view to completely transforming the production of castings;
- Agreement on specialisation and co-operation in the production of computer equipment. This agreement provides for the reciprocal delivery of equipment totalling 15 billion transferable rubles over the five-year (1981-1985) period. The aim is to adapt advanced electronic technologies (micro-computers) to the requirements of CMEA economies;
- A separate agreement covering the design and production of industrial robots and handling gear to be applied in various fields with a view to reducing manual labour;
- Agreement on the development of micro-electronics and the manufacture of electronic devices to improve the technological level of engineering products and automate the telecommunications sector, develop automated systems and improve the standard of manufactured consumer goods.

*Source*: Compiled from the Eastern literature.

Table 4 Chapter 4

AGREEMENTS ON SPECIALISATION AND CO-OPERATION IN PRODUCTION SIGNED BETWEEN THE USSR AND THE EAST EUROPEAN COUNTRIES AND VOLUME OF MUTUAL DELIVERIES ORIGINATING FROM THE AGREEMENTS

Number of Treaties (Agreements) for the Periods 1971-1975 and 1976-1980

| | 1971-1975 | | 1976-1980 | |
|---|---|---|---|---|
| | Total | Including machine building | Total | Including machine building |
| Multilateral Agreements | 27 | 19 | 99 | 74 |
| Bilateral Agreements | 72 | 57 | 132 | 98 |
| Including: | | | | |
| Bulgaria | 15 | 12 | 25 | 19 |
| Hungary | 16 | 12 | 34 | 23 |
| German Democratic Rep. | 18 | 16 | 25 | 22 |
| Poland | 12 | 10 | 22 | 18 |
| Romania | 2 | - | 8 | 4 |
| Czechoslovakia | 9 | 7 | 18 | 12 |
| Total multi- and bilateral agreements | 99 | 76 | 231 | 172 |

Table 4 (Chapter 4) contd

Volume of Mutual Deliveries of Specialised Products Between the USSR and Other CMEA Countiries
(million of rubles)

| Years | Export | | Import | |
|---|---|---|---|---|
| | Total | Including machines and equipment | Total | Including machines and equipment |
| 1971 | 196.6 | 196.6 | 348.6 | 348.6 |
| 1972 | 311.8 | 273.3 | 667.0 | 571.1 |
| 1973 | 433.9 | 377.8 | 934.5 | 839.4 |
| 1974 | 588.6 | 507.2 | 1 141.4 | 1 005.3 |
| 1975 | 683.4 | 559.4 | 1 551.1 | 1 380.3 |
| 1971-1975 | 2 214.3 | 1 914.3 | 4 642.6 | 4 145.0 |
| 1976 | 1 137.9 | 938.1 | 2 834.5 | 2 374.0 |
| 1977 | 1 388.7 | 1 125.0 | 3 384.6 | 2 843.0 |
| 1978 | 1 568.1 | 1 281.7 | 3 814.4 | 3 227.0 |
| 1979 | 1 627.0 | 1 277.3 | 4 144.2 | 3 577.1 |
| 1980 | 1 868.8 | 1 432.1 | 5 141.3 | 4 445.9 |
| 1976-1980 | 7 590.5 | 6 054.2 | 19 319.3 | 16 467.0 |

Note: Products of the radio-electronic industry are included in the cost of machines and equipment deliveries.

Source: Lev Afanasyev and Sergei Chenchikovski, "Socialist Economic Integration Viewed", Foreign Trade (English edition), No. 11, Moscow, November, 1981, p. 7.

Table 5 (Chapter 4)

COMPREHENSIVE PROGRAMME: AREAS OF CO-OPERATION IN PRODUCTION WHERE THE APPLICATION OF ADVANCED TECHNOLOGY WAS SEEMINGLY ENVISAGED

| Sectors | Branches (Eastern classification) | Areas or individual products |
|---|---|---|
| **Industry** | Fuel and energy | – Catalysts and additives for the oil-refining industry. |
| | Ferrous metallurgy | – Advanced technological processes and integrated automated systems for metallurgical products;<br>– Stainless and other high-quality steels. |
| | Non-ferrous metallurgy | – Advanced equipment for mineral extraction and refining;<br>– High quality solid-state materials to obtain high-purity metals for the radio-engineering and electronic industries. |
| | Machine-building | – Equipment for atomic electric power stations with water-cooled reactors and possible fast-neutron reactors;<br>– Advanced basic and auxiliary energy-producing equipment;<br>– Advanced minerals prospecting and drilling equipment;<br>– Basic mining equipment at world technical level;<br>– Programme controlled, automatic metal-cutting, forging and pressing equipment;<br>– New, highly productive metal-cutting machine tools;<br>– Highly productive equipment for chemical industry;<br>– Modern, programme-controlled equipment for the light industry (textile, footwear, woodworking, polygraphic, etc.);<br>– Advanced automated machinery and equipment for the food-industry and packaging of food-industry products;<br>– Advanced types of agricultural machinery, inclusive of tractors. |
| | Radio-engineering and electronics | – Modern electronic computers;<br>– Data control systems;<br>– Electronic solid-state devices and assemblies for micro-electronic circuits;<br>– Digital automated radio-electronic metering devices;<br>– Colour T.V. system;<br>– Apparatus for automatic air traffic control;<br>– Modern types of radio engineering products. |

Table 5 (Chapter 4) *contd*

| Sectors | Branches (Eastern classification) | Areas or individual products |
|---|---|---|
| | Chemicals and paper-pulp industry | – Aminoacid, vitamins and other feed additives for animal husbandry;<br>– New types of plastics;<br>– New types of chemical fibers;<br>– Advanced packing, wrapping and packaging materials (polymer, paper, cardboard, tin, glass containers, etc.). |
| **Agriculture and food industry** | Agriculture | – High quality seeds and hybrids for farm crops;<br>– Advanced types of construction facilities for animal husbandry products. |
| | Food industry | – Measures for automating the food industry. |

*Source:* D. Tosi, (Directorate-General for External Relations, Commission for the European Communities), *Trends of Technical-Scientific Co-operation, Co-operation in Production of High R&D-Intensive Products and Related Mutual Trade among the European CMEA Countries during the 1970-1981 period,* July 1984. Report prepared for the OECD.

Table 6 (Chapter 4)

INCREASE IN THE SCOPE OF MULTILATERAL SCIENTIFIC-TECHNICAL CO-OPERATION AMONG THE CMEA COUNTRIES (1)

| Item | 1966-1970 | 1971-1975 | 1976-1980 | 1981-1985 (Plan) (2) |
|---|---|---|---|---|
| Number of problems contained in the integrated plan of scientific-technical co-operation | 50 | 193 | 289 | 300 |
| Number of subjects contained in the individual problems | 183 | 706 | 1 330 | 1 400 |
| Number of self-contained subjects in the integrated plan | - | 102 | 69 | - |

Sources: 1. A. Shitkova, "Scientific-Technical Co-operation and Problems Concerning Its Improvement", in Plan-Conformity of the Socialist Economic Integration -- Problems and Prospects, Moscow, 1980, p. 40, in H. Zschiedrich and M. Perlzer, "Scientific-Technical Progress and Planned Intensification of Socialist Economic Integration", Deutsche Aussenpolitik (East Berlin), Vol. 26, No. 11, November, 1981, as given in Joint Publications Research Service (80171), East Europe Report, 24th February, 1982, p. 5.

2. L. Nolting, Integration of Science and Technology in CEMA, Foreign Economic Report No. 21, US Department of Commerce, Bureau of the Census, Washington, D.C., December 1983, p. 11. The figures are projections.

Table 7 (Chapter 4)

THE CO-ORDINATED PLAN OF MULTILATERAL INTEGRATION MEASURES FOR 1981-1985: S&T PRIORITY TASKS FOR DEVELOPMENT INCLUDED IN THE 13 PROJECTS ADOPTED

- MHD (magnetohydrodynamic) energy converters whose use is supposed to increase the efficiency of thermal power plants to 50-60 per cent and which, compared to the currently best power plants, is to facilitate a 20-25 per cent saving of organic fuels;
- New effective methods for the conversion of solar, chemical, wind, and geothermal energy into electrical, thermal, and mechanical energy as well as the development of the corresponding plants;
- Advanced technological processes and equipment for energy-intensive production sectors, the transportation industry and the communal economy as well as the development of less energy-consuming processes;
- Perfection of technological processes and equipment for the deep-cracking of petroleum which by 1985 is to make it possible to increase the yield of light petroleum products by 10 million tons;
- Technologies for the effective processing of coal into liquid, gaseous, and more easily transportable fuels;
- More effective corrosion protection for metals;
- Technological processes for powder metallurgy;
- Technological processes and equipment for the mechanisation and automation of transloading and storage processes.

*Source*: "Main Points of Emphasis of Multilateral Scientific-Technical Co-operation in CMEA", *AW-DDR-Aussenwirtschaft*, Vol. 10, No. 40, 29th September, 1982, as given in *Joint Publications Research Service* (82299), *East Europe Report*, 23rd November, 1982, p. 3.

Table 8 (Chapter 4)

EXAMPLES OF MOST IMPORTANT PROBLEMS FOR FUTURE S&T CO-OPERATION: PER THE 35TH SESSION OF THE CMEA

**I. Problems**[a]

- Development and use of control and regulation equipment for production processes, creation of a uniform and unified basis of electronic elements, of special materials and technological equipment for electronics and microelectronics, as well as the development and broad use of microprocessor techniques in the national economy;
- Development and use of new advanced technological processes and equipment for the energy sector which will guarantee the efficient employment and the economically justifiable expenditure of fuels and energy, particularly in energy-intensive production sectors and in the energy industry, as well as greater involvement of secondary resources in the economic cycle;
- Development and use of new highly-productive machine systems and equipment, including industrial robots, which will reduce the expenditure of manual labour in the national economy, especially in transportation, storage and auxiliary functions;
- Completion of existing and development of new technological processes and equipment for the more profound cracking of petroleum, for a significantly higher yield of light [clear] petroleum products and the more effective use of this raw material;
- Development and use of new, as well as perfection of existing, methods and technical means for effective processing of coal into liquid, gaseous, and more easily transportable solid fuels;
- Increase in livestock production on an industrial scale, including development of the fodder base and machine systems for complex mechanisation of agriculture (on the basis of highly-productive technologies) as well as for the efficient processing of agricultural food raw materials;
- Co-operation in the solution of these problems must, from the very beginning, include the entire science-technology-production-sales cycle.

**II. Agreements Signed (at the 35th and 36th Sessions of the CMEA) to Accomplish the Above**

- The government-level agreement on the creation of a uniform and unified basis for products of electronic technology, of technological special equipment, as well as semi-conductor materials and special raw materials for their production;
- The government-level agreement on specialisation and co-operation in the production of products of the micro-electronic component basis for computer technology equipment as well as materials and technological equipment for microelectronics;
- The general agreement on co-operation in the development and comprehensive use of microprocessor techniques in the national economy;
- The general agreement on the development of specialisation and co-operation in the production of industrial robots.

*a)* These, as well as other "highly important problems" are also contained in the document "Basic Directions of the CMEA Countries' Scientific and Technical Co-operation through 1990 and The Technical Economic Consequences Expected on this Basis". This would imply that they had not been resolved by the 35th session.

*Source*: See Table 7, *Ibid.*, pp.1-2.

Table 9 (Chapter 4)

INTRA-CMEA MULTILATERAL CO-OPERATION PROGRAMMES IN R&D, ACCORDING TO BRANCHES, INDUSTRIES AND PRODUCTS: PROGRAMME FOR THE 1981-1985 PERIOD AND BEYOND

| Sectors or Branches | Sub-branches, Industries and Products |
|---|---|
| *Machinery other than electrical* | Power generating machines, engines<br>= nuclear reactor equipment<br>= hydraulic equipment<br>– hydraulic equipment based on synthetic liquids<br>– components for the "Hydraulic and Pneumatic Unified System"<br>= equipment for hydroelectric stations |
| | Machines specialised for particular industries<br>= agricultural machinery<br>– multi-use advanced machinery<br>– rational processing of agricultural raw materials<br>= electrically propelled tractors<br>= machines and appliances for the rubber industries<br>– continuous vulcanization<br>= civil engineer and plant equipment<br>– excavating gazoduct trenches<br>= wood and wood-processing advanced machinery<br>= Coal-mining "programme-led" machinery<br>= oil and gas drilling machinery |
| | Machine tools for working metals<br>– automatic forging equipment<br>– forging/pressing automatic equipment<br>– universal advanced machine tools<br>= high precision metal-cutting machines<br>= machine-tools (automatic lines)<br>– machine-tools (automatic mechanisms)<br>– numerically programme-led metal-cutting machinery |
| | General industrial machinery and equipment<br>= fuel- and energy-saving equipment for machine tools<br>= programme-led manipulators<br>= lifting, handling, loading/unloading automatic machinery and equipment<br>= underground transportation equipment |
| | Office Machinery and automatic data-processing equipment<br>= office machines<br>– computers and other calculating equipment<br>– computer and electronic equipment<br>= automatic data-processing equipment |
| | Telecommunication apparatus and receivers<br>= television receivers<br>– "fourth generation" TV sets, parts and components<br>– TV sets with "perspective" images<br>– TV cables |

Table 9 (Chapter 4) *contd*

| Sectors or Branches | Sub-branches, Industries and Products |
|---|---|
| | = Telecommunication equipment<br>– "Unified system of ciphered information transmitters" |
| *Electrical machinery* | Electric power machinery<br>= equipment for converting sun, wind, thermal and other energy sources into electric energy<br>= equipment for distributing electricity<br>– equipment for high voltage electric lines<br>– electroceramic equipment<br>= electric apparatus for medical purposes<br>– X-ray apparatus<br>– electro medical apparatus<br>= electronic micro-circuits, semiconductor devices<br>– electronic sets, parts and components<br>– micro-processor systems<br>= other electrical machinery and apparatus<br>– advanced electrical machinery<br>– batteries, accumulators<br>– alternators<br>– flashing lights<br>– electro-magnets and parts<br>– condensors<br>– robots and programme-led manipulators |
| *Transport equipment* | Road vehicles<br>– containers<br>Other transport equipment<br>– railway dispatching equipment |
| *Other fields than machinery and equipment* | Non-electric medical equipment |
| *Chemicals and petrochemical products* | Catalyties<br>Drugs<br>Lacs<br>Pesticides<br>Fodder-yeasts from paraffin |

*Note*: The headings refer almost exclusively to Class 7, "Machinery and Transport Equipment", according to the standard SITC classification-2nd revision. A few headings from Class 5 (chemical products) and Class 8 (other manufactured products) have also been included. The first column contains the section-headings of the SITC, where for the sake of simplification, "Machinery other than electrical" subsumes several sections. In the second column, the headings with the initial capital letter indicate the SITC groups; the headings preceded by the (=) sign indicate, in principle, the SITC sub-groups; those preceded by the (–) sign indicate, in principle, the items, or individual products. Due to differences between the CMEA classification and SITC, it has not always been possible to transpose the original Soviet definition into the SITC heading. In such cases, the Soviet definition has simply been translated; for instance, "advanced electrical machinery".

*Source*: D. Tosi, *Multilteral and Bilateral Co-operation of the Eastern Countries in Research and Development within the CMEA Framework over the 1981-1985 Period and Beyond*, June, 1983. Report prepared for the OECD.

Table 10 (Chapter 4)

INTRA-CMEA BILATERAL CO-OPERATION AND SPECIALISATION IN "R&D" ACCORDING TO THE RESPONSIBLE COUNTRIES, IN 1981-1985 AND BEYOND

| Responsible Country | Sectors or Fields of Co-operation and Specialisation |
|---|---|
| Bulgaria | Electronic controlled manipulators<br>Convectors for electronic devices<br>Cables |
| German Democratic Republic | Micro-electronic components<br>Electrical and electronic computing machines<br>Microbiology equipment<br>Laser equipment IES/CIS/700 for welding techniques<br>Unified System for air-plasma metal cutting<br>Plastic fibres and materials |
| Czechoslovakia | Jets GN-1 for gazoduct welding<br>Equipment for protecting the environment from energy and chemical wastes<br>Nickel welding equipment<br>Super-conduction reels |
| Soviet Union | Computer technology<br>Machinery for in-depth and surface mineral drilling<br>Pneumatic valves |
| Poland | Transformers for multi-electrode machinery<br>Switchers for multi-electrode machinery<br>Other equipment for distributing electricity and for electrical circuits |
| Hungary | No detailed information available |
| Romania | No detailed information available |

*Source*: *Ibid.*

## NOTES AND REFERENCES

1. The Comprehensive Programme was, in essence, a complex plan for total reform of the CMEA. It foresaw improved co-ordination of Five-Year and long-term (15-20) plans, joint planning for the production of selected products, co-operation in long-term target programmes and in joint investment projects, joint scientific and technical research in certain economic branches and so forth. Perhaps most importantly, it also addressed itself to the issue of currency and financial relations. That is, the pricing system in intra-CMEA trade was to be improved by setting prices on an economic (presumably market) basis and by providing the conditions by which currency convertibility (hence multilaterality in trade) would eventually be introduced. In short, it was the basic blueprint mapping out the fundamental outlines for broadening co-operation – hence the evolution towards economic integration. For details see: "Kompleksnaya programme dal'ney-shego uglubleniya i sovershenstvovaniya sotrudnichestva i razvitiya sotsialisticheskoy ekonomi-cheskoy integratsii stran-chlenov SEV" (The Comprehensive Programme for the Further Intensification and Improvement of Co-operation and the Development of Socialist Economic Integration of the CMEA Member Countries), *Politizdat*, Moscow, 1971.
2. See for instance: Paul Marer and John Michael Montias (editors), *East European Integration and East-West Trade*, Indiana University Press, Bloomington, 1980; Zbigniew Fallenbuchl, Carl H. McMillan (editors), *Partners in East-West Economic Relations: The Determinants of Choice*, Pergamon Press, New York, 1980 and, Morris Bornstein, "East-West Economic Relations and Soviet-East European Economic Relations" in *Soviet Economy in a Time of Change*, Vol. 1, *op. cit.*
3. Some *Western* scholars associate integration with Soviet regional policy. Consequently, integration is said to imply "more joint decisions on the location of new capacity in relation to total demand within CMEA rather than with regard to national demand only." Hence, integration also means that "a larger share of any particular output crosses national borders"Shabad). Others focus on greater co-ordination of Member countries' national economic plans (Bornstein) or on attempts that are made to rationalise the regional structure of production by means of increased interaction among Member states. (McMillan) According to such concepts, inter-regional trade shares are not considered as being indicative of integration. See respectively: Theodore Shabad, "Soviet Regional Policy and CMEA Integration", in P. Marer and J.M. Montias, *op. cit.*, p. 224; M. Bornstein, *art.cit.*, p. 302 and C.H. McMillan, *art.cit.*, p. 186.

   Among *Eastern* scholars, differing definitional nuances are particularly interesting in that they reflect domestic policy proclivities. According to the socialist definition, integration is essentially a process encompassing the gamut of economic activities: country specialisation, joint industrial projects, increased intra-CMEA trade and so forth. However, in the pursuit of effective integration, the Soviet school places primary emphasis on the need to strengthen joint planning and increasing centralisation of production, although the role of the market is also acknowledged. In contrast, the Hungarian school of thought while recognising the need for common planning, accentuates market relations: increasing the economic autonomy of enterprises, increasing the role of money in economic transactions and so forth. For a discussion see Alfred Zauberman, *Topics in Trade Co-ordination of Planned Economies*, The MacMillan Press Ltd., London and Basingstoke, 1980.

4. For the Summit Declaration see, *CMEA: The Strength of Friendship and Co-operation, Documents and Materials of the Top-Level Economic Conference of the CMEA Member States, June 12th-14th, 1984*, Novosti Press Agency Publishing House, Moscow, 1984. For the Communiqué of the Havana meeting see, *Neues Deutschland*, 2nd November, 1984. For the 37th session of the CMEA see, *Neues Deutschland*, 21st October, 1983.
5. Czechoslovak Prime Minister L. Strougal, *Neues Deutschland*, 20th October, 1983, p. 4.
6. *Ibid.*
7. The degree to which the East European countries depend on Soviet oil deliveries varies – the German Democratic Republic and Bulgaria being most dependent and Romania and Hungary least. However, Romania pays for its imports from the Soviet Union in hard currency, a considerable burden given that country's financial situation.
8. See, for instance, N. Tikhonov (then Prime Minister) and N. Baybakov (then Chairman of Gosplan), *Neues Deutschland*, 19th October, 1983, p. 3.
9. N. Tikhonov, *Ibid.*
10. See Allen J. Lenz, *Controlling International Debt: Implications for East-West Trade*, US Department of Commerce, Office of Trade and Investment Analysis, Washington, D.C., September, 1983.
11. For instance, the fall-out from the "invasion" of Afghanistan, the declaration of martial law in Poland, problems relative to the issue of arms control and so forth. As has been noted, "the evolution of attitudes toward technology transfer has closely followed changes in the general political climate of East-West relations". For a more detailed discussion of the cyclical nature of East-West technology transfer see E. Zaleski and H. Wienert, *Technology Transfer between East and West*, *op. cit.*, Chapter 4.
12. *Pravda*, 15th October, 1982.
13. From the speech of W. Jaruzelski, *Neues Deutschland*, 19th October, 1983, p. 4. Subsequent discussions of the 37th CMEA session indicate that re-integrating Poland tightly into the bloc is seen, at least for the moment, as a long-term phenomenon. For instance, "tightening Poland's economic co-operation with CMEA countries is not a temporary trend created by current difficulties, but a permanent orientation. This reorientation (with some of Poland's economic ties also shifted to Third World countries) will continue throughout the 1986-1990 plan period." Interview with Deputy Premier Z. Medej, *Zycie Warszawy*, 15th November, 1983, as given in *Joint Publications Research Service*, (85002), *East Europe Report*, 22nd December, 1983, pp. 1 and 2.
14. See Academician O. Bogomolov, "SEV Ekonomicheskaya strategiya 80. ch. godov", *Kommunist*, No. 7, May, 1983. Bogomolov notes that during the 1976-1980 period, the role of intra-CMEA trade in covering the needs of the Member countries had greatly deteriorated. In 1980, the Western share in total imports of machinery and equipment for the chemical industry had exceeded 65 per cent, for the mining, metallurgical and petroleum industries 52 per cent and for metal working machinery, 49 per cent.
15. The transferable ruble is simply an accounting unit – an aspect discussed in subsection 3(c). Hungary (since 1975), Poland (since 1981) and Romania (since 1980) express the value of their trade in national currency. Bulgaria, Czechoslovakia, the German Democratic Republic and the Soviet Union use devisa (valuta) currency; the valuta ruble being equal to the transferable ruble.
16. The evolution of trade flows must, of course, be placed within the context of price patterns. That is, differences in the evolution of prices within and outside the CMEA area, preclude a rigorous assessment of the actual growth of trade relative to its geographical distribution. Different reporting procedures, which can lead to sizeable discrepancies in the "mirror" foreign trade statistics between Eastern and Western data sources, also tend to skew the picture.

17. The notion of Soviet credits or subsidies extended to the East European countries is receiving a good deal of attention in Western analyses – though there is little agreement to date as to what Soviet trade surpluses are "costing" the Soviet Union or the East European countries. For a recent discussion see, Michael Marrese & Jan Vanous, *Soviet Subsidization of Trade with Eastern Europe, A Soviet Perspective*, Institute of International Studies, University of California, Berkeley, 1983. For an earlier discussion of CMEA pricing and the relationship to Soviet-East European trade developments see: Martin J. Kohn, "Soviet-Eastern Europe Economic Relations, 1975-1978" and Raimund Dietz, "Price Changes in Soviet Trade with CMEA and the Rest of the World Since 1975", in *Soviet Economy in a Time of Change*, Vol. 1, *op. cit.*.

18. For a discussion of this, as well as the varying attitudes among the Eastern countries, see Franz-Lothar Altmann, *Die Westhandelspolitik der RGW(6)-Länder*, Working Paper No. 96, Osteuropa Institut, Munich, November, 1983.

19. *OECD, Foreign Trade Statistics.*

20. Otta Henys (Office of the Czechoslovak Government Presidium), "More Intensive Co-ordination of Economic Development of the CMEA Countries", *Nova Mysl*, No. 3, 1983, p. 56.

21. R. Nyers, "Tradition and Improvement in CMEA Co-operation", *Kozgazdasagi Szemle*, No. 4, 1982, as given in *Joint Publications Research Service* (81022), *East Europe Report*, 10th June, 1982, pp. 14 and 15.

22. This observation (inclusive of the quotation) is taken from *Ibid.*, pp. 11 and 15.

    While Hungarian enterprises certainly do have far more autonomy in decision-making matters, neither are their powers as great as could be inferred by the source cited here. For a detailed analysis of decision-making and its relevance to East-West technology transfer, see Paul Marer, *East-West Technology Transfer, Study of Hungary: 1968-1984, op. cit.*

23. For a discussion of these problems and their effect on CMEA integration see, for example: Alfred Zauberman, *op. cit.*, Anita Tiraspolsky, "L'énigme du prix des échanges à l'intérieur du CAEM", *Le courrier des pays de l'est*, No. 271, Mars, 1983 and Hermann Clement, *Preisbildung im Intra-RGW-Handel – Ein Überblick*, Working Paper Nr. 94, Osteuropa Institut, Munich, October, 1983.

24. Note that "intra-CMEA trade transactions are denominated in TRs after world market prices (expressed in internationally traded currencies) are converted into TRs on the basis of the official ruble-dollar exchange rate specified by the USSR". See Morris Bornstein, "Soviet East-European Economic Relations", in *East-West Relations and the Future of Eastern Europe* (edited by M. Bornstein, Z. Gitelman and W. Zimmerman), George Allen & Unwin, 1981, p. 118.

25. R. Nyers, *art. cit.*, p.5.

26. D. Ghermani, "Romania and CMEA", *Südosteurope*, No. 9, 1983, as given in *Joint Publications Research Service* (849112), *East Europe Report*, 9th December, 1983, p. 87.

27. See Jozef M. van Brabant, "The USSR and Socialist Economic Integration – A Comment", *Soviet Studies*, Vol. XXXVI, No. 1, January 1984, note 49.

28. A. Belovich, "The Co-ordination of the Investment Policy of the CMEA Member Countries", *Izvestiya Akademii Nauk SSSR. Seriya Ekonomicheskaya*, No. 1, Jan.-Feb., 1983, as given in *Joint Publications Research Service* (83815), *USSR Report*, 1st July, 1983, p. 50. Hungarian estimates give the following breakdown: Hungary – 4 per cent of all productive investments; the German Democratic Republic – 3 per cent ; Bulgaria – 2 per cent and Poland – 2.4 per cent. Cited by Marie Lavigne in "The Soviet Union Inside Comecon", *Soviet Studies*, Vol. XXXV, No. 2, April, 1983, p. 147.

29. See Cam Hudson, "Are CMEA Joint Investment Schemes being Downgraded", *Radio Free Europe Research*, RAD Background Report 258, Munich, 10th December, 1983, pp. 5-6. The author cites Hungarian sources.

30. Raimund Dietz, Ilse Grosser, *Comecon Energy Perspectives and the Long-Term Target Programmes*, Reprint Serie No. 58, Wiener Institut für Internationale Wirtschaftsvergleiche, November, 1981, p. 21.

31. *Ibid.*, p. 25.

32. L. Strougal, "Co-ordination of Economic Policy – A Qualitatively New Form of Co-operation: On the Results of the CMEA Economic Summit Conference", *Pravda*, 5th July, 1984, as given in *Joint Publications Research Service* (EEI-84-088), *East Europe Report*, 6th August, 1984, p.4.

33. V. Shastiko, "Production Co-operation at the Enterprise Level", *Voprosy Ekonomiki*, No. 1, 1984.

34. T. Bartoszewicz, "New Impetus Needed", *Rynki Zagraniczne*, 23rd October, 1984, as given in *Joint Publications Research Service* (EEI-85-002), *East Europe Report*, 7th January, 1985, pp. 1-2.

35. J. Nitz, P. Freiberg, "Problems of East-West Economic Relations", *IPW Forschungshefte* (East Berlin), Vol. 19, No. 3, 1984, as given in *Joint Publications Research Service* (EEI-85-019), *East Europe Report*, 16th February, 1985, p.7.

36. R. Nyers, *art. cit.*, p. 17.

37. See, for instance, L. Csaba, "The Place of CMEA in the World Economy", *Valosag*, No. 7, 1982, as given in *Joint Publications Research Service* (81721), *East Europe Report*, 8th September, 1982 and L. Csaba, "Reform and Pragmatism. Further Development of CMEA Integration", *Valosag*, No. 12, 1983, as given in *Joint Publications Research Service* (EEI-84-070), *East Europe Report*, 13th June, 1984. The author argues that "if the CMEA countries are to be able to withstand the competition in the world economy and to successfully adjust to it, then integration into the world economy will proceed parallel with regional integration". He also points out that the alternative of "turning inward is nothing new; it would merely mean an extension of the economic policy of import substitution that the CMEA countries have been pursuing for three decades".

38. O. Labetskiy, A. Rybakov, "Economic Incompatibility of CMEA Countries Refuted", *Obshchestvennyye Nauki*, No. 6, November-December, 1983, as given in *Joint Publications Research Service* (UIE-84-003), *USSR Report,* 24th January, 1984, p. 31.

39. J. Nitz, P. Freiberg, *art. cit.*

40. John Hannigan and Carl McMillan, "Joint Investments in Resource Development: Sectoral Approaches to Socialist Integration", in *East European Economic Assessment*, Part 2, Regional Assessments, Joint Economic Committee, Congress of the United States, Washington, D.C., July 10, 1981, p. 261.

41. L. Csaba, "The Place of CMEA in the World Economy", *art. cit.*, p. 39.

42. Raimund Dietz, Ilse Grosser, *op. cit.*, p. 34.

43. For a history of attempts made to co-ordinate national economic plans (as well as an analysis of the problems encountered) see, for instance, Z. Fallenbuchl, "East European Integration: COMECON", in *Reorientation and Commercial Relations of the Economies of Eastern Europe*, Joint Economic Committee, Congress of the United States, Washington D.C., August 16, 1974. See also, R. Dietz, Ilse Grosser, *op. cit.*

44. *The Council for Mutual Economic Assistance; 30 Years*, CMEA Secretariat, Moscow, 1979, p. 20.

45. 0. Rybakov, "The Problems of the Co-ordination of the National Economic Plans of the CMEA Member Countries", *Planovoye Khosyaystvo*, No. 11, November, 1982.

46. See, respectively: "Answers to Questions on Socialist Economic Integration", *West Berlin Presse Informationen*, 20th May, 1977, p. 20; N. Faddeyev, "10 Years of Actualising the Comprehensive Programme for the Further Intensification and Improvement of Co-operation and

Development of Socialist Economic Integration", *Ekonomicheskoye Sotrudnichestvo Stran-Chlenov SEV*, No. 3, 1981 and N. Ivanov and A. Loshchakov, "Co-operation of CMEA Countries in Development of the Material-Technical Basis of Socialism", *Voprosy Ekonomiki*, No. 3, 1983.

47. Section II of the Plan contains measures relative to production specialisation and co-operation to be implemented on a multilateral basis, while Section III of the Plan is devoted to scientific and technical co-operation. As given by S. Kolchin, "Socialist Economic Integration and its Statistical Study", *Vestnik Statistiki*, No. 10, 1980, as given in *Joint Publications Research Service* (77357), *USSR Report*, 10th February, 1981, pp. 15-16.

48. A. Belovich, *art. cit.*, p. 50.

49. See John Hannigan and Carl McMillan, *art. cit.*, p. 269.

50. For a detailed discussion of the manner in which these projects were financed, inclusive of the role of the International Investment Bank (IIB) and the International Bank for Economic Co-operation(IBEC) see Raimund Dietz, Ilse Grosser, *op. cit.*, and John Hannigan and Carl McMillan, *art. cit.*

51. See A. Belovich, "Co-ordinating Capital Investments Among CMEA Members", *Voprosy Ekonomiki*, No. 3, 1983.

52. N. Faddeyev, *art. cit.*

53. V. Kotouch (CMEA Secretariat), "The Co-ordination of Plans for 1986-1990", *Ekonomicheskoye Sotrudnichestvo Stran-Chlenov SEV*, No. 1, 1983.

54. See, for instance, N. Inozemtsev, "Co-ordination of the National Economic Plans of the CMEA Member countries: New Problems and Tasks", *Planovoye Khosyaystvo*, No. 10, 1984, O. Rybakov, "Long-range Strategy for Deepening Socialist Economic Integration of the CMEA Countries (On the Results of the CMEA Member Country Economic Conference at the Highest Level)", *Planovoye Khosyaystvo*, No. 8, 1984 and V. Kotouch, *art. cit.*

55. O. Rybakov, "The Problems of the Co-ordination of the National Economic Plans of the CMEA Member Countries", *art. cit.*

56. N. Inozemtsev, *art. cit.*

57. V. Kotouch, *art. cit.*

58. N. Inozemtsev, "Some Results of the Co-ordination of the State Plans of the USSR and other CMEA Member Countries for 1981-1985", *Ekonomicheskoye Sotrudnichestvo Stran-Chlenov SEV*, No. 6, 1981.

59. For details as to how these programmes are elaborated and administered (particularly the relationship between planning and the Target Programmes), see, for instance, Jozef M.van Brabant, "Target Programming: A New Instrument of Socialist Economic Integration", in *Jahrbuch der Wirtschaft Osteuropas*, Vol. 9, No. 1, 1981, and Raimund Dietz, Ilse Grosser, *op. cit.*

60. The information concerning all the programmes is taken from K. Starzyk, "Economic Co-operation with CMEA Countries Discussed", *Nowe Drogi*, No. 5, 1983, as given in *Joint Publications Research Service* (84150), *East Europe Report*, 19th August, 1983, pp. 2-3 and V.A. Golubov, *et.al.*, *The Foreign Economic Relations of CMEA Countries under the Conditions of Socialist Integration*, Kiev, 1983, as given in *Joint Publications Research Service* (UIE-84-006-L), *USSR Report*, 15th May, 1984.

61. For a detailed discussion of the LTPC, particularly within the context of CMEA energy developments and their implication for East-West Trade see, Raimund Dietz, Ilse Grosser, *op. cit.*

62. V. Moiseyenko, "Specialisation and Co-operation in the Machine-Building Industry – an Important Factor in Promoting CMEA Member Countries' Mutual Trade", *Foreign Trade* (English edition), No. 2, Moscow, 1979, pp. 6-7. The author is citing A.N. Kosygin speaking at the 32nd session of the CMEA.

63. N. Ivanov and, A. Loshchakov, *art. cit.*

64. The figure of 200 is given by Marie Lavigne (herself citing official sources). See M. Lavigne, *art. cit.*, p. 144. The figure of 120 is cited in *Dolgosrochnye Tselevye Programme Sotrudnichestvo Stran-Chlenov SEV*, Moscow, 1982, p. 40.

65. I. Dudinskiy, "Socialist Integration as a Factor in Intensifying the Economy", *Voprosy Ekonomiki*, No. 8, 1982.

66. A. Belichenko (Chairman of the Board of IIB), *Foreign Trade*, (English edition), No. 9, Moscow, September, 1983, p. 35 and *Dengi i Kredit*, No. 7, 1984, as given in *Joint Publications Research Service* (UIE-84-024), *USSR Report*, 20th December, 1984. The figure of 20 billion TRs budgeted for 1976-1980 is given by Jozef M. van Brabant, *art. cit.*, p.138 (note 49).

67. A. Belovich, "Co-ordinating Capital Investments Among CMEA Members", *art. cit.* Similar observations are made by Soviet authors. See, for instance, O. Rybakov, "The Problems of the Co-ordination of the National Economic Plans of the CMEA Member Countries", *art. cit.*

68. A. Belovich, "Co-ordinating Capital Investments Among CMEA Members", *art. cit.*

69. B. Dyakin, B. Kheyfets, "Comprehensive Program for Integration and the Effectiveness of Co-operation", *Sotsialisticheskiy Trud*, No. 3, Moscow, 1982, as given in *Joint Publications Research Service* (81894), *USSR Report*, 30th September, 1982, p. 41.

70. Yu. Medvedkov, "Programmes for Specialisation and Co-operation in Production up to 1990 and Some Questions of Organisation of Socialist Countries' Foreign Trade", *Foreign Trade* (English edition), No. 2, Moscow, 1979, p. 32.

71. P. Bagudin, Ye. Gavrilov, "Improving the Sectoral Structure of CMEA Member Countries' Industry", *Voprosy Ekonomiki*, No. 2, 1983.

72. N. Barzin, L. Bibik, "Strengthening the World Socialist System and Co-operation of the Socialist Countries", *Ekonomicheskiye Nauki*, No. 4, 1982, as given in *Joint Publications Research Service* (82167), *USSR Report*, 4th November, 1982, p. 8 and B. Vyazovov, "CMEA: 35 Years of Co-operation and Integration", *Foreign Trade* (English edition), No. 4, Moscow, 1984.

73. S. Kolchin, *art. cit.*

74. I. Kareyev, "Bilateral Programmes for Production Specialisation and Co-operation", *Voprosy Ekonomiki*, No. 5, 1981.

75. I. Portyannikov and N. Rakhutin, "Co-ordinating National Economic Plans of CMEA Member States for 1981-1985", *Voprosy Ekonomiki*, No. 9, 1981.

76. Martin J. Kohn, "Soviet-Eastern European Economic Relations, 1975-1978", *art. cit.*, p. 258.

77. I. Dudinskiy, *art. cit.* It should be noted, that there is considerable discrepancy in the Eastern literature as to the number of agreements actually concluded. We have cited the figure still given most frequently.

78. Various Eastern sources. It should be noted that data limitations are particularly apparent as regards 1983. Those figures which were found relative to country shares indicate rather sizeable country increases. Yet a Polish source (*Zicie Warszawy*, 3rd July, 1984, cited by A. Tiraspolsky, *Le courrier des pays de l'est*, No. 289, November, 1984, p.64) gives the same total output and country shares in 1983 as for 1982.

79. Angel Chaushev, "Foreign Trade in 1982: Rates, Structure and Trends", *Ekonomicheskoye Sotrudnichestvo Stran-Chlenov SEV*, No. 8, 1983.

80. S. Dulovets, "International Specialisation and Co-operation in the Industry of the CMEA Member Countries, "*Ekonomicheskoye Sotrudnichestvo Stran-Chlenov SEV*, No. 1, 1983.

81. Angel Chaushev, *art. cit.* In 1981, the chemical industry comprised 10.4 per cent of the total exports of specialised industrial output and other branches of industry, 7.1 per cent. See source cited in note (80) above.

82. Angel Chaushev, *art. cit.*

83. Communiqué of the 103rd session of the CMEA Executive Committee, *Neues Deutschland*, 22nd January, 1982.

84. L. Afanasyev, S. Chenchikovski, "Ten Years of Comprehensive Programme of Socialist Economic Integration and International Specialisation of Production", *Foreign Trade* (English edition), No. 11, Moscow, 1981, p. 4.

85. "Decisive Role of Machine Construction in Realisation of the Goals", *Aussenwirtschaft* (East Berlin), No. 40, 3rd October, 1984, as given in *Joint Publications Research Service* (EEI-85-029), *East Europe Report*, 12th March, 1985, p.2.

86. *Heti Vilaggazdasag*, No. 24, 16th June, 1984 as given in *Presseschau Ostwirtschaft*, No. 6, Österreichisches Ost-und Südosteuropa-Institut, 1984, p. 5.

87. *DDR-Aussenwirtshcaft*, Vol. 10, No. 49, 1st December, 1982, pp. 1-2, as given in *Joint Publications Research Service* (83030), *East Europe Report*, 8th March, 1983, p. 2.

88. M. Engerst, "New Steps in Economic Co-operation", *Horizont* (East Berlin), No. 5, 30th April, 1984, as given in *Joint Publications Research Service*) (EEI-84-067), *East Europe Report*, 7th June, 1984, pp.7-8.

89. "Special Tasks of Specialisation and Co-operation", *Figyelo*, Budapest, 14th April, 1983, as given in *Joint Publications Research Service* (83470), *East Europe Report*, 16th May, 1983, p. 5.

90. "Machines and Equipment-Dynamic Factor in Mutual Trade of CMEA Countries", *DDR-Aussenwirtschaft*, Vol. 10, No. 49, 1st December, 1982, as given in *Joint Publications Research Service* (83030), *East Europe Report*, 8th March, 1983, p. 4.

91. See excerpts of articles by Prof. K. Morgenstern and H. Tschanter in *Informationen*, No. 3, Bonn, February, 1982, as given in *Joint Publications Research Service* (80507), *East Europe Report*, 7th April, 1982, p. 16 and O. Rybakov, "Long-range Strategy for Deepening Socialist Economic Integration of the CMEA Member Countries", *art. cit.*

92. *Aussenwirtschaft*, No. 43, (East Berlin), 19th October, 1983, as given in *Joint Publications Research Service* (84832), *East Europe Report*, 28th November, 1983, p. 7.

93. N. Alekhin, "Production and Scientific-Technical Co-operation of CMEA Countries", *Voprosy Ekonomiki*, No. 10, 1982.

94. Prof. Morgenstern, in publication cited above, p. 16.

95. N. Alekhin, *art. cit.*

96. D. Tosi (Directorate-General for External Relations, Commission of the European Communities), *Trends of Technical-Scientific Co-operation, Co-operation in Production of High R&D Intensive Products and Related Mutual Trade among the European CMEA Countries during the 1970-1981 Period*, July, 1984. Report prepared for the OECD.

97. See Paul Marer, "Western Multinational Corporations in Eastern Europe and CMEA Integration", in Z. Fallenbuchl, C. McMillan, *Partners in East-West Economic Relations, the Determinants of Choice, op.cit.*

98. Louvan E. Nolting, *Integration of Science and Technology in CEMA*, Foreign Economic Report No. 21, US Department of Commerce, Bureau of the Census, Washington, D.C., December, 1983. See also Vladimir Sobell, *The Red Market: Industrial Co-operation and Specialisation in Comecon*, Gower Publishing Company Limited, 1984.

99. See "Kompleksnaya programma..." in *Politizdat, op. cit.*, pp. 58-60.

100. G. St.-Georgiev, "Scientific and Technical Integration Between CMEA Countries and the Participation of Bulgaria in It", *Planovo Stopanstvo*, No. 8, 1976 (the author is quoting from a Soviet source), as given in *Joint Publications Research Service* (68739), *Translations on Eastern Europe*, 8th March, 1977, p. 23. For details, see L. Nolting, *op. cit.*, Chapter 2.

101. See "Kompleksnaya programma..." in *Politizdat, op. cit.*, pp. 33-34.

102. D. Tosi, *Trends of Technical-Scientific Co-operation..., op. cit.*

103. See L. Nolting, *op. cit.*, p. 7.

104. *Ibid.*, p. 11.

105. *Ibid.*, p. 11.

106. See S. Kolchin, *art. cit.*, p. 17, and V.A. Golubov, *op. cit.*, pp. 41 and 106.

107. See "A Year of Fruitful Co-operation", *Ekonomicheskoye Sotrudnichestvo Stran-Chlenov SEV*, No. 10, 1983, and V.A. Brykin, *et.al.*, *The Effectiveness of Foreign Economic Ties: Urgent Problems*, Moscow, 1983, as given in *Joint Publications Research Service* (UIE-85-004-L), *USSR Report*, 11th February, 1985 and "Proceedings: At the 36th Session of the Council for Mutual Economic Assistance", *Ekonomicheskoye Sotrudnichestvo Stran-Chlenov SEV*, No. 9, 1982.

108. L. Nolting, *op. cit.*, p. 11.

109. V. Leont'ev, "Certain Problems in Improving Scientific and Technical Co-operation", *Ekonomicheskoye Sotrudnichestvo Stran-Chlenov SEV*, No. 4, 1983.

110. V.A. Brykin, *et. al.*, *op. cit.*, p. 27.

111. "Criticism and Bibliography: The Economic Co-operation of the CMEA Member Countries", *Politicheskoye Samoobrazovanyie*, No. 1, 1984, as given in *Joint Publications Research Service* (UIE-84-008), *USSR Report*, 5th April, 1984, p. 15. The 198 agreements concern the LTPCs.

112. N. Ivanov and A. Loshchakov, *art. cit.*

113. H. Zschiedrich and M. Perlzer, "Scientific-Technical Progress and Planned Intensification of Socialist Economic Integration" *Deutsche Aussenpolitik* (East Berlin), Vol. 26, No. 11, November, 1981, as given in *Joint Publications Research Service* (80171), *East Europe Report*, 24th February, 1982, p. 5.

114. N. Ivanov and A. Loshchakov, *art. cit.*

115. N. Faddeyev, *art. cit.*

116. V.A. Gnutov, "Financing of Scientific and Technical Research Carrried out by the CMEA Member Countries", *Finansy SSSR*, No. 5, 1983.

117. A. Leznik, "Effectiveness of International Production Specialisation", *Voprosy Ekonomiki*, No. 4, 1982. The author is quoting N.A. Tikhonov, speaking at the 35th session of the CMEA.

118. N. Alekhin, *art. cit.*

119. A. Filipenko, "Scientific and Technical Co-operation in the International Economic Complex of the CMEA Member Countries", *Ekonomika Sovetskoy Ukrainy*, No. 9, 1982, as given in *Joint Publications Research Service* (83139), *USSR Report*, 25th May, 1983, p. 18.

120. Speech by N. Tikhonov as given in *Neues Deutschland*, 19th October, 1983, p. 3.

121. D. Tosi, *Multilateral and Bilateral Co-operation of the Eastern Countries in Research and Development within the CMEA Framework Over the 1981-1985 Period and Beyond*, June, 1983. Report prepared for the OECD.

122. *Ibid.* The author points out that according to Western classifications, products which under intra-CMEA multilateral co-operation belong to large sub-branches "machines specialised for particular industries", "machine tools for working metals" or "general industrial machinery and equipment", fall, in principle, into the "low R&D intensive" class. However, as he also mentions, in the Eastern literature such products or groups of products are specified as "advanced", "programme-led", "high-precision" and so forth. In other words, they are considered to be "high R&D-intensive" in that they contain electronic components and other micro-processing equipment. The relative importance of "high", "medium" or "low" R&D-intensive products in East-West technology transfer is discussed in more detail in Part II of the study.

123. There is, of course, great variation among the individual Eastern countries. For instance, OECD exports to the German Democratic Republic and Romania, show a clearly rising trend for the

"high" and "moderate" R&D-intensive groups. For countries such as Czechoslovakia, the corresponding relative shares are declining. Nevertheless, the situation for the individual Eastern countries mirrors rather closely the overall pattern. In no country does the share of "high R&D-intensive OECD exports exceed 6 per cent, or the share of "low R&D-intensive OECD exports fall below 70 per cent: referring always to the total OECD exports of embodied technology to the relevant country. *Ibid.*

124. See L. Nolting, *op. cit.*, p. 39.

125. See V.A. Gnutov, *art. cit.*, V. Leont'ev, *art. cit.*, and A. Filipenko, *art. cit.*

126. E. Halizak, "Trends in International Division of Labour within the Socialist Countries", *Sprawy Miedzynarodowe*, No. 8, August 1981, as given in *Joint Publications Research Service* (82419), *East Europe Report*, 8th December, 1982, p. 9.

127. See Z. Tvrdik, "Stimulating Scientific and Technological Co-operation Among the CMEA Countries", *Voprosy Ekonomiki*, No. 5, 1982, V.A. Brykin, *et.al.*, *op.cit.*, and V.A. Golubov, *et.al.*, *op.cit.*

*Part II*

# EMBODIED TECHNOLOGY TRANSFER AND TOTAL TRADE

# INTRODUCTION

This part of the study examines embodied technology transfers (capital goods, together with a selection of intermediate goods with a significant technology content) between East (seven individual countries of the European CMEA) and West (total of 24 OECD countries). Chapter 5, section 1 describes the data bases built up to quantify the embodied technology element in East-West and total Eastern countries' trade flows. In section 2, the changing share of embodied technology is surveyed in relation to total Western exports, imports and trade balances over the period 1970-1983. Finally, changes in the contributions of various country groupings within OECD's export and import trade in capital and intermediate (technology) goods with the Eastern countries are summarised. The findings in section 2 are based upon OECD member countries' trade returns at current prices. Chapter 6 is broader in scope. First, the Eastern countries' own trade returns, converted into US dollars at current prices, are analyzed to illustrate the role of embodied technology (capital goods only) in their trade with the world and by region of origin (section 1). In section 2, intra-CMEA embodied technology flows are discussed. In section 3, a number of topics are analyzed (level of sophistication of Eastern embodied technology imports from the West, the movement of capital good imports relative to investment, their development and structure by type and, the sectoral and industrial branch structure of Eastern capital goods imports from East and West) in order to throw light on past and present strategies towards these imports.

The main findings of the two chapters are first, that the share of embodied technology, as defined in section 1 of Chapter 5, has averaged around three-quarters of total OECD exports to the East, but a much smaller share of OECD imports from the Eastern area. Second, variations in the former flow have been the main determinant of total OECD exports to the area. Third, capital goods have borne the main burden of East European balance of payments adjustments in the early 1980s. In the Soviet Union, though the growth of capital goods imports moderated in 1980/1981, expanding fuel exports played a major role in trade deficit reduction. Thus, while there has been a fairly general fall in the share of capital goods in Eastern Europe's embodied technology imports, it has been much less marked in the Soviet Union and indeed was partially reversed after 1981 in that country. The unchanged share of intermediate (technology) goods which characterized, in particular, Eastern Europe's technology import pattern, and that of the Soviet Union up to 1980, suggests a fourth conclusion: the import of Western capital goods may incorporate hidden costs by expanding demand for imports of more sophisticated intermediate goods. These are not readily available either from indigenous sources or from other Eastern countries. Fifth, overall strategy toward technology imports during the 1970s paid too little attention to export expansion. Not only was the share of capital goods imports for the production of tradeables surprisingly small, but the difficulties of maintaining or expanding the Eastern share of Western markets (except for fuel and a very limited selection of other goods) were clearly vastly underestimated. Finally, Western technology, notably capital goods, still accounts for a minority share of Eastern-imported embodied technology from all regions. When financial conditions permitted – in the 1970s – it nonetheless expanded rapidly – especially in those countries where it was low at the beginning of the period covered by this analysis (1970-1983). Subsequent cuts, necessary for the redressment of trade deficits in the late 1970s and early 1980s, were not pushed beyond a certain limit, and there is evidence that the Eastern countries as a whole seek to maintain a minimum share of Western technology in their imports and to expand it where hard currency availabilities permit.

*Chapter 5*

# EAST-WEST EMBODIED TECHNOLOGY TRANSFER: THE STATISTICAL EVIDENCE

## 1. The Data Bases

This section describes the set of trade data developed as an aid to quantify and assess the effects of Western technology exports to the six countries of Eastern Europe and the Soviet Union. A survey of the criteria according to which data on East-West and intra-CMEA embodied technology trade (mainly capital goods) were selected, and the main difficulties encountered in doing so, is given in sub-section *a).* In sub-section *b),* comparisons between the data taken from OECD and Eastern sources are made, and some of the shortcomings thereby revealed, are brought out.

### a) *General Principles of Data Selection*

It is not intended in this presentation to belabour the concept of "technology" – either by offering alternatives to definitions proposed by other authors in this field or by justifying in great detail those definitions which, implicitly or explicitly, were the starting point for the work described. Suffice it to say that we first distinguished between "embodied" and "disembodied" technology and in fact limit our comments to the former. In this part of the paper, we attempt to show the size and development of a certain relatively simply definable component of technology only, with a view to highlighting changes in its magnitude during the 1970s.

All we have done therefore, is to use two classes of traded goods which seem to us relevant to the technology transfer issue on *a priori* grounds. The first type is capital goods (investment goods). It should be noted that we do not distinguish between capital goods imported to remove more or less temporary capacity constraints and those which fill gaps due to specific, long-term shortcomings in domestic technology; all investment goods are retained for our purpose. The second type of goods we retained are intermediate goods with a significant technological content. Here we have, in fact, tried to distinguish between different classes of intermediate goods by selecting imports, the domestic production of which might be expected to have a high opportunity cost. We have done so, however, simply on the implicit assumption that it is goods with a significant technology content which have the highest opportunity costs. Again, while some of these imports of intermediate goods fill gaps in (or help to improve) domestic production technology, others simply supplement domestic output capacity. However, as in the case of investment goods, we make no systematic effort to distinguish between the two.

The selection of capital goods has been made both on the basis of OECD statistics of East-West trade and, secondly, on the basis of East European and Soviet national statistics. However, Eastern-reported data are insufficiently complete to enable the selection of intermediate goods with a significant technology content, so that this part of the exercise was limited to East-West flows based on OECD trade statistics.

A considerable effort has been made to gather together data on flows of Western capital goods to the Eastern countries from both OECD and CMEA country sources – that is, to look at essentially the same phenomenon from two different data sets. The reason is that, prior to this exercise, little systematic information was available on the breakdown of technology flows of any kind by economic sector or industrial branch end-users. The presentation of capital goods, but not intermediate goods flows, by end-users is a unique feature of the OECD exercise. To accomplish it, we first used total OECD trade flows with the seven Eastern countries individually, based on the UN Standard International Trade Classification (Revision I). The SITC is essentially a classification by type of good, which, as will be discussed later, leads to difficulties in allocating capital goods by end-uses. However, whatever its shortcomings in this respect, it is a more complete and consistent data set over time than the trade statistics of the Eastern countries themselves.

While retaining this advantage, but also to check on the breakdowns derived from these data when disaggregated by end-users, it was also considered desirable to perform as nearly as possible the same operation on the Eastern countries' own trade figures. Despite a number of omissions which will be discussed in section 2 below, the Eastern countries' trade statistics do contain a large number of capital goods items already classified by end-use. It was in any case thought desirable to exploit Eastern countries' own data in order to examine intra-CMEA trade in capital goods by end-use sector and branch – to examine their implications for East-West technology trade, to set the size of the latter into context and, where possible, to compare the pattern of supplies from the two sources. Since it was relatively simple to include data for capital goods trade with the world, it was possible, by substraction, also to obtain data on CMEA exports to and imports from the Rest of the World. In the case of Eastern capital goods imports, the Rest of the World means, effectively, the OECD countries (though other suppliers which are not included in the OECD aggregates – notably Yugoslavia – are not altogether negligible). CMEA countries' own import data thus provide an independent check on the OECD data in aggregate as well as by end-uses.

The Western data cover exports by 24 OECD countries (excluding Yugoslavia and trade between the Federal Republic of Germany and the German Democratic Republic). They also include imports by the OECD from the CMEA countries. As mentioned, both capital and intermediate technology-based goods are included. Capital goods are as specified in the United Nations Classification by Broad Economic Categories[1], and correspond to category 41 (machinery, other capital equipment except transport excluding parts and accessories) plus item 521 (industrial transport equipment). It should be noted that in both cases parts and accessories are excluded from capital goods. The BEC listing was slightly expanded to include pipeline, within which large diameter pipe for the transport of oil and gas is included, and also a few other items such as rails which are clearly also capital goods items.

Intermediate goods with a significant technology content are selected first from the BEC intermediate goods listing and second, according to a method of classifying output by industrial branches according to the R&D intensity of the producer branch[2]. The broad headings under which they were selected are: parts and accessories of machinery and equipment; ditto for transport equipment; paper manufactures; textile manufactures; synthetic fibres; special manufactures of leather, wood, glass and minerals; miscellaneous mineral manufactures; iron and steel; non-ferrous metals; organic and inorganic chemical elements; final chemical manufactures; chemical and plastic materials and, a very small miscellaneous component[3].

The selection of capital goods items made on the basis of the East European and Soviet country data is based upon a "control group" – SFTC (Standard Foreign Trade Classification of the CMEA) group 1 (engineering goods), for which a number of items have been added (pipes) and others subtracted (mainly a few intermediate goods such as ballbearings, cable and wire and metals, etc.). Investment goods imports from and exports to other CMEA countries and also to and from the Rest of the World were both extracted. This gives not only the possibility of checking OECD-reported exports to the East in total and by end-uses, as already mentioned, but also enables intra-CMEA shipments by country to be completed. Since the data cover both exports and imports, they enable "mirror" data to be used to supplement gaps in individual reporting countries' data on intra-CMEA trade.

End-use sectors and branches were specified according to the more or less similar national accounting and industrial classifications used by the Eastern countries. First, the total economy was divided between the "material" and "non-material" spheres – that is, between activities resulting in material output – and consumer services; notably medical and administrative services which are the main ones involving identifiable trade flows in capital goods. Within the material sphere, five end-use sectors were identified: industry, construction, agriculture, transport (including pipeline and ancillary equipment for the transport of oil and gas) and other "non-attributable" sectors. Within industry, there is some variation in the scope of the various branch classifications between Eastern countries, but nine branches (see the stub of Table 4 of this chapter) were selected. Differences in their coverage are summarized in the note to Table 3.

Both sets of statistics are presented in current US dollars. OECD data are, of course, reported in this currency. Eastern country trade flows with other Eastern countries in national currencies were converted into transferable rubles at official rates, and then into US dollars at the official ruble/dollar rate. For trade with the Rest of the World, the official national currency/dollar rate was used. (This involves some discrepancy in that the implied ruble/dollar cross-rates sometimes diverge from the Soviet ruble/dollar rate – see note 4 to Chapter 6.) It needs to be borne in mind that the valuta currencies on which these dollar values are generally based (but not in Hungary after 1975 or in Poland and Romania after 1980) are accounting units which bear no necessary relation to domestic price levels or to inter-branch and sectoral price relatives. Moreover, while the resulting dollar values for East-West transactions presumably reflect accurately the hard currency values in which they were, in most cases, originally designated, their value as an indicator of the value of intra-CMEA trade is more limited. Even though intra-CMEA trade prices are, in theory, fixed on the basis of lagged world market prices, a certain unquantifiable arbitrary element remains since it will not always have been possible for CMEA negotiatiors to find equivalent products traded on world markets for the purpose of price setting. Finally, it can be expected that many transactions are designated at prices reflecting particular circumstances determining the partner countries' bargaining positions for a given bilateral flow at a particular time. These also may diverge from world market prices.

In the attribution of OECD (SITC) data to Eastern end-uses there were, inevitably, many cases where instinct and experience rather than precise knowledge guided the attribution of a particular sector or branch. This was inevitable, since SITC is, essentially, a classification by type of good rather than by end-use. While some SITC end-uses are fairly obvious (the energy branch in the case of electricity generation equipment, the transport branch for railway engines or rolling stock and, within industry, textile machinery etc.), other items may be destined for a range of different end-users (e.g. structural parts; electrical or non-electrical machinery in general). In these cases, the approach used was one of cautious common sense – which does not, however, exclude the possibility of error. In the event, it was found possible to attribute over 90 per cent of investment goods flows to one or other sector of the material sphere (industry, construction, agriculture, transport and communications, trade) and the non-material sphere (office machinery, medical items). Within industry, branch attribution covered about 65 per cent of the total allocated to the industry sector. Of the industry residual, a quarter to a half was accounted for by SITC item 719.8 ("non-electrical machinery and equipment not elsewhere specifed").

Allocation problems were smaller in the case of East European and Soviet-reported figures which, as noted, contain many more categories than SITC relating trade directly to end-users[4]. However, other problems arise as a result of the very poor detailed commodity coverage of a number of CMEA countries: Bulgaria, where many items are shown in physical units only, and the German Democratic Republic and Romania for which only data at relatively high levels of aggregation are published. The practice of including certain items in some years but not in others is a misleading feature of the Soviet statistics which are otherwise among the most complete. In general, the Czechoslovak, Hungarian, Polish and Soviet data give the most complete and consistent end-use breakdowns. The figures reported by Eastern European countries as a whole (but not the Soviet Union or Czechoslovakia in recent years) do reveal a considerable degree of understatement of aggregate flows in trade with the Rest of the World. Where appropriate, OECD-reported "mirror" data were used to fill in some gaps in the coverage of trade with the Rest of the World.

With regard to intra-CMEA flows, these difficulties can also be resolved to some extent by filling obvious gaps in the end-use items in particular countries' reported figures by the use of "mirror" statistics. A serious deficiency is the existence of a substantial unexplained residual between the sum of capital goods inputs identified by country of origin and capital goods imports in total, which amounted in 1975 and 1983 to some $5.5 and 12 billion respectively – some 17 per cent of the total identified in 1975 – rising to 21 per cent in 1983 for the seven countries taken together (Table 3).

This description of the sources and methods used to create the data base already perhaps suffices to show that it can make no claim to completeness in measuring embodied technology flows at sector or branch levels of the economy. In addition, it omits disembodied technology of all kinds – patents, licences, other transfers of know-how via documentation or personnel training, etc.[5]. The data presented are also somewhat undifferentiated in terms of the various definitions even of embodied technology which have been offered by different scholars. In the East-West dimension, the OECD-reported figures do capture total Eastern capital goods purchases – as well as a selection of trade in intermediate goods as defined – though the data are incomplete given the omission of trade between the German Democratic Republic and the Federal Republic of Germany. The data sets based on Eastern-reported statistics also contain gaps. Their merit is that they exploit all statistics published by the Eastern countries as well as some Western data. Nevertheless, some inconsistencies remain.

### b) *The Quality of the Data*

#### i) *Aggregate West-to-East Capital Goods Trade Statistics*

As a first step in assessing the quality of the data sets described above, the OECD and CMEA country reported data on West-to-East flows are first compared (Table 1 of this chapter). It can be seen that the ratio between the total of Eastern countries' acquisitions of capital goods from the OECD area (OECD-reported export data) and Eastern countries' reported imports of such goods from the Rest of the World (i.e. from non-CMEA countries) falls from about 90 per cent in 1973, and 80 per cent in 1975 and 1978, to only 66 per cent in 1983. However, when the OECD data are adjusted to include spare parts and accessories (for all countries but Hungary and Czechoslovakia), and also exports of capital goods by Yugoslavia, the ratio improves considerably – and the adjusted OECD data give results ranging between plus or minus 10 per cent of the CMEA-reported totals in all years but 1983 (minus 16 per cent). In aggregate, therefore, capital goods imports from the Rest of the World and corresponding OECD exports can be reconciled reasonably well (though less so in 1983). However, this is due partly to offsetting discrepancie between Eastern countries.

In the first place, the Eastern country import data should yield higher figures because of the inclusion of imports by the German Democratic Republic from the Federal Republic of Germany. Second, as noted by Paul Marer in his investigation of mirror-image statistics in East-West trade, Eastern-recorded imports – particularly of engineering goods – may be bigger than OECD-recorded exports because "Soviet statistics include in imports the cost of disembodied technology and know-how associated with machinery purchases, which Western countries report as invisible exports"[6]. The same is presumably true of other countries using the SFTC or similar systems (but not, it should be recalled, SITC-based data for Czechoslovakia or for Hungary whose trade data also exclude such items). The correspondence in the seven country aggregate conceals, moreover, some substantial differences between OECD and Eastern country-reported data for the individual Eastern countries. In the first place, the adjusted OECD-reported data for the Soviet Union for years after 1970 are, as might be anticipated given the exclusion of disembodied technology, smaller than the Soviet-reported figures. (The former fell from 102 to 76 per cent of the latter between 1970 and 1983). But the converse is true for Eastern Europe as a whole and for all of the six countries individually except Czechoslovakia after 1975.

In part, this reflects gaps in East European data identifiable by origin and type of good. Data not so identifiable (henceforth referred to simply as the "unidentified" item) are sometimes large. This was so in the German Democratic Republic, where OECD and Eastern partner-country "mirror" data accounted for roughly two-thirds of the sector and branch detail within the "control group" of total imports of Western engineering goods. The OECD "mirror" data for the German Democratic Republic also exclude trade with the Federal Republic of Germany. It was also large in Romania, for which only partner-country data are available. In Bulgaria, the contribution of partner-country data was also high (about half). The contribution of mirror data in various years for Eastern exports and imports with the Rest of the World, and from other CMEA countries in intra-CMEA trade, is summarized in Table 2 of this chapter.

For most East European countries, the total sector and branch totals reported by them (purchases identified by region) appear to underestimate the actual purchases from OECD countries even when supplemented by OECD data ; though not for Hungary, Poland, or for Czechoslavakia in the earlier years.

These apparent discrepancies focus upon one of the principal shortcomings of the data set based on the Eastern countries' own trade returns. The relatively large size of the residual import of capital goods which cannot be identified by country of origin can be attributed, in the first place, to the method used to identify total capital goods imports. Since this starts from the extraction of the total imports of the engineering goods "control group" (adjusted, as noted earlier, to exclude non-capital goods items and to include pipeline)[7], it is basically the difference between this total and imports of capital goods identified by country of origin which gives the "unidentified" item shown in Table 3 and other tables.

This residual was large in all countries of Eastern Europe at the beginning of the 1970s – about 40 per cent of the total for all East European countries in 1973. But it was much higher in some countries (i.e. Bulgaria and Romania). It fell somewhat in most countries towards the middle of the decade (mainly because statistical reporting improved in some countries), especially in Hungary (19 per cent) and Poland (14 per cent) but remained high in the German Democratic Republic (40 per cent). The average in 1975 was 27 per cent for all six East European countries combined. By 1983, it had again risen in most countries – to 20-30 per cent in Czechoslovakia, Hungary and Poland and a somewhat uncertainly estimated 45-55 per cent in the German Democratic Republic and Romania. Taking into account the rough estimate made for these two countries, the six-country residual stood at 38 per cent in 1983.

In the Soviet Union, the unidentified residual was relatively small in the early 1970s (5-7 per cent), but roughly trebled toward the end of the decade. In the 1980s it settled around 10 per cent. Thus, in very summary terms, it can be seen that, over about one-third of East European imports of capital goods cannot be identified by country of origin. In Poland and the Soviet Union, the unidentified item amounted to a relatively much lower 10-20 per cent in the early 1980s.

In sum, while it is thus not possible fully to reconcile the OECD and Eastern-reported totals for capital goods imports of the Eastern countries, the two sets of figures taken together provide first, a reliable aggregate time series for Western exports to the East (the OECD data). Second, the Eastern-reported data provide at least a minimum basis for assessing the size of imports from the West relative to those from CMEA partner countries – and, as will be investigated in the following sub-section, useful information, broken down on a sector/industrial branch end-use basis, to check on and supplement the similarly disaggregated OECD data.

### ii) *Sectoral and Branch End-use Allocation: OECD and Eastern-reported Data*

Given the discrepancies at country levels between aggregate Eastern country- and OECD-reported data outlined in the previous sub-section, it is not surprising that there are some differences between the sectoral and industrial branch end-use allocations of Eastern capital goods imports from the West as revealed by the two data sets. The main source of discrepancies within the aggregates is that the OECD-reported data contain much bigger residuals which cannot be satisfactorily allocated either to sectors of the economy as a whole and, particularly, to branches within industry.

For reasons of space, comparisons of the two data sets are presented for only the four years 1975, 1980, 1982 and 1983. They are, however, sufficient to illustrate the general pattern (Table 4 of this chapter). The Table shows that for most Eastern countries the OECD data set points to a much smaller share of capital goods imports for the material sphere than do the Eastern-reported data. The broad magnitude of the sectoral end-user patterns revealed by the two data sets within the material sphere are broadly similar – though the CMEA figures tend to understate the share of industry and overstate that of transport and communications. OECD-reported data do not enable investment goods for the construction industry to be shown separately (all tractors, for instance, have been allocated to agriculture) whereas Eastern country data show that the sector took 2-6 per cent of imports for the material sphere of the economy in most years. This, and OECD's understatement of exports of agricultural equipment to the East, contributes to the residual unattributable by sector of about 10 per cent of capital goods imports in the total for the material sphere in the OECD data – which is generally below the CMEA residual.

Within industry, the overall magnitude of the share of imports by branch end-users shown by the two data sets are again broadly similar, but there are, nonetheless, some sizeable differences in some countries. This relates first to the fact that the OECD data contain a non-attributable residual of considerable importance – generally between 30-40 per cent of the total and even higher in Bulgaria[8]. It is thus not surprising that the share of most industrial end-users appears to be underestimated by the OECD-reported data. Given the apparent understatement of industry as a whole in the totals reported by CMEA countries, no detailed reconciliation by industrial end-users is possible. However, despite these demonstrable discrepancies, the relative orders of magnitude of the various industrial end-users can, as in the case of the sectoral allocations, be roughly identified and the changes over time discerned.

iii) *Intra-CMEA Capital Goods Trade Statistics (Eastern reported data)*

The data on intra-CMEA exchanges of capital goods is fairly complete despite the failure of Romania – and subsequently the German Democratic Republic – to produce detailed commodity breakdowns of foreign trade since the early 1970s. Improved coverage was attained for these and other countries by the use of "mirror" statistics for intra-CMEA exchanges. The general principle has been to retain an individual reporting country's own import or export data, except where discrepancies are particularly large, and to fill in gaps due to apparent substantial underestimates, or missing components, by inserting the sum of the other CMEA countries' exports to that country.

While it should be borne in mind that the end-use detail of imports excludes shipments from Romania to the German Democratic Republic and vice versa, the aggregate dimension of CMEA imports from other CMEA countries, as well as their composition by end-use, has been surprisingly well reconciled with the corresponding export data in several countries (Table 5 of this chapter). Leaving aside Romania, whose import figures after 1972 are in fact wholly derived from partner-country export data, discrepancies for Bulgaria, Poland and the Soviet Union fall within a range of 2-5 per cent. In Hungary, the much lower figure for recorded imports from 1975 onwards compared with the sum of partners' exports may result in part from exchange rate inconsistencies. Capital goods import data reported by Czechoslovakia and the German Democratic Republic are roughly 10 per cent in excess of imports calculated from partners' exports. Apart from the limited availability of data in the German Democratic Republic, there is no obvious explanation for this.

Since the data presented reflect the final result of "mirror" adjustments, a fair degree of consensus between the direct import and partner country export-derived data might have been expected. The coverage of the two sets of figures does check out surprisingly well given the known gaps in some countries' data – not only in aggregate but also in terms of end-use sectors and industrial branches.

## 2. East-West Embodied Technology Transfer and Total East-West Trade 1970-1983

The following sub-sections analyse East-West technology exchanges of a "productive" nature. That is, they cover first flows of capital goods and second intermediate (technology) goods. The intermediate goods retained for study are hereinafter referred to as "intermediate (technology) goods" as defined in sub-section 1*a*) above. In sub-section *a*) and *b*) following, the trends of these two items in OECD's exports to and imports from the Eastern countries are discussed in turn. In sub-section *c*), the contribution of these technology items to changes in the Eastern balances of trade with the OECD area in 1970-1983 is examined.

a) *Western Embodied Technology Exports to the East*

Western exports of capital and intermediate (technology) goods to the East amounted to nearly \$22 billion in 1983 – a substantial decline from \$26 billion in 1980 (as derived from Table 6). Although 30 per cent higher in value compared with their level in 1975 (\$19 billion), they were only 10 per cent higher in real terms than in that year (Table 1 of Chapter 8). The aggregate two-item value in current prices thus feel since 1980, but it rose in real terms by an average of 2 per cent annually between 1980-1983. There was a very big change in the share of the Soviet Union and Eastern Europe in the region's imports of embodied technology. Within the seven-country total, the Soviet Union accounted for 65 per cent in 1983, compared with 49 per cent in 1975 and 52 per cent as recently as 1980.

This strong rise in the Soviet share of total imports of technology items reflected, in the first place, large declines in such imports by Poland and Romania. It was also due to a very sharp upsurge in the value of Soviet capital goods imports in 1982. Within the Soviet technology (two-item) total, the share of capital goods, which fell substantially from well over to well under half between 1975 and 1980, rose to 55 per cent in 1983. For Eastern Europe, the capital goods share declined from 48 per cent of the total in 1975 to about 40 per cent in 1981-1983. Thus for most of the period since 1975, imports of capital goods have had less priority than intermediate (technology) goods – but this pattern was abruptly reversed by the Soviet Union in 1982 and 1983[9].

In the paragraphs which follow, it will be shown first that the ability or willingness of the Eastern countries to import technology has been affected more than other classes of imports by the trade adjustment policies described in Chapter 1 above. Second, that the technology imports of Eastern Europe were affected more than those of the Soviet Union even before the Soviet upsurge in 1982. Finally, that particularly in Easter Europe, it has been found generally easier since 1975 to cut back the growth of imports of Western capital goods than purchases of intermediate (technology) goods, which consist mainly of industrial inputs.

These observations, which are supported by the data in volume terms, enable two further underlying features of West-to-East technology flows since 1975 to be identified. First, the burden of economic re-adjustment after the mid-1970s was rather lighter for the Soviet Union than for Eastern Europe – principally because overall terms-of-trade movements favoured the former and resulted in a net transfer of resources of considerable proportions from both the West and from its (European) CMEA trade partners. Second, that the progressive and virtually uninterrupted rise in the *share* of intermediate (technology) goods in total embodied technology imports which occurred, particularly in Eastern Europe, can be interpreted as an unanticipated extra cost to the East of imports of Western capital goods. It also suggests that the Eastern countries have not succeeded in gearing up their domestic output of goods for further processing to the technical levels or the quantities necessary for supplying their new imported plant with appropriate domestically-produced inputs.

Movements in technology items relative to total OECD exports to the East since 1970 have shown great variation but, by the early 1980s, the pattern seems to have reverted to that of the late 1960s. In 1970, the two broad embodied technology items retained for study accounted together for around two-thirds of the OECD countries' total exports to the Eastern area. Just over half were capital goods. At the end of the period of fast expansion of technology imports, 1970-1975, the share of the two technology items amounted to nearly three-quarters of total OECD exports to the East, but had fallen to 61 per cent by 1983. However, while capital goods alone constituted a declining proportion of total exports by the West to the East – they fell from just over one-third in 1970 to little more than one-quarter by 1980 – intermediate (technology) goods retained a fairly constant share: just above one-third in 1970 and slightly more in 1980. This trend was maintained in Eastern Europe but not in the Soviet Union between 1980 and 1983.

The very high share of these technology items in Western exports to the East, which is not a new phenomenon, reflects partly differences in the level of development between the two areas, and partly the fact that the Eastern countries do not import other manufactures – for instance consumer goods – on the scale normal in non-socialist countries at a similar level of development. These high shares are easier to explain than their low share in Western imports from the East, discussed below. Moreover, while such "inter-industry" trade has tended to increase fast between other areas of the world, it has noticeably regressed in East-West trade over the 1970s (see sub-section *b)* of this section).

Capital goods accounted for a much higher proportion of the West's total exports to the Soviet Union in 1970 than to Eastern Europe (42.8 against 29.3 per cent). This reflected the capital-hungry nature of Soviet development due to its size and technical backwardness relative to several East European countries. Hard currency earning constraints at the beginning of the decade imposed a strict selectivity on Soviet imports to favour the growth of investment, paralleling the restriction of internal private consumer demand in favour of capital accumulation and large defence commitments. However, the share of capital goods in OECD exports to the Soviet Union fell somewhat up to 1975 and very sharply between 1975 and 1980. This partly reflected cutbacks in 1980 itself, but also the broadening of Soviet import options due to higher fuel earnings after the oil price rises of the mid- and late-1970s. The share rose fast after 1980. In contrast – mainly because of the massive rise in exports of capital goods to Poland – the share of those goods in OECD's exports to Eastern Europe rose considerably up to the middle of the 1970s but even the big decline thereafter left their share, unlike in 1970, only marginally lower than in the Soviet case. In contrast with capital goods, intermediate (technology) goods maintained a fairly stable 30 per cent share of all Western exports to the Soviet Union up to 1980, but their share declined thereafter. In Eastern Europe the corresponding share remained steady over the 1970s and also into the 1980s at well over one-third.

The changes in the share of technology in Western exports to the area need to be viewed in the light of changes in the share of other components, and notably agricultural products. The decline in the share of capital goods in Western exports to all the seven countries and to Eastern Europe between its peak in 1977 and 1981 reflects primarily the more than doubling in the share of OECD agricultural products over the decade. In the Soviet case especially, it also results from a doubling in the share of industrial commodities other than the two technology items and fuel – mainly, in fact, processed food. Given these changes in the composition of Western exports, the relatively constant share of intermediate (technology) goods in Eastern Europe is striking.

Thus, while exports of capital goods, and hence the two technology items together, lost priority in terms of total Eastern expenditures on Western products, intermediate (technology) goods generally gained ground within Western exports as a whole. In ranking import priorities, therefore, it can be seen that: greatest emphasis has been placed on feeding the population; maintaining supplies of industrial inputs, especially intermediate (technology) goods, has taken second place and, that supplying new technology in the form of imported capital goods has ranked third. These observations were also true for Czechoslovakia and Hungary, even though the share of agricultural products in total Western exports to these two countries declined, uniquely in the region, over the 1970s and in the early 1980s due to their own better agricultural performance relative to the Soviet Union.

As shown in Table 7 of this chapter, the Soviet Union is by far the West's biggest single Eastern market for capital goods. Its share was, however, rather small relative to its share in total output of the seven Eastern countries combined up to 1981 – the second lowest in the case of capital goods and smallest with regard to intermediate (technology) goods. However, in 1982 and 1983 the relative size of Soviet capital goods imports became for the first time almost commensurate with its contribution to the Eastern area's aggregate outputNMP). At the other extreme, Hungary was the country most open to Western capital goods exports relative to its size in 1983, followed by Bulgaria, Czechoslovakia, Poland (first in 1975), the Soviet Union, the German Democratic Republic and Romania. The order is similar (with the Soviet Union, however, in last place) with regard to intermediate goods. In both cases, the German Democratic Republic would certainly be placed higher if account were taken of exports by the Federal Republic of Germany to that country.

b) *Western Imports of Embodied Technology from the East*

By 1983, Western imports of Eastern capital goods stood at $1.1 billion and of intermediate (technology) goods at the much higher level of $5.4 billion. These were equivalent to 11 and 51 per cent respectively of the corresponding flows in the opposite direction and increased in value terms at only half and four-fifths the rates of respective West-to-East flows since 1970. In the case of capital goods, OECD's import growth was rather faster than its exports of these items to the East up to 1981. However, by 1983 – despite the acceleration in Soviet purchases of capital goods – growth rates of OECD imports of both capital and intermediate (technology) goods from the East declined more than the corresponding post-1980 Western export flow. Within the Eastern countries' technology export totals, the Soviet Union, following a steep decline after 1975, accounted for only 14 per cent of the region's capital goods exports to the West in 1983. While it exported over one-third of intermediate (technology) goods sold by the Eastern countries to the West, both figures are low compared with its weighting of about 72 per cent in total output of the European CMEA countries (Table 8 of this chapter).

The 9 percentage point decline between 1975 and 1983 in the Soviet share of OECD capital goods imports from the East reversed the rising trend apparent between 1970-1975. The biggest suppliers in 1983 were Poland and, despite the exclusion from the statistics of the Federal Republic of Germany's imports, the German Democratic Republic – followed by Czechoslovakia, Hungary (which nearly doubled its share since 1975), the Soviet Union, Romania and Bulgaria. Declines in absolute values since 1980 have been substantial. Even so, most countries' exports to OECD were higher than in 1975, the notable exception being Poland, where the fall was small, and the Soviet Union, whose exports of capital goods to the West were only two-thirds of their 1975 level. However, and perhaps more significantly, OECD's capital goods imports from the East in volume terms, declined by 20 per cent in aggregate between 1975 and 1983. The decline affected all Eastern countries but the German Democratic Republic, Hungary and Romania. No sign of any reversal of the overall trend is apparent between 1980 and 1983.

The most successful Eastern exporting country was Hungary, which more than doubled the absolute value of its capital goods exports to the West over the 1975-1983 period. Bulgaria and Romania did relatively well – though from low initial levels. In part, the decline registered by the seven countries as a whole was compensated by the rise in intra-CMEA trade in capital goods (Chapter 6 below) – which, in contrast, more than doubled in value terms over the period.

As indicated by the figures in Table 8 of this chapter, the most successful Eastern country for its size, relative to other Eastern countries, on the Western countries capital goods market in 1983 was Hungary, followed by Czechoslovakia, the German Democratic Republic,

Poland, Romania, Bulgaria and the Soviet Union. However, this excludes the capital goods imports of the Federal Republic of Germany from the German Democratic Republic. If allowance is made for this, then the German Democratic Republic would probably take first place by a considerable margin on Poland. As noted earlier, Hungary's position has also improved very quickly since 1975. This was due, above all, to a rapid expansion of the OECD's imports of Hungarian transport equipment, capital goods for the energy and miscellaneous engineering industries, and of office machinery.

In contrast to the position with regard to capital goods, the structure of OECD's intermediate (technology) goods imports from the East is characterized by the leading position of the Soviet Union in the total flow – about one-third and rising. While in 1983, Eastern Europe maintained shares in the total broadly similar to those in 1975, the German Democratic Republic and Hungary increased their shares while the other countries – particularly Poland and Romania – lost ground.

c) *Technology Transfer and the East-West Balance-of-Trade*

This section aims to show the effects of East-West embodied technology trade as defined on the total balance of East-West trade during the period 1970-1983, and in particular, its contribution to the cumulated total trade deficit over the period. It should be noted that the OECD trade figures on which this approach is based show exports f.o.b. and imports c.i.f. No adjustment has been made to the trade balances on this account (as a rule of thumb, such adjustment lowers the value of Western imports by about 7.5 per cent – i.e., it pushes up Western surpluses or reduces Western deficits). The first item treated here is *changes* in the balance of trade over the 14 years reviewed. These are not substantially affected by the use, or non-use, of estimated c.i.f./f.o.b. adjustment coefficients, but the point should be borne in mind in the subsequent discussion concerning the *levels* of cumulated trade deficits.

As shown above in Table 6 of this chapter, the Western balance on total East-West trade shifted from a small Western surplus of $327.1 million in 1970 to a very substantial one of $8 billion in 1975. The increase of about $8 billion was divided half-and-half between the Soviet Union ($4.05 billion) and Eastern Europe ($4.28 billion). Within the latter, Western trade with Poland alone was responsible for $2.50 billion of the rise in the Western surplus, followed, in descending order, by $0.62 for Bulgaria, $0.49 for Hungary, $0.18 for Czechoslovakia and Romania and $0.07 billion for the German Democratic Republic. Thus, between the two years some 80 per cent of the rise in the overall Eastern deficit was due to Poland and the Soviet Union alone – a figure which rises to 95 per cent if Bulgaria and Hungary are included.

Cumulated Western surpluses on total East-West trade amounted, on the f.o.b./c.i.f. figures presented in Table 9 of this chapter, to about $37 billion over the whole period 1970-1980. Using these data, the role of technology imports in the build-up of the cumulative trade deficit in the 1970s can be clearly charted relative to other commodity groups[10]. The figures indicate a very rapid rise of the total trade deficit in the first half of the 1970s – mainly in the year 1975 alone when, as noted above, the deficits of both the Soviet Union and Eastern Europe each amounted to $4 billion – equivalent to over 50 per cent of the total accumulated deficit of the seven countries during 1971-1975. Further annual deficits of $6.5, 4.6, 6.2 and 3.6 billion were reported for all Eastern countries taken together in the OECD trade returns for the four years 1976-1979 inclusive. But, apart from the year 1978, their size decreased substantially and progressively as the internal adjustment process within the Eastern economies, initiated in 1975, was further tightened in 1978-1979. In 1980 the deficit fell to only $0.4 billion. The cumulated trade deficits in 1976-1980 – though, at $21.3 billion, 40 per cent larger than registered in 1971-1975 – were in fact smaller in real terms.

The aggregate seven-country figures conceal a number of intra-country differences. First, the contribution of the Soviet Union to the cumulated deficit – half the total in 1971-1975 – fell to one-quarter in 1976-1980. Indeed, the Soviet cumulative deficit, at roughly $5.2 billion in both 1971-1975 and 1976-1980, remained virtually constant, while that of Eastern Europe rose from $10 to $16 billion between the two periods. The East European debt build-up not only started earlier but lasted longer and was bigger than the Soviet Union's. Within Eastern Europe, Poland alone accounted for 44 per cent of the cumulative deficit over 1971-1980 and somewhat less in the later period even though its cumulative level rose by a further 30 per cent over 1971-1975.

Throughout the 1970s, the cumulated Western surpluses on capital goods amounted to $68 billion, and to $49 billion for intermediate (technology) goods for the seven countries together i.e. three times bigger than the corresponding overall trade deficit. The structure of the cumulative overall trade deficit shows marked similarities between the Soviet Union and Eastern Europe in the early 1970s. In both, it coincided closely with the deficit on industrial goods as a whole. Within the latter, the Western surplus on the two technology items was, however, only partly offset by the Western deficit in other industrial goods – mostly fuels in the Soviet Union, and fuels (including re-exports of Soviet oil), processed food and other manufactures in Eastern Europe. The Western balance on agricultural exports was either at a manageable (Western) surplus level for the Soviet Union or negative in the case of Eastern Europe. But in 1976-1980 an increasing divergence became apparent. In these years, the Soviet Union converted the Western surplus on total industrial goods into a deficit. It more than covered the West's technology surplus – notably by big rises in its own sales of fuel, leaving something over to offset partially the West's by then large surplus on agricultural products. In contrast, over the 1976-1980 period the East European countries limited the growth of the Western technology surplus to a slower rate than the Soviet Union (it rose by 60 per cent compared with two-and-one-half-times in the Soviet Union), but were not able to offset it fully by the growth of other industrial exports. Quite simply, the scale of the East European technology deficit was disproportionate in comparison with the export earning possibilities of their other industrial goods. Moreover the previous Western deficit on agricultural produce with Eastern Europe was transformed into a sizeable ($5.5 billion) Western surplus.

The divergent trends became even more marked in 1981-1983 – the period in which the escalating foreign trade deficits of the Eastern countries with the West were brought under control. In the Soviet Union, the annualized rate of the Western surplus on technology trade continued to rise relative to 1976-1980 – though at a slower rate. Moreover, the previous Western deficit on industrial goods excluding technology and fuel was turned into surplus and the West's agricultural surplus doubled. But the overall Western surplus was eliminated; the key element was a further and very marked rise in the Western deficit on fuel imports.

In contrast, the annualised Western surplus on embodied technology trade with the East European countries was cut by nearly half in 1981-1983 – mainly by severe cuts in their imports of these items. The Western deficit on non-technology, non-fuel items ("other industrial goods" in Table 9) was also halved. The Western surplus on agriculture remained constant. As a result, the Western surplus on overall trade was eliminated in the three years taken together (there was, in fact, a Western deficit in 1982 and 1983). The point to note is that Eastern Europe was not able to raise its fuel (or other) exports to anywhere near the same extent as the Soviet Union. Hence overall balance on trade could only be achieved, in contrast with that country, by proportionately sharper cuts in embodied technology deficits.

These figures illustrate the relative success of the Soviet Union, and the relative failure of Eastern Europe, in their strategy of paying for expanded technology imports by raising exports of industrial goods (which, on this definition, includes fuel). In the Soviet Union, where the import strategy was concentrated on fuel export, gas- and oil-distribution technology having a direct export-facilitating role, the outcome has been totally satisfactory in this respect. Fuels accounted for 73 per cent of OECD imports from the Soviet Union, compared with 30 per cent in 1970. The gap between success and failure in the Soviet Union relative to Eastern Europe, has, of course, also come about largely because of the windfall gains which accrued to it as a result of the post-1973 and post-1978 oil price rises. The East European countries, on the contrary, were hit by terms of trade losses, as well as by sluggish demand in the West for their manufactures. But low demand on Western markets generally was only one reason for the failure of their strategy of paying for current technology with increased future sales of manufactures; most Eastern countries also experienced falls in the share of their products in total OECD imports. This issue is however reserved to Chapter 8 below.

d) *OECD Suppliers and Recipients of Embodied Technology in East-West Trade*

Although this study addresses itself principally to East-West trade in technology in the context of flows between the OECD as a whole and the Eastern countries, it was felt useful to present a short summary in the following paragraphs of the role of the major OECD country groupings in the supply of technology to the East.

In 1983, the EEC countries were still the biggest suppliers of capital goods to both Eastern Europe and the Soviet Union, as they had been throughout the 1970s (Table 10). They were also the biggest suppliers of intermediate (technology) goods. Nevertheless, the decade witnessed a doubling of Japan's share of OECD exports of capital goods to both the Soviet Union and Eastern Europe. The movement was mainly concentrated, due to Japan's originally small 1970 share of OECD's supplies to Eastern Europe, in the Soviet market, where Japan's share grew from a tenth in 1970 to a fifth in 1983. In the same years the EEC's share fell from just under two-thirds in 1970 to a half in 1980 and 1983. The share of the other North European countries also declined over the decade, but recovered somewhat in the 1980s when they accounted for about a quarter. North American supplies have been small (2-6 per cent) over most of the 1970s, though they rose very rapidly during the détente years (as seen mainly in 1973-1976) to over 10 per cent. Even so, they did not keep pace with Japan. The Southern European countries' share of the Soviet market rose but remained marginal (less than 1 per cent in 1983). The EEC countries' share in OECD capital goods supplies to Eastern Europe fell from just under three-quarters to well below two-thirds between 1970 and 1983. The share of 'other' North European countries (see note to Table 10) rose slightly – to one-quarter in 1983. But the Japanese share increased from below 2 per cent to 14 per cent over the same period. North America's share in the total was below 5 per cent in most years, but still rose from its minimal 2 per cent share in 1970.

With regard to intermediate (technology) goods, much less change in the pattern of supplies took place over the decade. The EEC countries suffered a fall in their 1970 two-thirds share of the East European market, but retained their share of the Soviet market – somewhat less, however, than the middle years of the 1970s. Japan gained on the other North European countries' shares and, in 1983, equalled their one-fifth share of the market. North America and Southern Europe throughout supplied less than 3 per cent of the East European market for these goods. On the Soviet market, North American suppliers overtook the South European countries' share, but remained less than 5 per cent in most years.

Despite their geographical propinquity, the EEC countries have very clearly been out-performed on the Eastern capital goods markets by Japanese suppliers, especially in the 1980s. This contrasts with the performance of the other North European countries, especially in Eastern Europe and, though to a lesser extent, the Soviet Union. Special financing arrangements and traditionally close trading ties between the Eastern countries and some European non-EEC OECD members may be partly responsible for this relatively good performance. Japan's success on Soviet markets may be connected with the shift of new Soviet industrial operations to the Central and Eastern area of the country, where Western Europe's geographical advantage may be less marked, (but this would not explain Japan's fast increasing share of the East European market). The efforts of the United States to restrict the export of technology to the Eastern countries has clearly distorted the geographical pattern of OECD exports to the CMEA area as a whole, resulting, on the Soviet market particularly, in a bigger share of all other supplier regions than would otherwise have been the case. The experience of the détente years 1973-1976, however, suggests that United States competition affected non-EEC North European countries more than suppliers in the EEC and Japan.

On the Eastern export side, the bulk of embodied technology goods has been sold on the EEC and other North European markets (70-80 per cent). The share of Eastern capital goods exports to North America however increased over the 1970s – both within Eastern Europe's and the Soviet Union's total sales, though the latter remains low (about 5 per cent). Southern Europe took 20-30 per cent of the total in 1983 – roughly the same as in 1970. Eastern Europe is a bigger supplier both of capital goods and intermediate (technology) goods to the OECD countries than the Soviet Union. The difference is particularly marked with respect to sales of capital goods to the EEC countries, where the Eastern European countries supplied more than ten times as much in value as did the Soviet Union. They also supplied above five times as much to Southern Europe. The discrepancy is much less marked for intermediate (technology) goods, but Eastern Europe's total sales are still twice as big as those of the Soviet Union.

Table 1 (Chapter 5)

COMPARISON OF OECD-REPORTED EXPORTS OF CAPITAL GOODS TO THE EASTERN COUNTRIES (A AND B) WITH EASTERN COUNTRY-REPORTED CAPITAL GOODS IMPORTS FROM THE REST OF THE WORLD (C)

($ billion and percentage differences)

| | 1970 | | | 1973 | | | 1975 | | | 1978 | | | 1980 | | | 1982 | | | 1983 | | |
|---|---|---|---|---|---|---|---|---|---|---|---|---|---|---|---|---|---|---|---|---|---|
| | A | B | C | A | B | C | A | B | C | A | B | C | A | B | C | A | B | C | A | B | C |
| Bulgaria | 0.09 | 0.12 | .. | 0.15 | 0.18 | 0.13 | 0.51 | 0.60 | 0.46 | 0.38 | 0.47 | 0.37 | 0.40 | 0.56 | 0.38 | 0.56 | 0.68 | 0.54 | 0.57 | 0.69 | 0.57 |
| - per cent of C | .. | .. | .. | 115 | 138 | .. | 111 | 130 | .. | 103 | 127 | .. | 105 | 147 | .. | 104 | 125 | .. | 100 | 121 | .. |
| Czechoslovakia | 0.26 | 0.28 | 0.28 | 0.39 | 0.40 | 0.40 | 0.64 | 0.67 | 0.68 | 0.85 | 0.89 | 0.98 | 0.85 | 0.92 | 1.49 | 0.63 | 0.74 | 0.99 | 0.56 | 0.66 | 0.96 |
| - per cent of C | 98 | 100 | .. | 98 | 100 | .. | 94 | 99 | .. | 87 | 91 | .. | 57 | 62 | .. | 64 | 75 | .. | 58 | 69 | .. |
| German D. Rep. (a) | 0.16 | 0.19 | 0.16 | 0.17 | 0.23 | 0.18 | 0.33 | 0.42 | 0.41 | 0.33 | 0.43 | 0.41 | 0.52 | 0.68 | 0.59 | 0.55 | 0.69 | 0.62 | 0.63 | 0.79 | 0.64 |
| - per cent of C | 100 | 119 | .. | 94 | 128 | .. | 80 | 102 | .. | 80 | 105 | .. | 88 | 115 | .. | 89 | 111 | .. | 98 | 123 | .. |
| Hungary | 0.13 | 0.13 | .. | 0.23 | 0.24 | 0.23 | 0.41 | 0.42 | 0.45 | 0.84 | 0.86 | 0.70 | 0.73 | 0.76 | 0.63 | 0.74 | 0.78 | 0.58 | 0.57 | 0.60 | 0.47 |
| - per cent of C | .. | .. | .. | 100 | 104 | .. | 91 | 93 | .. | 120 | 123 | .. | 116 | 121 | .. | 128 | 134 | .. | 121 | 128 | .. |
| Poland | 0.22 | 0.27 | 0.26 | 1.09 | 1.27 | 1.16 | 2.19 | 2.58 | 2.51 | 1.82 | 2.26 | 2.16 | 1.57 | 2.10 | 1.88 | 0.48 | 0.74 | 0.69 | 0.49 | 0.76 | 0.74 |
| - per cent of C | 81 | 104 | .. | 94 | 109 | .. | 87 | 103 | .. | 86 | 105 | .. | 84 | 112 | .. | 71 | 107 | .. | 66 | 102 | .. |
| Romania | 0.25 | 0.30 | 0.21 | 0.51 | 0.62 | 0.50 | 0.64 | 0.87 | 0.76 | 1.07 | 1.40 | 1.38 | 0.81 | 1.18 | 0.96 | 0.21 | 0.37 | 0.18 | 0.13 | 0.23 | 0.11 |
| - per cent of C | 119 | 143 | .. | 102 | 124 | .. | 84 | 114 | .. | 78 | 101 | .. | 84 | 123 | .. | 117 | 207 | .. | 118 | 208 | .. |
| Total, Eastern Europe (a) | 1.12 | 1.29 | (1.04) | 2.54 | 2.94 | 2.62 | 4.72 | 5.56 | 5.26 | 5.29 | 6.31 | 6.00 | 4.85 | 6.20 | 5.93 | 3.15 | 4.01 | 3.58 | 2.95 | 3.73 | 3.49 |
| - per cent of C | 108 | 124 | .. | 96 | 111 | .. | 90 | 106 | .. | 88 | 105 | .. | 82 | 105 | .. | 88 | 112 | .. | 85 | 107 | .. |
| Soviet Union | 1.10 | 1.32 | 1.30 | 1.89 | 2.15 | 2.43 | 4.93 | 5.81 | 6.75 | 6.54 | 7.90 | 8.78 | 5.72 | 7.32 | 8.82 | 7.26 | 9.25 | 11.26 | 7.57 | 9.64 | 12.43 |
| - per cent of C | 85 | 102 | .. | 78 | 88 | .. | 73 | 86 | .. | 74 | 90 | .. | 70 | 83 | .. | 65 | 83 | .. | 61 | 76 | .. |
| Total, Eastern Europe and Soviet Union (a) | 2.22 | 2.61 | (2.34) | 4.47 | 5.09 | 5.05 | 9.65 | 11.37 | 12.01 | 11.83 | 14.21 | 14.78 | 10.57 | 13.52 | 14.75 | 10.41 | 13.25 | 14.85 | 10.52 | 13.37 | 15.92 |
| - per cent of C | 83 | 112 | .. | 89 | 101 | .. | 80 | 95 | .. | 80 | 96 | .. | 72 | 92 | .. | 71 | 90 | .. | 66 | 84 | .. |

Notes: Column A (OECD data) excludes spare parts and accessories for investment goods items; column B (also OECD data) includes them, (except data for Hungary and Czechoslovakia which exclude these items from the totals in column C) plus the value of Yugoslav capital goods exports to all countries concerned. Components may not sum to totals due to totals to rounding.

a) Columns A and B exclude exports by the Federal Republic of Germany to the German Democratic Republic. Total for column C includes rough estimates for Bulgaria and Hungary in 1970.

Source: OECD East-West technology transfer data base (Eastern and Western statistics).

Table 2 (Chapter 5)

VALUE OF TRADE IN CAPITAL GOODS CLASSIFIABLE BY TECHNOLOGICAL CHARACTERISTICS OR SECTOR AND BRANCH END-USERS AND BY GEOGRAPHICAL AREAS
(percentage of the total reported value of trade in capital goods) (a)

| | Partner statistics not utilised | | | | | Partner statistics utilised | | | | |
|---|---|---|---|---|---|---|---|---|---|---|
| | 1970 | 1975 | 1980 | 1982 | 1983 | 1970 | 1975 | 1980 | 1982 | 1983 |
| Bulgaria | | | | | | | | | | |
| Exports | n.a. | 9.6 | 10.8 | 8.8 | 10.4 | n.a. | 94.6 | 97.9 | 89.5 | 92.8 |
| Imports | n.a. | 35.1 | 33.3 | 31.8 | 32.5 | n.a. | 68.0 | 67.6 | 66.2 | 62.3 |
| Czechoslovakia | | | | | | | | | | |
| Exports | 66.8 | 64.8 | 63.9 | 68.2 | 70.0 | 79.9 | 73.7 | 73.4 | 74.9 | 73.8 |
| Imports | 60.3 | 60.0 | 61.5 | 66.0 | 63.7 | 69.1 | 66.6 | 74.0 | 68.7 | 66.8 |
| German Democratic Republic | | | | | | | | | | |
| Exports | 40.0 | 39.9 | 35.3 | 37.1 | 39.7 | 69.4 | 71.4 | 68.6 | 64.9 | 76.9 |
| Imports | 12.8 | 16.0 | 13.6 | 10.3 | 10.1 | 45.7 | 60.4 | 51.2 | 47.4 | 45.6 |
| Hungary | | | | | | | | | | |
| Exports | 70.0 (b) | 97.8 | 91.3 | 88.1 | 87.1 | 82.0 | 97.9 | 91.4 | 89.1 | 88.7 |
| Imports | 49.8 (b) | 82.3 | 77.0 | 82.9 | 77.7 | 62.2 | 83.1 | 77.9 | 83.7 | 78.3 |
| Poland | | | | | | | | | | |
| Exports | 65.0 | 73.4 | 87.9 | 72.8 | 76.2 | 65.0 | 73.4 | 87.9 | 72.8 | 76.2 |
| Imports | 64.9 | 85.7 | 82.2 | 67.4 | 75.6 | 64.9 | 85.7 | 82.2 | 67.4 | 75.6 |
| Romania | | | | | | | | | | |
| Exports | 89.8 | n.a. | n.a. | n.a. | n.a. | 94.4 | 68.1 | 68.5 | 54.4 | 60.2 |
| Imports | 59.7 | n.a. | n.a. | n.a. | n.a. | 66.0 | 65.1 | 63.0 | 57.7 | 54.5 |
| Soviet Union | | | | | | | | | | |
| Exports | 95.7 | 92.8 | 91.0 | 88.2 | 88.0 | 97.2 | 93.9 | 92.7 | 91.7 | 91.7 |
| Imports | 92.5 | 93.7 | 84.0 | 87.3 | 88.5 | 94.9 | 95.4 | 87.1 | 89.3 | 90.6 |

Notes: a) Percentages are where possible, based on "capital goods" (engineering adjusted downward for non-capital items plus pipes). However, in certain cases the "control group" (engineering products) had to be used because the data needed for adjustment were not available.

b) 1971.

Source: OECD East-West technology transfer data base (Eastern statistics).

Table 3 (Chapter 5)

EASTERN-REPORTED CAPITAL GOODS IMPORTS OF EASTERN EUROPE AND THE SOVIET UNION BY REGION OF ORIGIN
($ million and per cent of importing country totals)

| | 1970 | | | | 1973 | | | | 1975 | | | |
|---|---|---|---|---|---|---|---|---|---|---|---|---|
| | From CMEA | From Rest of World | Uniden-tified | Total | From CMEA | From Rest of World | Uniden-tified | Total | From CMEA | From Rest of World | Uniden-tified | Total |
| Bulgaria | .. | .. | .. | .. | 932.4 | 133.2 | 472.4 | 1 538.0 | 1 141.4 | 456.5 | 752.4 | 2 350.3 |
| | .. | .. | .. | .. | 60.6 | 8.7 | 30.7 | 100.0 | 48.6 | 19.4 | 32.0 | 100.0 |
| Czechoslovakia | 552.0 | 279.9 | 371.6 | 1 203.5 | 1 037.6 | 404.8 | 806.2 | 2 248.6 | 1 526.9 | 677.0 | 1 106.5 | 3 310.4 |
| | 45.8 | 23.3 | 30.9 | 100.0 | 46.1 | 18.0 | 35.9 | 100.0 | 46.1 | 20.5 | 33.4 | 100.0 |
| German Democratic Republic | 620.2 | 156.3 | 923.6 | 1 700.0 | 1 150.0 | 178.6 | 1 331.2 | 2 659.8 | 1 751.0 | 406.7 | 1 415.6 | 3 573.4 |
| | 36.5 | 9.2 | 54.3 | 100.0 | 43.2 | 6.7 | 50.0 | 100.0 | 49.0 | 11.4 | 39.6 | 100.0 |
| | (a) | (a) | (a) | (a) | | | | | | | | |
| Hungary | 288.1 | 201.3 | 297.4 | 786.7 | 389.9 | 233.2 | 174.9 | 798.0 | 784.8 | 446.2 | 251.0 | 1 482.0 |
| | 36.6 | 25.6 | 37.8 | 100.0 | 48.9 | 29.2 | 21.9 | 100.0 | 52.9 | 30.1 | 16.9 | 100.0 |
| Poland | 648.6 | 257.1 | 489.2 | 1 394.9 | 1 431.6 | 1 163.8 | 859.5 | 3 454.8 | 1 739.1 | 2 513.3 | 711.1 | 4 963.5 |
| | 46.5 | 18.5 | 35.1 | 100.0 | 41.3 | 33.7 | 24.9 | 100.0 | 35.0 | 50.6 | 14.3 | 100.0 |
| Romania (b) | 281.1 | 211.4 | (297.4) | (789.9) | 518.3 | 503.4 | (406.4) | (1 428.0) | 533.0 | 760.5 | (560.4) | (1 853.9) |
| | 35.6 | 26.8 | 37.7 | 100.0 | 36.3 | 35.3 | 28.5 | 100.0 | 28.8 | 41.0 | 30.2 | 100.0 |
| | (c) | (c) | (c) | (c) | | | | | | | | |
| Total, Eastern Europe | (2 101.7) | (904.7) | (2 081.9) | (5 088.3) | 5 459.6 | 2 617.0 | 4 050.6 | 12 127.2 | 7 476.2 | 5260.1 | 4 797.3 | 17 533.6 |
| | (41.3) | (17.8) | (40.9) | 100.0 | 45.0 | 21.6 | 33.4 | 100.0 | 42.6 | 30.0 | 27.4 | 100.0 |
| Soviet Union | 2 750.4 | 1 303.9 | 218.4 | 4 272.7 | 4 522.5 | 2 433.9 | 514.8 | 7 471.1 | 6 514.3 | 6 746.7 | 641.8 | 13 902.8 |
| | 64.4 | 30.5 | 5.1 | 100.0 | 60.5 | 32.6 | 6.9 | 100.0 | 46.9 | 48.5 | 4.6 | 100.0 |
| | (c) | (c) | (c) | (c) | | | | | | | | |
| Total, Eastern Europe and the Soviet Union | (4 852.1) | (2 208.6) | (2 300.3) | (9 361.0) | 9 982.1 | 5 050.8 | 4 565.4 | 19 598.3 | 13 990.5 | 12 006.8 | 5 439.1 | 31 436.4 |
| | (51.8) | (23.6) | (24.6) | 100.0 | 50.9 | 25.9 | 23.2 | 100.0 | 44.4 | 38.2 | 17.4 | 100.0 |

Table 3 (Chapter 5) contd

| | 1978 | | | | 1980 | | | |
|---|---|---|---|---|---|---|---|---|
| | From CMEA | From Rest of World | Uniden-tified | Total | From CMEA | From Rest of World | Uniden-tified | Total |
| Bulgaria | 1 680.1 | 372.6 | 1 111.9 | 3 164.6 | 1 941.1 | 376.6 | 1 110.1 | 3 427.8 |
| | 53.1 | 11.8 | 35.1 | 100.0 | 56.6 | 11.0 | 32.4 | 100.0 |
| Czechoslovakia | 2 593.4 | 976.5 | 1 458.2 | 5 028.1 | 2 639.3 | 1 494.5 | 1 322.1 | 5 455.9 |
| | 51.6 | 19.4 | 29.0 | 100.0 | 48.4 | 27.4 | 24.2 | 100.0 |
| German Democratic Republic | 2 500.8 | 405.2 | 2 280.3 | 5 186.2 | 2 416.0 | 592.3 | 2 869.0 | 5 877.3 |
| | 48.2 | 7.8 | 44.0 | 100.0 | 41.1 | 10.1 | 48.8 | 100.0 |
| Hungary | 1 327.7 | 703.0 | 453.6 | 2 484.3 | 1 479.3 | 625.2 | 595.9 | 2 700.3 |
| | 53.4 | 28.3 | 18.3 | 100.0 | 54.8 | 23.2 | 22.1 | 100.0 |
| Poland | 2 756.4 | 2 162.7 | 1 075.5 | 5 994.6 | 3 274.4 | 1 878.6 | 1 115.2 | 6 268.2 |
| | 45.0 | 36.1 | 17.9 | 100.0 | 52.2 | 30.0 | 17.8 | 100.0 |
| Romania (b) | 1 021.1 | 1 377.6 | (1 091.3) | (3 490.0) | 1 086.0 | 962.9 | (1 202.5) | (3 251.4) |
| | 29.3 | 39.5 | 31.3 | 100.0 | 33.4 | 29.6 | 37.0 | 100.0 |
| Total, Eastern Europe | 11 879.4 | 5 997.5 | 7 470.9 | 25 347.7 | 12 836.0 | 5 930.1 | 8 214.8 | 26 980.8 |
| | 46.9 | 23.7 | 29.5 | 100.0 | 47.6 | 22.0 | 30.4 | 100.0 |
| Soviet Union | 9 630.2 | 8 783.1 | 3 924.9 | 22 338.2 | 12 430.0 | 8 816.1 | 3 138.3 | 24 384.4 |
| | 43.1 | 39.3 | 17.6 | 100.0 | 51.0 | 36.2 | 12.9 | 100.0 |
| Total, Eastern Europe and the Soviet Union | 21 509.6 | 14 780.6 | 11 395.8 | 47 686.0 | 25 266.0 | 14 746.1 | 11 353.1 | 51 365.2 |
| | 45.1 | 31.0 | 23.9 | 100.0 | 49.2 | 28.7 | 22.1 | 100.0 |

Table 3 (Chapter 5) contd

| | 1982 | | | | 1983 | | | |
|---|---|---|---|---|---|---|---|---|
| | From CMEA | From Rest of World | Uniden-tified | Total | From CMEA | From Rest of World | Uniden-tified | Total |
| Bulgaria | 2 026.2 | 538.3 | 1 310.0 | 3 874.4 | 2 054.8 | 565.3 | 1 551.0 | 4 171.1 |
| | 52.3 | 13.9 | 33.8 | 100.0 | 49.3 | 13.6 | 37.2 | 100.0 |
| Czechoslovakia | 2 548.1 | 986.2 | 1 612.0 | 5 146.3 | 2 703.4 | 958.4 | 1 819.5 | 5 481.3 |
| | 49.5 | 19.2 | 31.3 | 100.0 | 49.3 | 17.5 | 33.2 | 100.0 |
| German Democratic Republic | 2 470.2 | 623.7 | 3 429.5 | 6 523.4 | 2 253.3 | 640.3 | 3 456.0 | 6 349.6 |
| | 37.9 | 9.6 | 52.6 | 100.0 | 35.5 | 10.1 | 54.4 | 100.0 |
| Hungary | 1 511.8 | 578.5 | 405.9 | 2 496.2 | 1 510.9 | 466.7 | 548.9 | 2 526.5 |
| | 60.6 | 23.2 | 16.3 | 100.0 | 59.8 | 18.5 | 21.7 | 100.0 |
| Poland | 2 093.7 | 686.5 | 1 345.6 | 4 125.8 | 2 431.2 | 743.6 | 1 026.3 | 4 201.0 |
| | 50.7 | 16.6 | 32.6 | 100.0 | 57.9 | 17.7 | 24.4 | 100.0 |
| Romania (b) | 926.6 | 175.9 | (807.1) | (1 909.6) | 788.4 | 107.3 | (744.9) | (1 643.6) |
| | 48.5 | 9.2 | 42.3 | 100.0 | 48.3 | 6.5 | 45.5 | 100.0 |
| Total, Eastern Europe | 11 576.6 | 3 589.0 | 8 910.1 | 24 075.7 | 11 742.0 | 3 481.5 | 9 149.6 | 24 373.2 |
| | 48.1 | 14.9 | 37.0 | 100.0 | 48.2 | 14.3 | 37.5 | 100.0 |
| Soviet Union | 14 592.1 | 11 245.2 | 3 088.7 | 28 926.0 | 16 998.9 | 12 427.2 | 3 039.6 | 32 465.7 |
| | 50.4 | 38.9 | 10.7 | 100.0 | 52.4 | 38.3 | 9.4 | 100.0 |
| Total, Eastern Europe and the Soviet Union | 26 168.7 | 14 834.2 | 11 998.9 | 53 001.8 | 28 740.9 | 15 908.7 | 12 189.3 | 56 838.8 |
| | 49.4 | 28.0 | 22.6 | 100.0 | 50.6 | 28.0 | 21.4 | 100.0 |

Notes: The "Unidentified" item consists of the sum of those items reported as a total but not by country of origin.

a) 1971.
b) For Romania, "unidentified" item is a residual of total machinery and equipment imports plus pipes, and the sum of of imports from CMEA and the rest of the world.
c) "Total, Eastern Europe" and "Eastern Europe and the Soviet Union" exclude Bulgaria and Hungary in that year.

Source : OECD East-West technology transfer data base (Eastern statistics).

Table 4 (Chapter 5)

END-USE SECTOR AND INDUSTRIAL BRANCH DISTRIBUTION OF EASTERN COUNTRIES' CAPITAL GOODS IMPORTS FROM THE WEST: OECD AND CMEA REPORTING COUNTRY DATA COMPARED (a)
(percentage shares of respective totals and sub-totals; grand total in $ million)

| End-Users | Bulgaria | | | | | | | | Czechoslovakia | | | | | | | |
|---|---|---|---|---|---|---|---|---|---|---|---|---|---|---|---|---|
| | 1975 | | 1980 | | 1982 | | 1983 | | 1975 | | 1980 | | 1982 | | 1983 | |
| | OECD data | CMEA data | OECD data | CMEA data | OECD data | CMEA data | OECD data | CMEA data | OECD data | CMEA data | OECD data | CMEA data | OECD data | CMEA data | OECD data | CMEA data |
| **Material sphere** | | | | | | | | | | | | | | | | |
| Industry | 60 | 54 | 62 | 45 | 68 | 58 | 67 | 59 | 75 | 62 | 78 | 71 | 78 | 70 | 73 | 63 |
| Construction | - | 1 | - | 5 | - | 3 | - | 3 | - | 4 | - | 5 | - | 9 | - | 7 |
| Agriculture | 2 | 3 | 3 | 4 | 3 | 4 | 2 | 3 | 3 | 3 | 1 | 4 | 2 | 4 | 3 | 4 |
| Transport and communications (including gas and oil distribution) | 26 | 35 | 23 | 34 | 20 | 29 | 22 | 27 | 13 | 17 | 9 | 12 | 10 | 14 | 14 | 21 |
| Non-attributable | 12 | 7 | 12 | 12 | 9 | 7 | 9 | 9 | 10 | 13 | 11 | 9 | 9 | 3 | 10 | 4 |
| Total | 100 | 100 | 100 | 100 | 100 | 100 | 100 | 100 | 100 | 100 | 100 | 100 | 100 | 100 | 100 | 100 |
| -- per cent of grand total | 98 | 95 | 97 | 90 | 96 | 92 | 97 | 90 | 94 | 84 | 96 | 92 | 95 | 85 | 95 | 97 |
| **Industry** | | | | | | | | | | | | | | | | |
| Energy | 2 | 7 | 3 | 12 | 3 | 11 | 3 | 10 | 3 | 13 | 4 | 55 | 5 | 22 | 4 | 18 |
| Fuel | 5 | - | 6 | 1 | 7 | 1 | 3 | 1 | 7 | 7 | 7 | 3 | 7 | 5 | 9 | 6 |
| Metallurgy | 7 | 2 | 8 | 5 | 9 | 22 | 7 | 5 | 10 | 9 | 8 | 3 | 6 | 4 | 8 | 8 |
| Engineering and metal-working | 9 | 12 | 11 | 17 | 17 | 22 | 25 | 31 | 11 | 23 | 12 | 11 | 9 | 13 | 12 | 17 |
| Chemicals | 12 | 25 | 11 | 37 | 12 | 21 | 12 | 28 | 10 | 9 | 11 | 5 | 15 | 8 | 15 | 6 |
| Construction materials | - | 11 | - | 9 | - | 5 | - | 4 | - | 1 | - | 1 | - | - | - | 1 |
| Wood and paper | 6 | 7 | 2 | 2 | 3 | 4 | 5 | 7 | 8 | 11 | 18 | 12 | 10 | 32 | 6 | 26 |
| Light industry | 22 | 20 | 12 | 11 | 10 | 7 | 8 | 6 | 15 | 18 | 10 | 7 | 13 | 12 | 9 | 13 |
| Food | 9 | 17 | 5 | 6 | 4 | 6 | 4 | 8 | 6 | 8 | 5 | 4 | 4 | 3 | 6 | 5 |
| Non-attributable | 28 | - | 42 | - | 35 | - | 34 | - | 29 | - | 25 | - | 31 | - | 31 | - |
| Total | 100 | 100 | 100 | 100 | 100 | 100 | 100 | 100 | 100 | 100 | 100 | 100 | 100 | 100 | 100 | 100 |
| **Non-material sphere** | | | | | | | | | | | | | | | | |
| Administration and public health -- per cent of grand total | 2 | 15 | 3 | 10 | 4 | 8 | 3 | 10 | 6 | 16 | 4 | 8 | 5 | 15 | 5 | 13 |
| Grand total | 100 | 100 | 100 | 100 | 100 | 100 | 100 | 100 | 100 | 100 | 100 | 100 | 100 | 100 | 100 | 100 |
| -- in $ million | 510 | 457 | 400 | 377 | 550 | 538 | 572 | 565 | 632 | 677 | 844 | 1494 | 623 | 986 | 555 | 958 |

Table 4 (Chapter 5) contd

| End-Users | German Democratic Republic | | | | | | | | Hungary | | | | | | | |
|---|---|---|---|---|---|---|---|---|---|---|---|---|---|---|---|---|
| | 1975 | | 1980 | | 1982 | | 1983 | | 1975 | | 1980 | | 1982 | | 1983 | |
| | OECD data | CMEA data | OECD data | CMEA data | OECD data | CMEA data | OECD data | CMEA data | OECD data | CMEA data | OECD data | CMEA data | OECD data | CMEA data | OECD data | CMEA data |
| **Material sphere** | | | | | | | | | | | | | | | | |
| Industry | 50 | 47 | 70 | 53 | 73 | 55 | 79 | 63 | 65 | 42 | 68 | 41 | 68 | 42 | 67 | 37 |
| Construction | - | 9 | - | 5 | - | 1 | - | 2 | - | 1 | - | 3 | - | 2 | - | 1 |
| Agriculture | 2 | 2 | 2 | 2 | 1 | 2 | - | - | 9 | 6 | 8 | 8 | 8 | 7 | 10 | 15 |
| Transport and communications (including gas and oil distribution) | 42 | 37 | 20 | 23 | 17 | 3 | 14 | 17 | 19 | 39 | 16 | 30 | 15 | 32 | 14 | 29 |
| Non-attributable | 7 | 6 | 8 | 17 | 8 | 20 | 7 | 18 | 8 | 12 | 9 | 18 | 8 | 18 | 9 | 17 |
| Total | 100 | 100 | 100 | 100 | 100 | 100 | 100 | 100 | 100 | 100 | 100 | 100 | 100 | 100 | 100 | 100 |
| -- per cent of grand total | 98 | 86 | 98 | 80 | 98 | 80 | 97 | 77 | 95 | 89 | 96 | 87 | 95 | 87 | 95 | 83 |
| **Industry** | | | | | | | | | | | | | | | | |
| Energy | 10 | 14 | 5 | 14 | 4 | 11 | 5 | 8 | 6 | 17 | 4 | 16 | 4 | 12 | 5 | 16 |
| Fuel | 9 | 9 | 5 | 5 | 3 | 2 | 3 | 1 | 8 | 4 | 11 | 4 | 9 | 4 | 10 | 7 |
| Metallurgy | 3 | 2 | 7 | 8 | 8 | 8 | 18 | 24 | 4 | 8 | 6 | 16 | 4 | 5 | 3 | 6 |
| Engineering and metal-working | 8 | 11 | 14 | 23 | 10 | 15 | 10 | 16 | 6 | 13 | 8 | 17 | 4 | 15 | 7 | 18 |
| Chemicals | 14 | 9 | 10 | 8 | 18 | 25 | 13 | 18 | 17 | 16 | 15 | 9 | 16 | 23 | 19 | 15 |
| Construction materials | - | 1 | - | 8 | - | 9 | - | 13 | 1 | 1 | 1 | 3 | 1 | 2 | 1 | 2 |
| Wood and paper | - | 9 | - | 6 | .. | 3 | - | 3 | - | 15 | - | 5 | .. | 6 | - | 4 |
| Light industry | 30 | 40 | 18 | 25 | 15 | 23 | 12 | 16 | 20 | 16 | 13 | 13 | 19 | 19 | 13 | 20 |
| Food | 5 | 5 | 2 | 3 | 2 | 3 | 1 | 1 | 5 | 9 | 6 | 18 | 6 | 14 | 7 | 11 |
| Non-attributable | 21 | - | 39 | - | 40 | - | 38 | - | 34 | - | 37 | - | 36 | - | 36 | - |
| Total | 100 | 100 | 100 | 100 | 100 | 100 | 100 | 100 | 100 | 100 | 100 | 100 | 100 | 100 | 100 | 100 |
| **Non-material sphere** | | | | | | | | | | | | | | | | |
| Administration and public health -- per cent of grand total | 2 | 14 | 2 | 20 | 2 | 20 | 3 | 23 | 5 | 11 | 4 | 13 | 5 | 13 | 5 | 17 |
| Grand total | 100 | 100 | 100 | 100 | 100 | 100 | 100 | 100 | 100 | 100 | 100 | 100 | 100 | 100 | 100 | 100 |
| -- in $ million | 328 | 407 | 512 | 592 | 551 | 624 | 635 | 640 | 404 | 446(b) | 729 | 625(b) | 740 | 578(b) | 571 | 467 |

Table 4 (Chapter 5) contd

| End-Users | Poland | | | | | | | | Romania | | | | | | | |
|---|---|---|---|---|---|---|---|---|---|---|---|---|---|---|---|---|
| | 1975 | | 1980 | | 1982 | | 1983 | | 1975 | | 1980 | | 1982 | | 1983 | |
| | OECD data | CMEA data | OECD data | CMEA data | OECD data | CMEA data | OECD data | CMEA data | OECD data | CMEA data | OECD data | CMEA data | OECD data | CMEA data | OECD data | CMEA data |
| **Material sphere** | | | | | | | | | | | | | | | | |
| Industry | 58 | 51 | 59 | 52 | 71 | 57 | 69 | 51 | 63 | 61 | 68 | 63 | 64 | 43 | 66 | 56 |
| Construction | - | 3 | - | 3 | - | 3 | - | 3 | - | 3 | - | 4 | - | 5 | - | - |
| Agriculture | 2 | 3 | 1 | 2 | 1 | 3 | 1 | 2 | - | 4 | - | 4 | - | - | - | 1 |
| Transport and communications (including gas and oil distribution) | 31 | 33 | 32 | 31 | 20 | 27 | 20 | 28 | 29 | 28 | 23 | 23 | 25 | 40 | 27 | 38 |
| Non-attributable | 9 | 11 | 8 | 13 | 7 | 10 | 10 | 15 | 7 | 4 | 9 | 6 | 11 | 12 | 7 | 5 |
| Total | 100 | 100 | 100 | 100 | 100 | 100 | 100 | 100 | 100 | 100 | 100 | 100 | 100 | 100 | 100 | 100 |
| -- per cent of grand total | 98 | 97 | 98 | 95 | 96 | 94 | 94 | 93 | 97 | 93 | 97 | 90 | 96 | 80 | 96 | 81 |
| **Industry** | | | | | | | | | | | | | | | | |
| Energy | 4 | 7 | 4 | 8 | 6 | 15 | 5 | 19 | 6 | 13 | 5 | 12 | 4 | 26 | 4 | 7 |
| Fuel | 7 | 7 | 9 | 22 | 14 | 21 | 11 | 11 | 7 | 8 | 7 | 7 | 6 | 16 | 7 | 11 |
| Metallurgy | 11 | 14 | 9 | 9 | 9 | 6 | 6 | 6 | 10 | 15 | 8 | 15 | 5 | 9 | 7 | 11 |
| Engineering and metal-working | 14 | 19 | 12 | 26 | 25 | 27 | 12 | 23 | 18 | 17 | 23 | 23 | 8 | 17 | 27 | 46 |
| Chemicals | 11 | 16 | 11 | 18 | 10 | 17 | 11 | 15 | 12 | 26 | 11 | 25 | 10 | 9 | 7 | 7 |
| Construction materials | - | 7 | - | 1 | - | 2 | - | 3 | - | - | - | 1 | - | - | - | 1 |
| Wood and paper | 4 | 5 | 4 | 5 | 2 | 2 | 3 | 1 | 4 | 9 | 2 | 7 | 1 | 3 | 2 | 3 |
| Light industry | 14 | 12 | 8 | 5 | 6 | 4 | 14 | 14 | 10 | 8 | 10 | 9 | 8 | 18 | 8 | 12 |
| Food | 6 | 12 | 4 | 5 | 4 | 6 | 4 | 8 | 3 | 3 | 2 | 2 | 1 | 3 | 1 | 1 |
| Non-attributable | 29 | - | 39 | - | 25 | - | 34 | - | 30 | - | 33 | - | 57 | - | 36 | - |
| Total | 100 | 100 | 100 | 100 | 100 | 100 | 100 | 100 | 100 | 100 | 100 | 100 | 100 | 100 | 100 | 100 |
| **Non-material sphere** | | | | | | | | | | | | | | | | |
| Administration and public health | | | | | | | | | | | | | | | | |
| -- per cent of grand total | 2 | 3 | 2 | 5 | 4 | 6 | 6 | 7 | 3 | 7 | 3 | 10 | 4 | 20 | 4 | 19 |
| Grand total | 100 | 100 | 100 | 100 | 100 | 100 | 100 | 100 | 100 | 100 | 100 | 100 | 100 | 100 | 100 | 100 |
| -- in $ million | 2167 | 2513 | 1568 | 1879 | 484 | 687 | 492 | 744 | 632 | 761 | 799 | 963 | 213 | 176 | 127 | 107 |

Table 4 (Chapter 5) contd

| End-Users | Soviet Union | | | | | | | |
|---|---|---|---|---|---|---|---|---|
| | 1975 | | 1980 | | 1982 | | 1983 | |
| | OECD data | CMEA data | OECD data | CMEA data | OECD data | CMEA data | OECD data | CMEA data |
| Material sphere | | | | | | | | |
| Industry | 57 | 46 | 57 | 53 | 48 | 39 | 52 | 48 |
| Construction | - | 5 | - | 3 | - | 9 | - | 3 |
| Agriculture | 3 | 1 | 2 | 1 | 2 | 1 | 1 | 1 |
| Transport and communications (including gas and oil distribution) | 34 | 41 | 34 | 36 | 42 | 42 | 41 | 41 |
| Non-attributable | 7 | 8 | 7 | 7 | 8 | 10 | 6 | 6 |
| Total | 100 | 100 | 100 | 100 | 100 | 100 | 100 | 100 |
| -- per cent of grand total | 99 | 97 | 98 | 96 | 99 | 96 | 98 | 94 |
| Industry | | | | | | | | |
| Energy | 4 | 3 | 2 | 9 | 5 | 9 | 10 | 7 |
| Fuel | 5 | 12 | 5 | 13 | 16 | 19 | 11 | 30 |
| Metallurgy | 9 | 14 | 15 | 17 | 15 | 16 | 14 | 15 |
| Engineering and metal-working | 19 | 32 | 22 | 27 | 14 | 24 | 17 | 22 |
| Chemicals | 18 | 17 | 19 | 28 | 19 | 14 | 17 | 12 |
| Construction materials | - | 2 | - | 1 | - | 2 | - | 2 |
| Wood and paper | 3 | 6 | 2 | 6 | 3 | 7 | 3 | 5 |
| Light industry | 8 | 9 | 3 | 4 | 4 | 5 | 4 | 3 |
| Food | 3 | 4 | 4 | 6 | 3 | 5 | 3 | 5 |
| Non-attributable | 31 | - | 38 | - | 32 | - | 32 | - |
| Total | 100 | 100 | 100 | 100 | 100 | 100 | 100 | 100 |
| Non-material sphere | | | | | | | | |
| Administration and public health -- per cent of grand total | 1 | 3 | 2 | 4 | 1 | 4 | 2 | 6 |
| Grand total | 100 | 100 | 100 | 100 | 100 | 100 | 100 | 100 |
| -- in $ million | 4 918 | 6 747 | 5 713 | 8 816 | 7 262 | 11 245 | 7565 | 12 427 |

Notes: Chemical branch in Bulgaria, the German Democratic Republic and Hungary includes oil refining. Light industry includes textiles, clothing and footwear plus (except in the Soviet Union) glass, ceramics and printing. In the German Democratic Republic and Hungary it also officially includes wood and paper (shown separately in CMEA data above). Components may not sum to totals due to rounding. "CMEA data" column refers to data reported by the CMEA country concerned (i.e. not by the CMEA statistical services).

a) OECD data exclude spare parts and accessories and all shipments by Yugoslavia and trade between the Federal Republic of Germany and the German Democratic Republic.

b) For Hungary, imports of complete plant (valued at $55, 64 and 78 million in the years shown) are excluded from the percentage distribution by sector and branch (since these are not broken down by region of origin).

Source: OECD East-West technology transfer data base.

Table 5 (Chapter 5)

COMPARISON OF INTRA-CMEA IMPORTS AND EXPORTS OF CAPITAL GOODS (a)
(percentage shares of respective total and sub-totals; grand total in $ million)

| | Bulgaria | | | | | | | |
|---|---|---|---|---|---|---|---|---|
| | 1975 | | 1980 | | 1982 | | 1983 | |
| | Imports from CMEA | CMEA exports to country | Imports from CMEA | CMEA exports to country | Imports from CMEA | CMEA exports to country | Imports from CMEA | CMEA exports to country |
| **Material sphere** | | | | | | | | |
| Industry | 42 | 45 | 42 | 48 | 39 | 46 | 41 | 46 |
| Construction | 5 | 3 | 6 | 6 | 7 | 6 | 6 | 5 |
| Agriculture | 3 | 13 | 11 | 11 | 2 | 11 | 15 | 13 |
| Transport and communications (including gas and oil distribution) | 15 | 34 | 34 | 34 | 33 | 30 | 34 | 31 |
| Non-attributable | 5 | 5 | 4 | | 9 | 7 | 5 | 5 |
| Total | 100 | 100 | 100 | 100 | 100 | 100 | 100 | 100 |
| -- per cent of grand total | 93 | 94 | 95 | 90 | 94 | 94 | 95 | 95 |
| **Industry** | | | | | | | | |
| Energy | 34 | 30 | 33 | 31 | 23 | 33 | 29 | 40 |
| Fuel | 4 | 11 | 4 | 14 | 12 | 13 | 10 | 10 |
| Metallurgy | 9 | 7 | 14 | 12 | 10 | 7 | 11 | 9 |
| Engineering and metal-working | 13 | 11 | 17 | 15 | 23 | 18 | 19 | 16 |
| Chemicals | 14 | 16 | 10 | 7 | 12 | 10 | 12 | 9 |
| Construction materials | 8 | 7 | 8 | 6 | 6 | 4 | 5 | 3 |
| Wood and paper | 3 | 4 | 4 | 3 | 3 | 4 | 2 | 2 |
| Light industry | 11 | 9 | 6 | 6 | 8 | 8 | 7 | 8 |
| Food | 4 | 4 | 4 | 5 | 4 | 3 | 4 | 3 |
| Total | 100 | 100 | 100 | 100 | 100 | 100 | 100 | 100 |
| **Non-material sphere** | | | | | | | | |
| Administration and public health -- per cent of grand total | 7 | 6 | 5 | 10 | 6 | 6 | 5 | 5 |
| Grand total | 100 | 100 | 100 | 100 | 100 | 100 | 100 | 100 |
| -- in $ million | 1 141 | 1 165 | 1 941 | 1 969 | 2 026 | 2 209 | 2 055 | 2 250 |

Table 5 (Chapter 5) contd

| | Czechoslovakia | | | | | | | |
|---|---|---|---|---|---|---|---|---|
| | 1975 | | 1980 | | 1982 | | 1983 | |
| | Imports from CMEA | CMEA exports to country | Imports from CMEA | CMEA exports to country | Imports from CMEA | CMEA exports to country | Imports from CMEA | CMEA exports to country |
| **Material sphere** | | | | | | | | |
| Industry | 25 | 26 | 30 | 38 | 39 | 41 | 39 | 42 |
| Construction | 4 | 4 | 8 | 6 | 9 | 6 | 8 | 6 |
| Agriculture | 22 | 23 | 12 | 12 | 11 | 15 | 12 | 17 |
| Transport and communications (including gas and oil distribution) | 42 | 36 | 35 | 32 | 31 | 26 | 32 | 24 |
| Non-attributable | 7 | 11 | 15 | 12 | 9 | 12 | 10 | 11 |
| Total | 100 | 100 | 100 | 100 | 100 | 100 | 100 | 100 |
| -- per cent of grand total | 82 | 80 | 86 | 86 | 86 | 87 | 84 | 86 |
| **Industry** | | | | | | | | |
| Energy | 27 | 19 | 24 | 34 | 40 | 37 | 41 | 39 |
| Fuel | 8 | 9 | 9 | 8 | 9 | 9 | 7 | 7 |
| Metallurgy | 5 | 6 | 10 | 9 | 8 | 11 | 8 | 11 |
| Engineering and metal-working | 33 | 35 | 27 | 22 | 16 | 17 | 15 | 17 |
| Chemicals | 2 | 3 | 6 | 5 | 6 | 6 | 9 | 9 |
| Construction materials | - | 1 | 1 | - | 3 | 3 | 2 | 2 |
| Wood and paper | 2 | 2 | 4 | 4 | 3 | 3 | 1 | 1 |
| Light industry | 15 | 17 | 14 | 12 | 9 | 8 | 11 | 8 |
| Food | 7 | 8 | 5 | 6 | 6 | 6 | 5 | 6 |
| Total | 100 | 100 | 100 | 100 | 100 | 100 | 100 | 100 |
| **Non-material sphere** | | | | | | | | |
| Administration and public health -- per cent of grand total | 18 | 20 | 14 | 14 | 14 | 13 | 16 | 14 |
| Grand total | 100 | 100 | 100 | 100 | 100 | 100 | 100 | 100 |
| -- in $ million | 1 527 | 1 398 | 2 639 | 2 439 | 2 548 | 2 465 | 2 703 | 2 461 |

Table 5 (Chapter 5) contd

| | German Democratic Republic | | | | | | | |
|---|---|---|---|---|---|---|---|---|
| | 1975 | | 1980 | | 1982 | | 1983 | |
| | Imports from CMEA | CMEA exports to country | Imports from CMEA | CMEA exports to country | Imports from CMEA | CMEA exports to country | Imports from CMEA | CMEA exports to country |
| Material sphere | | | | | | | | |
| Industry | 32 | 36 | 33 | 34 | 32 | 37 | 36 | 40 |
| Construction | 4 | 4 | 4 | 4 | 3 | 3 | 3 | 3 |
| Agriculture | 8 | 8 | 13 | 11 | 14 | 12 | 14 | 11 |
| Transport and communications (including gas and oil distribution) | 36 | 41 | 39 | 40 | 32 | 38 | 32 | 35 |
| Non-attributable | 20 | 11 | 11 | 11 | 19 | 10 | 16 | 12 |
| Total | 100 | 100 | 100 | 100 | 100 | 100 | 100 | 100 |
| -- per cent of grand total | 85 | 84 | 92 | 91 | 91 | 89 | 90 | 89 |
| Industry | | | | | | | | |
| Energy | 32 | 29 | 39 | 39 | 48 | 49 | 48 | 48 |
| Fuel | 4 | 6 | 3 | 3 | 3 | 3 | 3 | 3 |
| Metallurgy | 11 | 11 | 10 | 10 | 8 | 8 | 7 | 7 |
| Engineering and metal-working | 22 | 22 | 22 | 22 | 19 | 17 | 23 | 23 |
| Chemicals | 9 | 10 | 5 | 5 | 3 | 3 | 3 | 3 |
| Construction materials | 3 | 3 | 1 | 1 | 2 | 2 | 1 | 1 |
| Wood and paper | 1 | 1 | 2 | 2 | 2 | 2 | 2 | 1 |
| Light industry | 13 | 12 | 12 | 12 | 10 | 10 | 9 | 8 |
| Food | 5 | 5 | 6 | 6 | 6 | 6 | 5 | 6 |
| Total | 100 | 100 | 100 | 100 | 100 | 100 | 100 | 100 |
| Non-material sphere | | | | | | | | |
| Administration and public health -- per cent of grand total | 5 | 6 | 8 | 9 | 9 | 11 | 10 | 11 |
| Grand total | 100 | 100 | 100 | 100 | 100 | 100 | 100 | 100 |
| -- in $ million | 1 751 | 1 549 | 2 416 | 2 356 | 2 470 | 2 112 | 2 253 | 2 059 |

Table 5 (Chapter 5) contd

| | Hungary | | | | | | | |
|---|---|---|---|---|---|---|---|---|
| | 1975 | | 1980 | | 1982 | | 1983 | |
| | Imports from CMEA | CMEA exports to country | Imports from CMEA | CMEA exports to country | Imports from CMEA | CMEA exports to country | Imports from CMEA | CMEA exports to country |
| Material sphere | | | | | | | | |
| Industry | 23 | 23 | 29 | 31 | 33 | 32 | 32 | 27 |
| Construction | 5 | 4 | 2 | 3 | 3 | 5 | 3 | 5 |
| Agriculture | 20 | 18 | 11 | 12 | 9 | 15 | 13 | 16 |
| Transports and communications (including gas and oil distribution) | 40 | 45 | 46 | 44 | 39 | 37 | 38 | 43 |
| Non-attributable | 13 | 11 | 12 | 11 | 12 | 10 | 14 | 10 |
| Total | 100 | 100 | 100 | 100 | 100 | 100 | 100 | 100 |
| -- per cent of grand total | 88 | 99 | 90 | 92 | 89 | 91 | 88 | 91 |
| Industry | | | | | | | | |
| Energy | 28 | 29 | 48 | 51 | 54 | 49 | 57 | 52 |
| Fuel | 4 | 8 | 7 | 8 | 5 | 9 | 4 | 7 |
| Metallurgy | 4 | 3 | 2 | 6 | 3 | 7 | 1 | 5 |
| Engineering and metal-working | 20 | 19 | 22 | 15 | 21 | 17 | 19 | 16 |
| Chemicals | 13 | 11 | 3 | 3 | 4 | 5 | 4 | 6 |
| Construction materials | 3 | 2 | 2 | 5 | 1 | 1 | 2 | 1 |
| Wood and paper | 2 | 2 | 1 | 1 | 1 | 1 | 1 | 1 |
| Light industry | 19 | 18 | 9 | 7 | 6 | 6 | 6 | 6 |
| Food | 6 | 7 | 6 | 5 | 5 | 5 | 6 | 6 |
| Total | 100 | 100 | 100 | 100 | 100 | 100 | 100 | 100 |
| Non-material sphere | | | | | | | | |
| Administration and public health -- per cent of grand total | 12 | 11 | 10 | 8 | 11 | 9 | 12 | 9 |
| Grand total | 100 | 100 | 100 | 100 | 100 | 100 | 100 | 100 |
| -- in $ million | 785 (b) | 913 | 1 479 (b) | 1 791 | 1 512 (b) | 1738 | 1 511 (b) | 1 905 |

Table 5 (Chapter 5) contd

| | Poland | | | | | | | |
|---|---|---|---|---|---|---|---|---|
| | 1975 | | 1980 | | 1982 | | 1983 | |
| | Imports from CMEA | CMEA exports to country | Imports from CMEA | CMEA exports to country | Imports from CMEA | CMEA exports to country | Imports from CMEA | CMEA exports to country |
| **Material sphere** | | | | | | | | |
| Industry | 35 | 36 | 29 | 28 | 25 | 28 | 28 | 26 |
| Construction | 4 | 4 | 5 | 5 | 3 | 4 | 4 | 4 |
| Agriculture | 12 | 12 | 15 | 15 | 18 | 15 | 16 | 13 |
| Transport and communications (including gas and oil distribution) | 34 | 33 | 40 | 41 | 42 | 43 | 39 | 46 |
| Non-attributable | 15 | 15 | 11 | 12 | 12 | 9 | 13 | 12 |
| Total | 100 | 100 | 100 | 100 | 100 | 100 | 100 | 100 |
| -- per cent of Grand total | 92 | 90 | 92 | 92 | 90 | 90 | 90 | 90 |
| **Industry** | | | | | | | | |
| Energy | 14 | 16 | 21 | 22 | 25 | 32 | 27 | 25 |
| Fuel | 13 | 13 | 11 | 14 | 12 | 14 | 8 | 12 |
| Metallurgy | 12 | 11 | 16 | 13 | 4 | 3 | 2 | 1 |
| Engineering and metal-working | 30 | 27 | 21 | 22 | 26 | 21 | 31 | 29 |
| Chemicals | 4 | 5 | 6 | 4 | 5 | 4 | 4 | 3 |
| Construction materials | 5 | 6 | 4 | 5 | 1 | 1 | 1 | 1 |
| Wood and paper | - | 1 | - | - | - | - | - | - |
| Light industry | 16 | 16 | 15 | 15 | 22 | 19 | 22 | 23 |
| Food | 6 | 6 | 5 | 5 | 6 | 5 | 5 | 5 |
| Total | 100 | 100 | 100 | 100 | 100 | 100 | 100 | 100 |
| **Non-material sphere** | | | | | | | | |
| Administration and public health -- per cent of grand total | 8 | 10 | 8 | 8 | 10 | 10 | 10 | 10 |
| Grand total | 100 | 100 | 100 | 100 | 100 | 100 | 100 | 100 |
| -- in $ million | 1 739 | 1 717 | 3 274 | 3 356 | 2 094 | 2 122 | 2 431 | 2 412 |

Table 5 (Chapter 5) contd

| | Romania | | | | | | | |
|---|---|---|---|---|---|---|---|---|
| | 1975 | | 1980 | | 1980 | | 1983 | |
| | Imports from CMEA | CMEA exports to country | Imports from CMEA | CMEA exports to country | Imports from CMEA | CMEA exports to country | Imports from CMEA | CMEA exports to country |
| **Material sphere** | | | | | | | | |
| Industry | 51 | 51 | 53 | 53 | 58 | 58 | 59 | 59 |
| Construction | 4 | 4 | 6 | 6 | 9 | 9 | 7 | 7 |
| Agriculture | 3 | 3 | 3 | 3 | 5 | 5 | 4 | 4 |
| Transports and communications (including gas and oil distribution) | 33 | 33 | 31 | 31 | 21 | 21 | 21 | 21 |
| Non-attributable | 10 | 10 | 7 | 7 | 7 | 7 | 9 | 9 |
| Total | 100 | 100 | 100 | 100 | 100 | 100 | 100 | 100 |
| -- per cent of grand total | 95 | 95 | 96 | 96 | 95 | 95 | 90 | 90 |
| **Industry** | | | | | | | | |
| Energy | 24 | 24 | 14 | 14 | 16 | 16 | 18 | 18 |
| Fuel | 7 | 6 | 10 | 10 | 9 | 9 | 12 | 12 |
| Metallurgy | 13 | 13 | 9 | 9 | 22 | 22 | 11 | 11 |
| Engineering and metal-working | 30 | 30 | 35 | 35 | 30 | 30 | 35 | 35 |
| Chemicals | 9 | 9 | 9 | 9 | 4 | 4 | 6 | 6 |
| Construction materials | - | - | - | - | 1 | 1 | 1 | 1 |
| Wood and paper | 2 | 2 | 3 | 3 | 2 | 2 | 2 | 2 |
| Light industry | 14 | 14 | 19 | 19 | 15 | 15 | 14 | 14 |
| Food | 2 | 2 | 1 | 1 | 2 | 2 | 1 | 1 |
| Total | 100 | 100 | 100 | 100 | 100 | 100 | 100 | 100 |
| **Non-material sphere** | | | | | | | | |
| Administration and public health -- per cent of grand total | 5 | 5 | 4 | 4 | 5 | 5 | 10 | 10 |
| Grand total | 100 | 100 | 100 | 100 | 100 | 100 | 100 | 100 |
| -- in $ million | 533 | 533 | 1 086 | 1 086 | 927 | 927 | 788 | 788 |

Table 5 (Chapter 5) contd

| | Soviet Union | | | | | | | |
|---|---|---|---|---|---|---|---|---|
| | 1975 | | 1980 | | 1982 | | 1983 | |
| | Imports from CMEA | CMEA exports to country | Imports from CMEA | CMEA exports to country | Imports from CMEA | CMEA exports to country | Imports from CMEA | CMEA exports to country |
| Material sphere | | | | | | | | |
| Industry | 33 | 33 | 40 | 39 | 41 | 41 | 42 | 41 |
| Construction | 1 | 1 | 2 | 1 | 2 | 2 | 2 | 2 |
| Agriculture | 8 | 7 | 10 | 8 | 9 | | 9 | 8 |
| Transport and communications (including gas & oil distribution) | 47 | 46 | 36 | 37 | 36 | 36 | 35 | 37 |
| Non-attributable | 11 | 13 | 13 | 15 | 12 | 12 | 12 | 12 |
| Total | 100 | 100 | 100 | 100 | 100 | 100 | 100 | 100 |
| -- per cent of grand total | 88 | 89 | 91 | 88 | 84 | 86 | 86 | 86 |
| Industry | | | | | | | | |
| Energy | 21 | 20 | 23 | 20 | 21 | 22 | 21 | 19 |
| Fuel | 6 | 6 | 10 | 10 | 10 | 9 | 11 | 10 |
| Metallurgy | 5 | 5 | 7 | 7 | 7 | 7 | 5 | 5 |
| Engineering and metal-working | 17 | 19 | 16 | 19 | 19 | 22 | 20 | 21 |
| Chemicals | 19 | 19 | 15 | 16 | 1 | 11 | 12 | 12 |
| Construction materials | 4 | 4 | 3 | 2 | 3 | 3 | 3 | 4 |
| Wood and paper | 2 | 1 | 2 | 1 | 11 | 1 | 1 | 1 |
| Light industry | 14 | 14 | 15 | 15 | 17 | 15 | 18 | 17 |
| Food | 11 | 11 | 10 | 9 | 10 | 10 | 10 | 11 |
| Total | 100 | 100 | 100 | 100 | 100 | 100 | 100 | 100 |
| Non-material sphere | | | | | | | | |
| Administration and public health -- per cent of grand total | 12 | 11 | 9 | 12 | 16 | 14 | 14 | 14 |
| Grand total | 100 | 100 | 100 | 100 | 100 | 100 | 100 | 100 |
| -- in $ million | 6 514 | 6 310 | 12 430 | 12 862 | 14 592 | 13 915 | 16 999 | 16 838 |

Notes: For coverage of industrial branches, see Note to Table 4 of this Chapter. Components may not sum to totals due to rounding.

a) "Imports from CMEA" are those reported by the importing country; "CMEA Exports to country" are the sum of the other six countries' exports to that country.

b) For Hungary, imports of complete plant (valued at $30, 193, 199 and 192 million in the years shown) are included in the totals in $ million but excluded from the percentage distribution by sector and branch.

Source: OECD East-West technology transfer data base.

Table 6 (Chapter 5)

EAST-WEST TRADE AND TRADE BALANCES: COMPOSITION BY SELECTED COMMODITY GROUPS
(components expressed as per cent of total exports/imports; total exports/imports and balances
(in $ million)

| | Bulgaria | | | | | Czechoslovakia | | | | | German Democratic Republic | | | | |
|---|---|---|---|---|---|---|---|---|---|---|---|---|---|---|---|
| | 1970 | 1975 | 1980 | 1982 | 1983 | 1970 | 1975 | 1980 | 1982 | 1983 | 1970 | 1975 | 1980 | 1982 | 1983 |
| OECD exports f.o.b | | | | | | | | | | | | | | | |
| Capital goods | 29.9 | 46.4 | 24.9 | 35.4 | 36.7 | 33.0 | 33.6 | 28.5 | 28.9 | 28.5 | 35.5 | 28.9 | 20.6 | 32.3 | 32.3 |
| Intermediate (technology) goods | 44.3 | 34.7 | 44.0 | 36.6 | 36.4 | 29.6 | 36.2 | 35.0 | 35.0 | 39.0 | 29.6 | 37.9 | 28.5 | 23.0 | 25.3 |
| Fuel | 1.0 | 0.5 | 0.4 | 1.0 | 1.1 | 0.5 | 1.4 | 1.0 | 1.3 | 1.1 | 3.5 | 0.1 | 0.7 | 0.1 | 0.5 |
| Other industrial goods | 20.6 | 15.2 | 22.9 | 22.3 | 22.0 | 25.2 | 22.8 | 21.4 | 24.0 | 24.3 | 18.8 | 24.1 | 24.3 | 24.0 | 19.4 |
| Total industrial goods | 95.8 | 96.8 | 92.2 | 95.3 | 96.1 | 88.3 | 94.1 | 85.9 | 89.2 | 92.9 | 87.4 | 91.0 | 74.1 | 79.4 | 77.4 |
| Agricultural goods | 4.2 | 3.2 | 7.8 | 4.7 | 3.9 | 11.7 | 5.9 | 14.1 | 10.8 | 7.1 | 12.6 | 9.0 | 25.9 | 20.6 | 22.6 |
| Total exports | 327.7 | 1099.4 | 1606.9 | 1552.9 | 1557.6 | 784.6 | 1878.3 | 2965.4 | 2157.1 | 1945.7 | 438.5 | 1134.5 | 2490.3 | 1705.7 | 1966.6 |
| OECD imports c.i.f. | | | | | | | | | | | | | | | |
| Capital goods | 6.6 | 8.5 | 8.7 | 6.9 | 5.2 | 15.2 | 12.8 | 10.6 | 9.0 | 8.6 | 20.2 | 18.3 | 16.5 | 12.0 | 9.6 |
| Intermediate (technology) goods | 23.1 | 27.3 | 22.1 | 27.9 | 27.4 | 27.1 | 24.4 | 25.6 | 28.2 | 27.8 | 35.3 | 29.8 | 30.0 | 31.0 | 31.5 |
| Fuel | 0.5 | 2.0 | 31.5 | 23.7 | 25.7 | 5.3 | 9.0 | 14.3 | 14.2 | 14.8 | 5.1 | 5.9 | 20.2 | 30.6 | 34.4 |
| Other industrial goods | 34.4 | 37.3 | 26.5 | 29.1 | 30.3 | 44.0 | 48.2 | 42.9 | 42.4 | 43.3 | 32.3 | 36.5 | 29.4 | 23.6 | 22.2 |
| Total industrial goods | 64.6 | 75.0 | 88.7 | 87.6 | 88.6 | 91.6 | 94.4 | 93.4 | 93.7 | 94.5 | 92.9 | 90.6 | 96.0 | 97.2 | 97.6 |
| Agricultural goods | 35.4 | 25.0 | 11.3 | 12.4 | 11.4 | 8.4 | 5.6 | 6.6 | 6.3 | 5.5 | 7.1 | 9.4 | 4.0 | 2.8 | 2.4 |
| Total imports | 239.9 | 392.3 | 994.0 | 805.5 | 729.2 | 723.5 | 1642.1 | 3215.1 | 2677.3 | 2621.3 | 409.9 | 1039.6 | 2102.8 | 2385.8 | 2408.6 |
| OECD balance of trade | | | | | | | | | | | | | | | |
| Capital goods | 82.1 | 476.2 | 314.1 | 494.6 | 533.9 | 149.3 | 421.8 | 501.9 | 382.6 | 328.8 | 73.0 | 137.2 | 165.7 | 263.5 | 404.1 |
| Intermediate (technology) goods | 89.6 | 274.8 | 487.7 | 343.8 | 366.5 | 36.0 | 280.0 | 214.8 | 1.9 | 31.0 | -15.2 | 119.9 | 78.5 | -348.4 | -260.8 |
| Fuel | 2.3 | -2.3 | -306.7 | -175.3 | -170.0 | -34.4 | -120.3 | -429.2 | -351.4 | -366.3 | -5.4 | -60.4 | -406.7 | -728.1 | -819.0 |
| Other industrial goods | -15.0 | 20.7 | 104.8 | 111.6 | 121.3 | -120.7 | -363.4 | -743.6 | -618.4 | -662.5 | -49.7 | -106.9 | -12.0 | -152.1 | -151.9 |
| Total industrial goods | 159.0 | 769.5 | 600.0 | 774.7 | 851.7 | 30.2 | 218.1 | -456.1 | -585.3 | -669.0 | 2.6 | 89.8 | -174.4 | -965.0 | -827.6 |
| Agricultural goods | -71.2 | -62.4 | 12.9 | -27.3 | -23.3 | 30.9 | 18.0 | 206.5 | 65.1 | -6.6 | 26.0 | 5.1 | 561.9 | 284.9 | 385.6 |
| Total balance of trade | 87.8 | 707.1 | 612.9 | 747.4 | 828.4 | 61.1 | 236.1 | -249.7 | -520.2 | -675.6 | 28.6 | 95.0 | 387.5 | -680.1 | -442.0 |

Table 6 (Chapter 5) contd

| | Hungary | | | | | Poland | | | | | Romania | | | | |
|---|---|---|---|---|---|---|---|---|---|---|---|---|---|---|---|
| | 1970 | 1975 | 1980 | 1982 | 1983 | 1970 | 1975 | 1980 | 1982 | 1983 | 1970 | 1975 | 1980 | 1982 | 1983 |
| **OECD exports f.o.b** | | | | | | | | | | | | | | | |
| Capital goods | 19.3 | 22.1 | 22.3 | 25.5 | 22.1 | 24.2 | 39.4 | 24.2 | 15.0 | 17.0 | 34.9 | 31.7 | 20.6 | 12.8 | 9.9 |
| Intermediate (technology) goods | 42.2 | 48.0 | 46.9 | 45.4 | 45.8 | 37.0 | 33.2 | 31.7 | 30.1 | 32.9 | 38.6 | 41.7 | 37.2 | 39.3 | 39.8 |
| Fuel (including coal) | 0.4 | 0.4 | 0.5 | 0.5 | 0.6 | 1.1 | 0.7 | 0.8 | 0.9 | 0.8 | 0.5 | 1.3 | 2.4 | 1.7 | 2.2 |
| Other industrial goods | 30.8 | 26.6 | 25.4 | 26.4 | 28.9 | 24.8 | 15.4 | 19.5 | 32.2 | 31.5 | 20.1 | 16.1 | 25.6 | 33.2 | 35.9 |
| Total industrial goods | 92.8 | 97.1 | 95.1 | 97.8 | 97.5 | 87.1 | 88.7 | 76.2 | 78.1 | 82.2 | 94.2 | 90.8 | 85.8 | 86.9 | 87.9 |
| Agricultural goods | 7.2 | 2.9 | 4.9 | 2.2 | 2.5 | 12.9 | 11.3 | 23.8 | 21.9 | 17.8 | 5.8 | 9.2 | 14.2 | 13.1 | 12.1 |
| Total exports | 625.3 | 1831.8 | 3268.3 | 2866.1 | 2579.8 | 882.4 | 5499.4 | 6478.6 | 3218.2 | 2890.4 | 703.3 | 1992.7 | 3887.6 | 1670.0 | 1278.4 |
| **OECD imports c.i.f.** | | | | | | | | | | | | | | | |
| Capital goods | 4.5 | 6.0 | 7.4 | 7.6 | 5.7 | 4.7 | 8.4 | 8.0 | 7.6 | 7.2 | 4.8 | 5.9 | 5.8 | 6.1 | 4.4 |
| Intermediate (technology) goods | 19.7 | 18.7 | 23.0 | 23.5 | 22.4 | 17.3 | 14.3 | 23.3 | 21.8 | 19.4 | 21.5 | 17.8 | 18.6 | 20.1 | 19.8 |
| Fuel (including coal) | 2.6 | 1.9 | 5.9 | 5.4 | 9.1 | 21.3 | 36.7 | 23.8 | 25.7 | 25.5 | 10.3 | 22.6 | 38.6 | 34.2 | 41.1 |
| Other industrial goods | 46.4 | 57.6 | 54.4 | 52.6 | 53.0 | 43.9 | 34.0 | 37.2 | 36.1 | 38.1 | 43.4 | 48.0 | 34.2 | 36.7 | 32.3 |
| Total industrial goods | 73.2 | 84.2 | 90.7 | 89.0 | 90.2 | 87.3 | 93.3 | 92.4 | 91.2 | 90.2 | 80.1 | 94.4 | 97.3 | 97.2 | 97.6 |
| Agricultural goods | 26.8 | 15.8 | 9.3 | 11.0 | 9.8 | 12.7 | 6.7 | 7.6 | 8.8 | 9.8 | 19.9 | 5.6 | 2.7 | 2.8 | 2.4 |
| Total imports | 536.0 | 1250.6 | 2821.2 | 2304.2 | 2338.4 | 1061.4 | 3173.7 | 5471.4 | 3316.3 | 3263.2 | 553.5 | 1662.3 | 3504.9 | 2618.4 | 2805.2 |
| **OECD balance of trade** | | | | | | | | | | | | | | | |
| Capital goods | 96.8 | 328.7 | 521.5 | 555.0 | 437.9 | 163.4 | 1901.4 | 1128.4 | 230.6 | 255.4 | 218.8 | 533.8 | 595.1 | 53.5 | 3.8 |
| Intermediate (technology) goods | 158.7 | 645.9 | 883.2 | 759.4 | 658.1 | 142.7 | 1368.3 | 782.0 | 246.0 | 320.1 | 152.5 | 535.0 | 794.7 | 128.2 | -45.5 |
| Fuel (including coal) | -11.6 | -16.9 | -149.8 | -107.6 | -197.2 | -216.1 | -1123.5 | -1255.0 | -825.9 | -810.0 | -53.4 | -351.1 | -1259.3 | -867.0 | -1125.8 |
| Other industrial goods | -56.0 | -232.1 | -704.2 | -454.2 | -493.2 | -248.0 | -231.4 | -775.8 | -162.6 | -334.4 | -99.1 | -476.8 | -206.2 | -407.1 | -447.9 |
| Total industrial goods | 187.9 | 725.6 | 550.7 | 752.6 | 405.5 | -158.0 | 1914.8 | -120.4 | -511.9 | -568.9 | 218.8 | 240.9 | -75.8 | -1092.4 | -1615.3 |
| Agricultural goods | -98.7 | -144.4 | -103.6 | -190.8 | -164.2 | -21.0 | 410.8 | 1127.5 | 413.8 | 196.1 | -69.1 | 89.6 | 458.4 | 144.0 | 88.6 |
| Total balance of trade | 89.3 | 581.2 | 447.1 | 561.9 | 241.4 | -179.0 | 2325.6 | 1007.2 | -98.1 | -98.1 | 149.8 | 330.5 | 382.7 | -948.3 | -1526.8 |

Table 6 (Chapter 5) contd

| | Total Six | | | | | Soviet Union | | | | | Total Seven | | | | |
|---|---|---|---|---|---|---|---|---|---|---|---|---|---|---|---|
| | 1970 | 1975 | 1890 | 1982 | 1983 | 1970 | 1975 | 1980 | 1982 | 1983 | 1970 | 1975 | 1980 | 1982 | 1983 |
| **OECD exports f.o.b** | | | | | | | | | | | | | | | |
| Capital goods | 29.1 | 34.8 | 23.4 | 23.9 | 24.2 | 42.6 | 39.2 | 26.5 | 31.5 | 33.5 | 34.6 | 36.9 | 25.0 | 28.7 | 30.2 |
| Intermediate (technology) goods | 36.4 | 37.4 | 36.2 | 35.2 | 36.5 | 30.3 | 33.7 | 35.3 | 28.5 | 27.4 | 33.9 | 35.6 | 35.7 | 30.9 | 30.6 |
| Fuel (including coal) | 1.0 | 0.8 | 1.0 | 0.9 | 0.9 | 0.2 | 0.3 | 0.5 | 1.2 | 0.7 | 0.7 | 0.6 | 0.8 | 1.1 | 0.8 |
| Other industrial goods | 23.9 | 18.8 | 22.7 | 27.5 | 27.1 | 18.9 | 11.0 | 20.6 | 19.4 | 20.2 | 21.9 | 15.0 | 21.6 | 22.3 | 22.6 |
| Total industrial goods | 90.4 | 91.8 | 83.4 | 87.5 | 88.7 | 92.0 | 84.2 | 82.9 | 80.6 | 81.9 | 91.1 | 88.1 | 83.1 | 83.1 | 84.3 |
| Agricultural goods | 9.6 | 8.2 | 16.6 | 12.5 | 11.3 | 8.0 | 15.8 | 17.1 | 19.4 | 18.1 | 8.9 | 11.9 | 16.9 | 16.9 | 15.7 |
| Total exports | 3761.8 | 13436.1 | 20697.1 | 13170.0 | 12218.6 | 2643.5 | 12546.8 | 21586.9 | 23056.6 | 22589.9 | 6405.3 | 25982.9 | 42284.0 | 36226.5 | 34808.5 |
| **OECD imports c.i.f.** | | | | | | | | | | | | | | | |
| Capital goods | 8.8 | 9.5 | 9.0 | 8.3 | 7.0 | 3.0 | 3.0 | 1.5 | 0.7 | 0.6 | 6.4 | 6.4 | 4.7 | 3.4 | 2.9 |
| Intermediate (technology) goods | 22.8 | 19.7 | 23.5 | 24.9 | 24.0 | 18.5 | 10.6 | 10.8 | 7.7 | 8.3 | 21.0 | 15.3 | 16.3 | 13.9 | 14.0 |
| Fuel (including coal) | 10.1 | 19.4 | 22.2 | 22.5 | 25.4 | 30.1 | 49.1 | 62.3 | 72.8 | 73.2 | 18.5 | 33.7 | 44.9 | 54.7 | 55.8 |
| Other industrial goods | 42.2 | 42.7 | 38.8 | 37.6 | 37.3 | 29.4 | 22.1 | 18.8 | 13.8 | 13.4 | 36.9 | 32.8 | 27.5 | 22.4 | 22.1 |
| Total industrial goods | 84.0 | 91.4 | 93.5 | 93.2 | 93.6 | 81.1 | 84.8 | 93.4 | 94.9 | 95.6 | 82.8 | 88.2 | 93.4 | 94.3 | 94.9 |
| Agricultural goods | 16.0 | 8.6 | 6.5 | 6.8 | 6.4 | 18.9 | 15.2 | 6.6 | 5.1 | 4.4 | 17.2 | 11.8 | 6.6 | 5.7 | 5.1 |
| Total imports | 3524.3 | 9160.6 | 18109.5 | 14107.4 | 14166.0 | 2553.9 | 8493.7 | 23788.5 | 25059.9 | 24663.3 | 6078.2 | 17654.4 | 41898.0 | 39167.3 | 38829.2 |
| **OECD balance of trade** | | | | | | | | | | | | | | | |
| Capital goods | 783.4 | 3799.1 | 3226.7 | 1979.8 | 1963.8 | 1048.7 | 4665.5 | 5361.3 | 7088.0 | 7408.1 | 1832.1 | 8464.6 | 8587.9 | 9067.7 | 9371.9 |
| Intermediate (technology) goods | 564.3 | 3224.0 | 3240.9 | 1131.0 | 1069.3 | 327.8 | 3333.8 | 5042.1 | 4650.7 | 4149.6 | 892.1 | 6557.8 | 8283.0 | 5781.7 | 5218.9 |
| Fuel (including coal) | -318.6 | -1674.5 | -3806.7 | -3055.2 | -3488.3 | -764.9 | -4134.7 | -1470.3 | -1796.5 | -1789.1 | -1083.5 | -5809.2 | -1850.9 | -2102.0 | -2138.0 |
| Other industrial goods | -588.5 | -1389.9 | -2336.9 | -1682.8 | -1968.5 | -250.4 | -497.7 | -26.6 | 1010.7 | 1259.3 | -838.8 | -1887.6 | -2363.5 | -672.1 | -709.3 |
| Total industrial goods | 440.6 | 3958.7 | 324.0 | -1627.2 | -2423.6 | 361.3 | 3366.9 | -4325.9 | -5215.7 | -5074.4 | 801.9 | 7325.7 | -4001.9 | -6842.9 | -7498.0 |
| Agricultural goods | -203.1 | 316.7 | 2263.6 | 689.8 | 476.3 | -271.7 | 686.2 | 2124.3 | 3212.3 | 3001.0 | -474.8 | 1002.9 | 4388.0 | 3902.1 | 3477.3 |
| Total balance of trade | 237.5 | 4275.5 | 2587.6 | -937.4 | -1947.4 | 89.6 | 4053.1 | -2201.6 | -2003.3 | -2073.4 | 327.1 | 8328.5 | 386.0 | -2940.8 | -4020.8 |

Note: For trade balance panel, "+" = OECD surplus, "-" OECD deficit. Note that this panel is based on f.o.b. export and c.i.f. import values and the trade balance in OECD's favour (or the OECD trade deficit) is thus understated (overstated). As in all following Tables based on Western statistics, data exclude trade between the Federal Republic of Germany and the German Democratic Republic.

Source: OECD East-West technology transfer data base (Western statistics).

Table 7 (Chapter 5)

WESTERN TECHNOLOGY EXPORTS TO THE EAST AND OUTPUT (NMP):
SHARES OF EASTERN COUNTRIES IN RESPECTIVE SEVEN-COUNTRY TOTALS,
1970, 1975, 1980, 1982 and 1983
(per cent)

| | Western capital goods exports | | | | | Western intermediate (technology) goods exports | | | | | NMP | |
|---|---|---|---|---|---|---|---|---|---|---|---|---|
| | 1970 | 1975 | 1980 | 1982 | 1983 | 1970 | 1975 | 1980 | 1982 | 1983 | 1970 | 1982 |
| Bulgaria | 4.4 | 5.3 | 3.8 | 5.3 | 5.4 | 6.7 | 4.1 | 4.7 | 5.1 | 5.3 | 1.9 | 2.3 |
| Czechoslovakia | 11.7 | 6.6 | 8.0 | 6.0 | 5.3 | 10.7 | 7.3 | 6.9 | 6.7 | 7.1 | 5.0 | 4.6 |
| German Democratic Republic | 7.0 | 3.4 | 4.8 | 5.3 | 6.0 | 6.0 | 4.6 | 4.7 | 3.5 | 4.7 | 6.1 | 6.0 |
| Hungary | 5.4 | 4.2 | 6.9 | 7.0 | 5.4 | 12.2 | 9.5 | 10.1 | 11.6 | 11.1 | 2.7 | 2.6 |
| Poland | 9.6 | 22.6 | 14.8 | 4.6 | 4.7 | 15.0 | 19.7 | 13.6 | 8.6 | 8.9 | 9.3 | 7.5 |
| Romania | 11.1 | 6.6 | 7.6 | 2.0 | 1.2 | 12.5 | 9.0 | 9.6 | 5.8 | 4.8 | 3.8 | 5.5 |
| Total six | 49.2 | 48.7 | 45.9 | 30.3 | 28.1 | 63.1 | 54.3 | 49.6 | 41.4 | 41.9 | 28.8 | 28.5 |
| Soviet Union | 50.8 | 51.3 | 54.1 | 69.7 | 71.9 | 36.9 | 45.7 | 50.4 | 58.6 | 58.6 | 71.2 | 71.5 |
| Total seven | 100.0 | 100.0 | 100.0 | 100.0 | 100.0 | 100.0 | 100.0 | 100.0 | 100.0 | 100.0 | 100.0 | 100.0 |

Source: Trade data: OECD East-West technology transfer data base (Western statistics). NMP weights are those used by the Economic Commission for Europe.

Table 8 (Chapter 5)

WESTERN TECHNOLOGY IMPORTS FROM THE EAST AND OUTPUT (NMP): SHARES OF EASTERN COUNTRIES IN RESPECTIVE SEVEN-COUNTRY TOTALS, 1970, 1975, 1980, 1982 and 1983 (per cent)

| | Western capital goods imports | | | | | Western intermediate (technology) goods imports | | | | | NMP | |
|---|---|---|---|---|---|---|---|---|---|---|---|---|
| | 1970 | 1975 | 1980 | 1982 | 1983 | 1970 | 1975 | 1980 | 1982 | 1983 | 1970 | 1982 |
| Bulgaria | 4.1 | 3.0 | 4.4 | 4.1 | 3.3 | 4.3 | 4.0 | 3.2 | 4.1 | 3.7 | 1.9 | 2.3 |
| Czechoslovakia | 28.4 | 18.6 | 17.3 | 17.9 | 19.8 | 15.4 | 14.8 | 12.1 | 13.9 | 13.4 | 5.0 | 4.6 |
| German Democratic Republic | 21.4 | 16.9 | 17.5 | 21.4 | 20.2 | 11.3 | 11.5 | 9.2 | 13.6 | 13.9 | 6.1 | 6.0 |
| Hungary | 6.2 | 6.7 | 10.5 | 13.0 | 11.6 | 8.2 | 8.6 | 9.5 | 10.0 | 9.6 | 2.7 | 2.6 |
| Poland | 13.0 | 23.6 | 22.2 | 18.8 | 20.6 | 14.4 | 16.8 | 18.7 | 13.3 | 11.6 | 9.3 | 7.5 |
| Romania | 6.9 | 8.7 | 10.3 | 11.9 | 10.8 | 9.3 | 11.0 | 9.6 | 9.7 | 10.2 | 3.8 | 5.5 |
| Total six | 80.0 | 77.5 | 82.2 | 87.1 | 86.3 | 63.0 | 66.7 | 62.3 | 64.6 | 62.4 | 28.8 | 28.5 |
| Soviet Union | 20.0 | 22.5 | 17.8 | 12.9 | 13.7 | 37.0 | 33.3 | 37.7 | 35.4 | 37.6 | 71.2 | 71.5 |
| Total seven | 100.0 | 100.0 | 100.0 | 100.0 | 100.0 | 100.0 | 100.0 | 100.0 | 100.0 | 100.0 | 100.0 | 100.0 |

Source: Trade data: OECD East-West technology transfer data base (Western statistics). NMP weights are those used by the Economic Commission for Europe.

Table 9 (Chapter 5)

COMPOSITION OF CUMULATIVE OECD SURPLUSES ON EAST-WEST TRADE BY SELECTED COMMODITY GROUPS 1971-1983
($ billion)

| | 1971-1975 | 1976-1980 | Total | 1981-1983 |
|---|---|---|---|---|
| Soviet Union | | | | |
| Capital goods | 10.5 | 29.0 | 39.5 | 19.8 |
| Intermediate (technology) goods | 7.0 | 17.4 | 24.4 | 13.5 |
| Fuel | -10.9 | -44.2 | -55.1 | -52.0 |
| Other industrial goods | -2.3 | -4.1 | -6.5 | 3.9 |
| Total industrial goods | 4.3 | -1.9 | 2.4 | -14.7 |
| Agricultural goods | 0.9 | 7.4 | 8.3 | 9.1 |
| Total | 5.2 | 5.5 | 10.7 | -5.7 |
| Eastern Europe | | | | |
| Capital goods | 10.7 | 17.4 | 28.1 | 6.2 |
| Intermediate (technology) goods | 9.3 | 15.0 | 24.3 | 4.2 |
| Fuel | -4.4 | -12.3 | -16.7 | -9.8 |
| Other industrial goods | -5.3 | -9.8 | -15.1 | -4.9 |
| Total industrial goods | 10.3 | 10.3 | 20.6 | -4.3 |
| Agricultural goods | -0.1 | 5.5 | 5.4 | 3.0 |
| Total | 10.2 | 15.8 | 26.0 | -1.2 |
| Total, Seven Eastern Countries | | | | |
| Capital goods | 21.2 | 46.4 | 67.6 | 26.1 |
| Intermediate (technology) goods | 16.3 | 32.5 | 48.7 | 17.7 |
| Fuel | -15.2 | -56.5 | -71.8 | -61.8 |
| Other industrial goods | -7.6 | -14.0 | -21.6 | -1.0 |
| Total industrial goods | 14.6 | 8.4 | 23.0 | -19.0 |
| Agricultural goods | 0.8 | 12.9 | 13.7 | 12.1 |
| Total | 15.4 | 21.3 | 36.7 | -6.9 |

Note: Symbol "-" = Western deficit. The Table is based on Western exports f.o.b. and Western imports c.i.f.; hence Western surpluses tend to be understated and deficits overstated.

Source: OECD East-West technology transfer data base (Western statistics).

Table 10 (Chapter 5)

SHARE OF OECD COUNTRY GROUPINGS IN OECD EXPORTS
OF EMBODIED TECHNOLOGY TO THE EAST
(per cent)

| | Capital goods | | | | Intermediate (technology) goods | | | |
|---|---|---|---|---|---|---|---|---|
| | 1970 | 1975 | 1982 | 1983 | 1970 | 1975 | 1982 | 1983 |
| Eastern Europe | | | | | | | | |
| EEC 9 | 73.0 | 68.9 | 62.2 | 58.5 | 67.2 | 69.9 | 68.0 | 67.6 |
| N. Europe (a) | 22.8 | 20.2 | 25.1 | 23.8 | 22.4 | 20.7 | 22.8 | 22.6 |
| Japan | 1.8 | 5.7 | 8.3 | 13.9 | 5.7 | 5.3 | 4.2 | 5.2 |
| N. American | 2.1 | 4.3 | 3.4 | 3.0 | 2.0 | 1.7 | 2.5 | 2.3 |
| S. Europe (b) | 0.2 | 0.8 | 0.9 | 0.7 | 2.8 | 2.3 | 2.5 | 2.5 |
| Oceania (c) | - | - | - | - | - | - | - | - |
| Total | 100.0 | 100.0 | 100.0 | 100.0 | 100.0 | 100.0 | 100.0 | 100.0 |
| Soviet Union | | | | | | | | |
| EEC 9 | 62.5 | 57.6 | 38.6 | 50.7 | 55.5 | 62.7 | 52.9 | 56.5 |
| N. Europe (a) | 23.6 | 16.4 | 26.0 | 27.7 | 23.8 | 16.3 | 19.1 | 17.3 |
| Japan | 10.0 | 14.7 | 32.2 | 19.0 | 15.8 | 17.7 | 19.1 | 17.3 |
| N. America | 3.9 | 11.2 | 2.6 | 1.9 | 4.9 | 2.5 | 6.6 | 5.8 |
| S. Europe | - | 0.1 | 0.5 | 0.9 | - | 0.8 | 2.3 | 3.1 |
| Oceania (c) | - | - | - | - | - | 0.1 | - | - |
| Total | 100.0 | 100.0 | 100.0 | 100.0 | 100.0 | 100.0 | 100.0 | 100.0 |

Notes: a) Iceland, Norway, Switzerland, Finland, Austria and Sweden.
b) Greece, Portugal, Spain and Turkey.
c) Australia and New Zealand.

Source: OECD East-West technology transfer data base (Western statistics).

## NOTES AND REFERENCES

1. *Statistical Papers*, Series M, No.53, United Nations, New York, 1971.
2. Regina Kelly, "Alternative Measurements of Technology-Intensive Trade", United States Department of Commerce, Office of Economic Research, Paper OER/ER-17, Washington, D.C., September 1976. The listings therein were slightly expanded – notably to cover all metals.
3. The BEC intermediate goods categories from which this selection was made are 111: primary food and beverages mainly for industry; 121: processed food and beverages mainly for industry; 2: industrial supplies not elsewhere specified; 31: primary fuels and lubricants; 322: other fuels and lubricants; 42: parts and accessories of machinery and other capital equipment (except transport) and 53: parts and accessories of transport equipment.
4. The Standard Foreign Trade Classification (SFTC) of the Council for Mutual Economic Assistance was supposed to be introduced in all CMEA member-countries from January 1971. In fact only three countries (Bulgaria, Romania and the Soviet Union) did so. Hungary, Poland and the German Democratic Republic follow their own classification systems – which have, however, many points in common with SFTC – while Czechoslovakia publishes trade data according to SITC.
5. However, Eastern-reported data do include some purchases of know-how and licensing fees in cases where such costs are included in the price of plant bought on a package deal basis. This cannot be quantified [see section 2*a)*].
6. Paul Marer, "Towards a Solution of the Mirror Statistics Puzzle in East-West Commerce", in Friedrich Levcik (ed.), *Studien über Wirtschaftsvergleiche*, Springer Verlag, Vienna/New York, 1978.
7. Except for Romania, where insufficient geographical or commodity detail is available.
8. For the seven countries taken together, SITC item 719.8 alone (miscellaneous non-electrical machinery and equipment) accounted for 52 per cent of the total of SITC items not allocated by branch in 1970, falling to 46 per cent in 1975, 38 per cent in 1980 and 28 per cent in 1983.
9. The share of embodied technology in total Western exports to the East peaked in Eastern Europe in 1975 and in 1977 in the Soviet Union and has declined steadily up to 1981 (except for 1977 in Eastern Europe).
10. Adjusted to take account of the fact that Western imports are expressed *c.i.f.* in the Table, the cumulated trade deficits in 1971-1980 would total about $50 billion, out of a net indebtedness to the West of $68 billion at the end of 1980 (Table 8 of Chapter 1). See also *Economic Survey of Europe in 1980*, *op. cit.*, p. 182. The difference between the total debt and accumulated trade deficits represents the contribution of total debt service, net negative invisible transactions – the latter rising from small amounts in 1971-1975 (cumulative total estimated at about $2 billion) to much larger sums in 1976-1980 (cumulative total of over $10 billion) and, changes in reserves and residual debt built up during the 1960s, offset by gold sales and net transactions with the developing countries.

*Chapter 6*

# EASTERN COUNTRIES' TOTAL TRADE IN EMBODIED TECHNOLOGY

As described in Chapter 5 (section 1), the OECD East-West technology transfer statistics include data based on the Eastern countries' own trade returns. These cover trade in capital goods between the Eastern countries themselves and trade with the Rest of the World. In section, 1 below, the role of this trade is set in the context of transactions reported by the Eastern countries themselves by region of the world and in relation to their total trade. Section 2 discusses the principal characteristics of intra-Eastern area trade in capital goods. In section 3, a number of topics are examined with the aim of illuminating the Eastern countries' recent strategies towards embodied technology imports, notably capital goods. Section 2 is based on both Eastern and OECD-reported statistics.

## 1. CMEA Capital Goods Trade by Region

### a) *Eastern Imports of Capital Goods*

As shown in Table 3 of Chapter 5, purchases of capital goods from the world by the Eastern countries totalled $57 billion in 1983. The deceleration in the growth of these imports during the period after 1975 mainly reflected the slowdown, followed by an absolute decline, in the growth of imports from Western countries[1]. However, this deceleration gathered pace in Eastern Europe in the early 1980s as the fairly high rates of growth previously maintained in capital goods imports from within the Eastern area also slowed down – the level of such imports by Eastern Europe as a whole even declining in absolute terms between 1980 and 1983. The rise registered between 1980 and 1983 for the region as a whole owes most to Soviet imports from Eastern Europe, though Soviet imports from the West also increased quite strongly.

Eastern imports of capital goods identified as originating within the CMEA region accounted for over half the total in 1970, declining to about 40 per cent in 1975, the year in which the West's identified share reached its highest level (nearly 40 per cent)[2]. In 1983 the CMEA share was over half. Imports from Eastern partners took a particularly big share of the total for Bulgaria and the Soviet Union up to the early 1970s (around 60 per cent). However, even in 1975, only in Poland and Romania did the Eastern share fall significantly below 40 per cent. By the early 1980s the distribution by region of origin had, in any case, reverted to the earlier 1970 pattern – with the notable exception of the Soviet Union where the Eastern share of capital goods imports was just over half (in 1983) compared with 60 per cent in the early 1970s.

These movements generally mirror the movement in the share of Western capital goods in the total (the third element being the "unidentified" item – see below). The rapid build-up over 1970-1975 described in Chapter 5 increased the identified Western ("Rest of the World") share in total Eastern capital goods imports from just under a quarter in the former year to nearly 40 per cent in the latter. Though the West's share in the total was still low in some countries (Bulgaria, Czechoslovakia and the German Democratic Republic), in others (Poland, Romania and the Soviet Union) it attained virtually half the total. By 1980, the previously rising trend of the Western share over 1970-1975 had been largely reversed, and by 1983, completely so in Eastern Europe where it halved. In the Soviet case, however, the change was less abrupt and by 1983, the Western share had fallen by about a fifth from the peak years of the mid-1970s (to just under 40 per cent).

The generally smaller share of the West in most countries and in most years except in the mid-1970s needs to be qualified in certain respects. The high share of the "unidentified" item in all years (in Eastern Europe, it was well over one-third of the total at the beginning and end of the period, and about one-quarter in the mid-1970s, but, due to its much smaller size in the Soviet Union, it was 10-15 points less for the region as a whole) leaves a considerable margin for possibly substantial revision to the percentage share estimates of Eastern relative dependence on Western and other Eastern partners[3]. However, as can be seen from the detail given in note 3, adjustments taking into account the OECD-reported data do not, in fact, generally affect the *trends* described. The West's share would rise, for the Eastern region as a whole, by 3-4 percentage points in each year shown, the share of the unidentified item falling by the same amount. The implication is that the bulk of the unidentified import item consists of unreported imports within the Eastern area, plus perhaps some from non-OECD countries included in the Rest of the World.

For Czechoslovakia since 1975, and for the Soviet Union virtually throughout the period, the apparent over-reporting of imports of Western capital goods in the two countries' own trade figures gives rise to a contrary problem. The figures suggest strongly that commodities registered as imports from the West by Czechoslovak and Soviet data, are not so registered in the corresponding Western export statistics. How far this discrepancy consists of re-exports of Western goods to these two countries by countries outside the OECD area, but reported by the two countries in terms of imports from the Western exporter of origin, is not possible to state. Part of such "middleman trade" may also consist of re-exports by other East European countries to Czechoslovakia and the Soviet Union. However, the combined under-reporting of the five East European countries excluding Czechoslovakia amounted to only $0.7-0.8 billion in each of the years 1980-1983, compared with a much bigger $2 billion over-reporting by Czechoslovakia and the Soviet Union in the same years. Hence, if the Soviet and Czech discrepancies do reflect "middleman" trade, the latter appears to have been conducted largely via countries outside Eastern Europe.

A final reservation concerns the reliability of the dollar valuation of Eastern trade figures. It will be recalled that the figures presented are based on Eastern countries' statistics, revalued into dollars. Eastern currencies are not convertible – not least with regard to other Eastern currencies – and published exchange rates are used for accounting but not for multilateral clearing purposes. Indeed, Eastern countries' exchange rates *vis à vis* the ruble are often not reflected in cross-rates in terms of other Eastern currencies or with the dollar (or other Western currencies)[4]. In addition, as noted in Chapter 5 (section 1), intra-Eastern area transaction prices of such non-homogenous commodities as investment goods in general, although theoretically based on average world market prices over the five years preceding the transaction year, may in practice bear little relationship to their price on world markets even assuming that the items traded within the Eastern area and world markets are similar.

Moreover, the introduction after 1975 of the use of the CMEA's five-year sliding average system means that the lags after that year are different from the previous ones. Prices were formerly based on world market prices at the start of the corresponding plan period (i.e. in prices of about 1970 for the years 1971-1975). Before 1976, therefore, the dollar 'value' of intra-area exchanges should have corresponded roughly with their volume in 1970 dollar prices – which means that their subsequent levels and growth rates are probably understated in the Eastern-reported 'value' figures. Between 1975 and 1980, the 'lagged' system of intra-CMEA pricing may have interacted with the continuing but decelerating rate of change in world market and East-West prices to produce a growing degree of under-pricing of goods traded between Eastern countries. Thus, the observed fall in the share of Western capital goods in total imports of this kind may have been bigger than those shown in Tables 1-5 of this chapter. Since 1980, world market prices, and thus, probably East-West capital goods trade prices, have fallen. Hence, after 1980 the real share of Western goods may, conversely, have declined by less (or, in the case of the Soviet Union, increased by more) relative to the totals shown. It is not known how far these technicalities affect the trend of development described.

Leaving aside possible discrepancies of this nature, the Eastern data show that the already high share of the Soviet Union in the area's total capital goods imports from the world during the 1970s rose further into the 1980s – from just under to well over one-half (Table 1 of Chapter 6). This reflects both the absolute decline in East European purchases and also the very large jump in Soviet imports of capital goods from the West in 1982. The Soviet upsurge gave rise to an abrupt break in the Eastern country pattern of import demand for Western capital goods. Before 1982, Eastern Europe as a whole took more than half of the seven countries' imports of Western capital goods – very large in relation to its contribution of roughly one-quarter of the seven countries' combined NMP. But the rise in the Soviet share of the region's imports of Western capital goods in 1982 reduced Eastern Europe's share to one-quarter. As seen in section 2 below, the Soviet import share in intra-Eastern area exchanges also rose quite substantially. Eastern Europe's import share of intra-area exchanges consequently also fell. The movements in West-to-East and intra-Eastern area trade reflect the combined effects of Eastern Europe's increasingly energetic attempts to solve its Western debt repayment problems on one hand, and also the need for Eastern Europe to export more to (and import less from) the Soviet Union in order to offset the growth of their fuel deficits with that country.

As in East-West capital goods trade, the East European countries have seen a fall in the share of capital goods imports in total trade since 1975 (Table 2 of Chapter 6). Although the fall was marginal in Eastern Europe's intra-Eastern area exchanges up to 1980, it gathered pace after that year. Recalling that the fall in the share of capital goods in total imports from the West was due probably to the maintenance at high levels of intermediate (technology) imports, it is of interest that, up to 1980 at least, this does not seem to have been the case in intra-area trade[5]. However, by 1983, the downward trend of capital goods in East European imports as a whole extended to intra-Eastern area trade also. The movement owed most to Poland, but occurred in all countries except the Soviet Union. The latter country, in contrast, considerably expanded its imports of capital goods from Eastern Europe. Weak investment performance, except in the Soviet Union and Bulgaria, continues to exert a depressing influence on imports of capital goods from all areas in a majority of the East European countries.

The trade figures in fact show that a series of very large cutbacks in absolute terms was imposed on Eastern imports of Western capital goods, beginning in about 1978. But rapid growth (averaging some 12 per cent annually in value terms) did take place in intra-Eastern area capital goods trade between 1975-1980. This more than offset the steep fall in purchases from the OECD countries. At global level, therefore, efforts were made for a time to substitute Eastern for Western-produced capital goods.

The large apparent upturn in Soviet (and Bulgarian) investment which accompanied the faster growth of imports of Western capital goods in 1982, suggests that export prospects for Western capital goods on Eastern markets are determined largely by the trend of investment, as well as hard currency availability. The Soviet Union has been the best placed of all the seven countries, given its ability to increase fuel exports for hard currency. Its future ability to do so, and the prospects for the East European countries to increase sales of their principal export commodities, will be one determining factor in the future development of East-West technology trade. But unless this is accompanied by favourable, or at least neutral, terms of trade developments which would end unrequited transfers of real resources abroad, investment restrictions will continue to limit the prospects for an upturn in Western capital good exports to this market. The point is taken up again in section 2 *b*) of this chapter.

### b) *Eastern Capital Goods Exports*

Eastern exports of capital goods in total, within the Eastern area and to the Rest of the World contained in the OECD's Eastern-reported data bases do not coincide with the coverage of the corresponding import statistics in one important respect. Imports from the Rest of the World do, in practice (even taking into account the discrepancies noted in Chapter 5, section 2), serve to indicate West-to-East flows, since no country outside the OECD or the Eastern area itself is an important capital goods exporter. The data on Eastern exports of capital goods to the Rest of the World cover exports to the same geographical areas, but the developing countries (as well as other socialist countries outside the European CMEA area) are substantial markets for Eastern capital goods. Thus, no correspondence can be expected, as was the case for Eastern imports (Table 1 of Chapter 5), between exports by the Eastern countries to the Rest of the World, as derived from their own trade statistics, and the OECD countries' import statistics. No independent check on the Eastern-reported statistics underlying this flow is, therefore, possible.

As in the case of Eastern imports, a substantial residual item remains unidentified by region of destination. It was rather smaller than for imports (about a quarter of the total) throughout the period 1970-1980, but it rose to around one-third by 1983 (Table 3 of this chapter). The close correspondence of intra-Eastern area imports and exports (Table 5 of Chapter 5) suggests the working assumption, retained below, that most of this residual consists of trade with the Rest of the World although it could also mean that a proportion of intra-Eastern area trade in capital goods is left unreported in sector/branch detail both in the corresponding individual country import and export data.

In order to improve the data shown in Table 3 on the principal regional destinations of Eastern exports of capital goods, OECD-reported data have been compared with Eastern-reported statistics on exports to the Rest of the World – i.e. to non-European CMEA countries (Table 4 of this chapter). Differences between them should indicate the approximate order of magnitude of capital goods to non-European CMEA, non-OECD countries, i.e. mainly the developing countries (Table 4 of Chapter 6). It should be stressed that these orders of magnitude are only approximations[6].

In 1983, the Eastern countries as a whole exported some $51 billion of capital goods – not far below their total imports of these items in the same year of about $52 billion (Table 3 of this chapter and Table 3 of Chapter 5). As already discussed in Chapter 5, large trade deficits on embodied technology, including capital goods flows, were recorded in East-West trade during the 1970s and, though at a much more moderate rate, into the 1980s. However, they were increasingly offset by compensating growth of exports to the developing countries. Such exports, as evidenced by Table 4 of this chapter, apparently more than doubled between 1975 and 1980, and continued to increase fast even thereafter. A similar, if not quite so buoyant, development was observed in the growth of the "unidentified" item – which may contain a considerable element of arms shipments to the developing countries[7].

The country pattern of Eastern exports confirms, as for imports up to 1981, the low share of the Soviet Union in the total relative to its size (Table 5 of this chapter). Though it has traditionally been the biggest single supplier to Eastern Europe (as will be seen in section 2 of this chapter), the Soviet Union accounted for only a fifth of total CMEA exports of capital goods to the world in 1983. Its contribution to the CMEA's aggregate NMP was about three-quarters. However, despite the low Soviet share in exports to the West it remained the biggest Eastern supplier to the Rest of the World, supplying nearly half of total Eastern capital goods sales to the developing countries. Perhaps surprisingly, the Soviet Union accounted for only about 10 per cent of the unidentified items in 1983. The other big suppliers to the world as a whole were the German Democratic Republic and Czechoslovakia. The former country is also a particularly important supplier of capital goods to the CMEA countries. (This offsets the low share shown for its exports to the Rest of the World, which also reflects to a large extent data shortcomings[8].) The data show that Poland and Czechoslovakia together supply the bulk (about half) of the seven countries' exports of capital goods to the Rest of the World and roughly the same share to the developing countries.

Given their small size and level of development, the apparently high value of exports of capital goods by Bulgaria and Romania to unidentified destinations are also worthy of comment. This is due to the limited commodity detail reported by these countries and the consequently high "unidentified" item in their total trade – even though their data have been supplemented by Eastern and OECD partner countries' mirror statistics. This is also true for the German Democratic Republic. It seems likely that these three countries' capital goods exports to the Rest of the World and to the developing countries are understated in Tables 3-5 of this chapter, whereas a substantial part of the "high unidentified" item does, in reality, represent shipments not only to OECD and to other CMEA countries, but also to the developing countries.

Despite this source of statistical error which affects the Eastern area totals, the importance of the developing countries in the Eastern countries' capital goods exports is indicated from the fact that they appear to have consistently accounted for a large and rising share of the seven countries' total identified exports to non-CMEA destinations (79 per cent in 1975, and 90 per cent in 1983, according to the rough estimates given in Table 4 of this chapter). They also accounted for not far short of one-fifth of the East's total capital good sales to the world. This rises to nearly one-half if the "unidentified" item is added. Thus the developing-country market's importance has grown while exports to the OECD countries slowed down. Clearly, part of the impact of imported Western technology on the Eastern countries' export potential could thus have manifested itself in East-South trade in capital goods.

The growth of the Eastern region's exports of capital goods to all regions was relatively high in 1980-1983 (about 4-5 per cent annually in current dollar terms) but was much less buoyant than growth in 1970-1975 and, to a lesser extent, in 1975-1980. The region's capital goods fared particularly badly in Western countries due to lack of competitivity. Redressment measures to reduce debt adversely affected the Eastern countries' two principal markets: the Eastern countries themselves and also the developing countries. As a result, the share of investment goods in exports as a whole fell between 1975-1980 (though not in intra-Eastern area trade) and again into the 1980s in all (including unidentified) regions (Table 6 of this chapter). Present prospects do not indicate any great improvements in the immediate future. Higher growth in the West from 1983/1984 raised investment demand in that region, but the decline in the competitivity of Eastern capital goods in Western markets apparent since 1975, has not been reversed (see section 1 *c*) of Chapter 8 below). Continuing low investment is likely in the debt-encumbered developing countries. This, and slow investment growth within the Eastern region, have not provided a big stimulus to Eastern capital goods exports within the context of the 1981-1985 Five-Year Plans.

## 2. Intra-CMEA Trade in Capital Goods

The value of intra-Eastern area trade in capital goods amounted to about $25 billion in 1980 and $29 billion in 1983 – compared with total capital goods imports of about $57 billion from all suppliers in the latter year (Table 3 of Chapter 5). This compares with about $15 billion from the West since 1980. As noted in the previous section, intra-Eastern area exchanges grew substantially faster than Eastern imports of capital goods from the West during 1975-1980 – by over 80 per cent, compared with 20 per cent in value terms respectively. This was an abrupt reversal of the relationship between the growth of the two flows in the early 1970s. As noted above, between 1980 and 1983 there was also a marked slowdown in intra-Eastern area exchanges; in volume terms, the slowdown is likely to have been even more apparent. This largely reflected absolute reductions in Polish and Romanian imports of capital goods. These were, however, offset by big increases in Soviet imports of capital goods from Eastern Europe. In the case of Poland, imports from the Soviet Union fell by about a half and Romanian imports by a quarter, both countries reporting a somewhat smaller decline in imports from Eastern Europe. Even so, the other four countries of Eastern Europe registered only minor increases in their imports from CMEA partners.

Soviet capital goods imports from Eastern Europe rose by just over 14 per cent between 1980 and 1983. The rise in its imports from the CMEA area was marginal in 1981-1982 compared with the over 40 per cent rise in its imports from the OECD countries. Nevertheless, it marks a further step in the past shift away from mutual trade between the six countries of Eastern Europe in favour of increased supplies to satisfy Soviet needs. The Soviet Union now takes two-thirds of all capital goods exported by Eastern Europe to other Eastern countries.

The matrices shown in Table 7 of this chapter indicate the value of each country's imports from all other countries of the Eastern region, the right-hand vertical column showing the total imports of each Eastern country and the intra-area import totals, while the bottom line shows each country's exports and the intra-area export totals. As this Table shows, 68 per cent of East European exports to other Eastern countries were sold to the Soviet Union in 1983 (nearly $17 billion out of $25 billion). This is substantially higher than in 1975 and 1980 (59 and 62 per cent respectively – Table 8). The diversion of capital goods exports from other East European partners to the Soviet Union is a clear reflection of Eastern Europe's high fuel and raw material bills with the Soviet Union. Conversely Eastern Europe takes a smaller share of a smaller absolute level of Eastern-produced capital goods from the Soviet Union.

The large size of the Soviet market for Eastern Europe's capital goods relative even to its own substantial exports to the six countries is clearly brought out in the country ranking of suppliers in the total of intra-Eastern area exchanges shown in Table 7. The German Democratic Republic takes first place followed by Czechoslovakia, which has now replaced the Soviet Union as the second biggest supplier of capital goods within the area as a whole. In fact, the value of Soviet exports to other Eastern countries in 1983 was only two-thirds that of the German Democratic Republic and 85 per cent of that of Czechoslovakia. These countries together alone supplied over half of both Soviet and Eastern European purchases from within the region in that year.

By 1983, Poland had been replaced as the third biggest East European supplier to Eastern Europe – as well as to the Soviet Union – by Czechoslovakia. The Polish economic and balance of payments crisis is the special and well-known reason for this. Of more interest is the performance of the other countries of the area. Bulgaria appears to have lost some of its share of East European capital goods supplies to the Soviet Union, while Romania's contribution to both East European and Soviet imports of Eastern origin also declined. In general, the process of intra-Eastern regional integration appears to have favoured Czechoslovakia (in contrast to its performance on Western markets) and Hungary, where shares in both East European and the Soviet markets increased between 1975 and 1983.

The Soviet Union is the biggest single Eastern capital goods exporter to Eastern Europe. In sharp contrast to its low share in Eastern exports to the West and to the Rest of the World, the Soviet Union, despite a flagging East European market for investment goods, supplied over one-third of Eastern Europe's total imports from other countries of the area in 1983 – $4 billion out of $12 billion. Despite this, the Soviet Union still buys over four times as many capital goods from Eastern Europe as it sells to the six countries (in 1983, some $17 billion compared with $4 billion).

To summarize the general trend of intra-Eastern area trade: a large rise in exports of capital goods to the Soviet Union by the East European countries after 1980 was accompanied by absolute declines in East European capital goods imports from each other and from the Soviet Union. They affected mainly the six countries' imports from the Soviet Union, but the absolute level of the six countries' mutual trade in these items also declined.

This change in the trend reflects first and foremost the East European countries' overall resource constraints. The consequent cutback in import growth has fallen mainly on imports of capital goods relative to other classes of imports. Thus, the same trend noted for 1975-1980 with regard to imports from the West (Chapter 5) subsequently appeared, with some lag, also in intra-CMEA trade.

Soviet imports of capital goods from both East and West showed a substantial upturn by 1982. The contrast between Eastern Europe and the Soviet Union is very striking. Soviet investment in the early 1980s grew at higher rates than expected in the Five Year Plan, and much higher than in Eastern Europe. However, the 10 per cent rise in its capital goods imports from Eastern Europe is relatively modest compared with the rise in its imports from the West. How far this reflects relative Soviet preference for Western rather than Eastern goods in general cannot be verified, given the rather narrow range of capital goods on which the increment in imports from the West was concentrated (pipeline, energy equipment, etc). Certainly, however, a substantial surplus on trade with the West in 1981-1983 left the Soviet Union ample room for choice. In contrast, supply constraints in Eastern Europe (notably the persistence of problems of producing equipment attractive to the now more demanding Soviet market without sacrificing possible hard currency earnings) may have helped to limit the growth of the six countries' sales to the Soviet Union.

Sector and branch patterns of intra-Eastern area exchanges of capital goods will be dealt with in more detail in the context of strategies determining technology transfers (sub-section *d)* below). However, it is of interest to note here that for both Eastern Europe and the Soviet Union, nearly half of capital goods supplies traded within the Eastern area in 1982 consisted of transport equipment and machinery for the fuel and energy industries. It was somewhat lower – about one-third – in the case of Romania (Table 5 of Chapter 5). In general, the above shares taken together have remained constant since 1975, a decline in the share of transport items in a few countries having been offset by higher fuel and energy equipment imports. Thus, an important segment of total intra-Eastern area capital goods trade has been preempted on a continuous basis by a rather narrow range of basic capital goods for the fuel-energy branches and for transportation. Hence, despite the moves to expand economic integration initiated at the beginning of the 1970s (Chapter 4 above), the end-use pattern of capital goods imports in this respect has remained essentially unchanged. Put another way, past moves to increase further the degree of integration (however defined) can have affected, at most, only just over one-half of intra-Eastern area capital goods trade (i.e. those other than equipment for fuel/energy and transport). It may be, of course, that the attainment of integration is best achieved by a continuation of past intra-area trading patterns. But the size and stability of this narrow range of items within a general slowdown of capital goods exchanges is certainly conclusive evidence against the proposition that any fundamental changes have taken place, of a kind which are officially considered necessary to bring about longer-term changes in intra-Eastern area economic relationships. This conclusion is of some significance. Expanded growth of exchanges of the means of production, and in particular expansion of the commodity range of such goods, could be taken as one of the more important objectives of CMEA programmes introduced during the 1970s. These programmes to promote exchanges of know-how, joint research, production specialization in the output of standard machinery types and the like have been the subject of much fanfare in Eastern-country and CMEA policy statements over the last ten years. While further analysis of these questions is beyond the scope of this sub-section, the data presented do provide factual confirmation of the limited progress actually made, as noted in Part I, Chapter 4 of this study.

## 3. Strategy of Eastern Technology Acquisitions

Technology trade, as one aspect of international trade, is subject to interpretation within the framework of conventional trade theory based on comparative advantage. However, a number of special circumstances are present in the case of Eastern countries' trade in general, and with regard to technology trade in particular. They are of both a technical and socio-political nature. In the first place, the lack of price formation systems reflecting real relative scarcities (real supply/demand relationships) in the Eastern countries hampers rational resource allocation in general, and also the rational allocation of total resources to foreign trade as a whole, to expenditures on different classes of import, and finally, within the technology import complex, on the priority accorded to imports for different end-use sectors and branches. This "price veil" also precludes much *ex post* analysis of import flows which have actually occurred. To carry out tests of the economic rationality of such flows would involve re-pricing of sectoral/branch output, investment and labour remuneration series. In the second place, the essence of a planned economy is that short-run criteria of economic management can sometimes be ignored in favour of longer-term strategies and priorities (which may or may not be justified in the light of actual developments)[9]. Also, socio-economic priorities in force at various times (for example overall policy stances towards private as against public consumption, or towards consumption as a whole as against investment) lead to decisions which may or may not appear rational within a given time frame but which have economic consequences which may need to be judged against a longer-term perspective[10].

In order to reduce the evaluation of technology transfer strategies of the Eastern countries to manageable proportions, the present part of the study aims to approach the subject from a number of different directions, basing conclusions on the observed facts of technology transfer to and within the Eastern countries. The rest of this section hence analyses a number of areas which shed light – albeit in some cases rather obliquely – on various aspects of Eastern countries' acquisition strategies. In sub-section *a)* the distribution of Eastern technology import according to its level of sophistication is discussed. In sub-section *b)*, these flows are examined in more detail by type – that is according to the categories of machinery and equipment actually imported. Sub-section *c)* attempts to set changes in the magnitude of Eastern imports of Western capital goods in context by examining changes in the share of imported capital goods in total investment. Finally, the sector and branch end-use destinations of imports from the OECD and from other CMEA countries are first described and compared, and an attempt is made on this basis to assess the underlying features of technology import strategies vis à vis exports on one hand and production for domestic consumption on the other [sub-section *d)*].

It should again be emphasized that this discussion on strategies takes as its starting point the observed facts of technology transfer. Little discussion of technology acquisition has been published in the Eastern countries themselves, and high-level policy statements tend to ignore the subject except for pronouncements in very general terms. In the absence of clear statements of macro-economic strategy towards the acquisition of non-CMEA technology, details can only be guessed at, and, success or failure in those terms cannot be accurately judged. Moreover, it should be borne in mind that the "strategies" revealed by the above approach may well have been modified during the period under review in the light of actual outcomes – since such outcomes may have been different from what was originally expected. Nevertheless, the identification of actual changes in the pattern and trend of technology trade relative to various domestic and overall external trade indicators provides a framework within which the impact of the technology imports which actually took place is assessed in Chapter 8 below.

### a) *Level of Sophistication of Eastern Imports of Western Technology*

As noted in the Preface to this study, a major problem in approaching an assessment of embodied technology tranfers is how to define technology. Should the definition be limited to equipment for producing high technology products (in which case, we need to define "high technology" also)? Should the definition be extended to cover high technology products themselves or – as mostly in this study – should we narrow the definition to cover transfers of productive capacity (capital goods) only?

Given the high share of embodied technology in total East-West trade, which results from the broad definition (i.e. including intermediate (technology) goods) given in section 1 of Chapter 5, it is of interest first to consider one of the narrower definitions available. One such study classifies 28 items (distributed among SITC items 7 and 8 at SITC 5-digit level) as "high technology"[11]. The criterion retained is that they "are likely to contain products embodying world best practice in critical technologies". These items were found to account for about 9 per cent of the exports of 17 industrialized Western countries to Eastern Europe and the Soviet Union in 1981 – somewhat lower than in 1970 (14 per cent). The above study suggests first, that the shares of high technology items in industrialized Western countries' exports to communist countries "do not differ markedly from the respective shares in industrialized Western exports to the world", and also that, referring to the period up to 1978, "while the total volume of industrialized Western exports to the world as a whole and to communist countries increased dramatically in the 1970s, there was no discernible trend suggesting that industrialized Western high-technology products exports to communist countries were increasing as a portion of total industrialized Western exports to communist countries. More recently the volume of industrialized Western exports to Eastern Europe and the Soviet Union has declined".

This provides a useful summary foreword to the paragraphs which follow. These attempt to supplement the description of global embodied technology (capital and selected intermediate goods) flows from West to East retained for this study by differentiating embodied technology imported by the Eastern countries from the West according to the R&D-intensity of the producing branch. The classification applies mainly to capital and intermediate goods, but some consumer goods and passenger motor vehicles are also included. The aggregated data were broken down into three groups: flows from industrial branches classified as high- medium- and low-R&D intensive respectively (Table 9 of this chapter)[12].

High R&D-intensive items as defined in the source noted above accounted for 2.7 per cent of total exports of Western embodied technology to the Eastern countries in 1983, and only 1.7 per cent of exports of all goods. Thus, the OECD definition of "high R&D-intensive items" is considerably narrower than "high-technology" as defined in the US study quoted earlier. The share of moderately R&D-intensive goods was much higher at one-fifth of the total, whereas the bulk of OECD embodied technology exports as defined in Chapter 5, section 1 above is classed as low R&D-intensive – over three-quarters. Thus, as far as can be ascertained from the trade figures, embodied technology exports of the OECD to the Eastern countries were only to a small extent products of the most R&D-intensive industries in 1983. Overall, the broad pattern of distribution of technology imported by Eastern countries from the West by level of sophistication in 1983 was broadly similar throughout the 1970s for the seven countries taken together. There was, however, some evolution over time and some differences in country patterns and movements.

First, the seven countries as a whole, and for most Eastern countries individually, there was a slow but steady rise in high R&D intensive items relative to the total over the 1970s and into the 1980s. Nevertheless, their share in total Western technology exports to the Soviet Union in 1983 amounted to only 1.8 per cent – considerably less than the 4.3 per cent average for Eastern Europe. The item's share was as high as 10.7 per cent for Romania and over 4 per cent for Bulgaria, Czechoslovakia and Hungary in the same year, and its share in the remaining countries was still high compared with the Soviet Union. The increase for Romania is of comparatively recent date and appears to reflect, above all, a near doubling of Western sales of mining and fuel extraction machinery between 1979 and 1983. The West's total technology exports to that country moderated and subsequently declined – by more than any other Eastern country – over those years, indicating that in the process of these cut-backs preference was given by Romania to high R&D-intensive items. This was also true, but to a much smaller extent, for Poland.

Second, restrictions on the growth of Eastern imports of Western embodied technology have also been accompanied by rises in the share of moderately R&D-intensive technology. This accounted for a lower share in Western embodied technology exports to the Soviet Union than to any of the East European countries. Within Eastern Europe, Czechoslovakia, Hungary and Poland received a higher than average share of these items in 1983 – though in Poland this was a new phenomenon.

Third, the Soviet Union has imported low R&D-intensive technology to a substantially greater extent than the Eastern European countries. The Soviet shares of high and medium R&D-intensive goods in the total were both lower than in Eastern Europe in all the years under consideration. The Soviet share of low R&D-intensive items ranged between 80-85 per cent, compared with 71-78 per cent for the East European countries, throughout 1970-1983. The share of low R&D-intensive technology has generally tended to fall during the crisis years which followed the mid-1970s – by 5-6 percentage points in the Soviet Union and Eastern Europe. The German Democratic Republic and Hungary were unique in increasing the share of low R&D-intensive technology – which remained however, one of the lowest in the area in 1983 as in 1970. In the other countries of Eastern Europe the reduction was very marked since 1978, but for the Soviet Union, in contrast, it was partially reversed after that year.

The overall fall in the share of low R&D-intensive items indicates a pattern of increasing selectivity, mentioned earlier, which appears to be accompanying the cutbacks in capital goods import growth throughout the region. This general observation, reinforced by particularly large rises in high and medium R&D-intensive technology in these countries where the overall decline in imports of Western embodied technology was biggest (Poland, Romania) strongly suggests an overall policy designed to protect imports of more sophisticated types of imports in times of financial stringency.

Nevertheless, the relative stability of the overall pattern of embodied technology imports of different R&D intensity throughout the 1970s and early 1980s, and notably the still preponderant shares of low-to-medium R&D-intensive technology, suggests the following comments. First, the general strategy towards the level of sophistication of technology acquisitions does not seem inappropriate for countries in which the general level of technology is considered to be lower than in OECD countries. The products of low-to-medium R&D industries in Western countries may well be a considerable improvement on the domestic product, or may fill a gap in the range of output of domestic industry more cheaply than could be done by creating a new production facility. This finding, which suggests some complementary role of Western technology, is supported by recent econometric work (see Chapter 8, concluding paragraphs of section 2 *a)* below). Second, the lower the level of technology imported, the less is it likely to be disruptive. High technology plant frequently requires high quality inputs. These may not be readily called forth in the context of Eastern management systems which still place excessive stress on quantity rather than quality. On the assumption that the capital goods component of even low R&D-intensive technology (by Western standards) causes such disruptions, this may indeed be one explanation for the rapid growth of intermediate (technology) imports from the OECD countries noted in Chapter 5 above. Third, though the generalized decline in the share of low R&D-intensive imports in the total is not dramatic, it is consistent with the hypothesis that more sophisticated technology items are currently accorded a higher priority than the less sophisticated. This is not very surprising even in the light of the foregoing – if only because such items are harder to replace from domestic or other CMEA country suppliers.

### b) *Eastern Countries' Investment and their Capital Goods Imports from the West*

As a further step in assessing strategies toward technology acquisition during the 1970s and early 1980s, the evolution of general economic policies over the period should be recalled (Chapter 1 above). The Eastern countries began in the 1970s with total trade in approximate equilibrium with all regions, and ended it with debts totalling around $70 billion – equivalent to about half of their total export effort to all regions of the world in 1980 and to about 85 per cent of their exports outside the European CMEA area in that year. Their initial foreign exchange equilibrium position, as well as the increasing willingness of Western institutions to finance Eastern technology deals, essentially permitted domestic savings to be supplemented by external resources, and to increase the share of foreign technology in total investment. High output growth had already been achieved in 1966-1970. Plans for 1971-1975 also envisaged buoyant expansion of output, though a number of countries, aware of the slower growth in active population, were already stressing the importance of labour productivity improvement. The need to accelerate the substitution of labour by capital clearly indicated not only further growth in investment, but also the use of more capital-intensive production processes.

The size of the volume increase in capital goods imports in the majority of countries was, in fact, considerably faster than the growth of investment up to the middle of the decade. There was thus a substantial rise in the share of imported OECD-produced capital goods in total investment. It is not possible to make reliable estimates for most Eastern countries of the absolute share of Western capital goods in total investment (or, in total investment in machinery and equipment). However, on the basis of the few data produced by the seven countries, together with such Western estimates as are available, it seems unlikely that the share of Western capital goods in total investment rose above about 5-10 per cent in 1975-1977, the peak or near-peak year of such imports in most Eastern countries. In the Soviet Union, the share was probably considerably lower[13]. Thus Table 10 of this chapter, which is based on the volume growth of Western capital goods exports relative to the volume of investment expenditures, only shows *changes* in the share of Western capital goods in investment.

The figures indicate that the 1970 whole-economy share approximately doubled by 1975 in Bulgaria and Poland, rose by 50 per cent in the Soviet Union, and by about 20 per cent in Hungary. The share apparently fell in Czechoslovakia (which registered high investment growth but exercised, in contrast to the 1960s, a cautious capital goods import policy in the 1970s), in the German Democratic Republic (excluding imports from the Federal Republic) and in Romania (where investment growth itself was extremely high). Except in Czechoslovakia, the German Democratic Republic and Poland, the rise in the share of imported Western capital goods in the industry sector's investment was higher than for the economy as a whole. In some of these countries, however, fast-growing allocations to transport and communications – including oil and gas distribution – thus produced a bigger rise in the share of the material sphere than in industry (Czechoslovakia, the German Democratic Republic and Hungary) between 1970 and 1975.

By 1980, the cutbacks in the growth of Western capital goods exports relative to investment can be seen to have reduced the share of imported capital goods in the investment of all countries but Hungary. Indeed, in Bulgaria, the German Democratic Republic and Romania, their share fell by between one-third and two-thirds of the 1970 level for the total economy, and, except in Romania, the cuts for all countries were most drastic in industry. This took place, moreover, against a background of decelerating investment activity – particularly in 1979/1980. Thus, in the second half of the 1970s, there was a general and very marked reversal of strategy towards expanding the share of Western capital goods imports in investment in all countries in which it had been implemented in the early 1970s.

By 1983, the declining trend had intensified further in Poland – despite massive investment cuts – and in Romania. In the Soviet Union it was strongly reversed. Indeed in the latter country, imports of investment goods were again edging close to the share of investments they held in 1975. A reversal of the trend was also apparent in some other countries. But whereas it took place against a rise in investment over 1980 levels in both Bulgaria and the Soviet Union, the upward change in countries such as the German Democratic Republic and Hungary was due partly to declining investment levels. In Czechoslovakia and Romania, where investment also fell, imports of Western investment were cut back even more sharply and their share in investment continued to decline.

These latest figures can be interpreted first as illustrating that some Eastern countries appear unwilling to cut back the share of Western goods in investment to below a certain minimum level even when investment as a whole is falling and, indeed, may try to increase it (the German Democratic Republic and Hungary). Moreover, where investment levels have bottomed out, or have begun to recover, the opportunity will be taken to reverse past declines in that share (Bulgaria, the Soviet Union).

Czechoslovakia, Poland and Romania provide important exceptions. In Poland, which accounted for one-third of the total cumulated deficits on the regions's East-West balance of trade in 1971-1980, the consequent readjustment has necessitated particularly stringent import cuts. In Romania, cumulated balance of trade deficits were also high relative to its share of regional NMP – though not as high as for Poland. Moreover, the heavy dependence of its export earnings on fuel (40 per cent of its total exports to the OECD countries in 1981-1983) have been largely offset by import expenditures on crude oil from OPEC countries (rather than, as in the case of other CMEA countries by lower-priced supplies from the Soviet Union). Its process of trade adjustment with the OECD has had to take these hard currency costs into account. Its surplus on OECD trade in fact contributed nearly one half of the total surplus recorded by the Eastern area with OECD countries in 1981-1983 ($3 billion out of $6.9 billion). Cuts in capital goods imports from OECD countries were, for this reason, much more far-reaching than those of any other country of the area.

Czechoslovakia presents a contrast to the crisis situation of these two countries. Like Romania its share in the region's imports of capital goods were very high (12 per cent) compared with its share of the region's NMP and investment (about 5 per cent) in 1970. And although the share of OECD-supplied capital goods in Czechoslovakia's total investment is now only half what it was in 1970, its share of the region's imports of these goods in 1981-1983 was still rather higher (about 6 per cent) than its share in the whole region's NMP and investment (about 4½ per cent) in 1981-1983. Thus developments in Czechoslovakia reflect a reduction in its relatively high dependence at the beginning of the 1970s. It is not inconsistent with the hypothesis of a "minimum level", since Czechoslovakia does not appear to have reached that level as yet.

Thus with the exception of Poland and Romania, where post-1980 balance of payments readjustments have been particularly onerous, the evidence suggests that Eastern planners acknowledge the desirability of maintaining a minimum share of OECD capital goods in total investment. The recent rise in the share to levels above those of 1970 and, in some cases, to levels close to or above those of 1975, suggest that the import of capital goods from the West fell close to the desirable level in some countries at the height of the readjustment period. It is likely therefore to turn upwards when economic conditions permit and, in some countries, has done so already.

### c) *Eastern Capital Goods Imports from the West by Type*

Within the general strategies – and changes in strategies – towards the role of capital good imports relative to investment, the composition of capital goods imports by type reveals a number of features on the role such imports were expected to play in growth strategy.

As indicated in a previous OECD paper[14], OECD exports of capital goods can be broken down according to three broad categories by type (Table 11 of this chapter). The first covers capital goods for investment in sectors producing non-tradeable goods (i.e. goods which are not generally exported or imported). This includes power and power distribution, transport machinery and equipment (including oil and gas distribution), construction, communications, office machines, medical apparatus, heating and cooling of buildings. The second consists of imports of capital goods which make a direct contribution to output (engineering and metal-working machinery and equipment, agricultural machinery, furnaces and gas generators, machinery and tools for plastics and wood, machinery for special industries, non-electrical machinery and equipment – SITC item 719.8). This can be called "product-creating technology". The third consists of capital goods which make only an indirect contribution to output (communications, mechanical handling and storage equipment, electrical control gear, instruments and structural parts). This type of technology can be called "output-assisting technology".

The Table shows that at the beginning of the decade, some 75-80 per cent of Soviet and East European technology imports from the West was destined for use in direct production activities (tradeables and non-tradeables) and more than half of this sub-total (50 per cent in the Soviet Union and 59 per cent in Eastern Europe) was for the production of tradeable goods. These proportions remained roughly similar over the period to 1983 in Eastern Europe. But in the Soviet Union, the share of imports of capital goods for the production of non-tradeables rose strongly, while those for the production of tradeables fell. The bulk of other capital goods imports from the West was intended for output-assisting technology. About one-fifth of all imports of Western capital goods over the years 1970-1983 has, in other words, been destined for activity with no direct trading potential (output-assisting capital goods plus the residual).

Gas and oil distribution equipment, included among capital goods for the production of non-tradeables, has, however, had a very real effect on export capacity at least in the Soviet Union. Thus, in 1970 the greater part of Soviet capital goods imports was tied to activities with a potential, indeed a successful, export role (taking oil and gas distribution equipment together with capital goods for the production of tradeables). In Eastern Europe, where apart from transit facilities, oil and gas distribution equipment is tied to domestic rather than export requirements, the proportion is rather lower – around 41 per cent, which refers to tradeable product-creating technology alone.

It is noteworthy that while there was no basic difference in the overall pattern of Western capital goods imports from the West between the two markets in 1970, a good deal of differentiation has since developed. The distribution of East European capital goods imports has stayed roughly constant. But the Soviet profile has changed very considerably, the share of capital goods for the production of tradeables having in fact halved in 1983 compared with 1970. In terms of export capacity, this has been largely, but not completely, offset by a near doubling of the share of oil and gas distribution equipment in the total between 1970 and 1983.

The finer breakdown of Eastern capital good imports from the West by type (Table 12 of this chapter) does not add much to the above discussion, except to provide a quick reference to the broad classes of goods traded. It shows a general decline in the share of transport equipment (excluding oil and gas distribution) throughout the 1970s in both Eastern Europe and the Soviet Union. There has also been a fall in the share of capital good imports for special (mainly consumer) industries – notably in Eastern Europe where they accounted for nearly one-fifth of the total in 1970. The share of oil and gas distribution equipment has declined in Eastern Europe, but, as noted earlier, it rose fast in the Soviet Union. Machine tools, etc., which were important in 1970, fell to a low share of Soviet imports of capital goods but have increased their importance in East European imports. A similar divergence is apparent for the 'other' item (mainly machinery and mechanical appliances of various kinds).

That the acquisition of embodied productive technology was the most important element in Eastern strategies is clear. However, apart from statements of general intent, and such detail as exists on the buy-back commitments entered into by Western firms under such arrangements, it is not possible to pin down the *extent* to which the general policy of purchasing Western productive technology was undertaken specifically to promote production for export[15]. A considerable share of imports of technology in general was effectively bought on credit. The stress laid on compensation agreements, in order to ensure future export earnings to repay these credits, confirm an element of export promotion in Eastern technology acquisition strategy which is well documented in business literature as well as in Eastern official statements. The bulk of these concerned the exploitation of Soviet oil and gas. In fact, as will be shown in section 2 of this chapter, the impact of technology imports on manufactured goods export performance was, in practice, limited. The fact that more than half of Western capital goods purchased were for activities which do not have a potential export role to play, already indicates that the increase of export capacity was only one element in capital goods import strategies. The data presented thus indicate that the purchase of Western capital goods to improve domestic performance was perhaps the main objective. However, the need to service and liquidate past debt by export expansion has probably played an increasingly important role in recent years, especially in Eastern Europe where the share of capital goods acquired for production of tradeables has risen since 1975. This orientation is likely to prevail in the future.

d) *Eastern Capital Goods Imports from the West by End-use Sector and Branch*

Table 4 of Chapter 5 shows the sector and industrial end-use branch distribution of Eastern capital goods imports from the West in 1975, 1980 and 1983. The sector and branch breakdowns are given as far as possible in terms of individual Eastern countries' own national accounts and industrial branch classifications – which are similar but not identical between Eastern countries. In particular, it should be borne in mind that in three countries – Bulgaria, the German Democratic Republic and Hungary – oil refining is included in the chemical sector whereas in the other countries it is included under fuels (in two of the Tables which follow, oil refining and extraction equipment imports are, nonetheless, shown separately). Definitions of the coverage of sectors and branches are given in the footnote to Table 4 of Chapter 5.

The Table referred to above presents end-use breakdowns based on both OECD and Eastern-reported data [see description in Chapter 5, section 1*a)* and section 1*b)*]. While, as noted, the two breakdowns cannot be fully reconciled, their joint presentation enables shortcomings in one to be supplemented by better information in the other in the paragraphs which follow.

The Table shows that Eastern country capital goods imports from the West have been biased very heavily towards the industrial sector in the 1970s (between half and three-quarters of the total), in line with the pattern of domestic investment allocations. Although the service sectors of the economy accounted for 20-30 per cent of total investment, the items shown in the Table for the non-material sphere – principally medical and office equipment – have accounted for only 2-7 per cent (OECD data). The Eastern data show a similarly small share.

Within the material sphere, industry and transport equipment have accounted for the bulk of the 85-90 per cent of total Eastern imports of Western capital goods, the unallocated residual has accounted for 6-8 per cent of the total. The 85-90 per cent combined share of industry and transport (including oil and gas distribution equipment) was allocated between them roughly on a 2:1 basis according to OECD-reported data.

Industry has varied between a half and three-quarters of the total for the material sphere (the CMEA-reported data tend to understate this sector). While the share of transport (including communications) is generally higher in the case of CMEA-reported East-West trade data, its joint share combined with industry remains preponderant, as in OECD data. Of the total allocated to transport and communications, a share of between one-quarter and one-half has been spent on gas and oil distribution – notably large-diameter pipes, compressors and ancillary equipment. In other words, 10-20 per cent of capital goods for the material sphere of the economy has been expended on pipeline equipment in Eastern Europe (20 per cent in Romania and 10-15 per cent in the other East European countries). The share was as high as one-quarter in the Soviet Union in 1976-1983.

Overall patterns did not, in most cases, vary significantly between the first and second halves of the 1970s or in the early 1980s. A partial exception was the Soviet Union, where rises in purchases of pipeline equipment towards the end of the decade pushed industry below its two-thirds share of capital goods imports for the material sphere in the early 1970s. There was a further and substantial jump in imports of pipeline equipment by 1983, and a corresponding fall in industry's share. Of the other sectors, apart from negligible purchases of construction machinery, only agricultural capital goods have been purchased in any quantity by the Eastern countries from the West. But they did not rise above 2-3 per cent of the material sphere total except in the single case of Hungary – where they regularly accounted for 8-9 per cent of that total.

Within industry, the biggest single imported capital goods item, which averages about 30 per cent of the industry total, is the unallocated item according to OECD data. About half of this, falling to about a quarter in more recent years, consists of SITC 719.8 – "machinery and mechanical appliances n.e.s.". This contains a very wide range of goods, many of which – even if their value were known – could not be precisely allocated by branch. The share in capital goods imports of chemicals, engineering and also fuels in the Western-reported data is consistently lower than shown in the Eastern-reported statistics. Most of item 719.8 is thus probably distributed among these branches. Because of the incompleteness of OECD statistics on this score, it is likely that the Eastern-reported data give a more accurate impression of the branch distribution within industry. This confirms what is, despite omissions, also indicated by the OECD-reported data – namely, the predominant share of the engineering and chemical items in most countries in most years. Their combined share of the industry total in 1983, for instance, was between one-third and one-half except in Czechoslovakia where it was about one-fifth[16].

In most countries of Eastern Europe (except Bulgaria) the share of chemicals has increased in the 1980s, following some fall in the late 1970s. The preceding decline reflected cuts in domestic investment in the petroleum feedstock- and energy-intensive branch, as oil shortages began to bite. Largely because of this, engineering has been the biggest importing branch in most countries in recent years. The engineering branch, as the principal engine of domestic technology change might, however, have been expected to absorb an even higher share of its capital goods imports than it actually has. However, only in Poland and the Soviet Union has it consistently accounted for more than a quarter of all industrial capital goods imports from the West.

In Table 13 of this chapter, the Eastern-reported distribution of capital goods for industrial branch end-users in Table 4 of Chapter 5 has been re-arranged by broad industrial divisions. The "fuel and energy" division is self-explanatory. The "supply" division covers output by the metallurgical, engineering, chemicals and building material branches. The "consumer" division consists of wood and paper, light industry, and food processing. The Table shows that in all countries but the German Democratic Republic and Hungary, there has been a considerable shift in priorities towards Western capital goods imports for the fuel/energy branches between 1975 and 1983. In most countries (especially Czechoslovakia and the Soviet Union) energy equipment imports were particularly buoyant, in others (Poland, Romania and again the Soviet Union) oil extraction or refining equipment has grown strongly. The striking change is illustrated by the fact that whereas in 1975 the division took 7-23 per cent of the total for industry, by 1983 the shares had risen to within a range of 10-42 per cent.

The Table also shows that the supply industries in 1983 continued, as in the past, to account for around half of imported Western capital goods for industrial end-users, and the consumer goods industries for about one-quarter. The exceptions are Czechoslovakia – reflecting the rise, already mentioned but not confirmed by OECD sources, in equipment for the wood and paper industries in 1983 – and Hungary which has always relied heavily on Western capital goods for its imports of equipment for the food and textile export industries. Seen from this aspect, the main capital goods importing industries are those which contribute to the provision of industrial semi-finished goods. They are, in some cases, important export industries – notably chemicals. The engineering industry, while a big potential exporter, has not attained great success on Western markets at least since 1975 [see Chapter 8, Section 1*c)* below]. Perhaps because of this, no clear region-wide trend has emerged with regard to changes in its relative importance as a capital goods importer between 1975 and 1983. In general, the supply division tended to increase its share up to 1980 but levelled out thereafter. The consumer division lost shares in all countries between 1975 and 1980, but had recovered in some up to 1983.

If any conclusions can be drawn about the strategy of capital goods imports from the West on the basis of the above movements, they are first, that the oil price shock introduced a new element into those strategies in most countries of the area. Equipment for the fuel and in some countries the energy sectors have gained substantially in priority. In a number of countries the movement was headed by imports for energy production. This itself is a departure of some magnitude given the high degree of intra-CMEA self-sufficiency, based largely on Soviet technology, which existed up to the mid-1970s. Second, for the region as a whole, there appears to have been a general fall in the already rather low priority accorded to Western capital goods to the consumer industries. This applies particularly (except for Czechoslovakia and Hungary) to those countries where overall dependence on Western equipment was relatively high. Third, the supply industries, which also account for the highest and fastest growing share of Eastern investments, have retained a high degree of priority in imports of Western capital goods. It is these sectors which provide the bulk of indigenous embodied technology – notably the investment capacity-creating engineering branch – and where one of the most competitive export branches (chemicals) is located. Even so, it is perhaps surprising, as already mentioned with regard to the engineering branch, that the priority accorded to this division as a whole in the procurement of Western capital goods has not in fact been higher[17].

Table 5 of Chapter 5 summarizes the sector and branch distribution of intra-CMEA capital goods trade by country. It can be seen that the material sphere predominates, but imports for the non-material sphere are still somewhat higher than in imports from the West. Within the material sphere, the share of transport and communications is actually bigger than that of industry in several East European countries (the German Democratic Republic, Hungary and Poland). In general, the share of industry is lower than in capital goods imports from the West. Exchanges of agricultural and construction equipment are however higher, agriculture accounting for 10-20 per cent or even more except in Romania. Construction equipment accounts for 5-10 per cent. Both these sectors are, as already indicated, negligible in most Eastern countries' capital goods imports from the West.

Within industry, the main differences in 1983 were the high share of equipment for the energy and fuel branches in imports from the East as compared with the West, although the difference has diminished in recent years – especially in Bulgaria, Romania and the

Soviet Union (Table 14). Conversely, the supply branches – metallurgy, engineering, chemicals and construction materials – took a lower share in imports from the East. The share of the chemical industry is particularly low in supplies from the East. The consumer branches accounted for a lower share in total industrial capital goods imports from the East than from the West in several countries but not in the seven-country total. Moreover, their share in capital goods imports generally declined further after 1980 – especially in Bulgaria and the German Democratic Republic in which these branches accounted for particularly high shares in imports from the West in 1975. The overall results for the region are strongly affected by the low share of the Soviet food industry in that country's capital goods imports from the West relative to its imports from the East.

The actual extent of, and changes in, the relative degree of dependence of the Eastern countries on Eastern and Western industrial capital goods suppliers can, in principle, be assessed on the basis of comparisons of the values of imports from the two areas. Given the reservations expressed earlier concerning the valuation of the two flows (section 1 above), as well as data inconsistencies and omissions including the large size of the unidentified item, the results presented in Table 15 of this chapter (showing the share of CMEA suppliers in total identified branch imports of capital goods), should be interpreted with some caution.

Tables 4 and 5 of Chapter 6 indicate that industrial capital goods as a whole accounted for a higher share in total capital goods imported from the West than in those imported from the East. It is of interest that in 1975, on the basis of the figures presented (i.e excluding imports unidentified by origin), the seven countries as a whole apparently imported in fact, more than half of their total industrial capital goods requirements from the West (54 per cent in Eastern Europe and 62 per cent in the Soviet Union). By 1983, the Western share was still around 40 per cent (30 per cent only in Eastern Europe and 48 per cent in the Soviet Union). Because of previously discussed pricing discrepancies, the declines in these shares are, however, probably understated in the Table (see section 1*a)* of this chapter). A decline in the Western share certainly did take place. Even so, Western countries apparently remained important sources of capital goods for the oil extraction, metallurgical, chemical and wood and paper branches. However, by 1983, some of the branches most heavily dependent on Western suppliers in 1975 – notably oil extraction and mining, engineering, the light and food industries – show a considerable rise in the share of Eastern suppliers. In several of these branches, the "substitution" which apparently occurred was accompanied by falls or slow growth in overall imports of capital goods for the branches concerned. This was the case particularly in chemicals and metallurgy. Moreover, it should be recalled that, since 1980, there was no growth at all in Eastern Europe's imports of capital goods in total. For the Soviet Union, where growth in imports from the East was moderate and from the West very fast, substantial dependence on Western suppliers still apparently remains with regard to equipment for oil extraction equipment, engineering, chemicals and the wood and paper branches.

## 4. General Conclusions

A major question is the extent to which the Eastern countries have shown any sign of a coherent strategy towards their imports of Western technology. The breakdown, based on *the type* of capital goods imported contained in Table 11 of this chapter, suggests that several elements are present. Given the roughly 40 per cent imported for the production of tradeable goods in Eastern Europe, export development can be expected to have benefitted. This was clearly a less marked characteristic of the Soviet Union with regard to most manufacturing sectors. By 1983, the share devoted to tradeable goods production was only half that in Eastern Europe. There was, however, a very significant degree of Soviet concentration on pipeline construction for developing exports of crude and refined fuels.

The end-use sector/branch breakdown confirms and elaborates these findings. In all countries, chemicals have been the single biggest or second biggest branch importer – disproportionately big, in fact, relative to the branch's share of industrial investment. Fuel refining – sometimes included with chemicals and sometimes within fuel depending on the country's individual classification method (see note to Table 4 of Chapter 5) – has also been a sizeable branch importer. A similar correlation within industrial branch technology import shares can be discerned for some countries where food or some light industry elements have increased their shares of Western markets, or which have grown faster than average exports [Chapter 8, section 1*c)*]. These correlations, while far from perfect, do suggest that a significant element of export development has entered into Eastern strategies of technology acquisition in recent years.

In contrast, the high share of industry in total Eastern imports of Western capital goods, has been accompanied by substantial losses of shares in OECD markets for non-fuel industrial goods as a whole. The bulk of purchases of Western technology has thus – whether this was foreseen or not – been devoted to the development of output for domestic consumption or for exports to markets other than the OECD area. Up to 1980, at least, there was, moreover, no appreciable success in substituting indigenous products for goods imported from the West. "Successes" thereafter reflected generalised cuts in imports to reduce trade deficits and, the resulting cuts in industrial supplies, were costly in terms of output slowdowns.

These findings are consistent with the analysis given in section 3*a)* above concerning the level of sophistication of Western technology imported. The very low share of high R&D intensive items and the correspondingly high share of low R&D intensive items (70-80 per cent), is perfectly consistent with a strategy of integrating supplies of Western technology with a wide range of domestic activities capable of absorbing them. This presupposes that the level of technology embodied in the imported items is not too far above existing industry-wide standards. That it may, nonetheless, have been too high is indicated by the parallel increases in imports of Western intermediate (technology) goods.

In short, no evidence of any single strategy towards imports of Western technology can be derived from actual developments over 1970-1983. Export promotion played some role, but was concentrated on a narrow range of activity. The large share of low R&D intensive technology suggests an "incremental" approach to raising economic performance, rather than the creation of high technology industrial nodes. An obvious reservation here is the degree to which known high technology items have been reserved mainly for military or space uses. These activities are obviously not reported as such in the Eastern statistical sources and constitute a possible special case which must be excluded from the foregoing. The existence of a group of capital goods imports unidentified by end-user or region of origin, and noted in section 1*a)* above, is recalled in this connection.

Table 1 (Chapter 6)

EASTERN COUNTRIES CAPITAL GOODS IMPORTS, BY REGION OF ORIGIN
(per cent of totals)

| | 1975 | | | | 1980 | | | | 1982 | | | | 1983 | | | |
|---|---|---|---|---|---|---|---|---|---|---|---|---|---|---|---|---|
| | From : | | | | From : | | | | From : | | | | From : | | | |
| | East | Rest of World | Uniden-tified | Total | East | Rest of World | Uniden-tified | Total | East | Rest of World | Uniden-tified | Total | East | Rest of World | Uniden-tified | Total |
| Bulgaria | 8.1 | 3.8 | 13.7 | 7.5 | 7.7 | 2.6 | 9.8 | 6.7 | 7.7 | 3.6 | 10.9 | 7.3 | 7.1 | 3.6 | 12.7 | 7.3 |
| Czechoslovakia | 10.9 | 5.6 | 20.2 | 10.5 | 10.2 | 10.0 | 11.6 | 10.6 | 9.7 | 6.6 | 13.4 | 9.7 | 9.4 | 6.0 | 14.9 | 9.6 |
| German Democratic Republic | 12.5 | 3.4 | 25.9 | 11.3 | 9.6 | 4.0 | 25.3 | 11.4 | 9.4 | 4.2 | 28.6 | 12.3 | 7.8 | 4.0 | 28.4 | 11.2 |
| Hungary | 5.6 | 3.7 | 5.1 | 4.7 | 5.9 | 4.2 | 5.2 | 5.3 | 5.8 | 3.9 | 3.4 | 4.7 | 5.3 | 2.9 | 4.5 | 4.4 |
| Poland | 12.5 | 20.9 | 13.1 | 15.8 | 13.1 | 12.7 | 9.8 | 12.2 | 8.0 | 4.6 | 11.2 | 7.8 | 8.5 | 4.7 | 8.4 | 7.4 |
| Romania | 3.8 | 6.3 | 10.3 | 5.9 | 4.3 | 6.5 | 10.6 | 6.3 | 3.5 | 1.2 | 6.7 | 3.6 | 2.7 | 0.7 | 6.1 | 2.9 |
| Eastern Europe | 53.4 | 43.8 | 88.2 | 55.9 | 50.8 | 40.0 | 72.4 | 52.5 | 44.2 | 24.2 | 74.3 | 45.4 | 40.9 | 21.9 | 75.1 | 42.9 |
| Soviet Union | 46.6 | 56.2 | 11.8 | 44.2 | 49.2 | 60.0 | 27.6 | 47.5 | 55.8 | 75.8 | 25.7 | 54.6 | 59.1 | 78.1 | 24.9 | 57.1 |
| Eastern Europe and the Soviet Union | 100.0 | 100.0 | 100.0 | 100.0 | 100.0 | 100.0 | 100.0 | 100.0 | 100.0 | 100.0 | 100.0 | 100.0 | 100.0 | 100.0 | 100.0 | 100.0 |

Note: The "Unidentified" item consists of the sum of those items reported as a total for all regions but not by country of origin.

Source: OECD East-West technology transfer data base (Eastern statistics).

Table 2 (Chapter 6)

EASTERN COUNTRIES' IMPORTS OF CAPITAL GOODS COMPARED WITH TOTAL IMPORTS, BY REGION OF ORIGIN
(components expressed as per cent of total import from region shown; totals in $ million)

| | 1970 | | | 1975 | | | 1980 | | | 1982 | | | 1983 | | |
|---|---|---|---|---|---|---|---|---|---|---|---|---|---|---|---|
| | East | Rest of World | World | East | Rest of World | World | East | Rest of World | World | East | Rest of World | World | East | Rest of World | World |
| Bulgaria | | | | | | | | | | | | | | | |
| Identified capital goods | (a) 38.8 | (a) 16.5 | (a) 33.2 | 30.8 | 27.0 | 29.6 | 26.7 | 15.8 | 24.0 | 23.7 | 18.0 | 22.2 | 21.8 | 19.6 | 21.3 |
| Unidentified capital goods | - | - | 14.7 | - | - | 13.9 | - | - | 11.5 | - | - | 11.4 | - | - | 12.6 |
| Total imports ($ million) | 2 405 | 805 | 3 210 | 3 713 | 1 690 | 5 403 | 7 285 | 2 382 | 9 666 | 8 551 | 2 989 | 11 540 | 9 409 | 2 880 | 12 289 |
| Czechoslovakia | | | | | | | | | | | | | | | |
| Identified capital goods | 23.7 | 20.6 | 22.5 | 26.1 | 20.9 | 24.3 | 26.9 | 27.9 | 27.2 | 23.8 | 20.9 | 22.5 | 22.7 | 21.5 | 22.4 |
| Unidentified capital goods | - | - | 10.1 | - | - | 12.2 | - | - | 8.7 | - | - | 10.2 | - | - | 11.1 |
| Total imports ($ million) | 2 333 | 1 362 | 3 695 | 5 855 | 3 233 | 9 088 | 9 827 | 5 356 | 15 183 | 10 725 | 4 729 | 15 454 | 11 916 | 4 453 | 16 369 |
| German Dem. Rep. | | | | | | | | | | | | | | | |
| Identified capital goods | 19.4 | 9.4 | 16.0 | 24.7 | 9.7 | 19.1 | 21.4 | 7.6 | 15.8 | 19.1 | 8.6 | 15.3 | 16.8 | 7.9 | 13.4 |
| Unidentified capital goods | - | - | 19.1 | - | - | 12.5 | - | - | 15.1 | - | - | 17.0 | - | - | 16.1 |
| Total imports ($ million) | 3 191 | 1 656 | 4 847 | 7 090 | 4 202 | 11 292 | 11 265 | 7 815 | 19 080 | 12 947 | 7 245 | 20 196 | 13 448 | 8 076 | 21 524 |
| Hungary | | | | | | | | | | | | | | | |
| Identified capital goods | (b) 15.3 | (b) 18.1 | (b) 16.4 | 19.1 | 16.7 | 18.2 | 19.2 | 12.8 | 16.7 | 18.1 | 12.6 | 16.2 | 17.0 | 10.1 | 14.7 |
| Unidentified capital goods | - | - | 9.9 | - | - | 3.7 | - | - | 4.7 | - | - | 3.1 | - | - | 4.1 |
| Total imports ($ million) | 1 881 | 1 109 | 2 990 | 4 101 | 2 669 | 6 770 | 7 692 | 4 876 | 12 568 | 8 348 | 4 586 | 12 934 | 8 865 | 4 615 | 13 480 |
| Poland | | | | | | | | | | | | | | | |
| Identified capital goods | 27.5 | 20.6 | 25.1 | 31.9 | 35.4 | 33.9 | 32.5 | 20.8 | 27.0 | 21.3 | 16.3 | 19.8 | 21.4 | 17.5 | 20.4 |
| Unidentified capital goods | - | - | 13.6 | - | - | 5.7 | - | - | 5.8 | - | - | 9.6 | - | - | 6.6 |
| Total imports ($ million) | 2 361 | 1 247 | 3 608 | 5 453 | 7 092 | 12 545 | 10 085 | 9 031 | 19 116 | 9 818 | 4 245 | 14 063 | 11 333 | 4 254 | 15 587 |

Table 2 (Chapter 6) contd

| | 1970 | | | 1975 | | | 1980 | | | 1982 | | | 1983 | | |
|---|---|---|---|---|---|---|---|---|---|---|---|---|---|---|---|
| | East | Rest of World | World | East | Rest of World | World | East | Rest of World | World | East | Rest of World | World | East | Rest of World | World |
| Romania | | | | | | | | | | | | | | | |
| Identified capital goods | 29.8 | 20.7 | 25.1 | 27.1 | 22.5 | 24.2 | 21.5 | 11.8 | 15.5 | 30.1 | 3.4 | 13.3 | 24.0 | 2.5 | 11.7 |
| Unidentified capital goods | - | - | 15.1 | - | - | 10.5 | - | - | 9.1 | - | - | 9.7 | - | - | 9.8 |
| Total imports ($ million) | 942 | 1 019 | 1 961 | 1 965 | 3 376 | 5 341 | 5 052 | 8 148 | 13 200 | 3 079 | 5 245 | 8 324 | 3 279 | 4 366 | 7 645 |
| Eastern Europe | | | | | | | | | | | | | | | |
| Identified capital goods | 23.8 | 17.1 | 21.3 | 26.5 | 23.8 | 25.3 | 24.9 | 15.8 | 21.1 | 21.6 | 12.4 | 18.4 | 20.2 | 12.1 | 17.5 |
| Unidentified capital goods | - | - | 14.7 | - | - | 9.6 | - | - | 9.2 | - | - | 10 8 | - | - | 10.5 |
| Total imports ($ million) | 8 827 | 5 284 | 14 111 | 28 177 | 22 262 | 50 439 | 51 206 | 37 608 | 88 814 | 53 472 | 29 043 | 82 515 | 58 250 | 28 644 | 86 894 |
| Soviet Union | | | | | | | | | | | | | | | |
| Identified capital goods | 41.5 | 25.6 | 34.6 | 41.6 | 31.7 | 35.9 | 42.2 | 22.6 | 31.0 | 43.6 | 25.4 | 33.3 | 45.9 | 28.7 | 36.7 |
| Unidentified capital goods | - | - | 1.9 | - | - | 1.7 | - | - | 4.6 | - | - | 4.0 | - | - | 3.8 |
| Total imports ($ million) | 6 634 | 5 098 | 11 732 | 15 667 | 21 272 | 36 939 | 29 428 | 39 094 | 68 552 | 33 489 | 44 180 | 77 669 | 37 055 | 43 151 | 80 206 |
| Total, Eastern countries | | | | | | | | | | | | | | | |
| Identified capital goods | 31.4 | 21.2 | 27.3 | 31.9 | 27.7 | 29.8 | 31.3 | 19.3 | 25.4 | 30.1 | 20.3 | 25.6 | 30.1 | 22.2 | 26.7 |
| Unidentified capital goods | - | - | 8.9 | - | - | 6.3 | - | - | 8.5 | - | - | 7.5 | - | - | 7.3 |
| Total imports ($ million) | 15 461 | 10 382 | 25 843 | 43 844 | 43 534 | 87 378 | 80 634 | 76 702 | 157 336 | 86 961 | 73 223 | 160 184 | 95 305 | 71 795 | 167 100 |

Notes: Total Eastern Europe and Total Eastern countries for 1970 exclude Bulgaria and Hungary. The "Unidentified" item consists of the sum of those items reported as a total for all regions but not by country of origin.

a) 1973.
b) 1971.

Source: OECD East-West technology transfer data base (Eastern statistics).

Table 3 (Chapter 6)

EASTERN COUNTRIES' CAPITAL GOODS EXPORTS, BY REGION OF DESTINATION
($ billion)

| | 1970 | | | | 1975 | | | | 1980 | | | | 1982 | | | | 1983 | | | |
|---|---|---|---|---|---|---|---|---|---|---|---|---|---|---|---|---|---|---|---|---|
| | To : East | Rest of World | Unidentified | Total | To : East | Rest of World | Unidentified | Total | To : East | Rest of World | Unidentified | Total | To : East | Rest of World | Unidentified | Total | To : East | Rest of World | Unidentified | Total |
| Bulgaria | (a) 0.83 | (a) 0.32 | (a) 0.17 | (a) 1.33 | 1.26 | 0.50 | 0.10 | 1.87 | 2.79 | 1.50 | 2.13 | 4.38 | 2.88 | 1.92 | 3.02 | 5.36 | 3.49 | 1.96 | 0.42 | 5.88 |
| Czechoslovakia | 1.25 | 0.29 | 0.39 | 1.93 | 2.44 | 0.57 | 1.08 | 4.11 | 4.52 | 1.05 | 2.01 | 7.59 | 4.75 | 1.43 | 2.07 | 8.25 | 5.43 | 1.26 | 2.38 | 9.07 |
| German Democratic Republic | 1.50 | 0.16 | 0.73 | 2.39 | 3.30 | 0.38 | 1.48 | 5.16 | 5.92 | 0.74 | 3.05 | 9.71 | 6.27 | 0.58 | 3.70 | 10.55 | 7.18 | 1.46 | 2.59 | 11.23 |
| Hungary | (b) 0.29 | (b) 0.20 | (b) 0.30 | (b) 0.79 | 1.32 | 0.38 | 0.04 | 1.74 | 2.76 | 0.56 | 0.31 | 3.62 | 2.80 | 0.70 | 0.42 | 3.92 | 3.21 | 0.54 | 0.48 | 4.23 |
| Poland | 0.62 | 0.25 | 0.46 | 1.33 | 1.86 | 0.87 | 0.98 | 3.71 | 3.97 | 1.30 | 0.73 | 6.00 | 3.72 | 1.09 | 1.79 | 6.61 | 4.36 | 1.03 | 1.69 | 7.07 |
| Romania | 0.28 | 0.15 | 0.03 | 0.45 | 0.60 | 0.41 | 0.47 | 1.49 | 1.02 | 0.91 | 1.22 | 3.16 | 0.99 | 0.70 | 1.42 | 3.11 | 1.23 | 0.56 | 1.18 | 2.96 |
| Eastern Europe | (c) 3.64 | (c) 0.85 | (c) 1.61 | (c) 6.10 | 10.79 | 3.12 | 4.16 | 18.08 | 21.00 | 6.06 | 7.41 | 34.45 | 21.40 | 6.42 | 9.98 | 37.80 | 24.91 | 6.80 | 8.73 | 40.44 |
| Soviet Union | 0.98 | 0.59 | 1.03 | 2.60 | 2.13 | 2.39 | 0.34 | 5.56 | 5.21 | 4.88 | 0.79 | 10.88 | 4.32 | 5.05 | 0.84 | 10.21 | 4.10 | 5.45 | 0.86 | 10.42 |
| Seven Eastern countries | (c) 4.62 | (c) 1.44 | (c) 2.64 | (c) 8.70 | 13.82 | 5.51 | 4.50 | 23.63 | 26.20 | 10.94 | 8.20 | 45.35 | 25.72 | 11.47 | 10.82 | 48.00 | 29.01 | 12.25 | 9.60 | 50.86 |

Notes: The "Unidentified" item consists of the sum of those items reported as a total for all regions but not by country of destination.

a) 1973.
b) 1971.
c) Excluding Bulgaria and Hungary.

Source: OECD East-West technology transfer data base (Eastern statistics).

Table 4 (Chapter 6)

EASTERN EXPORTS OF CAPITAL GOODS TO THE REST OF THE WORLD(a)
($ million)

| | 1975 | | | 1980 | | | 1982 | | | 1983 | | |
|---|---|---|---|---|---|---|---|---|---|---|---|---|
| | To Rest of World | of which: | | To Rest of World | of which: | | To Rest of World | of which: | | To Rest of World | of which: | |
| | Total | To OECD | To "other" countries | Total | To OECD | To "other" countries | Total | To OECD | To "other" countries | Total | To OECD | To "other" countries |
| Bulgaria | 504 | 32 | 472 | 1 502 | 86 | 1 416 | 1 917 | 55 | 1 862 | 1 965 | 38 | 1 927 |
| Czechoslovakia | 575 | 210 | 365 | 1 048 | 343 | 705 | 1 430 | 241 | 1 189 | 1 255 | 226 | 1 029 |
| German Democratic Republic | 380 | 192 | 188 | 739 | 348 | 391 | 578 | 289 | 289 | 1 459 | 231 | 1 228 |
| Hungary | 383 | 76 | 307 | 556 | 208 | 348 | 700 | 176 | 524 | 538 | 133 | 405 |
| Poland | 866 | 267 | 599 | 1 302 | 443 | 859 | 1 093 | 254 | 839 | 1 027 | 236 | 791 |
| Romania | 410 | 99 | 311 | 914 | 205 | 709 | 700 | 160 | 540 | 555 | 123 | 432 |
| Total Eastern Europe | 3 117 | 876 | 2 241 | 6 062 | 1 626 | 4 428 | 6 418 | 1 175 | 5 343 | 6 800 | 987 | 5 813 |
| Soviet Union | 2 388 | 254 | 2 134 | 4 881 | 351 | 4 529 | 5 050 | 174 | 4 876 | 5 451 | 157 | 5 294 |
| Total Eastern Europe and the Soviet Union | 5 505 | 1 130 | 4 375 | 10 943 | 1 977 | 8 966 | 11 468 | 1 349 | 10 219 | 12 251 | 1 145 | 11 106 |

Note: National data have been augmented by the use of mirror statistics of OECD imports only. For this reason the Table partially understates exports of the Eastern countries which do not publish detailed partner trade statistics -- especially Bulgaria, the German Democratic Republic and Romania -- which would allow their exports to the Rest of the World in total, and especially to "other" countries, to be calculated. "Other" countries refer in practice to the developing countries, including those of the non-European CMEA (Cuba, Mongolia and the Socialist Republic of Vietnam).

a) Exports to World less exports to other Eastern countries and to unidentified destinations as shown in Table 3 of Chapter 6.

Source: OECD East-West technology transfer data base; 1st column of each year is based on Eastern statistics and 2nd column of each year is based on Western Statistics. The third column is a residual of the first two and is subject to a considerable margin of error.

Table 5 (Chapter 6)

EASTERN COUNTRY SHARES IN EASTERN CAPITAL GOODS EXPORTS
(per cent)

| | 1975 | | | | 1980 | | | | 1982 | | | | 1983 | | | |
|---|---|---|---|---|---|---|---|---|---|---|---|---|---|---|---|---|
| | From: | | | | From: | | | | From: | | | | From: | | | |
| | East | Rest of World | Uniden-tified | Total | East | Rest of World | Uniden-tified | Total | East | Rest of World | Uniden-tified | Total | East | Rest of World | Uniden-tified | Total |
| Bulgaria | 9.3 | 9.2 | 2.2 | 7.9 | 10.7 | 13.7 | 1.1 | 9.7 | 11.2 | 16.7 | 5.2 | 11.2 | 12.0 | 16.0 | 4.4 | 11.6 |
| Czechoslovakia | 17.9 | 10.4 | 24.0 | 17.4 | 17.3 | 19.6 | 24.6 | 16.7 | 18.5 | 12.5 | 19.2 | 17.2 | 18.7 | 10.2 | 24.8 | 17.8 |
| German Democratic Republic | 24.2 | 6.9 | 32.8 | 21.8 | 22.6 | 6.8 | 37.2 | 21.4 | 24.4 | 5.0 | 34.2 | 22.0 | 24.8 | 11.9 | 27.0 | 22.1 |
| Hungary | 9.7 | 7.0 | 0.8 | 7.4 | 10.5 | 5.1 | 3.8 | 8.0 | 10.9 | 6.1 | 3.9 | 8.2 | 11.1 | 4.4 | 5.0 | 8.3 |
| Poland | 13.7 | 15.7 | 22.0 | 15.7 | 15.2 | 11.9 | 8.9 | 13.2 | 14.5 | 9.5 | 16.6 | 13.8 | 15.0 | 8.4 | 17.6 | 13.9 |
| Romania | 4.4 | 7.4 | 10.6 | 6.3 | 4 0 | 8.4 | 14.9 | 7.0 | 4.0 | 6.1 | 13.1 | 6.5 | 4.2 | 4.5 | 12.3 | 5.8 |
| Eastern Europe | 79.2 | 56.6 | 92.4 | 76.5 | 80.1 | 55.4 | 90.3 | 76.0 | 83.2 | 56.0 | 92.2 | 78.7 | 85.9 | 55.5 | 91.0 | 79.5 |
| Soviet Union | 20.8 | 43.4 | 7.6 | 23.5 | 19.9 | 44.6 | 9.7 | 24.9 | 16.8 | 44.0 | 7.8 | 21.3 | 14.1 | 44.5 | 9.0 | 20.5 |
| Seven Eastern countries | 100.0 | 100.0 | 100.0 | 100.0 | 100.0 | 100.0 | 100.0 | 100.0 | 100.0 | 100.0 | 100.0 | 100.0 | 100.0 | 100.0 | 100.0 | 100.0 |

Note: The "Unidentified" item consists of those items reported as a total for all regions but not by country of destination.

Source: OECD East-West technology transfer data base (Eastern statistics).

Table 6 (Chapter 6)

EASTERN COUNTRIES' EXPORTS OF CAPITAL GOODS COMPARED WITH TOTAL EXPORTS, BY REGION OF DESTINATION
(components expressed as per cent of total export to regions shown; totals in $ million)

| | 1970 | | | 1975 | | | 1980 | | | 1982 | | | 1983 | | |
|---|---|---|---|---|---|---|---|---|---|---|---|---|---|---|---|
| | East | Rest of World | World | East | Rest of World | World | East | Rest of World | World | East | Rest of World | World | East | Rest of World | World |
| **Bulgaria** | | | | | | | | | | | | | | | |
| Identified capital goods | (a) 33.9 | (a) 41.2 | (a) 35.6 | 36.1 | 42.4 | 37.7 | 40.4 | 43.1 | 41.3 | 37.2 | 52.0 | 31.0 | 39.4 | 58.3 | 44.9 |
| Unidentified capital goods | - | - | 5.6 | - | - | 2.1 | - | - | 0.8 | - | - | 4.9 | - | - | 3.5 |
| Total exports ($ million) | 2 460 | 779 | 3 239 | 3 498 | 1 189 | 4 687 | 6 901 | 3 486 | 10 387 | 7 749 | 3 686 | 11 439 | 8 872 | 3 265 | 12 137 |
| **Czechoslovakia** | | | | | | | | | | | | | | | |
| Identified capital goods | 51.5 | 21.3 | 40.7 | 44.6 | 19.9 | 36.1 | 47.8 | 19.2 | 37.3 | 45.6 | 27.4 | 39.5 | 48.6 | 23.6 | 40.5 |
| Unidentified capital goods | - | - | 10.2 | - | - | 13.1 | - | - | 13.5 | - | - | 13.3 | - | - | 14.4 |
| Total exports ($ million) | 2 435 | 1 358 | 3 793 | 5 469 | 2 891 | 8 360 | 9 461 | 5 465 | 14 926 | 10 422 | 5 218 | 15 640 | 11 188 | 5 312 | 16 500 |
| **German Dem. Rep.** | | | | | | | | | | | | | | | |
| Identified capital goods | 47.8 | 11.1 | 36.2 | 48.2 | 11.7 | 36.5 | 53.6 | 11.8 | 38.5 | 48.6 | 6.6 | 31.5 | 50.6 | 15.2 | 36.3 |
| Unidentified capital goods | - | - | 15.8 | - | - | 14.6 | - | - | 17.6 | - | - | 17.0 | - | - | 10.9 |
| Total exports ($ million) | 3 127 | 1 454 | 4 581 | 6 850 | 3 239 | 10 089 | 11 052 | 6 258 | 17 311 | 12 897 | 8 846 | 21 743 | 14 182 | 9 611 | 23 793 |
| **Hungary** | | | | | | | | | | | | | | | |
| Identified capital goods | (b) 29.0 | (b) 6.7 | (b) 21.1 | 32.0 | 19.4 | 29.8 | 35.0 | 15.6 | 28.6 | 31.5 | 19.3 | 28.0 | 34.6 | 13.3 | 28.8 |
| Unidentified capital goods | - | - | 4.6 | - | - | 0.6 | - | | 6.4 | - | - | 3.4 | - | - | 3.5 |
| Total exports ($ million) | 1 616 | 885 | 2 501 | 3 753 | 1 966 | 5 719 | 7 876 | 3 727 | 11 603 | 8 853 | 3 630 | 12 483 | 9 276 | 4 029 | 13 305 |
| **Poland** | | | | | | | | | | | | | | | |
| Identified capital goods | 28.7 | 17.6 | 24.3 | 32.0 | 19.4 | 26.5 | 44.6 | 16.0 | 31.0 | 41.1 | 19.3 | 32.7 | 41.2 | 18.3 | 33.3 |
| Unidentified capital goods | - | - | 13.1 | - | - | 9.6 | - | - | 4.3 | - | - | 12.2 | - | - | 10.4 |
| Total exports ($ million) | 2 145 | 1 403 | 3 548 | 5 821 | 4 468 | 10 289 | 8 902 | 8 119 | 17 021 | 9 059 | 5 676 | 14 735 | 10 593 | 5 605 | 16 198 |

Table 6 (Chapter 6) contd

| | 1970 | | | 1975 | | | 1980 | | | 1982 | | | 1983 | | |
|---|---|---|---|---|---|---|---|---|---|---|---|---|---|---|---|
| | East | Rest of World | World | East | Rest of World | World | East | Rest of World | World | East | Rest of World | World | East | Rest of World | World |
| Romania | | | | | | | | | | | | | | | |
| Identified capital goods | 30.0 | 16.0 | 23.0 | 29.5 | 12.4 | 18.9 | 24.1 | 12.8 | 17.0 | 30.7 | 10.2 | 16.7 | 35.8 | 8.2 | 17.5 |
| Unidentified capital goods | - | - | 1.4 | - | - | 8.9 | - | - | 10.7 | - | - | 14.0 | - | - | 11.6 |
| Total exports ($ million) | 924 | 927 | 1 851 | 2 041 | 3 300 | 5 341 | 4 244 | 7 157 | 11 401 | 3 236 | 6 887 | 10 123 | 3 419 | 6 744 | 10 163 |
| Total, Eastern Europe | | | | | | | | | | | | | | | |
| Identified capital goods | 42.2 | 16.5 | 32.6 | 40.1 | 16.2 | 30.9 | 43.4 | 17.4 | 32.7 | 41.0 | 18.9 | 32.3 | 43.3 | 19.7 | 34.4 |
| Unidentified capital goods | - | - | 12.0 | - | - | 10.6 | - | - | 9.0 | - | - | 11.6 | - | - | 9.5 |
| Total exports ($ million) | 8 631 | 5 142 | 13 773 | 27 432 | 17 053 | 44 485 | 48 436 | 34 820 | 82 650 | 52 215 | 33 947 | 86 162 | 57 531 | 34 566 | 92 096 |
| Soviet Union | | | | | | | | | | | | | | | |
| Identified capital goods | 18.5 | 21.1 | 19.7 | 17.2 | 14.2 | 15.7 | 16.1 | 11.0 | 13.1 | 11.9 | 10.2 | 10.8 | 10.5 | 10.5 | 10.5 |
| Unidentified capital goods | - | - | 5.7 | - | - | 1.0 | - | - | 1.0 | - | - | 1.0 | - | - | 1.0 |
| Total exports ($ million) | 6 758 | 6 042 | 12 800 | 16 435 | 16 846 | 33 281 | 32 239 | 44 253 | 76 492 | 36 205 | 50 764 | 86 969 | 39 241 | 52 145 | 91 386 |
| Total, Eastern countries | | | | | | | | | | | | | | | |
| Identified capital goods | 31.8 | 19.0 | 26.4 | 31.5 | 15.2 | 24.4 | 32.5 | 13.9 | 23.3 | 29.1 | 13.5 | 21.4 | 30.0 | 14.1 | 22.5 |
| Unidentified capital goods | - | - | 6.5 | - | - | 6.5 | - | - | 5.2 | - | - | 6.2 | - | - | 5.2 |
| Total exports ($ million) | 15 389 | 11 184 | 26 573 | 43 867 | 33 899 | 77 766 | 80 675 | 78 465 | 159 142 | 88 420 | 84 707 | 173 129 | 96 772 | 86 711 | 183 482 |

Notes: Total, Eastern Europe and Total Eastern countries for 1970 exclude Bulgaria and Hungary. The "Unidentified" item consists of the sum of those items reported as a total for all regions but not by country of destination.
a) 1973.
b) 1971.

Source: OECD East-West technology transfer data base (Eastern statistics).

Table 7 (Chapter 6)

MATRICES OF INTRA-EASTERN AREA CAPITAL GOODS TRADE, 1975 AND 1983(a)
($ million)

| Suppliers / Importers | Bulgaria | Czechoslovakia | German Democratic Republic | Hungary | Poland | Romania | Total Eastern Europe | Soviet Union | Total Eastern and Soviet Union |
|---|---|---|---|---|---|---|---|---|---|
| **1975** | | | | | | | | | |
| Bulgaria | - | 96 | 151 | 26 | 128 | 19 | 420 | 721 | 1 141 |
| Czechoslovakia | 76 | - | 519 | 149 | 293 | 73 | 1 109 | 418 | 1 527 |
| German Democratic Republic | 46 | 355 | - | 170 | 471 | 32 | 1 075 | 676 | 1 751 |
| Hungary (b) | 41 | 110 | 193 | - | 77 | 63 | 483 | 271 | 754 |
| Poland | 115 | 347 | 508 | 70 | - | 73 | 1 112 | 626 | 1 743 |
| Romania | 6 | 98 | 69 | 73 | 45 | - | 334 | 244 | 533 |
| Total Eastern Europe (b) | 284 | 1 006 | 1 442 | 488 | 1 014 | 260 | 4 435 | 2 956 | 7 449 |
| Soviet Union | 979 | 1 346 | 1 914 | 776 | 1 159 | 341 | 6 514 | - | 6 514 |
| Total Eastern Europe and Soviet Union (b) | 1 263 | 2 350 | 3 355 | 1 264 | 2 173 | 647 | 11 049 | 2 956 | 14 004 |
| **1983** | | | | | | | | | |
| Bulgaria | - | 246 | 358 | 106 | 256 | 17 | 983 | 1 072 | 2 026 |
| Czechoslovakia | 218 | - | 887 | 330 | 471 | 120 | 2 026 | 678 | 2 703 |
| German Democratic Republic | 102 | 664 | - | 282 | 318 | 25 | 1 391 | 863 | 2 253 |
| Hungary (b) | 80 | 222 | 420 | - | 124 | 63 | 909 | 4 0 | 1 319 |
| Poland | 158 | 601 | 653 | 256 | - | 128 | 1 796 | 635 | 2 431 |
| Romania | 15 | 109 | 132 | 67 | 148 | - | 471 | 317 | 788 |
| Total Eastern Europe (b) | 572 | 1 842 | 2 451 | 1 042 | 1 317 | 353 | 7 576 | 3 973 | 11 550 |
| Soviet Union | 2 867 | 3 612 | 4 744 | 2 296 | 2 607 | 873 | 16 999 | - | 16 999 |
| Total Eastern Europe and Soviet Union (b) | 3 439 | 5 454 | 7 195 | 3 338 | 3 924 | 1 226 | 24 575 | 3 973 | 28 549 |

Notes: Since data for both the German Democratic Republic and Romania, whose statistics are lacking, are constructed from partner mirror statistics, flows between these two countries are greatly understated in the matrix.
a) Based on import data.
b) Excluding Hungary's imports of complete plant.

Source: OECD East-West technology transfer data base (Eastern statistics).

Table 8 (Chapter 6)

RESPECTIVE SHARES OF EASTERN EUROPE AND THE SOVIET UNION IN THEIR MUTUAL TRADE IN CAPITAL GOODS IN 1975, 1980, 1982 AND 1983

| | Eastern Europe Imports from Soviet Union (per cent of Eastern European imports from 7 Eastern countries) | | | | Soviet Union Imports from Eastern Europe (per cent of 7 Eastern countries imports from Eastern Europe) | | | |
|---|---|---|---|---|---|---|---|---|
| | 1975 | 1980 | 1982 | 1983 | 1975 | 1980 | 1982 | 1983 |
| Bulgaria | 63.2 | 58.2 | 49.5 | 52.1 | 77.5 | 76.4 | 81.1 | 83.4 |
| Czechoslovakia | 27.4 | 31.9 | 24.7 | 25.1 | 57.2 | 59.8 | 63.7 | 66.2 |
| German Democratic Republic | 38.6 | 37.5 | 41.9 | 38.3 | 57.0 | 60.5 | 67.3 | 65.9 |
| Hungary (a) | 35.9 | 30.7 | 32.8 | 31.1 | 61.3 | 60.3 | 66.5 | 68.8 |
| Poland | 36.0 | 41.1 | 30.2 | 26.1 | 53.3 | 63.5 | 59.4 | 66.4 |
| Romania | 42.2 | 33.5 | 34.3 | 40.2 | 52.8 | 43.5 | 56.9 | 71.2 |
| Total Eastern Europe (a) | 39.5 | 39.3 | 35.6 | 34.4 | 59.0 | 61.5 | 66.3 | 69.2 |
| Total Eastern Europe and Soviet Union (a) | (b) 21.2 | (b) 19.9 | (b) 15.7 | (b) 13.9 | (c) 46.5 | (c) 49.2 | (c) 55.9 | (c) 59.5 |

Notes: Based on import matrices in Table 7 of Chapter 6.

a) Excluding Hungary's imports of complete plant.
b) Including Soviet imports from Eastern Europe (i.e. Eastern Europe's share of total intra-area trade).
c) Including Eastern European imports from the Soviet Union (i.e. Soviet Union's share of total intra-area trade).

Source: OECD East-West technology transfer data base (Eastern statistics).

Table 9 (Chapter 6)

OECD EXPORTS OF TECHNOLOGY AND TECHNOLOGY BASED-PRODUCTS TO THE EASTERN COUNTRIES (CLASSIFIED BY R&D INTENSITY)
(percentage of total embodied technology exports to the East)

| | 1970 | 1975 | 1980 | 1982 | 1983 |
|---|---|---|---|---|---|
| Highly R&D-Intensive | | | | | |
| Bulgaria | 2.30 | 2.89 | 3.72 | 4.34 | 4.64 |
| Czechoslovakia | 3.63 | 3.37 | 3.35 | 4.08 | 4.21 |
| German Democratic Republic | 1.94 | 1.36 | 3.25 | 3.93 | 2.13 |
| Hungary | 3.88 | 3.50 | 4.53 | 4.22 | 4.22 |
| Poland | 2.48 | 1.82 | 1.83 | 2.71 | 3.26 |
| Romania | 4.12 | 4.31 | 6.20 | 8.33 | 10.70 |
| Total excluding the Soviet Union | 3.20 | 2.69 | 3.65 | 4.32 | 4.32 |
| Soviet Union | 1.97 | 1.31 | 1.53 | 1.47 | 1.80 |
| Total including the Soviet Union | 2.67 | 2.05 | 2.58 | 2.56 | 2.73 |
| Moderately R&D-Intensive | | | | | |
| Bulgaria | 20.77 | 14.39 | 20.29 | 19.33 | 19.87 |
| Czechoslovakia | 20.66 | 25.04 | 24.14 | 24.99 | 26.75 |
| German Democratic Republic | 13.68 | 20.14 | 23.41 | 19.78 | 18.62 |
| Hungary | 28.52 | 27.83 | 24.86 | 25.08 | 24.36 |
| Poland | 21.73 | 15.20 | 22.56 | 23.38 | 29.03 |
| Romania | 16.14 | 15.66 | 18.00 | 21.05 | 24.60 |
| Total excluding the Soviet Union | 20.39 | 18.59 | 22.29 | 24.61 | 24.29 |
| Soviet Union | 16.11 | 13.70 | 19.15 | 16.19 | 17.92 |
| Total including the Soviet Union | 18.55 | 16.32 | 20.70 | 19.40 | 20.25 |
| Low R&D-Intensive | | | | | |
| Bulgaria | 76.93 | 82.72 | 75.99 | 76.34 | 75.49 |
| Czechoslovakia | 75.71 | 71.60 | 72.51 | 70.93 | 69.03 |
| German Democratic Republic | 84.37 | 78.50 | 73.33 | 76.29 | 79.24 |
| Hungary | 67.61 | 68.67 | 70.61 | 70.69 | 71.42 |
| Poland | 75.79 | 82.98 | 75.61 | 64.91 | 67.61 |
| Romania | 79.74 | 80.03 | 75.80 | 70.61 | 64.70 |
| Total excluding the Soviet Union | 76.40 | 78.72 | 74.06 | 71.06 | 71.49 |
| Soviet Union | 81.91 | 84.99 | 79.32 | 82.34 | 80.28 |
| Total including the Soviet Union | 78.78 | 81.63 | 76.73 | 78.04 | 77.02 |

<u>Source</u>: OECD East-West technology transfer data base (Western statistics).

Table 10 (Chapter 6)

RELATIONSHIP BETWEEN INVESTMENT GOODS EXPORTS FROM THE OECD TO EASTERN COUNTRIES AND GROSS FIXED INVESTMENT IN THESE COUNTRIES (a)
(Indices, 1970 = 100)

| | Industry | | | | Material sphere | | | | Total economy | | | |
|---|---|---|---|---|---|---|---|---|---|---|---|---|
| | 1975 | 1980 | 1982 | 1983 | 1975 | 1980 | 1982 | 1983 | 1975 | 1980 | 1982 | 1983 |
| Bulgaria | 173 | 66 | 95 | 101 | 174 | 72 | 110 | 108 | 170 | 70 | 99 | 105 |
| Czechoslovakia | 78 | 50 | 49 | 43 | 81 | 53 | 52 | 47 | 81 | 56 | 55 | 51 |
| German Democratic Republic (b) | 60 | 65 | 84 | 102 | 86 | 76 | 99 | 113 | 81 | 71 | 91 | 108 |
| Hungary | 123 | 123 | 158 | 128 | 128 | 131 | 166 | 142 | 118 | 121 | 155 | 129 |
| Poland | 149 | 97 | 65 | 62 | 201 | 116 | 64 | 62 | 218 | 114 | 60 | 59 |
| Romania | 63 | 32 | 12 | 7 | 74 | 38 | 14 | 8 | 75 | 40 | 15 | 8 |
| Soviet Union | 162 | 95 | 118 | 124 | 150 | 98 | 131 | 138 | 158 | 105 | 139 | 147 |

Notes:

a) Volume index of OECD investment goods exports to each country divided by the volume index of the country's total gross fixed capital formation.
b) Excluding exports from the Federal Republic of Germany to the German Democratic Republic.

Source: National Statistics; OECD East-West technology transfer data base (Western volume statistics).

Table 11 (Chapter 6)

EAST EUROPEAN IMPORTS FROM THE WEST OF CAPITAL GOODS FOR NON-TRADEABLES, PRODUCT-CREATING CAPITAL GOODS AND OUTPUT-ASSISTING CAPITAL GOODS(a)
(per cent of total capital goods imports in current prices)

| | 1970 | 1975 | 1980 | 1982 | 1983 |
|---|---|---|---|---|---|
| Eastern Europe | | | | | |
| Non-tradeable product-creating technology | 30.2 | 41.2 | 38.1 | 34.7 | 34.5 |
| -- whereof gas and oil distribution | 11.2 | 15.5 | 13.3 | 13.5 | 12.3 |
| transport | 5.6 | 12.0 | 9.5 | 4.6 | 4.7 |
| Tradeable product-creating technology (a) | 43.2 | 37.7 | 39.5 | 41.2 | 41.1 |
| Output-assisting technology (a) | 19.7 | 16.6 | 18.4 | 18.7 | 19.0 |
| Residual | 6.9 | 4.5 | 4.1 | 5.5 | 5.5 |
| Soviet Union | | | | | |
| Non-tradeable product-creating technology | 40.9 | 45.9 | 45.9 | 59.5 | 58.5 |
| -- whereof gas and oil distribution | 16.1 | 19.6 | 25.1 | 29.9 | 25.1 |
| transport | 17.3 | 16.2 | 10.2 | 14.6 | 17.5 |
| Tradeable product-creating technology (a) | 40.9 | 36.0 | 33.0 | 21.5 | 21.5 |
| Output-assisting technology (a) | 14.1 | 13.0 | 15.6 | 15.5 | 15.9 |
| Residual | 4.1 | 5.1 | 5.4 | 3.5 | 4.1 |

Note: a) Product-creating technology represented by capital goods contributing directly to output (exportables). Output-assisting technology represented by capital goods indirectly contributing to output. For actual classification see text.

Source: OECD East-West technology transfer data base (Western statistics).

Table 12 (Chapter 6)

EASTERN CAPITAL GOODS IMPORTS FROM THE WEST, BY TYPE
(per cent of total)

| Machinery/ equipment type | 1970 | 1975 | 1980 | 1982 | 1983 |
|---|---|---|---|---|---|
| Eastern Europe | | | | | |
| Power, power distribution equipment | 7.5 | 7.2 | 8.1 | 8.5 | 9.1 |
| Oil, gas distribution | 11.2 | 15.5 | 13.3 | 13.5 | 12.3 |
| Transport (excluding oil and gas distribution) | 5.6 | 12.0 | 9.5 | 4.6 | 4.7 |
| Machine tools, metal processing and metallurgy | 9.3 | 9.4 | 10.6 | 9.9 | 12.2 |
| Construction, mining etc. | 3.9 | 4.7 | 5.1 | 4.8 | 4.6 |
| Furnaces, gas generators | 3.0 | 3.4 | 3.6 | 3.3 | 3.0 |
| Machinery and tools for wood, plastics | 2.4 | 2.5 | 1.9 | 2.4 | 2.6 |
| Electronics, communications equipment | 5.9 | 3.5 | 3.1 | 3.1 | 3.4 |
| - computers | 3.9 | 2.0 | 1.4 | 1.5 | 1.7 |
| Machinery for mainly consumer industries (a) | 18.8 | 13.2 | 12.3 | 12.9 | 11.0 |
| Mechanical handling, storage | 4.2 | 5.3 | 5.7 | 5.0 | 5.3 |
| Instruments | 5.7 | 3.8 | 5.4 | 5.7 | 6.3 |
| Structural parts | 2.4 | 3.0 | 2.6 | 3.5 | 2.3 |
| Other (b) | 20.1 | 16.5 | 18.8 | 22.8 | 23.2 |
| Total | 100.0 | 100.0 | 100.0 | 100.0 | 100.0 |
| Soviet Union | | | | | |
| Power, power distribution equipment | 4.0 | 5.1 | 6.1 | 5.6 | 9.1 |
| Oil, gas distribution | 16.1 | 19.6 | 25.1 | 29.9 | 25.1 |
| Transport (excluding oil and gas distribution) | 17.3 | 16.2 | 10.2 | 14.6 | 17.5 |
| Machine tools, metal processing and metallurgy | 17.0 | 12.9 | 13.7 | 7.6 | 9.3 |
| Construction, mining, etc. | 2.7 | 4.3 | 3.2 | 8.2 | 5.6 |
| Furnaces, gas generators | 4.7 | 2.5 | 2.4 | 1.7 | 1.8 |
| Machinery and tools for wood, plastics | 1.2 | 1.6 | 0.9 | 0.9 | 0.9 |
| Electronics, communications equipment | 3.4 | 1.6 | 2.2 | 1.9 | 2.3 |
| - computers | 1.7 | 0.9 | 1.1 | 0.8 | 0.9 |
| Machinery for mainly consumer industries (a) | 7.6 | 7.0 | 5.9 | 4.8 | 5.0 |
| Mechanical handling, storage | 4.1 | 5.5 | 5.7 | 6.8 | 5.2 |
| Instruments | 4.7 | 3.1 | 4.6 | 3.1 | 3.6 |
| Structural parts | 0.8 | 1.7 | 2.4 | 3.3 | 4.5 |
| Other (b) | 16.4 | 18.9 | 17.6 | 11.6 | 10.1 |
| Total | 100.0 | 100.0 | 100.0 | 100.0 | 100.0 |

Notes:

a) For textile, leather, paper, pulp, food, glass and printing industries.

b) Largely SITC item 719.8 "machinery and mechanical appliances n.e.s".

Source: OECD East-West technology transfer data base (Western statistics).

Table 13 (Chapter 6)

EASTERN COUNTRY CAPITAL GOODS IMPORTS FROM THE WEST IDENTIFIED BY REGION OF ORIGIN, BY BROAD INDUSTRIAL END-USE DIVISIONS
(per cent of total industrial capital goods imports from the West)

| | Fuel and energy (a) | | | | Supply industries (b) | | | | Consumer industries(c) | | | |
|---|---|---|---|---|---|---|---|---|---|---|---|---|
| | 1975 | 1980 | 1982 | 1983 | 1975 | 1980 | 1982 | 1983 | 1975 | 1980 | 1982 | 1983 |
| Bulgaria | 7 | 13 | 13 | 11 | 50 | 68 | 70 | 67 | 43 | 19 | 18 | 22 |
| Czechoslovakia | 21 | 58 | 27 | 24 | 43 | 20 | 26 | 32 | 36 | 22 | 47 | 43 |
| German Democratic Republic | 23 | 19 | 13 | 10 | 23 | 46 | 57 | 71 | 54 | 34 | 30 | 20 |
| Hungary | 21 | 19 | 16 | 23 | 39 | 45 | 47 | 41 | 40 | 36 | 28 | 36 |
| Poland | 14 | 30 | 36 | 30 | 56 | 55 | 52 | 46 | 31 | 15 | 12 | 23 |
| Romania | 21 | 18 | 42 | 19 | 59 | 64 | 35 | 65 | 20 | 18 | 23 | 16 |
| Soviet Union | 15 | 22 | 28 | 37 | 65 | 63 | 56 | 50 | 19 | 16 | 17 | 12 |

Notes:
a) Fuel and energy branches.
b) Metallurgy, engineering, chemicals, construction materials branches.
c) Wood, paper, light and food branches.

Source: OECD East-West technology transfer data base (Eastern statistics).

Table 14 (Chapter 6)

EASTERN CAPITAL GOODS IMPORTS FROM THE CMEA COUNTRIES IDENTIFIED BY REGION OF ORIGIN, BY BROAD INDUSTRIAL END-USE DIVISIONS
(per cent of total industrial capital goods imports from the East)

| | Fuel and energy (a) | | | | Supply industries (b) | | | | Consumer industries(c) | | | |
|---|---|---|---|---|---|---|---|---|---|---|---|---|
| | 1975 | 1980 | 1982 | 1983 | 1975 | 1980 | 1982 | 1983 | 1975 | 1980 | 1982 | 1983 |
| Bulgaria | 38 | 37 | 34 | 39 | 44 | 49 | 51 | 48 | 18 | 14 | 14 | 13 |
| Czechoslovakia | 35 | 33 | 49 | 48 | 41 | 44 | 33 | 35 | 24 | 24 | 18 | 18 |
| German Democratic Republic | 35 | 42 | 51 | 50 | 46 | 38 | 31 | 34 | 19 | 21 | 18 | 16 |
| Hungary | 33 | 56 | 59 | 63 | 40 | 28 | 29 | 25 | 27 | 16 | 11 | 11 |
| Poland | 27 | 32 | 37 | 35 | 51 | 47 | 36 | 38 | 22 | 21 | 28 | 27 |
| Romania | 30 | 24 | 25 | 30 | 52 | 52 | 57 | 53 | 18 | 23 | 18 | 17 |
| Soviet Union | 28 | 33 | 31 | 33 | 45 | 41 | 41 | 39 | 27 | 26 | 28 | 29 |

Notes:
a) Fuel and energy branches.
b) Metallurgy, engineering, chemicals and construction materials branches.
c) Wood, paper, light and food branches.

Source: OECD East-West technology transfer data base (Eastern statistics).

Table 15 (Chapter 6)

SHARE OF WESTERN SUPPLIERS IN INDUSTRIAL CAPITAL GOODS IMPORTS
IDENTIFIED BY REGION OF ORIGIN, BY END-USE BRANCHES
(per cent of total)

| | Eastern Europe | | | | Soviet Union | | | |
|---|---|---|---|---|---|---|---|---|
| | 1975 | 1980 | 1982 | 1983 | 1975 | 1980 | 1982 | 1983 |
| Energy | 30 | 40 | 18 | 15 | 20 | 28 | 25 | 24 |
| Minining | 60 | 60 | 36 | 31 | 90 | 73 | 55 | 45 |
| Oil extraction | 24 | 19 | 8 | 8 | 81 | 69 | 68 | 81 |
| Oil refining | - | - | 68 | 74 | 56 | 18 | 11 | 39 |
| Metallurgy | 60 | 37 | 30 | 37 | 82 | 49 | 64 | 74 |
| Engineering | 48 | 39 | 29 | 30 | 75 | 62 | 51 | 51 |
| Chemicals | 73 | 64 | 56 | 47 | 59 | 65 | 52 | 48 |
| Construction materials | 57 | 35 | 36 | 45 | 38 | 31 | 33 | 38 |
| Wood, paper | 84 | 70 | 76 | 77 | 84 | 78 | 84 | 85 |
| Light industry | 55 | 34 | 33 | 33 | 51 | 16 | 14 | 11 |
| Food | 69 | 43 | 36 | 36 | 37 | 37 | 32 | 29 |
| Total | 54 | 43 | 32 | 30 | 62 | 50 | 45 | 48 |

Source: OECD East-West technology transfer data base (Eastern statistics).

## NOTES AND REFERENCES

1. Strictly speaking, the world other than the CMEA countries themselves. As shown in Chapter 5, section 1*b)* and Table 1, the great bulk of such imports are from OECD suppliers.
2. For 1970, import data for two countries – Bulgaria and Hungary – are lacking.
3. The basic facts of apparently substantial under-reporting by some Eastern European countries could involve a simple adjustment – viz. a reduction in the share of the unidentified item and its transfer to the Western component. The discrepancies between Eastern and Western-reported figures given in Table 1 of Chapter 5 suggest that for Eastern Europe as a whole, about 8 per cent of the unidentified item of $4 billion in 1973 (Table 3 of Chapter 5) originated in the West – but that this would mainly affect the figures for Bulgaria, the German Democratic Republic and Romania (where the West's share in total capital goods imports might rise, in the first two cases, from 12 and 9 per cent respectively to the still low figures of 17 and 12 per cent). For Poland and Romania, the shares of 34 and 35 per cent would rise to 37 and 43 per cent respectively. By 1983, East European under-reporting relative to OECD data had risen to 25 per cent of the unidentified item. Statistical under-reporting of West-to-East flows also appears in Hungary after 1975, while the under-reporting by Romania in 1973 had increased to the point of complete unreliability by 1983, but has almost disappeared for Poland. This was partially offset by the build-up of an apparent over-reporting of West-to-East flows by Czechoslovakia after 1975. Similar adjustments to the "unidentified" item for the year 1983, would raise the share of the Western component from 14 to 17 per cent for Bulgaria; from 10 to 12 per cent for the German Democratic Republic; from 19 to 24 per cent in Hungary; and from only 7 to 14 per cent for Romania. For Eastern Europe as a whole, the West's share would rise by about 2 percentage points in all years shown.
4. For instance, the official Soviet official ruble/dollar rate in 1982 was 0.743. The ruble/dollar exchange rate implied by the Bulgarian leva/dollar and leva/ruble rates was 0.732; and respectively, via the Czechoslovakian Crown, 0.762; via the German Democratic Republic Mark, 0.74; via the Hungarian Forint, 1.466; via the Polish Zloty, 1.247; and via the Romanian Leu, 1.00.
5. *Economic Survey of Europe in 1981*, *op. cit.*, Table 3.6.4.
6. However, the discussion of discrepancies found between "mirror" statistics of OECD-reported capital goods exports to the East, and Eastern-reported data on the same flow (see Table 1 of Chapter 5) should be borne in mind. It is also recalled that the figures for the German Democratic Republic's exports to unidentified countries are exaggerated because of the inclusion of its trade with the Federal Republic of Germany from the OECD-reported statistics. It should also be noted that OECD data on imports from the Eastern countries used in Table 3 of this chapter are on a c.i.f. basis, whereas the Eastern reported data are f.o.b.
7. This cannot be proved. Table 5 of Chapter 5 roughly confirms the quality of reported intra-CMEA export and import figures, and the size of OECD flows are known. Hence, it is possible that the

exports to other countries shown in Table 4 and the "unidentified" item shown in Table 3 both represent Eastern exports to developing countries. However, the possibility that some part of the unidentified item consists of shipments not registered in either the Eastern-reported data on exports to other CMEA countries, or in Eastern-reported figures on imports from other CMEA countries, cannot be ruled out.

8. The Bulgarian, German Democratic Republic and Romanian data understate exports to the Rest of the World as a whole since "mirror" data from OECD imports only were used to fill gaps in those countries' reported data ; discrepancies due to this are included in the unidentified item for these countries. See also note to Table 4 of this chapter.

9. In the absence of the facts on which a rational Eastern price system could be reconstructed, an alternative would be to use Western analogues – a procedure which would itself contain wide margins of error.

10. For instance, the decision of the German Democratic Republic to invest heavily in lignite extraction in the late 1950s-early 1960s which to some seemed less rational at the time than now.

11. John A. Martens, *Quantification of Western Exports of High Technology to Communist Countries*, US Department of Commerce, Washington, D.C., February 1983.

12. It also includes motor cars and a few consumer items, which are of negligible importance in Western exports to the East. The criteria for the selection of R&D intensity categories is explained in two publications by Regina Kelly: "Alternative Measures of Technology-intensive Trade" and "The Impact of Innovation on International Trade Patterns", US Department of Commerce, Office of Economic Research, Papers OER/ER-17 and ER-24, Washington D.C., September 1976 and December 1977, respectively. In addition, the second category in these papers ("low technology-intensive" items) were further sub-divided into "medium" and "low R&D"-intensive technology items. Details of this extended classification are to be found in the paper entitled "Technology Transfers to Developing Countries, Statistical Paper No. 2, Technology Transfer and Trade", OECD working document, Paris, 14 November 1978. The listings therein were subsequently slightly expanded to cover all metals.

13. There are few reliable estimates as to the actual share of imports of Western capital goods in Eastern investment expenditures. Official statistics of capital goods imports on a price basis supposedly comparable with domestic investment price levels indicate that in Hungary, the share of equipment imported from the dollar area rose from 9.6 per cent in 1976 to 9.8 per cent in 1980, and to 10.7 per cent in 1982; the corresponding figures for imports from the ruble area are 18.5, 17.3 and 16.1 (*Statiztikai Evkönyv*, various issues). In Poland, the corresponding figure for 1982 was 22.3 per cent for imports from all sources and 20.6 per cent in 1983 (*Rocznik Statystyczny* 1984, pp. 175 and 353). In Romania, where foreign trade has been supposedly valued in domestic prices since 1980, the share of machinery and equipment imported from all countries in total investment was 22.1 per cent in 1976-1980, 12.5 per cent in 1982 and 9.6 per cent in 1983 (*Anuarul Statistic* 1984, p. 191). For the Soviet Union, Philip Hanson has estimated Western equipment shares of 0.8 to 2.4, or alternatively 1.6-4.8 per cent of Soviet annual capital formation between 1955 and 1975 ("Soviet Machinery Imports", *Survey*, Vol. 23, No. 2, 1978, pp. 112-126 and "Technology Transfers to the Soviet Union", *Survey*, Vol.24, No. 2, 1977-1978, pp. 73-104).

14. Zdenek Drabek, "Western Embodied Technology and its Sectoral Impact on the Economies of Eastern Europe", OECD working document, Paris, 1981.

15. According to OECD data, up to the end of 1976, buy-back commitments by Western firms falling due in 1976-1980 were estimated at $10 billion for the Soviet Union – equivalent to some 13 per cent of average annual Soviet exports to the West in 1976-1980 – and to $4.5 billion for Eastern Europe (about 6 per cent of their average annual exports to the West in the same years).

16. The low share of Czechoslovakia is certainly connected with the growing degree of "overreporting" in national statistics vis-à-vis those of the OECD since 1975. The Czechoslovak-reported industry

total is much bigger than the OECD statistics suggest. Very large imports of capital goods for the energy industry in 1980 and for the wood and paper branches in 1982-1983, tend to reduce the Czechoslovak-reported shares of other components (particularly chemicals and engineering) – but this is not confirmed from OECD data.

17. In the two years 1975 and 1980, the chemical industry received "priority" (defined as having registered a share of industrial capital goods imported from the West higher than its share in industrial investment) in Bulgaria, Czechoslovakia, Poland, Romania and the Soviet Union. The engineering branch received priority in this sense only in Czechoslovakia (in 1975), Poland (in 1980) and the Soviet Union.

*Part III*

# EASTERN POLICIES AND THE ACQUISITION OF WESTERN TECHNOLOGY

*Chapter 7*

# TECHNOLOGY TRANSFER: FORMS OF CO-OPERATION

## COMMENT

For most of the Eastern countries hard currency constraints – hence the adoption of a more cautious policy towards the purchase of Western technology – explain essentially their interest in pursuing forms of industrial co-operation which will continue to provide access to Western technology, make provision for financial support and will assure some form of market sharing. In short, technology transfer mechanisms which are essentially complements to, or substitutes for, foreign trade (as classically defined) are being favoured by the Eastern countries.

The basic objective of this chapter is to highlight the individual Eastern country's approach to technology transfer, generally the overriding objective in seeking co-operation.

Section I is devoted to a general discussion of the major forms of co-operation. Particular attention is paid here to those types of inter-firm agreements which are being singled out by the CMEA countries as being particularly well-suited to meeting their objectives and hence likely to be most vigorously pursued by them. However, it must be noted at the outset that data limitations preclude an in-depth treatment. It is selective and must be regarded as being essentially illustrative. Section II briefly sums up the primary motivations of the Eastern and Western partners for pursuing industrial co-operation. Section III describes the individual Eastern country's participation in East-West industrial co-operation. Finally, Section IV draws some conclusions about the evolution of East-West industrial co-operation, its effectiveness in terms of meeting anticipated objectives and its likely future development.

## *Section I.* FORMS OF CO-OPERATION

### A. INTER-GOVERNMENTAL AGREEMENTS

Virtually all Western countries have concluded long-term inter-governmental agreements with the Soviet Union and the East European countries. These agreements, supplemented in most cases by annual protocols, provide for commercial, economic, industrial and scientific and technical co-operation.

According to the United Nations, at the end of August 1984, 336 bilateral trade and economic co-operation agreements were in force. Of these, 116 were trade agreements, 196 related to economic, industrial and technological co-operation and 24 encompassed both trade and economic co-operation. On the Eastern side it is the Soviet Union which has concluded the largest number (80), while on the Western side it is Finland and France (25 each)[1].

While the majority of them provide for economic co-operation, scientific and technical co-operation agreements also account for a large share of the total. The forms of co-operation undertaken within the framework of such agreements are multiple: ranging from simple exchange of visits and technical information to joint R&D work – inclusive of co-operation in the application of results ensuing from it.

In most cases, inter-governmental agreements provide for the establishment of mixed or joint commissions. Their purpose is to formulate guidelines for co-operation and to periodically review the operation and implementation of an agreement. The CMEA countries, particularly the Soviet Union, regard the commissions as being an especially useful instrument: either for developing economic relations or for furthering existing business relations with the Western countries – aspects frequently mentioned in the Eastern literature. The agreements either include a comprehensive list specifying those fields in which co-operation is to be developed, or such lists are subsequently established by the mixed commissions. Some agreements also specify the mode of co-operation – delivery of plant, licences and documentation, co-operation in production, and so forth.

The agreements are becoming increasingly complex: in terms of the fields covered; in that they now frequently include activities which had generally been negotiated on an *ad hoc* basis (for instance, tripartite co-operation or financing facilities for large scale, long-term compensation projects) and, because they often contain provisions which are designed to facilitate the expansion of trade and co-operation. Finally, since the mid-1970s the time frame of the agreements has been expanded and most now cover at least a decade while some agreements remain in force for 25 years[2].

It is clearly not feasible, within the confines of this study, to analyse inter-governmental co-operation agreements. They are so numerous and complex (and publicly available information about their effectiveness so scarce), that to assess their importance or impact calls for a separate (and different) research effort.

In any event, they are essentially enabling agreements and relatively few commercial contracts are concluded within this framework. These agreements do provide official support for co-operation at the firm or enterprise level. However, while they, therefore, facilitate such co-operation, they do not predicate it. It is also the opinion of many Western experts that the number of existing agreements has a purely relative significance, since they constitute primarily an expression of "goodwill" on the part of Western governments. Finally, the degree of interest in concluding such agreements is not uniform among Western governments and is highly dependent on the prevailing political climate.

Nevertheless, seen from the Eastern perspective, inter-governmental agreements are important precisely because "they are not confined to traditional trade agreements, do cover extremely diverse forms of co-operation for the long-term and facilitate the conclusion of commercial contracts and industrial co-operation agreements"[3].

## B. INTER-FIRM CO-OPERATION (EWIC)

### 1. Concept and Definitional Problems

There is no internationally agreed-to definition of East-West industrial (inter-firm) co-operation – EWIC. In part, this is because the concept encompasses a wide range of activities and forms which frequently overlap (see Annex Tables 2 and 3 of this chapter)[4].

Many EWIC agreements formally drawn up to cover a particular type of co-operation evolve into another (for example, joint marketing on Western or third markets), not foreseen at the time the original agreement was signed. This makes it difficult to classify, let alone define, the arrangement. It is also due to the fact that it is almost impossible to clearly demark the boundary between an industrial co-operation activity and other forms of East-West economic/market relationships. For instance, East-West subcontracting arrangements, often initiated by a Western firm to fulfill counter-purchase obligations, frequently evolve into more complex forms of co-production. Though not necessarily covered by a formal inter-enterprise agreement, they constitute, nevertheless, an important form of EWIC. Annex Table 1 (of this chapter) which gives a glossary of terms frequently used, also illustrates the difficulty in establishing a clear-cut boundary between EWIC and commercial arrangements.

Data limitations likewise contribute to the fact that any quantitative assessment made of industrial co-operation can only be approximate. First, data are frequently given in terms of the number of contracts or agreements concluded. However, they do not, necessarily, indicate whether or not an agreement was actually implemented, if (and when) terminated or renewed. Available data are, therefore, cumulative. This implies, in turn, double counting[5]. Few Western countries maintain records of industrial co-operation agreements which compounds the "counting problem". Second, data gathered through various surveys which have been undertaken are not strictly comparative – either because different definitions were used or because the sample of firms surveyed is too varied and small to be regarded as being representative. Also, surveys generally do not specify the dates when agreements take effect or expire. Thus, they indicate "flows" but not the entire "stock" of agreements in force. This also implies that the impact of technology transfer cannot be precisely assessed. Finally, many EWIC agreements do not attain their announced objectives, although they do lead to the exchange of information, personnel and know-how. Even though such agreements eventually become dormant they are, nevertheless, listed as being active – by both sides. Presumably this is done for the sake of appearance as well as to keep open channels for possible future exchanges.

Measuring industrial co-operation in value terms is also difficult. Little is generally known about the *content* of an industrial co-operation agreement because of secrecy maintained by the parties involved in a particular agreement. Also, the financial value of an industrial co-operation contract is rarely indicated. In any event, given the diversified nature of industrial co-operation agreements, how is one to specify the "value" of an agreement in a uniform manner?

Though bearing in mind that no unambiguous definition of industrial co-operation has been generally accepted (and the implications this has for conclusions drawn about it), the definition currently employed by the United Nations Economic Commission for Europe (ECE) will be adopted for the purpose of this analysis. Namely:

> "a contractual economic relationship between two or more enterprises of different nationalities, extending over a longer period, whereby a community of interests is established for the purpose of complementary activities relating to the supply of licences and equipment, development of new technologies, the exchange of information on and the use of those technologies, production and marketing with provision for the settlement in kind of whole or part of the obligations arising from co-operation activities"[6].

We have adopted this definition for a variety of reasons. First of all, it is the most convenient, in that it was negotiated and generally agreed-to by both the Eastern and Western side. It also captures the evolutionary character of East-West industrial co-operation and singles out aspects of co-operation which are currently emphasised by the Eastern countries. For instance, it is evident from the definition that it encompasses, in effect, some form of countertrade. While Western firms (or governments) do not consider countertrade as constituting a form of EWIC, it is clear that many Eastern countries, particularly the Soviet Union, consider it as being an intrinsic part of EWIC. Furthermore, the definition includes transfer modes which have been the subject of a number of analytical surveys. While such surveys cannot, for the reasons mentioned above, furnish documentary evidence of trends, they do indicate general developments and illustrate Eastern policies regarding industrial co-operation.

Finally, the very concept of co-operation as well as the modes included in the definition, are considered by some Western analysts as being the most effective in that they enable the Eastern countries to not only gain access to Western technology, but to use it. For instance, co-operation is extended beyond the production phase, the sale of embodied technology (plant or equipment) is supplemented by provision made for their further development and the transfer of know-how (inclusive of training), as well as provision of technical assistance, are implicitly stipulated in the definition[7].

## 2. Volume of East-West Industrial Co-operation Agreements

East-West industrial co-operation developed rapidly in the 1960s, especially during the first half of the 1970s. The United Nations has estimated that some 1 000 industrial co-operation agreements existed at the end of 1975[8].

To what extent EWIC has expanded since 1975 is difficult to evaluate. Table 1 of Chapter 7 implies that industrial co-operation (at least measured by the number of agreements concluded) continued to increase during 1975-1980. According to the UN, some 1 500 inter-firm co-operation agreements existed by 1983[9]. However, since the UN apparently does not delete agreements which have either ended or become dormant, this figure is probably inflated. This is also suggested by Tables 8-14 of the chapter which include only those agreements which were concluded during 1980-1984 and said to be in effect. According to this data, 758 agreements were signed between 1980-1984. Even though it is not known whether all agreements have been identified (or how many have led to actual co-operation), this total would seem to be more realistic. Furthermore, Tables 8-14 indicate a downward trend of EWIC growth over the 1980-1984 period for all countries except the Soviet Union.

Whatever the quantitative evolution of industrial co-operation may be, it can be assumed that the number of EWIC agreements is small compared with the total number of commercial arrangements. Furthermore, the UN points out that industrial co-operation "still represents a marginal share in East-West trade, of not more than 5-7 per cent"[10].

## 3. Type of Agreements

The most comprehensive published information on the development of industrial co-operation agreements is provided by the United Nations' (the ECE) surveys. However, the UN, itself, points out the limitations of their surveys. Namely, the overall value of contracts concluded is not known and, since the sample only covers a part of existing activities in industrial co-operation, it cannot be representative: in terms of country participation or industrial coverage. Classification of contracts by type of activity and by industry is somewhat arbitrary, since the various types of agreements overlap – a data shortcoming already pointed out. Furthermore, it should be recalled that while many agreements express the intent to co-operate, they frequently do not lead to actual co-operation. Since, as mentioned previously, the actual content of an agreement is rarely divulged, it is also difficult to assess the level of technology transferred.

Bearing in mind these shortcomings, we shall, nevertheless, take the latest UN survey (which now comprises 814 contracts, excluding Yugoslavia) as the basis for illustrating general developments in East-West industrial co-operation. We have done so primarily because it *is* the most recent survey currently available. Furthermore, juxtaposed to the earlier surveys, it does allow one to discern certain overall trends. For instance, that the CMEA countries continue to seek co-operation in roughly the same types of EWIC arrangements and industrial branches, though the percentage share varies as does the distribution of industrial co-operation contracts by country (see Tables 2 and 3 of Chapter 7).

Also, the survey encompasses those co-operative arrangements currently favoured by the Eastern countries. These forms clearly illustrate the change in emphasis away from such "classical" arrangements as the straightforward import of machinery and equipment or the purchase of turnkey plants, to more complex and varied types of industrial co-operation which allow for the acquisition of a combination of embodied and disembodied technology on other than a simple "purchase type" basis.

The definition of industrial co-operation adopted by the ECE explicitly mirrors this change. As is noted, the types of EWIC arrangements now most commonly practiced, all entail, "as a rule, the creation of a community of interests between parties which extends over a long period...and provide a stable basis for this continuous interaction such as could never be established through conventional trade"[11].

a) *Transfer Channels*

i) *Co-production and Specialisation*

Co-production based on the reciprocal supply of goods and services and sometimes according to the partners' respective patterns of specialisation, continues to be the most widespread form of co-operation[12]. It accounts for roughly 44 per cent of all EWIC agreements registered and is the preferred form for the majority of the Eastern countries (see Tables 2, 4 and 5 of Chapter 7).

It is interesting to note that the breakdown of co-production and specialisation agreements varies greatly between the Soviet Union on the one hand, and the East European countries, on the other. For instance, the Soviet Union has signed mainly agreements which involve R&D. The East European countries, in particular, Bulgaria and Czechoslovakia, appear to be primarily interested in co-production.

For the Western countries, the survey found that there is a clear preference for this type of co-operative agreement by firms in six of the nine countries and that the percentage share continues to be highest for Swedish firms – see Table 6 of Chapter 7.

As far as Sweden is concerned, it is noteworthy that an earlier survey of 53 Swedish firms involved in EWIC agreements did not lead to the conclusion that this type of co-operation was the most frequent. Rather, the survey indicated that "Swedish firms more often deliver production equipment in addition to know-how than other Western firms"[13].

Relative to the United Kingdom, the UN survey indicates that some 29 per cent of contracts concluded by British firms involved co-production and specialisation. It might also be remarked that a survey conducted of 10 British engineering companies with Eastern counterparts found no incidence of this type of co-operative agreement[14]. While the author acknowledges that this conclusion could be due to a different (and more confined) interpretation of the concept of co-operation and specialisation than is used by the ECE, it does underscore the selective and tenuous nature of the available data base – in general.

In any event, co-production and specialisation is currently by far the most important form of co-operation – in terms of the number of agreements concluded and their scope. As mentioned, many sub-contracting agreements eventually develop into co-production. Furthermore, countertrade arrangements are usually easier to undertake within the framework of a co-production agreement.

Co-production and specialisation is also generally considered to be the most complex form of co-operation. For one thing, it provides for sustained involvement of the Western partner in production procedures based on Western inputs of equipment: notably components and parts. As was pointed out in Chapter 4, inability of the Eastern countries to produce sufficient components and parts (in terms of quantity and quality) is a major weakness in these countries' production processes. As was also mentioned, the CMEA countries (particularly the Soviet Union) cite this lack as a major factor in hindering effective intra-CMEA production specialisation. Access to such Western technologies is clearly an important input in building up indigenous production capacity.

Also, in addition to supplying Western technology, these agreements are rarely restricted to the exchange of parts or products. Rather, since the Western partner is interested in receiving counter-products which meet Western quality standards, the agreements often include such ancillary delivery as know-how, documentation, designs, drawings and so forth. This, it is believed, frequently provides the impetus for subsequent wide-ranging co-operation in R&D[15].

Furthermore, co-production and specialisation agreements generally provide for co-operation in marketing: the Eastern firm having exclusive right to sell jointly-made products on CMEA markets while the Western firm handles sales in Western markets. This could benefit the East European countries as far as intra-CMEA trade is concerned since, as was noted in Chapter 4, their problem is to increase not only the quantity but the quality of products to be exported to the Soviet Union.

From the Eastern perspective, the desirability of concluding such agreements is most frequently linked to the fact that they can lower production costs through longer production runs and improve utilisation of installed capacity. Another advantage often cited in the Eastern literature is that access to Western designs and assistance in product adaptation help to modernise domestic production structures, thereby also upgrading the quality of output. Since co-production and specialisation is rather widespread in the machinery and electronics branches, it can be assumed that co-production agreements are regarded as an important input to modernising the engineering sector. Improving output of this sector, in the interest of expanding exports to both Eastern and Western markets, is a priority in the development strategy adopted by the Eastern countries (see Chapter 2).

ii) *Delivery of Plant or Equipment*

In the literature, the supply of plant or equipment is frequently used interchangeably with the supply of turnkey plants. The UN surveys include only those agreements which are based on repayment (at least partially) in products – in essence, deliveries on the basis of buy-back arrangements. This element is most clearly reflected by the Soviet Union's concept of EWIC which explicitly links the delivery of plant or equipment with countertrade – in the form of buy-back.

Turning then to the latest UN survey, delivery of plant or equipment is the second most widely practiced form of EWIC, representing 16.5 per cent of the sample of contracts analysed (see Table 2 of Chapter 7). While for the German Democratic Republic it accounts for the highest share of that country's agreements, it is the Soviet Union and Poland which have concluded the largest number of such contracts (see Tables 4 and 5 of Chapter 7). According to the UN, nearly one-half of all contracts for delivery of plant or equipment is concentrated in the chemicals industry.

On the Western side, the survey found that Japanese firms were principally involved in co-operation based on delivery of plant or equipment followed by France (see Table 6 of Chapter 7). Previous UN surveys have also placed these two countries in the lead. However, while during the 1980-1984 period the percentage share has dropped for both, the decline in the case of Japanese firms seems to have been very sharp. In 1980, delivery of plant or equipment had accounted for 40.5 per cent of total contracts. The equivalent figure for French firms was some 27 per cent.

According to the UN surveys, the share of agreements (in the total number of contracts) providing for delivery of plant or equipment was more or less stagnant during the 1979-1982 period, but has diminished since 1983 (see Table 2 of Chapter 7). Nevertheless, it would be incorrect to assume that this form of co-operation is, therefore, less advantageous than production and specialisation agreements. First, the supply of technology (notably, industrial technology) is an explicit component of provision of plants or equipment. Furthermore, the agreements include the furnishing of such related services as training and help in the start-up of plant operation, particularly in the early stages. Also (as is the case for production and specialisation agreements), co-operation frequently extends beyond the delivery and assembly of plants and includes exchange of technical data, joint research to improve products, the supply of licences and joint marketing[16].

It is these aspects which have led some Western analysts to conclude that industrial co-operation is a more effective means of providing the Eastern countries with Western technology and know-how than the provision of capital goods: seen from the viewpoint of furthering the economic development of the CMEA countries.

The financing aspect is a particularly attractive feature of agreements for delivery of plant or equipment. That is, repayment in products manufactured eases the financing burden of the Eastern partner. The development of co-operation based on the supply of plants or equipment with payment in resultant or related products is a relatively recent development. It has led to the progressive conclusion of large-scale, long-term contracts between Western and Eastern firms based on compensation payment.

iii) *Supply of Licences*

The supply of licences and/or know-how – sometimes also special equipment – as well as parts of the final product (in varying percentages) with payment in resultant products manufactured under the licence supplied, constitutes the third most prominent type of EWIC agreement. Based on the 1984 UN survey, such agreements accounted for 13 per cent of the total contracts concluded (see Table 2 of Chapter 7).

Hungary is the most active participant in licencing agreements (see Tables 4 and 5 of Chapter 7). According to the survey, licencing contracts are, in general, most widespread in the transport equipment and machinery sectors and in the light and chemicals industries.

Relative to the Western countries, the survey found that Austrian firms were predominantly involved with licensing agreements followed by firms from the United States and the United Kingdom (see Table 6 of Chapter 7). It might be noted that the technological intensity (or level) of the licences supplied is not known – hence the number of licencing agreements concluded by Western firms cannot be judged on qualitative terms.

Although the limitations of the data provided by the UN surveys have already been mentioned, it is necessary to underline that they are particularly striking in the case of licensing agreements. Here the "counting problem" is especially apparent. The reason for this is that licences are frequently supplied as part of supply of plants or joint venture contracts and are not recorded separately. Furthermore, licence agreements are registered differently from country to country (if registration is required at all) and statistics do not give a reliable record of valid licence agreements.

The downtrend in licensing agreements suggested by Table 2 of Chapter 7 may, therefore, be misleading. The above-mentioned may also have a bearing on the statistics given for Western country participation. For instance, previous UN surveys indicate wide fluctuations even for as recent a period as 1980-1984. For example, Italian firms were hardly engaged in licencing agreements in 1980 (5 per cent). In 1982, such agreements represented 29 per cent of contracts concluded, whereas in 1984, participation had dropped to 9.3 per cent. Similar fluctuations apply to US and UK firms.

It might also be noted that it is particularly difficult to accurately assess licence trade in value terms. In part, this is due to the payment method. Only a small portion of payment is made by lump sum. Since payment is primarily in the form of royalties calculated as a percentage of sales of licenced goods, the commercial value of a licence agreement can be evaluated only after it has expired[17]. Western countries generally include payments and receipts owing under licence contracts in their statistics. The Eastern countries, on the other hand, do not publish statistics on their balance of payments. This also makes it difficult to value licence agreements[18].

Nevertheless, licencing agreements are generally regarded as being among the most important forms of EWIC. Licences are granted for various purposes and involve two basic types of agreements.

The simplest and most straightforward of these are "pure" or disembodied licence agreements. These agreements involve only the sale of the results of R&D activities, patented or unpatented, and do not provide for the transfer of know-how or information. Nevertheless, they are significant in that they are frequently the first step towards more complex forms of co-operation.

The second type is basically package deals. That is, a licencing arrangement may be part of an industrial co-operation agreement involving patent licences accompanied by imports of certain capital goods, know-how licences, technical assistance, construction of turnkey plants and facilities, and a trademark licence[19]. This type is clearly preferred by the Eastern countries and, as the latest UN survey indicates, agreements which supply licences and/or know-how, as well as special equipment, are the most prevalent: a phenomenon also illustrated by the earlier surveys.

It might be useful to recall here that such components of an EWIC agreement as training, the supply of know-how or technical assistance have been most closely associated with effective technology transfer. Since the flow of licences is predominantly in the West-East direction, this underscores the importance attached by the Eastern countries to concluding "package" licencing agreements. This preference is also suggested by the survey of 13 EWIC agreements signed by ten British engineering companies. This survey led to the conclusion that a typical industrial co-operation agreement "usually consisted of the sale of a licence with the provision of associated know-how, together with the sale of relevant critical components, especially during the start-up phase"[20].

Apart from the acquisition of know-how and technical assistance, financial considerations also explain the CMEA countries' interest in pursuing licencing agreements. As is evident from the ECE's definition, payment arrangements for licence fees and associated equipment provided by the Western partner consist of counter-deliveries of the licenced product[21]. In many respects these agreements are, therefore, "self-financing". It might be noted that the recent trend toward payment of Western licences under countertrade arrangements has changed the pattern of licence sales in that large firms (multi-national corporations) are primarily in a position to enter into such deals, leaving the smaller owners of proprietary technology at a disadvantage.

Attention should also be drawn to the fact that increasing insistence of the Eastern countries to include provisions in licencing contracts which tie the fee for a licence to *export* of products manufactured under the licence, constitutes a major change from previous practices where contracts provided for the payment of a lump sum or royalties based on *domestic production* under a licence.

Since licencing agreements provide access to Western production standards, it is thought (by some Eastern countries) that they "allow more efficient production methods to be used for export items, improve competitiveness, raise quality, consolidate a country's position and help it to appear in new sectors on the world market"[22]. Western licences are also viewed as benefiting intra-CMEA co-operation: either in terms of developing regional specialisation of a particular line or, by means of market extension, facilitating the acquisition of technology-intensive products which a particular country may be unable to purchase or only with great difficulty[23].

iv) *Joint Ventures*

According to the latest UN survey, joint ventures now account for a growing share of EWIC agreements – some 13 per cent of total contracts (see Table 2 of Chapter 7). It should be noted however, that the data given in the Table, do not accurately reflect the real weight of these agreements, since no distinction is made between joint ventures located in the East and those located in the West. While joint ventures predominate in Romania (58 per cent of that country's total number of contracts), it is Hungary which has concluded by far the largest number of joint venture agreements (see Tables 4 and 5 of Chapter 7). According to the survey, joint ventures relate primarily to the chemical industry – taking the Eastern countries as a whole.

As far as the Western countries are concerned, the UN survey ascertained that UK firms concluded the largest number of joint venture agreements and that this type of EWIC arrangement was also the principal form of co-operation for British firms. Joint venture agreements are also well represented by Swiss, Italian and US firms (see Table 6 of Chapter 7).

Joint ventures are a form of business relationship which involves pooling of assets (capital equipment, licences, technical and organisational facilities and so forth), joint management, profit and risk sharing according to a commonly-agreed to formula and, provide for joint marketing, servicing and production[24]. Such related activities as joint R&D are also foreseen under them[25]. Joint ventures may be of an equity or non-equity type, although both entail risk and profit-sharing. Presently direct foreign investment is permitted in Bulgaria, Czechoslovakia, Hungary, Poland and Romania.

As in all EWIC arrangements, joint venture agreements strive to establish long-term relationships between the Eastern and the Western partner. However, they are unique in that joint ventures legally institutionalise this co-operative relationship[26]. It is this aspect which makes these agreements the most developed form of EWIC, distinguishes them from all other co-operative arrangements and makes them, at least potentially, the most attractive form of co-operation. It might also be noted that virtually all forms of industrial co-operation can take place within the framework of an East-West joint venture. In essence, joint ventures are, therefore, a framework for EWIC and not just another form of East-West industrial co-operation.

Furthermore, on the basis of current trends, it would appear that joint ventures have led to new and more innovative forms of co-operation, particularly in the manufacturing sector. It is said, for instance, that some joint ventures which have been concluded recently have led to the transfer of more efficient Western manufacturing technology, co-operation has, in a number of instances, been extended to include production of newly-developed designs not yet manufactured by either partner and, some agreements now provide consultancy and advisory services for product development. In the trading sector, some new developments have also become apparent. All joint ventures located in the East provide for co-operation in marketing. However, in the case of Hungary, jointly-owned trading companies have been established since the early 1980s. They are designed to not only distribute goods produced within the framework of a joint venture agreement, but also products of other manufacturers[27].

On the Western side, among the perceived advantages of entering into a joint venture agreement are that they provide a vehicle for penetrating Eastern markets, hence the possibility of promoting sales. The likelihood of obtaining a reasonable return on the capital invested is another motivation often cited by company executives. This expectation is linked to the fact that joint ventures call for joint management, thereby assuring a better control over the investment.

From the Eastern perspective, motivations differ somewhat according to the location of the joint venture. Those domiciled in the East are primarily established for manufacturing purposes. They are seen as a means of importing Western industrial and new non-standardised technology (without foreign currency outlay) which is not generally available through normal market transactions. The Eastern partner can also obtain technology which may not normally be sold or leased[28]. The very co-operative format of a joint venture agreement likewise gives the Eastern partner access to Western management and marketing know-how.

Despite revisions in, or additions to, existing legislation undertaken by all the countries to make investment more attractive to a Western firm, it may be presumed from the small number of joint ventures actually located in the East, with the notable exception of Hungary, that the above-mentioned expectations are not being met: at least to the degree desired by the Eastern countries.

Joint ventures located in the West differ from those domiciled in the East. They are of far longer duration, more diversified institutional forms are employed (equity or non-equity, with or without majority control) and they are used together with direct investment in the form of wholly-owned subsidiaries[29]. Their basic function, seen from the Eastern perspective, is to promote Eastern exports to the West and to improve the Eastern countries' marketing facilities. Given the debt problem and the need, therefore, to increase exports to the West, it is evident that hard currency considerations stand behind the considerable effort that has been made to increase the number of joint ventures or mixed companies located in the West.

v) *Tripartite (Third Country) Co-operation*

According to the UN definition, tripartite agreements are a form of co-operation in which an Eastern and Western enterprise jointly carry out a project (setting up of industrial complexes, plant assembly, prospecting of natural resources, provision of marketing services and so forth) in a third (usually developing) country, with the active participation of the latter. This definition is very restrictive, since those tripartite (TIC) deals where the host developing country is not active, are not included in the UN analysis. This precludes an accurate assessment of their actual significance.

Furthermore, in view of the fact that since 1980 the UN no longer treats tripartite co-operation as a separate category (joint tendering or joint projects are now included in it), it is difficult to gauge the quantitative importance of tripartite co-operation *per se*. Taking this shortcoming into account, the latest UN survey indicates that this type of co-operation represents 10.4 per cent of the agreements analysed (see Table 2 of Chapter 7). While it is significant for the majority of the East European countries, it is Hungary and the Soviet Union which account for the highest shares in the total number of contracts concluded (see Tables 4 and 5 of Chapter 7). According to the UN, tripartite co-operation is most widely represented in the electrical equipment industry (power stations) in particular, and generally speaking, in the energy sector.

On the Western side, the UN survey indicates that Japanese and Swiss firms participate most actively (see Table 6 of Chapter 7).

It seems useful to briefly point out the findings of another survey concerned with tripartite co-operation (up to the 1979 period), since these do not coincide with those of the UN[30]. According to this analysis, Hungary was indeed among the most active participants (during 1976-1979 – though second to Poland), while the Soviet Union's participation was minimal – 10 contracts out of 88 concluded. It might be noted that the Soviet Union has now signed a large number of protocol agreements indicating that it is more interested in pursuing this type of co-operation. On the Western side, Germany accounted for the highest number of agreements, while those concluded by Japanese and Swiss firms were on the lower end of the scale. Estimates by the same author (taken from his data bank) for the more recent period reflect a similar hierarchy of Eastern and Western participation in TIC.

It is not possible, on the basis of available surveys, to assess the overall significance of tripartite co-operation in volume or value terms. Nevertheless, it appears that such agreements are becoming more complex. For instance, they are said to have evolved into consortium-type arrangements (opposed to mere subcontracting) with each partner assuming equal contractual rights, projects now occasionally provide for joint management (previously never the case), some involve equity ownership or, co-operation extends beyond the construction phase[31].

For the Western partner, among the perceived advantages of participating in tripartite or third country agreements most frequently cited in available surveys are: access to new markets; the possibility of expanding sales within the CMEA area proper because of co-operating with Eastern enterprises; willingness of the Eastern partner to accept payment in resultant product for which the Western enterprise may have no use and, the relatively low cost of Eastern labour for construction of plants and installations in third countries.

On the Eastern side, market considerations (particularly hard currency earnings) and cost-sharing are also important motivations. However, the overriding purpose for seeking tripartite co-operation is evidently the possibility of acquiring and using Western technology and complementary Western equipment – hence also assimilating know-how and, perhaps equally important, gaining expertise in constructing and managing turnkey plants[32].

b) *Industrial Sectors*

With regard to the distribution of industrial contracts by industry (see Table 3 of Chapter 7), the latest UN survey notes that industrial co-operation continues to be concentrated in those industries for which technological progress is of particular importance. This emphasis is not surprising, given the desire of the Eastern countries to raise the technical level of their production structures – a policy which, as noted in Chapter 2, is a component part of their development strategy as articulated in the Five-Year Plans.

The Table indicates that the chemical industry dominates, followed by mechanical engineering (including machine tools), the electrical and electronics industries, transport equipment and metallurgy. Perhaps more interesting than this "global" distribution, is the industrial share relative to the various countries. According to this breakdown, contracts concluded for the mechanical engineering sector (including machine tools) predominate for the majority of the East European countries, while for the Soviet Union and Romania, it is the chemical industry which is in the leading position[33].

It might be noted that the data given in Tables 8-14 of the chapter, substantiate the leading position of the chemical industry. However, it also appears that far fewer agreements are concluded for such key sectors as mechanical engineering or the electrical and electronics industries than is suggested by the UN survey. This would imply that the sectors in which EWIC actually occurs do not necessarily (indeed, apparently do not) coincide with the sectors given top priority by the Eastern countries – as far as the allocation of investment resources is concerned. This could be due to the fact that Eastern planners are unable to attract Western firms to co-operate in these industries/sectors because they show no record of successful co-operation, are more oriented towards CMEA than Western markets (hence do not generate sufficient hard currency required to purchase equipment needed for co-production) or, because they stand a good chance of becoming serious competitors on Western markets.

c) *Country Participation*

On the *Eastern side*, Hungary continues to account for the largest number of co-operation contracts concluded (291 agreements or 36.0 per cent – see Table 7 of Chapter 7). The Soviet Union is in second place with 243 agreements or 30 per cent of the total number of co-operation agreements concluded. This percentage share has varied very little during the 1979-1984 period. It is very small given the size of the USSR's economy and its trade involvement with the Western countries. According to the survey, Poland has maintained its position in third place, although its share has continuously (and not surprisingly) dropped since 1980. Tables 8-14 of the chapter also indicate that Hungary and the Soviet Union have concluded the largest number of agreements – during the 1980-1984 period. However, Poland is not in third place having concluded fewer agreements than Bulgaria and Czechoslovakia – only three in 1984.

On the *Western side*, Table 6 of Chapter 7 indicates that German firms are clearly in first place, followed by firms from the United States and Italy. The surveys undertaken for the years 1980-1983, also place these three countries in the lead and the percentage shares vary very little. The same observation applies to the remaining countries included in the survey.

Tables 8-14 corroborate the UN's conclusion that German firms are the most active participants in EWIC. However, the data show that the US has dropped far down the list and, contrary to the ranking of the UN survey, Italy, Austria, France and the United Kingdom follow Germany – in this order.

## *Section II.* MOTIVES FOR CO-OPERATION

As is evident from the previous section, East-West industrial co-operation has various purposes. The most consequential of these are the potential establishment of long-term contractual relationships and the direct link that is made between co-operation and production. While industrial co-operation presupposes a mutuality of interests, it is apparent that Eastern and Western motivations in concluding a co-operation agreement frequently differ. It should also be noted that there is a growing link between industrial co-operation (in all its forms) and various types of countertrade. This link has implications for the economics of EWIC – comparative advantages, return to scale and so forth.

It must also be borne in mind that the CMEA countries, taken as a whole, do not have a common strategy towards EWIC, though there are similarities in approach. The policy adopted towards industrial co-operation, as well as the reasons for entering into a particular co-operative arrangement, differ from country to country. These differences are related to the degree of openess towards the West – in general. They are also the natural outcome of systemic differences, as far as centralisation of decision-making is concerned.

Relative to the West, it is also obvious that there is no commonly-shared policy toward, or interest in, pursuing co-operation with the East. In general, Western country participation in industrial co-operation is greatly influenced by political attitudes towards the CMEA countries. Furthermore, the economic interests of a firm in pursuing EWIC can (and sometimes do) run counter to the strategic considerations/concerns of their nation states. As far as the individual firms are concerned, their attitudes toward or interest in pursuing industrial co-operation are shaped by such factors as the general economic climate, past experience and current assessments made of the potential advantages thought to ensue from co-operation. Participation in EWIC is, therefore, conditioned by the degree to which a Western firm is willing to co-operate, the form co-operation will take, its duration, the means of financing it, and so forth.

The motivations briefly noted below are those which are most frequently cited in the general Eastern and Western literature and in surveys and case studies concerning EWIC. They should be regarded as being essentially illustrative of the basic considerations of the partners involved.

## A. THE EASTERN PERSPECTIVE

### 1. Acquisition of Western Technology

Gaining access to Western technology, inclusive of production know-how, seems to be the primary motivation. It reflects the belief of the Eastern countries that EWIC arrangements are instrumental in increasing their production capacity (as well as diversifying and extending the range of products produced) and in improving the overall efficiency of production. It is said, for instance, that industrial co-operation with Western firms accelerates introduction of improved equipment and technologies into production, raising thereby the technical level and quality of output, modernising existing production capacity and, since co-operation shortens the time needed to solve various important scientific and technical problems, the efficiency of R&D outlays are increased. A corollary consideration is that industrial co-operation provides access to new technologies more quickly and at a lower cost. Access to Western technology is also seen as an input in furthering intra-CMEA specialisation and production, in that it enables the CMEA countries to expand the mutual exchange of goods which conform to world standards[34].

### 2. Export and Marketing Considerations

Export promotion, due to balance of payments considerations, has become an increasingly strong motivation. In the Eastern literature, industrial co-operation with Western firms has been directly linked to the possibility of the Eastern countries to expand their market potential[35]. That is, the very improvement of production structures made possible by having access to Western technology, is seen by the Eastern countries as providing the possibility for changing their export structures. Notably, to increase the share of manufactured goods in their exports to the West. The actual provision made in EWIC arrangements for furnishing technical assistance and know-how, as well as the ability to learn Western marketing skills through long-term and close association with a Western firm, are also believed to lead to increased export potential.

This perspective has been substantiated by various surveys. For instance, the survey of British engineering companies engaged in EWIC established that the Eastern partner "wished to arrange for the production of an item which had the capability to be competitive in the international market, and thereby generate foreign currency earnings". It is also pointed out that this requirement appeared to have been met in almost every case, "although restrictions were sometimes placed on the region into which the East European customer was allowed to export"[36].

A corollary consideration is the expectation that through industrial co-operation hard currency imports of a particular product, or group of products, will be reduced. Furthermore, it is evident from the types of co-operation pursued (for instance joint ventures or third-country agreements) that the Eastern countries hope to gain marketing channels with a minimum of investment.

### 3. Financing

Closely linked to the above, are hard currency considerations. As noted in the previous section, an intrinsic feature of EWIC arrangements is their essentially "self-financing" nature. For instance, licences and know-how fees are covered by counter-deliveries, as is equipment supplied. Thus, directly or indirectly, EWIC agreements provide for payment in resultant product.

Eastern (and, in particular, Soviet) authors, frequently link the advantage of concluding an industrial co-operation agreement to the compensatory payment mechanism. In the case of the Soviet Union, ability to launch such large-scale projects as enlarging the CMEA's network of petroleum and gas pipelines or developing new raw material deposits, has been directly associated to Western financial, material and technical assistance made available on the basis of compensation payment[37].

## B. THE WESTERN PERSPECTIVE[38]

### 1. Marketing Considerations

Among the most frequently cited considerations here are gaining entry into an Eastern market (or, keeping a presence on such markets) and export promotion. That is, to acquire a market for sales. In any event, given the Eastern countries' attempt to minimise hard currency outlays, the conclusion of an EWIC agreement (by means of a compensation arrangement) is often the only means of marketing a product. While Western firms generally prefer to sell their products, they frequently do conclude a co-operation agreement rather than loosing a potential market. Some company executives also point out that once an agreement has been signed, export sales progressively increase.

Furthermore, through industrial co-operation Western firms can increase sales of complementary products – particularly components and parts. For some Western firms, industrial co-operation is viewed as simply providing a sales outlet for products which could not be sold on Western markets.

Industrial co-operation agreements are also regarded as instruments for extending an initial market entry. That is, access to one East European market is believed to provide access to markets of the other CMEA countries. Finally, co-operation agreements are thought to restrict or limit competition from other Western firms.

### 2. Cost and Financial Advantages

One aspect here is that Western firms can expand their own production capacity with a limited initial investment outlay. Though less frequently the case (and often confined to specific sectors), industrial co-operation also gives a Western firm access to low-cost materials and labour. This allows a Western firm to perform labour intensive, value-added production for profitable re-sale on Western markets. However, it should be noted that cost advantages are most often realised in agreements which involve sub-contracting.

Western firms are also interested in concluding co-operation agreements because they can extend the lifetime of a product. That is, products are transferred which are less easily sold on Western markets – either because of market saturation or because they are obsolescent. The cost advantages of such sales are evident.

Though infrequent and highly dependent on the general technological development of a particular Eastern country, some Western firms seek industrial co-operation in order to use the Eastern partner's R&D potential. For instance, some company executives believe that the Eastern countries possess excellent basic research, but are unable to apply it for specific manufacturing purposes. An industrial co-operation contract gives the Western firm access to it. The firm's ability to apply it leads, in turn, to highly profitable production. Also of interest to a Western firm is the fact that the costs of investment, research and design for the production of goods for which the demand on Western markets is relatively small, can be attributed to the Eastern partner.

The ability to profitably exploit the commercial use of a licence sold within the framework of industrial co-operation, is also a motive often cited by Western firms for seeking an EWIC arrangement. Nevertheless, it can be anticipated that as an increasing share of licence fees are covered by counter-deliveries, this potential source of hard currency earnings will diminish.

## *Section III.* PARTICIPATION IN EWIC

## A. BULGARIA

### 1. Involvement In and Policy Towards EWIC

Compared to Hungary, Bulgaria's participation in EWIC is relatively modest. This is partially related to Bulgaria's extreme trade orientation towards the CMEA countries and also to apparent difficulties in reconciling the requirements of industrial co-operation with a highly centralised economic system.

However, reforms undertaken to make the planning system more flexible, to grant enterprises a greater role in the formulation of production plans and to give them more freedom in establishing closer (and more direct) links with foreign firms, indicate that Bulgaria intends to become more active in pursuing EWIC.

For instance, the Decree (No. 535) promulgated by Bulgaria's Council of State in March 1980, was essentially aimed at reducing legal restrictions which had hindered co-operation with foreign firms. It also permitted the creation of joint ventures. The strengthening of the New Economic Mechanism in January 1982, constituted a further step in the direction of facilitating contact of Western firms with their Bulgarian counterparts. It is said, for instance, that "better co-ordination of national industry with world market requirements, is the main task in the sphere of foreign economic relations for the 1981-1985 Plan period and for the period ending in 1990. In line with the New Economic Mechanism, Bulgaria places great stress on setting up foreign trade companies which deepen industry's foreign trade ties to the greatest degree"[39].

Recent changes in Bulgaria's foreign trade system (Decree No. 2 of the Council of Ministers, adopted in January, 1984) are likewise intended to expand co-operation with Western firms. Relative to Bulgarian economic organisations, special attention is to be paid to the development of long-term production, technical and commercial co-operation with Western firms. Furthermore, the creation of small and medium-sized enterprises continues to be encouraged – with the expectation that such enterprises will attract foreign investment. As is noted, legislation adopted in July 1984 (Decree No. 3) is intended to create more favourable conditions for them, facilitating thereby co-operation with Western partners.

From the Bulgarian perspective, EWIC agreements are regarded as instruments for meeting overall production goals and for raising the technical level of output. Industrial co-operation is also sought in the interest of furthering Bulgaria's competitive position within the CMEA for those products in which it specialises. For instance, it is pointed out that "co-operation with Bulgarian organisations and the British firm Perkins for manufacturing motor loaders with a fork-type grapple has served as an additional factor in furthering Bulgaria's specialisation within the CMEA"[40].

As is the case for all of the Eastern countries, data as to the number of EWIC agreements concluded conflict. The UN survey indicates that Bulgarian enterprises accounted for 44 co-operation agreements of the sample analysed. Some sources claim that Bulgaria has 58 active co-operation agreements, while those used to draw up Table 8 (of Chapter 7), establish that 87 agreements were concluded over the 1980-1984 period. An overall assessment of exports ensuing from industrial co-operation was not found in the literature consulted.

## 2. Preferred Sectors and Partners in Industrial Co-operation

According to the latest UN survey, industrial co-operation takes place primarily in mechanical engineering and machine tools, followed by the light and electrical and electronic industries (see Table 3 of Chapter 7).

A somewhat different picture emerges from those agreements which have been recorded in the general literature. As Table 8 of Chapter 7 indicates, it is the light industry (notably textiles) which predominates, followed by chemicals. It is also interesting to note that the majority of them are "multi-sector" agreements. In addition to providing for co-operation in a number of branches (within the framework of one agreement), these agreements often include the exchange of scientific and technical information, the provision of a licence for further production, joint R&D and tripartite co-operation. It is presumably the wide scope of these agreements which makes them particularly appealing to Bulgaria.

As far as the Western partners are concerned, Bulgaria co-operates primarily with the West European countries (notably the EEC members) and Japan. Co-operation continues to be most extensive with German firms. Japanese firms seem to be the second most important partner, followed by Italy, and the United Kingdom. The sectoral distribution and type of agreement concluded is given in Table 8 of the chapter.

## 3. Forms of Co-operation

### a) *Co-production and Specialisation*

According to the latest UN survey, co-production is the preferred form, accounting for 50 per cent of the total number of contracts concluded – see Table 2 of Chapter 7. Insofar as the agreements listed in Table 8 of the chapter do include those aspects generally associated with co-production agreements, it would appear that this form of industrial co-operation is indeed the most prevalent. It may also be noted that according to Bulgarian data, co-production represented some 8 per cent of Bulgaria's total exports in 1981[41].

### b) *Joint Ventures*

In its endeavour to strengthen economic co-operation with the Western countries, Bulgaria intends to pay particular attention to the development of joint venture agreements. From the viewpoint of Bulgarian specialists, this type of industrial co-operation is best suited for forging close links between industry and foreign trade. Furthermore, such agreements are intended to expand exports, improve product quality and indigenous R&D, as well as introducing modern management methods.

It might also be mentioned that the joint venture form of co-operation (because the Western partner has direct control) generally encourages more complex forms of technology transfer: an aspect which has undoubtedly played an important role in the attempt made on the part of all those countries permitting foreign equity investment to encourage this type of industrial co-operation.

Bulgaria's interest in attracting Western capital through joint equity ventures is clear from the fact that it has the most liberal joint venture legislation. For instance, joint ventures may be established in all sectors of the economy and no limits are placed on the Western partner's share in the equity holding of a joint venture nor on the repatriation of profits. Bulgaria's joint venture legislation is also unique in that it includes various tax incentives and allows the use of both national currency and convertible currency accounts for the same company. Procedures for the removal of import duties for parts and materials used in export products are simple. Legislation also permits both an ordinary partnership (an equity with no legal personality) and an incorporated association. The partner in the latter arrangement also retains title to its investment in machinery, equipment and installations[42].

Despite the considerable effort that is thus being made to attract Western capital and to gain access to modern industrial technology, the number of joint companies located in Bulgaria is very modest. Information as to the actual number currently in operation conflicts. According to the general literature consulted, it appears that Bulgaria has concluded 9 joint venture agreements during the 1980-1984 period (see Table 8 of the chapter). As the Table also indicates, these agreements span a variety of sectors.

In the Bulgarian literature particular importance appears to be attached to those agreements concluded with Japan – notably Fanuk Machinex Ltd. and Atlas-Engineering.

The former is reported to have expanded its activities to provide engineering assistance to Bulgarian enterprises that manufacture industrial equipment under licence from Japanese partners. Atlas-Engineering is said to be the first joint company in the engineering sector. It is designed to further the construction of heavy machinery.

As mentioned previously, a main attraction for the East European countries for establishing joint or mixed companies in the West, is that they allow the Eastern partner easier entry into highly competitive markets.

Bulgaria's intention to expand its exports on Western markets can be seen from the fact that the number of joint ventures operating in the West has significantly expanded since 1976 – see Table 1 of Chapter 7. Estimates of the number of joint ventures presently operating on Western markets vary between 32-40. The majority are in the sphere of chemistry and pharmaceuticals, followed by electronics, electrical and mechanical engineering and the light industry. Western partners include German (clearly in the lead), French, Austrian, British, Japanese, Swiss, Belgian, Italian and Finnish firms[43]. It is said that in 1983, Bulgarian joint ventures accounted for 40 per cent of all Bulgarian exports to the EEC[44].

According to the latest UN survey, joint venturing accounts for 18 per cent of Bulgaria's total agreements and is the second most prevalent form of EWIC – see Table 2 of Chapter 7.

c) *Delivery of Plant or Equipment*

Though apparently still an important form of co-operation, this type of arrangement appears to be declining. The UN data also indicate that Bulgaria's share in this type of co-operative agreement is minimal (see Tables 2 and 5 of Chapter 7). Table 8 of the chapter indicates that plant construction is virtually non-existent – one agreement having been recorded in 1984.

d) *Licencing Agreements*

The latest UN survey indicates that licencing agreements represent a small share in the total number of contracts concluded (see Table 2 of Chapter 7). Contrary to the UN findings, licencing agreements appear to be more prevalent than joint venture agreements (18 and 9 respectively – see Table 8 of the chapter). Those recorded for the 1980-1984 period indicate that most concern the light industry and that German firms are the predominant partner.

## 4. Other Technology Transfer Channels

As noted, *licencing* is a particularly effective technology transfer mechanism: a consideration that has clearly been incorporated into Bulgaria's development strategy as articulated in the 1981-1985 Five-Year Plan. Bulgaria's intention to continue to purchase licences and know-how has been made explicit (see Chapter 2). Based on the guidelines governing Bulgaria's economic development for the 1986-1990 Five-Year Plan period, it would seem that licence trade is to expand. This expansion is linked to the intent to modernise and re-equip existing production structures in order to increase exports.

Whether this policy can, in fact, be realised, is difficult to assess. Hard data on Bulgaria's involvement in licence trade is scarce, although it appears that Bulgaria is a modest purchaser of Western licences.

## B. CZECHOSLOVAKIA

### 1. Involvement in and Policy Towards EWIC

Until recently, and in contrast to most of the other CMEA countries, Czechoslovakia's attitude towards co-operation with Western firms was negative. The scope or concept of EWIC, as defined by Czech authorities, was also much narrower. For instance, co-operation on the basis of licencing agreements, compensation arrangements or technical assistance were not permitted.

By mid-1976, Czechoslovakia had some 40 active co-operation agreements with Western firms. This represented some 4.7 per cent of the total agreements concluded between Eastern and Western firms – placing Czechoslovakia at the low end of the scale. Overall trade turnover ensuing from co-operation with Western firms was barely 1 per cent[45].

Export considerations as well as concerns related to furthering economic development are among the factors which explain Czechoslovakia's growing interest in EWIC. As was underlined by Czechoslovakia's Prime Minister, "realisation of our long-term objectives requires developing new forms of collaboration both with the capitalist and developing countries. To date, production co-operation is used insufficiently. This form of division of labour is becoming an objective necessity in the world, but our share in it is lower than that of the other socialist countries"[46]. Consequently, and as was pointed out in Chapter 2, increased industrial and scientific and technical co-operation, is now a component part of Czechoslovakia's economic development strategy. In other words, more active participation in EWIC is currently viewed as being indispensable for raising the efficiency of production and, consequently, increasing export capability.

In order to facilitate co-operation with Western firms, existing legislation governing trade has been made more flexible. The essential purpose of the Decrees passed since 1980, is to provide financial incentives to enterprises to conclude co-operation agreements with Western firms. For instance, foreign exchange accounts have been set up to assist the execution of agreements and hard currency surpluses generated by an agreement can be partially retained by an enterprise to finance unplanned imports. The administrative procedures to be followed in order to gain permission for entering into an agreement with Western firms have also been eased[47].

However, whether such changes will prove to be effective is questionable. A number of key obstacles which hinder the official desire to expand EWIC remain. Among these is the strict nature of the planning system itself which does not give enterprises sufficient flexibility in dealing with their Western counterparts. That is, the innate reluctance of many enterprises to become more closely involved in EWIC, is compounded by the need to juggle scarce labour and capital resources in order to meet both mandatory plan targets (i.e. plan fulfilment) and the obligations of a co-operation agreement[48].

Nevertheless, Czechoslovakia has manifestly adopted a more positive attitude towards EWIC and the scope of industrial co-operation has been enlarged. For instance, EWIC now encompasses licencing agreements, joint R&D and delivery of technological equipment for production co-operation. Compensation deliveries ensuing from new production are also allowed. Furthermore, the desirability of becoming involved in more complex forms of co-operation – including arrangements which entail third (notably, developing country) markets – is increasingly stressed.

According to the latest UN survey, Czechoslovakia accounts for 69 co-operation agreements. Table 9 of the chapter records 40 agreements for the 1980-1984 period. It might also be noted that the official literature claims that over 100 co-operation agreements were in effect by 1984 – though this figure includes agreements with all non-socialist countries[49].

In any event, and despite Czechoslovakia's increased participation in EWIC since the mid-1970s, exports resulting from co-operation arrangements in 1983 are said to still account for less than 1 per cent of total exports to the developed market economy countries[50].

## 2. Preferred Sectors and Partners in Industrial Co-operation

Based on the agreements recorded by the UN, it is mechanical engineering (inclusive of machine tools) which accounts for the largest share of agreements, followed by electronics and electrical equipment combined – see Table 3 of Chapter 7. The preponderance of deals in the field of general engineering products also emerges from those 77 agreements registered in 1982[51]. In the engineering field, co-operation is primarily aimed at furthering the production of machine tools, construction machinery, electrothermic equipment, hydraulic components, precision instruments, insulation glass and farm equipment.

The agreements given in Table 9 of the chapter indicate a rather more widespread distribution. It might also be noted that the chemical industry, which is placed at the low-end of the scale by the UN survey, appears to account for the largest share of the agreements recorded for the 1980-1984 period. The multi-sector agreements listed in Table 9 concern primarily co-operation in plant construction.

As far as Western participation is concerned, the official literature states that German firms are Czechoslovakia's primary partners (currently accounting for 46 per cent of all agreements), followed by Austria (12 per cent), the Netherlands, Sweden and France[52]. According to Table 9 of Chapter 7, it would appear that Austria is the leading partner, followed by France, Germany and Italy – firms from the latter two countries having concluded the same number of agreements. The discrepancy between our findings and those in the official literature, could be due to methodological and definitional problems noted earlier (see section B.1. above). That is to say, we have only taken into account those agreements which are said to be in effect and intended to involve both parties in long-term co-operation.

## 3. Forms of Co-operation

### a) *Co-production and Specialisation*

Co-production is the form most clearly preferred: a preference which emerges from the UN survey (see Table 2 of Chapter 7) and is explicitly stated in the official literature. It is said, for instance, that "from the Czechoslovak concept of co-operation, the predominant objective is [to seek] production co-operation; particularly joint production of parts, components, assemblies and final products – as well as follow-up to co-operative production. That is, parallel finishing or final assembly by the co-operating partners"[53].It is also noted that co-production is the most advantageous form of co-operation since it is self-financing, thereby enabling Czechoslovak enterprises to increase needed imports without resorting to Central Funds[54]. While joint production *per se* does not appear to be as prevalent as indicated by the UN survey, the agreements recorded for the 1980-1984 period (see Table 9 of the chapter) do indicate that the majority involve co-operation in production.

b) *Joint Ventures*

Joint ventures with foreign equity investment have only been permitted since mid-1985 and, at the time of witing, none had been established in Czechoslovakia. Joint venturing in the West is developing slowly (see Table 1 of Chapter 7). At the end of 1981, Czechoslovakia is reported to have invested in 32 Western firms. Like its Eastern partners, investments are designed for export promotion[55]. Based on the UN survey, Czechoslovakia's participation in this form of co-operation has diminished since 1980 and is currently stagnant (see Table 2 of Chapter 7). It might also be noted that no new joint venture agreements appear to have been concluded during the 1980-1984 period (see Table 9 of the chapter).

c) *Delivery of Plant or Equipment*

Compared to co-operation in production, all other forms of co-operation are negligible. This is particularly evident in agreements for delivery of plant or equipment. According to the UN surveys, no such agreements were concluded until 1981. Since then, their share in the total number of contracts concluded has been very modest (see Tables 2 and 5 of Chapter 7). As noted earlier, delivery of plant or equipment involves essentially buy-back arrangements. The fact that Czechoslovakia is frequently unable to meet counter-delivery obligations (in terms of quantity or quality), could explain why this form of co-operation has not progressed.

d) *Licensing Agreements*

While representing a much smaller share of the total contracts registered by the UN than, for instance, co-production, licencing is the second most prevalent form of co-operation for Czechoslovakia. However, its share is very modest compared to that of Hungary (see Tables 2 and 5 of Chapter 7). According to the UN, the majority of the agreements are accounted for by the transport equipment sector, followed by the food and agriculture and chemical industries and the electrical equipment sector. Due to the few licencing agreements recorded for the 1980-1984 period, it is not possible to confirm the conclusions reached by the UN.

## 4. Other Technology Transfer Channels

Licencing (whether on a purchase basis or within the framework of industrial co-operation) has always been considered by Czechoslovakia as one of the most effective forms of technology transfer. As is frequently mentioned, "by purchasing a licence, the purchaser gains access to top technology, which finally results in high-quality products, capable of competing in world markets"[56].

In the past, Czechoslovakia's approach towards licencing was passive. That is, licences were bought for domestic production purposes (import substitution). This is said to have been the case in the 300 licencing contracts concluded by Polytechnica (the State Organisation responsible for the sale and purchase of licences) between 1970-1980. Since 1980, licences have been primarily purchased to modernise those industrial branches whose production is export-oriented[57].

The chemical industry has the highest priority (in particular for crude oil processing, petro-chemicals, synthetic fibres and organic dyes). Other areas for which licence purchases are to be expanded include general and heavy engineering and metal-working, as well as the automotive industry[58]. The emphasis placed on the engineering and metallurgy branches is related to the fact that Czechoslovakia is to expand the construction of nuclear power plants, as that country is to become a major supplier of nuclear power engineering within the CMEA area. Recent licences purchased from the UK and France for the Skoda works have a dual purpose: to further the production of turbines and to modernise the automotive industry. The latter is essential for Czechoslovakia in order to increase exports of cars and trucks to its CMEA partners and to regain its competitive position on Western markets.

The importance which Czechoslovakia attaches to licencing, as one means of increasing its export potential, is evident from those measures (taken and recommended) designed to both increase the number of licences purchased and to improve their effectiveness. Among these are: that greater control over the quality of licenses purchased is being exercised; enterprises which do not meet production targets foreseen by a licence will be penalised; hard currency for the purchase of licences will be granted only if proposals include estimates of inputs needed to make the licence work and, only licences which are not import-intensive and which make use of domestic raw materials are to be acquired[59].

As of 1985, procedures for financing technology purchases from the West will be eased and, as noted in Chapter 2, more foreign exchange will be made available to enterprises to buy licences. According to the official literature, for the 1986-1990 plan period, licences will be bought primarily for the metallurgical and heavy engineering sectors.

## C. GERMAN DEMOCRATIC REPUBLIC

### 1. Involvement in and Policy Towards EWIC

Since 1973, the German Democratic Republic has concluded long-term co-operation agreements with virtually every Western country – notably OECD Member countries. According to the UN, a total of 32 were in force by the end of August 1984, with the majority providing for economic, industrial and scientific co-operation. In principle, this would seem to indicate interest in establishing closer economic relations with the West. Indeed, the desirability of maintaining long-term economic ties with the Western countries is supported in the official literature and has been formally upheld at recent sessions of the CMEA.

In fact, the German Democratic Republic's involvement in EWIC continues to be very modest. For instance, at the end of 1976, only 32 agreements had been concluded, representing some 3 per cent of the total – see Table 1 of Chapter 7. The Table likewise indicates that the percentage share in the total number of agreements had only risen to 4 per cent by the end of 1980. The latest UN survey also ascertains that participation in EWIC seems to be stagnant. That is, only 4 per cent of the total number of contracts analysed (i.e. 33 agreements) were accounted for by the German Democratic Republic – see Table 7 of the chapter. Earlier surveys conducted by the UN indicate that this percentage share has varied very little. Table 10 of Chapter 7 indicates that 31 agreements were concluded during 1980-1984.

It should be noted that concrete information concerning the German Democratic Republic's attitude towards EWIC is virtually non-existent. Nevertheless, based on recent trends, one can surmise that its involvement in EWIC will remain negligible. For instance, countertrade seems to be becoming the preferred mechanism in the German Democratic Republic's dealings with Western firms, and authorities are becoming increasingly demanding about its modalities. As has been noted, Western firms are currently being met by countertrade demands ranging as high as 50-100 per cent of the contract value. At the same time, the range of goods offered in exchange is being reduced[60]. This is hardly an inducement to Western firms to engage in co-operation agreements.

Furthermore, the German Democratic Republic appears to prefer framework agreements. Such agreements fix the amount of exchanges on an annual basis, provide for countertrade, industrial co-operation and also contain provisions for third market operations. Examples of agreements which contain one or all of these elements are those signed with Voest-Alpine AG (1984), Chemie Linz AG (1984), Ciba-Geigy (Switzerland, 1981) and Shell Petroleum (UK, 1980). It can also be assumed that the very rise in countertrade demands has led some Western firms to conclude framework agreements as a means of achieving some measure of flexibility in fulfilling their countertrade obligations.

Since the German Democratic Republic seeks co-operation primarily for those industrial sectors which are in urgent need of modernisation in order to retain current export markets (for instance, the agreement signed with Volkswagen in 1984 whereby the German Democratic Republic will produce engines under the VW licence enabling the production of cars which meet EEC anti-pollution standards), one can also assume that this narrow "sectoral scope" will limit the general expansion of EWIC agreements.

Access to EEC markets and, in particular, the German Democratic Republic's special relationship with Germany – in terms of credit availability and the fact that Germany presently accounts for over 50 per cent of its trade with the OECD area – could also explain lack of interest in pursuing EWIC. That is, vitally needed Western technology can be obtained from Germany with less pressure on the German Democratic Republic's hard currency account. Finally, the German Democratic Republic's special role as a supplier of advanced technology to the USSR also conditions its participation in EWIC – in terms of the industries (sectors) concerned and the type of co-operation.

## 2. Preferred Sectors and Partners in Industrial Co-operation

According to the UN survey, the transport equipment and mechanical engineering (including machine tools) industries account for the highest shares of the total number of agreements concluded – see Table 3 of Chapter 7. A somewhat different picture emerges from Table 10 of the chapter which shows that the chemical industry predominates. The little information available in the general and official literature indicates that co-operation agreements involving third countries are being more frequently sought, particularly for the chemical and petro-chemical industries. However, the number of such agreements appears to be limited.

Relative to Western participation, Table 10 indicates that French firms are the most actively involved in industrial co-operation, closely followed by German, Italian, Japanese and Austrian firms – the latter three having an equal share in the number of contracts concluded.

## 3. Forms of Co-operation

### a) *Co-production and Specialisation*

The UN survey shows that co-production is not particularly prevalent – see Tables 3 and 5 of Chapter 7. Table 10 of the chapter also indicates that joint production, as such, is virtually non-existent and that agreements involving some form of co-operation and development also appear to be negligible.

b) *Joint Ventures*

Joint ventures permitting equity investment are not allowed. The number of joint companies operating in the West is modest: 23 at the end of 1981[61]. Based on Table 10, it appears that only one joint venture (involving marketing in third countries) was concluded – in 1982. According to the UN survey, participation in joint venturing is minimal (see Tables 2 and 5 of Chapter 7).

c) *Delivery of Plant or Equipment*

Delivery of plant or equipment seems to be the preferred form of industrial co-operation, accounting for some 55 per cent of the total number of contracts registered by the UN. However, the UN's breakdown by country shares, shows that the German Democratic Republic is in fourth place (see Tables 2 and 5 of Chapter 7). It would appear from Table 10 of the chapter that only one agreement falling into this category was concluded during 1980-1984.

d) *Licencing Agreements*

Based on the UN survey, licencing agreements are also not a significant form of co-operation – an assessment which also emerges from Table 10 of the chapter. According to the UN, those agreements which have been concluded are primarily accounted for by the light industry and the transport equipment sector. Data available in the general literature are too sketchy to verify this conclusion.

### 4. Other Technology Transfer Channels

Given hard currency considerations, the purchase of *turnkey* plants, has been relatively high. According to available information, 8 such contracts were concluded during 1980-1984. Of these, six pertained to the chemical and petro-chemical sector (two with Japanese and one each with Austrian, German, French and Italian firms). One turnkey contract was concluded with Germany to construct a metallurgical complex and one with Italy for the manufacture of mineral wool.

During 1970-1978 *licencing* seems to have played a rather significant role in the acquisition of Western technology: 422 licencing contracts were concluded during that period. Licences were primarily purchased for chemicals and pharmaceuticals, heavy machinery, machine-building (metal-working, printing and food processing machinery), metallurgy, electrotechnics, electronics and shipbuilding[62]. Global figures for the recent period could not be found. Available data in the literature suggest that licences are primarily being purchased for the chemicals and petro-chemicals, construction, machinery and metal-working branches.

## D. HUNGARY

### 1. Involvement in and Policy Towards EWIC

Among the East European countries Hungary continues to be the most active participant in EWIC. This active participation is generally associated with its more open and less bureaucratic approach towards industrial co-operation and, in particular, with the liberal legislation governing industrial co-operation with Western firms. The most salient feature here is that Hungarian enterprises can establish direct links with their Western counterparts.

The series of economic reforms undertaken since 1981 are also believed to have favoured the development of EWIC. For instance, as part of the objective to decentralise decision-making (relative to production), new small and medium-sized enterprises have been set up. These enterprises are authorised to engage directly in foreign trade activities and can conclude co-operation agreements without the participation of a specialised foreign trade organisation. According to Hungarian commentators, the establishment of such enterprises is propitious for furthering EWIC since they tend to be more active in concluding co-operation agreements than large enterprises. Some Swedish experts familiar with EWIC have pointed out that those co-operation agreements concluded with small and medium-sized firms, have been the most successful (for both parties). Furthermore, they note that this is the "best end" of co-operation and a manner of co-operating that is unique to Hungary.

Hungary's concept of EWIC appears to coincide closely with that adopted by the ECE. It is interesting to note however, that while Hungary strongly favours the conclusion of industrial co-operation agreements with Western firms, official trade policy stresses that counter-purchase transactions are to be kept at a minimum. In other words, industrial co-operation is not be equated with commercial counter-purchase practices[63].

Hungary's attitude towards EWIC is made very explicit in the official literature, although differences in objectives to be achieved emerge between central authorities and the enterprises.

Relative to the former, export promotion (particularly exports in non-ruble trade) is the overriding motivation and it is directly linked to balance of payments considerations. While this aspect also plays a role for the enterprises, their primary objective in seeking co-operation is to increase overall output (particularly of new products), to raise the technical level of their products and to have access to new technology[64]. Assistance in applying acquired technology and cost savings in developing new technologies are also advantages attributed to industrial co-operation. Finally, capital shortages in fully meeting domestic production demands and the desire to obtain maximum profit for products exported within the framework of industrial co-operation, are also reasons why enterprises are interested in concluding EWIC agreements[65].

Current discussion in the Hungarian literature suggests that, in future, more emphasis will be placed on pursuing those forms of co-operation which require less investment and are also less import-intensive. Singled out among such forms are co-production and product-sharing arrangements which increase utilisation of existing capacities and which more fully exploit equipment and know-how purchased from the West, marketing-related co-operative agreements (notably, those arrangements involving third-markets) and joint equity ventures. Regarding the latter, particular emphasis is placed on seeking joint ventures which include provisions for third-market operations[66].

Turning then to Hungary's participation in EWIC, the latest UN survey establishes that Hungarian enterprises accounted for some 36 per cent of the total number of contracts – or, 291 agreements (see Table 7 of Chapter 7). Table 11 of the chapter ascertains that Hungary concluded 294 agreements over the 1980-1984 period.

Despite Hungary's active involvement in EWIC, such co-operation seems to have made only a modest contribution to Hungary's trade with the West. For 1981, it has been estimated that 7-8 per cent of Hungary's total exports to the West can be attributed to industrial co-operation. A more recent estimate places the share at 4 per cent – although it is noted that this probably is on the low side. The reason for this is that (according to Hungarian officials)

Hungarian published figures do not include trade ensuing from production agreements involving third markets[67]. It may also be noted that export shares are significantly higher in certain sectors. It has been estimated, for instance, that some 20 per cent of total exports in the engineering sector alone were made possible by co-operation[68].

## 2. Preferred Sectors and Partners in Industrial Co-operation

According to the latest UN survey, the largest share of co-operation agreements is accounted for by the electrical and electronics industries combined. The chemical industry has the second largest share, closely followed by mechanical engineering (if machine tools are included) – see Table 3 of Chapter 7. A somewhat different picture emerges from Table 11 of the chapter which indicates that the chemical industry predominates. However, if one sums up the agreements which correspond to the UN's light industry category, it is this industry which is then in first place. Furthermore, the Table also shows that electronics and electrical equipment are at the lower end of the scale.

Germany is clearly most actively involved in industrial co-operation. Based on the agreements listed in Table 11, it appears that co-operation agreements concluded with German firms are widespread throughout industry, although the majority pertain to chemicals and petro-chemicals, construction and the engineering sector, machine tools, machinery, electrical equipment and electronics.

Austrian firms are Hungary's second most active partners (primarily construction, electronics, consumer goods and the automotive branch) followed by the US (mainly agriculture and equipment). French, Italian and British firms follow closely behind the US – with roughly the same number of agreements. A relatively high concentration on the chemical and petro-chemical industries is noticeable in the case of Italy. As far as French firms are concerned, no particular pattern emerges from Table 11, while in the case of British firms, agreements pertaining to electrical equipment and electronics seem to be somewhat in the lead.

## 3. Forms of Co-operation

### a) *Co-production and Specialisation*

According to the UN survey, co-production is the preferred form of co-operation and, among the East European countries, Hungary's share of this type of agreement is also the highest (see Tables 2 and 5 of Chapter 7). The UN's findings are supported by the agreements listed in Table 11, particularly if those elements usually associated with co-production (co-operation in production, development, marketing and so forth) are taken into account.

### b) *Joint Ventures*

Joint ventures with foreign equity investment (maximum 49 per cent foreign participation) have been permitted since 1972. In order to encourage this form of co-operation, Hungarian legislation governing joint ventures has been progressively liberalised. For instance, as of 1977, joint manufacture between local enterprises and foreign firms is permitted. Preferential tax treatment has also been accorded to joint companies and Western firms allowed to hold a majority share in those companies not involved in production.

In 1982, a further revision came into effect. It provided for the establishment of duty-free zones for joint ventures. This Decree allows the creation of "offshore" manufacturing bases.

Duty on raw materials and component imports is not paid if the end-product is re-exported. Import tariffs will only be applied if the product is sold within Hungary. Joint ventures in the new custom-free zones will also be exempt from many Hungarian regulations concerning wages, prices and investment. In essence, joint ventures operating within these duty-free zones will be treated as foreign companies operating on Hungarian territory.

Hungarian legislation permits four types of joint ventures: unlimited liability partnerships, joint stock companies, companies with limited liability and joint enterprises. Joint ventures may now be established for production, trade and service activities. As is the case for those countries permitting foreign equity investment, profits may be transferred abroad. However, in the case of Hungary, this is only allowed if the joint venture has convertible currency available[69].

In order to promote joint ventures, an information bureau – designed to provide Western companies with information concerning the modalities for establishing joint equity companies – was established by the Hungarian Chamber of Commerce in 1984. Furthermore, in 1985, Hungary signed the first joint venture (with Schwarzkopf GmbH, Germany) which permitted the foreign partner to hold 51 per cent of the shares.

The effort being made to promote this form of co-operation reflects the Hungarian viewpoint that joint ventures are potentially the most effective form of co-operation. Among the advantages most frequently cited by Hungarian commentators are: that they are a means of getting around capital shortages; provide easier access to Western technology and production techniques; are most likely to attract Western firms (because such arrangements involve less risk and are reasonably flexible) and, that they extend marketing (export) possibilities – not only to Western, but to third country markets.

Whether Hungarian expectations will be met – both in terms of expanding this form of co-operation or meeting the desired objectives – is very much an open question. The modest expansion of joint ventures domiciled in Hungary, relative to the evident effort made to further this form of industrial co-operation, is linked to a number of factors.

On the Western side, interest in establishing joint ventures seems to have been dampened by the fact that negotiations for concluding them are frequently very long. Low or little profitability ensuing from them, seems to be another constraint.

On the Hungarian side, the obstacle most frequently mnentioned is that enterprises find it excessively difficult to sell their products on Western markets. It is also noted that the procedures for obtaining permission to enter into a joint venture agreement are too complicated and that enterprises are not sufficiently rewarded (in terms of financial incentives) – relative to the effort required to establish a joint venture[70].

Concerning the number of joint ventures actually established, Table 11 of Chapter 7 indicates that 33 such agreements were concluded over the 1980-1984 period – the majority of them with German and Austrian firms. According to the UN survey, it is Hungary's third most prevalent form of co-operation and Hungary's share (in the total number of agreements) is the highest of all the Eastern countries (see Tables 2 and 5 of the chapter). It might also be noted that only one "offshore" agreement (with Denmark) has been concluded.

In contrast to the relatively slow expansion of joint ventures located in Hungary, those domiciled in the West have grown rather rapidly. At the end of 1981, 68 had already been established. The majority of these are involved with import and market-related functions[71].

c) *Delivery of Plant or Equipment*

The UN survey establishes that delivery of plant or equipment is minimal (and appears to be declining since 1980), although with the exception of Poland, Hungary's share is the highest among the East European countries (see Tables 2 and 5 of Chapter 7). The UN's findings that this form of co-operation is not very pronounced in Hungary is also confirmed by Table 11 (using the category which comes closest to the UN definition – plant construction).

d) *Licencing Agreements*

Licencing agreements account for the second largest category of the agreements analysed by the United Nations. The survey also indicates that Hungary had the highest share of the total number of licencing agreements concluded by the Eastern countries (see Tables 2 and 5 of Chapter 7) – a fact that was also established by earlier surveys carried out by the UN.

According to UN data, the light industry accounts for the majority of the agreements which have been concluded, followed by the chemical industry, food and agriculture and mechanical engineering. The agreements recorded for the 1980-1984 period, corroborate that licencing is the second most important form of co-operation (see Table 11 of the chapter). German firms have clearly been the most active participants. However, no clear-cut pattern – in terms of sectoral distribution – is evident, though the chemical industry is somewhat in the lead.

According to official sources, Hungary's clear preference is for those agreements which provide most-favoured licence clauses, though cross-licencing, where applicable, is also welcomed[72].

## 4. Other Technology Transfer Channels

Hungary's participation in *licence trade* is considered to be modest. For the 1970-1983 period, cumulative licence and know-how purchases are said to have accounted for only 3 per cent of domestic R&D expenditures[73]. It is interesting to note that the share of the Western countries in the total number of licences purchased by Hungarian enterprises has been consistently high – fluctuating between 86-90 per cent during 1976-1983. Germany has been the principal country from which licences were bought[74].

Licences purchased (and used) during 1975-1979, were primarily aimed at developing machine-building and the light, chemical and food industries[75]. A similar trend applies to licences acquired by the end of 1983, although the share of the food-processing industry has decreased.

While Hungary has increased its licence purchases from the West since 1980, the export ratio was still only 37 per cent in 1983 (the same as in 1976), although the proportion of convertible currency exports (as a percentage of total export sales of licence-based products) was 68 per cent. Furthermore, relatively few enterprises buy licences, purchases are limited primarily to manufacturing and the share of products manufactured on the basis of licences remains small – some 6 per cent in 1983[76].

Apart from export restrictions and the evident reluctance of Western firms to sell know-how of their latest developments, internal factors also provide an explanation for this. For one thing, lengthy bureaucratic procedures to obtain permission (and foreign exchange) to buy a licence lead to considerable time delays between the purchase and application of a licence. They also serve as a disincentive to enterprises to attempt to acquire them. It is also noted that "research institutions are not interested in adapting foreign licences, because this does not count as creative work. Also, there is no connection between research and licence purchase policy. As a result, licences are frequently purchased for which there is no realistic chance of further development"[77]. Systemic difficulties, as well as hard currency shortages which prevent the purchase of complementary inputs needed to effectively exploit a licence, are among the major internal reasons which have prevented Hungary from significantly improving its competitive position on Western markets.

*Scientific and technological co-operation* within the framework of multi-sector agreements is relatively prevalent. The agreements listed in the general literature do not indicate any particular pattern, as far as sectoral distribution or Western firm participation is concerned. However, the majority of them involve know-how and licence transfers, delivery and installation of plant or production equipment and marketing co-operation in third markets.

The supply of *turnkey plants* is negligible. Available information suggests that two were actually acquired during 1980-1984.

## E. POLAND

### 1. Involvement in and Policy Towards EWIC

Export promotion was – and remains – Poland's primary objective in pursuing EWIC. Nevertheless, it is interesting to note that equal emphasis is now also being placed on effective assimilation of modern technologies and this aspect is included in the guidelines currently governing industrial co-operation. In essence, close and continuous association with the supplier of technology is seen as a means of procuring assistance in the implementation of technology acquired through industrial co-operation[78].

This viewpoint clearly reflects Poland's past, and primarily negative, experience with EWIC. That is, inability to effectively assimilate Western technology was a major factor in the collapse of its import-led growth strategy. Furthermore, most agreements which Poland had concluded during the early 1970s were short-term and, for the most part, did not entail real co-production or "two-way advanced co-operation". In other words, they did not lead to the establishment of long-term linkages – the essence of effective industrial co-operation. Nor, in most instances, did these arrangements involve technology transfer of any real significance: with the exception of those agreements concluded with the United States[79].

Since 1982, numerous laws and directives have been introduced designed to reform the economic mechanism. Re-establishment of market equilibrium and the resumption of international co-operation ties, are among the goals to be achieved by these reforms[80]. Consequently, enterprises were to be given more autonomy in their dealings with foreign firms. Hard currency needed by them for the production of goods for export and for meeting the obligations of an industrial co-operation agreement, was to be made more easily available and, the number of enterprises and organisations allowed to engage in foreign trade, was to be increased.

Furthermore, a greater number of small and medium-sized enterprises (considered to be very important for furthering economic co-operation with Western firms), were to be created. Such enterprises were also to be given greater autonomy. For instance, they would be allowed to maintain foreign exchange accounts and to engage directly in foreign trade, without the intermediary of foreign trade organisations[81].

Though not explicitly stated in the official literature, it can be assumed that one purpose was to establish long-term co-operative links with Western firms. Another, was to counteract the adverse effects which the drastic cuts in the investment programme (undertaken as of 1979) had on industrial co-operation. These cuts prevented Poland from meeting supply commitments to its Western partners. That is, components, parts, machinery and equipment, the dominant proportion of goods that were to be produced (and exported) under the industrial co-operation agreements concluded with Western firms, could not be delivered[82]. Consequently, many of the agreements which Poland had entered into were simply allowed to lapse.

Whether the reforms that have been undertaken will be effective, in terms of meeting the above-mentioned objectives, is questionable. It is the opinion of many Western analysts that they are, in fact, a mere expression of intent. Economic management remains tightly centralised. Whether the ability to retain a portion of hard currency earnings ensuing from exports will be a sufficient incentive for enterprises to pursue industrial co-operation, is also questionable. In any event, since the Ministry of Foreign Trade retains the authority to intervene in the policies of individual enterprises engaged in foreign trade, foreign exchange allocations remain, in effect, under central control. Furthermore, the Ministry of Foreign Trade also decides the percentage of hard currency export earnings which the enterprises are allowed to keep and, since hard currency is frequently not available, it can be assumed that the degree of freedom enjoyed by enterprises will continue to be severely constrained.

Current discussions in the Polish literature emphasise that the traditional forms of EWIC have, by and large, not met Polish expectations – as far as export expansion is concerned. Balance of payments considerations (or, as some Polish authors put it, the search for new sources of financing the country's external trade) are clearly behind the viewpoint that, in future, greater attention should be paid to furthering the joint venture form of co-operation as well as outright compensation arrangements[83].

However, the extent to which Poland can – or actually wishes to – expand EWIC, is difficult to assess and official attitudes are clearly divided. The measures taken to favour the expansion of industrial co-operation, indicate Poland's desire to maintain links with the West and reflect recognition of continuing need for Western technology in order to meet export goals. On the other hand, the developments which led to the collapse of the import-led growth strategy, has strengthened the position of those officials advocating closer links with CMEA partners. Disillusionment with EWIC and with Western technology (irrespective of the fact that, given the overly centralised command system of planning and management, Western technology imports could not meet expectations), have also contributed to the current policy of turning increasingly towards the Eastern countries: notably, the Soviet Union.

Turning then to Poland's involvement in EWIC, available data indicate that its participation has progressively diminished. According to official statistics, Poland's share in EWIC dropped from some 26 per cent in 1974-1975 to 19 per cent in 1976-1977[84]. Official estimates for subsequent years could not be found in the literature consulted. However, based on data of earlier UN surveys, it can be concluded that Poland's share in the total number of agreements analysed declined by some 23 per cent during the 1979-1984 period.

As noted earlier (see Section I B above), the latest UN survey ascertained that Poland presently accounts for some 12 per cent of the co-operation contracts analysed (i.e. 96 agreements – see Table 7 of the chapter). Table 12 of Chapter 7, indicates that Poland concluded 34 agreements over the 1980-1984 period.

Exports ensuing from EWIC arrangements are modest. In 1975, they represented only 2.4 per cent of total exports, although the share from the electrical and engineering branches was 25 per cent – of total exports[85]. While at the end of 1980, some 3 per cent of total Polish exports were the outcome of industrial co-operation, this share had fallen to 1.3 per cent in 1983[86]. Perhaps more disturbing, given the emphasis placed on co-production agreements, is that their share in total exports has remained virtually stagnant. This share was 3 per cent in 1975 and 3.7 per cent in 1981[87].

## 2. Preferred Sectors and Partners in Industrial Co-operation

According to the latest UN survey, mechanical engineering continues to account for the largest share of contracts concluded. It is followed by the chemical industry and by electronics and electrical equipment (taken together) – see Table 3 of Chapter 7. A somewhat different picture emerges from Table 12 of the chapter which indicates that the chemical industry is in the predominant position, while the share accounted for by electronics and electrical equipment is minimal. The Table also shows that German firms are most actively engaged in industrial co-operation with Poland, followed by firms from the United States, Italy, Japan and the United Kingdom – the latter four sharing almost the same number of agreements.

As far as Germany is concerned, the official literature points out that while German firms are indeed Poland's most active partners, co-operation with them has declined substantially since 1976. Furthermore, it is noted that the majority of the agreements concluded prior to 1979 (a total of 113) were of short-term duration and due to expire by 1984. It is presumably this development – together with the fact that only 11 new agreements (with German firms) appear to have been concluded during 1980-1984 – which has led to the expression of official concern relative to the future of industrial co-operation between Germany and Poland[88].

## 3. Forms of Co-operation

### a) *Co-production and Specialisation*

According to the latest UN survey, co-production and specialisation continues to be the most prevalent form of co-operation though Poland's share is not particularly high (see Tables 2 and 5 of Chapter 7). The agreements listed in Table 12 indicate that joint production *per se* is minimal. However, the majority of the agreements do involve co-operation in production.

It might be noted in this regard, that Poland's preference for agreements which entail co-production is clearly stated in the official literature. In fact, it is mentioned that the Decree issued in 1971 (the first of various measures undertaken to expand EWIC), was intended to enhance primarily this form of co-operation. Co-production was to be "developed above all in those industrial sectors which facilitate modernisation of domestic production structures and intensify trade with the outside world: that is, engineering"[89]. However, that this form of co-operation has clearly not progressed as desired, is also noted. It is said, for instance, that "out of the nearly 100 co-production agreements signed by Western firms by 1982, only some 15-20 can be considered to be real co-production agreements"[90].

### b) *Joint Ventures*

Foreign investment in Poland has been allowed since 1976, though largely limited to small-scale industry and to consumer services.

Since 1979, joint ventures permitting foreign equity investment (a maximum of up to 49 per cent) have been admitted. The sphere of joint venture activity was also enlarged, though it is still primarily confined to production and trading activities. Joint ventures are not permitted in such key industries as mining, raw materials, engineering, chemicals, electronics or fuels.

Nevertheless, in order to make the establishment of joint ventures more attractive to Western firms, their duration was extended to 10-15 years. Furthermore, the restriction on repatriation of profits (such repatriation had been limited to 9 per cent of the foreign partner's investment, except in cases where at least 50 per cent of the company's turnover originated from hard currency sales abroad), was waived. Various incentives were also provided to Polish enterprises to encourage the development of joint venture agreements. Among them are the establishment of hard currency accounts: either for financing investments in Poland or for financing export-import transactions – including sales in Poland for convertible currency.

Since 1982, further incentives for the creation of joint ventures were provided, with the evident intent to attract hard currency inflows. For instance, joint ventures may now be established for 20 years. Foreign or mixed enterprises, with unlimited equity participation, are now permitted in the light industry. The foreign partner is now also allowed to repatriate 50 per cent of its share of a joint venture's foreign trade surpluses.

Under the 1982 Law, the foreign investor is permitted to establish either a wholly-owned company or one with a local partner. However, in contrast to Hungary, Poland only permits the establishment of joint-stock companies and companies with limited liability – the latter type being favoured. Also, before a foreign partner can establish a mixed company in the small industry sector, it must make a substantial deposit in convertible currency[91]. At the end of 1982, some 250 private foreign companies were operating in Poland; involving primarily German, British, Austrian, US, French and Italian firms. Joint turnover amounted to some 17.5 million zloty and these companies exported goods worth $8 million[92]. Attracting foreign capital and increasing exports are clearly Poland's motivation in encouraging this form of co-operation.

As regards the number of equity joint ventures actually established in Poland, their number appears to be very modest – three were recorded during 1980-1984 (see Table 12 of the chapter). In contrast, joint ventures located in the West have expanded rapidly – see Table 1 of Chapter 7. According to official sources, at the end of 1982, 93 were active. Of these the majority were located in Germany (19), followed by the United Kingdom (11), Austria and France (5 each). It is said that 42 per cent of Polish exports were made possible through these mixed companies. However, their contribution to Poland's imports was only 14 per cent[93].

c) *Delivery of Plant or Equipment*

Based on the UN survey, delivery of plant or equipment is the second most prevalent form of co-operation and Poland appears to also command the largest share of this type of contract (see Tables 2 and 5 of Chapter 7). According to Table 12 of the chapter, agreements involving plant construction (the category which most closely approximates the UN's classification) has not been prominent during the 1980-1984 period.

d) *Licencing Agreements*

Licencing agreements played an extremely important role in Poland's "new development strategy" introduced at the beginning of the 1970s. That is, such agreements (as well as the direct purchase of foreign licences) were seen as a vital source for modernising entire industries or branches narrowing, thereby, the existing technological gap. It was also assumed that, in the long run, this would lead to export promotion[94]. During 1971-1975, Poland concluded some 283 licencing agreements, though between 1976-1980, only 128 new agreements were signed. Agreements were concentrated on machinery and equipment, transport equipment, electrical equipment, chemicals, electronics and precision instruments. This sectoral distribution mirrored the policy goal of modernising key sectors of the economy.

In terms of the number of agreements concluded, German firms were clearly in the lead (some 100 of the total), followed by firms from France, the United States and Great Britain. However, in value terms, France had the largest share, followed by the United States, Italy and Great Britain.

Despite the large number of agreements that were concluded up to 1980, Poland's inability to attain the policy objectives to be achieved by them, is officially linked to restrictive clauses included in the agreements. That is, frequently (and particularly in "key industries") application was restricted to a specific enterprise. Furthermore, the majority of the agreements (80 per cent) contained either total or partial restrictions on exports produced under a licence, and 75 per cent of the agreements permitted exports only to CMEA countries[95]. However, it is also officially stressed that inability to effectively assimilate technology and know-how acquired by means of licencing agreements, was equally decisive in preventing Poland from achieving the given policy objectives[96].

According to the latest UN survey, licencing agreements continue to account for a rather large proportion of Poland's industrial co-operation agreements – see Tables 2 and 5 of Chapter 7. Nevertheless, the UN surveys also indicate that, since 1979, Poland's share in the total number of licencing agreements registered has progressively diminished. It might also be noted, that the number of licencing agreements actually concluded during 1980-1984 is very modest (see Table 12 of the chapter).

## 4. Other Technology Transfer Channels

The *purchase of licences* was considered to be an especially advantageous means of acquiring technology; in terms of "giving immediate access to a higher level of technology at lower cost and because they did not require acceptance of foreign control and interference with the decision-making process"[97].

During the 1970-1980 period, a total of 428 licences were purchased,– primarily from Germany (always in the lead), the United Kingdom, France and the United States. Priority was given to the engineering and chemical industries. Presumably, they were deemed to be best suited for meeting the objective of "creating entirely new industries and industrial branches based on imported technologies and techniques"[98]. While licence imports fluctuated during the entire period (in value and in terms of numbers purchased), most were acquired during 1975-1976, after which they declined rapidly.

Given Poland's development strategy, it is very significant that licence purchases far exceeded the absorptive capacity of the economy. During 1974-1976, only 97 (out of 256) licences were actually utilised. In 1980, only 15 licences were put into operation. The highest proportion of delays were precisely in those sectors for which most licences had been purchased. That licences had, therefore, a limited impact on the development of the Polish economy, can be seen from the fact that during the entire period, exports of products made under a licence amounted to only about 5 per cent of total exports[99].

Poland's current financial situation has led to the adoption of a very restrictive policy towards licence imports and few licence purchases have been recorded since 1981. Those listed in the official literature suggest that licences have been acquired primarily for the chemical industry.

The purchase of *turn-key* plants has always been rare and this continues to be the case. During the 1971-1980 period, they represented roughly 7 per cent of total technology imports from the Western countries[100].

The purchase of *technical information*, rather prevalent during 1976-1980, seems also to have been largely abandoned. Between 1976-1980 Poland had spent some $6 million for such information: primarily for expanding machinery and equipment output[101].

## F. ROMANIA

### 1. Involvement in and Policy Towards EWIC

Based on the latest UN survey, Romania's involvement in EWIC is not extensive. Table 7 of Chapter 7 shows that its share in the total number of contracts analysed was some 5 per cent (or 38 agreements), placing that country in second to last position. UN data also indicate, that Romania's share had progressively diminished over the 1979-1983 period. According to official figures, Romanian enterprises had concluded more than 100 industrial co-operation agreements with Western firms by 1982[102]. For the mid-seventies, a figure of some 250 agreements was given. While Romania's definition of industrial co-operation is very broad, this would, nevertheless, confirm the general downtrend noted by the UN.

Table 13 of the chapter also indicates that Romania's participation in EWIC is minimal. Apparently only 19 agreements were, in fact, concluded during 1980-1984 and no agreements for 1984 could be traced in the sources consulted. Furthermore, according to this count, Romania is least involved in industrial co-operation.

The official literature continues to support the need for, and utility of, co-operating with the Western countries. Such co-operation is considered to be especially useful for: expanding existing energy resources; developing alternate sources of energy (notably, atomic energy) and, for gaining access to advanced technologies designed to expand output (and exports) of such key sectors as chemicals, machine construction and electronics[103]. The official literature also indicates that Romania is particularly interested in agreements which involve long-term co-operation in production (notably in the area of manufactures), arrangements which entail co-operation on third markets and joint ventures[104].

Nevertheless, current developments lead to the conclusion that Romania's involvement in EWIC will, in all likelihood, decrease even further. For instance – and as mentioned in Chapters 2 and 4 – Romania intends to turn increasingly towards its CMEA partners, particularly in the sphere of S&T development. In order to meet energy and raw materials needs, closer trading and co-operative relationships (notably with the Soviet Union) are also foreseen. Hard currency constraints, continual restructuring of Ministries and FTOs (foreign trade organisations), lack of flexibility on the part of enterprises and extraordinarily high countertrade demands (up to 200 per cent) are also likely to constrain the development of EWIC.

Balance of payments considerations have likewise led to the adoption of a series of measures designed to re-establish external equilibrium. These measures are unlikely to encourage the expansion of industrial co-operation as far as Western firms or Romanian enterprises are concerned. Thus, the Council of State Decree (No. 276) of July 1979, stipulates that a co-operation arrangement can only be undertaken if imports are balanced by exports. Long-term contracts for countertrade deliveries, particularly of machinery, are to be encouraged. Imports of machinery and equipment are to be allowed primarily by means of co-operation in exchange for products from the mechanical engineering sector. Such rigid stipulations, particularly as far as Western counter deliveries are concerned, could well explain why delivery of plant or equipment accounts for a very small share of Romania's industrial co-operation agreements.

As far as Romanian enterprises are concerned, the requirements of Art. No. 12 of the above-mentioned Decree (in effect as of January, 1981) are hardly conducive for their seeking co-operation arrangements. Namely, all foreign exchange liabilities arising from a co-operation agreement (imports, credits and so forth), must be met from their own foreign exchange earnings. At the same time, enterprises must try to obtain additional earnings by over-fulfilling their export plan[105]. Romania's financial difficulties, hence the need to maintain its strict consolidation policy, makes it highly unlikely that these requirements (designed to "rationalise imports and to improve the efficiency of economic co-operation") will be eased.

## 2. Preferred Sectors and Partners in Industrial Co-operation

According to the UN, the chemical industry and transport equipment occupy first place, as regards the number of industrial co-operation contracts concluded: each accounting for some 34 per cent of the total – see Table 3 of Chapter 7. Based on Table 13 of the chapter, it appears that the chemical industry is in first position (although the number of agreements recorded is small), while the UN's category of transport equipment does not appear to be well represented.

As far as Western participation is concerned, Italian and US firms are Romania's predominant partners, although no new agreements with the latter appear to have been signed since 1980. Based on the agreements signed with Italian firms, it seems that co-operation with them is particularly useful – in terms of Romania's development strategy (see Chapter 2). For instance, the agreement signed with Nuove Pignone in 1981, provides for mutual sale of each other's product lines, including plant, machinery and equipment for the chemical and petro-chemical, natural gas and nuclear power industries. The Italian firm is to make its entire production programme available, including turbines, compressors, pumps, valves and drive and control systems. The agreement with FATA (signed in 1982) covers development of an automated line for producing cast-steel engine blocks. Very significant for Romania is the agreement, signed in 1981, with Ansaldo Impianti Nucleari (together with the US's General Electric Company) to construct two nuclear power stations in Romania, scheduled for start-up in 1987.

It is also interesting to note, that some Western firms seem to have concluded agreements simply to ensure overdue payments from Romania. This is the case, for instance, in the agreements concluded in 1982 by ICI, Shell International and Courtaulds (all British firms). These firms have apparently agreed to increase their imports of Romanian machine tools, chemical and petro-chemical products.

## 3. Forms of Co-operation

### a) *Co-production and Specialisation*

According to the UN survey, this form of industrial co-operation is not prevalent – see Tables 2 and 5 of Chapter 7. Based on Table 13 of the chapter, it appears that only one joint production agreement was concluded during 1980-1984, although three of the agreements recorded do involve co-operation in production.

b) *Joint Ventures*

Romania has permitted joint equity investment since 1971. In 1972, two additional Decrees were adopted which fixed the legal regulations governing joint ventures. Since then, legislation has not been modified, in any significant manner.

Joint ventures are permitted in industrial and agricultural production, construction, transport, tourism, trade and research. The official literature makes it quite clear, however, that gaining access to Western technology (for the purpose of economic development) and export promotion (particularly of industrial goods) are Romania's principal motivations for seeking this form of co-operation. As regards the former, Romania's joint venture legislation specifically notes that "the constitution of joint companies in the Socialist Republic of Romania has in view (*inter alia*)...to accelerate the introduction of certain modern technologies..."[106].

Romanian legislation appears to be flexible. For instance, such aspects as management of a mixed company or choice of joint venture type (joint stock or limited liability company – the two forms permitted in Romania), is determined by the participating partners. Profit sharing, proportional to its capital investment (determined by the company's management), is permitted for the foreign partner. The foreign partner may also transfer its portion of profits abroad – after payment of the profits tax.

In reality, Romanian joint venture laws are very restrictive and seem to have the net effect of hindering the very policy objectives to be attained by joint ventures. For instance, equity participation of the Western partner is limited to 49 per cent. Profits are subject to a 30 per cent tax, with an additional 10 per cent for those profits not reinvested in Romania. The requirement that all foreign trade transactions be carried out in hard currency, limits the flexibility of a joint venture. Also, regardless of the form of the joint venture, the partners do not retain ownership over their investments and retain only those rights (relative to assets and liabilities) which result from their capacity as a shareholder.

Furthermore, imported products are liable to customs duty, as well as, to fiscal and other domestic dues. The net effect of this is that the final cost of a product is increased: hence competitiveness on foreign markets decreased. For the Romanian partner, joint venture legislation also stipulates that all payments in foreign currency of the joint companies shall only be made out of their own available foreign currency or from loans[107]. As is noted, one outcome of this requirement is that a joint company must already be earning foreign exchange from a portion of its exports in order to be able to import parts or materials from its Western partner. This is clearly impossible, if the company has not yet started production and has already exhausted its convertible currency loans in equipping its manufacturing plant[108].

In the case of Romania, it is particularly difficult to obtain accurate information concerning industrial co-operation. This is most striking as regards joint ventures. For instance, the latest UN survey indicates that joint ventures are Romania's prevalent form of co-operation, though its share in the total number of agreements is well below that of Hungary (see Tables 2 and 5 of Chapter 7). Data as to the number of joint ventures established conflict. Most sources cite 9 such agreements – concluded by mid-1983. Table 13 of the chapter indicates that none were signed during 1980-1984.

Perhaps more significant is that only four joint equity ventures are said to be currently active. These include: ROM-CONTROL DATA (with the US-Control Data Corporation) for the manufacture and marketing of computer peripheral equipment; RESITA-RENK (Germany-Renk. A.G.) for the production and sale of gear wheels, speed reduction gears and couplings; RIFIL (Italy-Romalfa) for the production and sale of synthetic (acrylic) fibres and OLTCIT (France-Citroën), for the manufacture and marketing of small passenger cars and replacement parts.

The latter appears however, to be on the verge of collapse. The reasons for this are manifold. Among them are Romanian inefficiency and management problems which have delayed production to the point that the car was outdated before it was exported. Under the terms of the agreement, Romania was to purchase parts and supplies for hard currency (amounting to roughly 30 per cent of the total price of the car), while Citroën was to purchase no less than 50 per cent of the total output for sales on Western markets. Inability to pay for parts and components caused French banks to freeze credits. Furthermore, production targets (originally set at 130 000 cars annually and reduced to some 60 000 by the time production started – 2 years behind schedule) clearly over-estimated market demand and Romanian production capacity. The ultimate outcome of this venture is uncertain, as negotiations, mutually "acceptable" to the parties involved, are on-going.

On the other hand, Romania has been more successful in establishing joint ventures in the West – see Table 1 of Chapter 7. At the end of 1981, Romania is reported to have invested in 32 Western firms, predominantly located in Germany and France. Like its East European partners, the majority entail importing and some related marketing functions[109].

c) *Delivery of Plant or Equipment*

As mentioned, agreements concerning delivery of plant or equipment remain insignificant (see Tables 2 and 5 of Chapter 7). According to Table 13 of the chapter, two agreements involving plant construction were concluded during 1980-1984.

d) *Licencing Agreements*

UN data indicate that licencing agreements are also minor: both in terms of the number of contracts concluded and Romania's share in the total number of agreements – see Tables 2 and 5 of Chapter 7. According to UN data, licencing agreements pertain exclusively to transport equipment. Table 13 of the chapter indicates that only two licencing agreements were concluded during 1980-1984: one for the automotive branch and one for the chemical and petrochemical industry.

## 4. Other Technology Transfer Channels

Romania's *licence purchases* are not significant. Since Romania does not include licence trade in its published statistics, participation in such trade can only be assessed on the basis of other Eastern sources. According to Polish statistics, for instance, during the mid-sixties Romania was among the leading purchasers of Western licences (in value terms), with purchases rapidly declining thereafter[110]. Western data as to the number of licences actually acquired by Romania are not available; for this or subsequent periods.

Export considerations are, not surprisingly, at the heart of the Decree issued in 1980 which defines Romania's procedures applicable to licence imports. These procedures have been simplified and made more uniform. At the same time, the Ministry of Foreign Trade "may refuse licence imports from countries which maintain or apply discriminatory quantitative restrictions on Romanian exports"[111].

According to official (and other Eastern) sources, licences have primarily been purchased for the chemical, engineering and light industries, as well as, for the metallurgy sector. Principal suppliers have been Germany, France, the United Kingdom and Switzerland. Japan, Sweden and Belgium are also said to have been important suppliers of licences[112].

From the Romanian perspective, licences are clearly seen as an important technology transfer channel and acknowledged to have played a vital role in the development of key sectors of the economy. Licence procurements are also considered to have been a key component for the conclusion of a number of co-operation agreements and, indispensable for reaching the objectives of joint venture co-operation[113]. Relative to the former, it is interesting to note that the agreement between a British company and Romania for the production of aircraft engines was made possible because of the willingness of the British company to supply a licence (together with requisite documentation, training and technical assistance) which gives Romania the right to manufacture a significant proportion of components and to assemble the engine[114].

One of the main objectives of current Romanian policy is to reconstruct its aircraft industry; chiefly on the basis of Western expertise. A milestone in achieving this objective is the licence supplied by British Aerospace (in 1982) for the production of the BAC 1-11 aircraft. Romania's ultimate objective is to manufacture the aircraft on its own. Given marketing difficulties in the West, it can be assumed that exports will be primarily for third world markets.

Romania's automotive industry has been developed almost entirely thanks to licences supplied by Western firms. Production of Dacia cars is on the basis of the licence from Renault. The seemingly ill-fated OLTCIT joint venture, also provided for production under a licence from Citroën. In addition, Romania has purchased licences for components. These include a licence for piston ring production from Goetz (Germany), electrical equipment from Dusseal (France), fuel equipment from Bosch (Germany) and suspension rings from Armstrong Equipment (UK).

The Romanian truck industry has also relied heavily on Western licences for production and for components and assemblies. For instance, a licence has been purchased from M.A.N. (Germany) for the manufacture of engines and truck propeller shafts and from Zahnrad Fabrik (Germany) for steering equipment.

## G. SOVIET UNION

### 1. Involvement in and Policy Towards EWIC

Based on the latest UN survey, the Soviet Union, with 243 agreements (30 per cent of the total number of contracts analysed) is the second largest participant in EWIC – see Table 7 of Chapter 7. Table 14 of the chapter, indicates that the Soviet Union had concluded 254 agreements during 1980-1984. Nevertheless, while as mentioned earlier, the Soviet Union's involvement in EWIC is modest relative to the size of the economy, the agreements concluded by the USSR are large-scale in nature; hence, in value terms, they far exceed those of Hungary. It might also be noted, that the Soviet Union's participation in industrial co-operation has steadily increased as of 1982, while the number of agreements concluded by Hungary during 1982-1984, has been relatively stagnant (see Tables 11 and 14 of Chapter 7).

Financing capability (which has enabled the Soviet Union to directly import materials, parts, machinery and equipment needed for production) explains, in part, the Soviet Union's relatively low participation in EWIC. The prevalence of, and preference for compensation arrangements, also accounts for this.

Furthermore, and in contrast to the East European countries, the Soviet Union has focused its co-operative efforts on realising joint industrial projects; particularly large scale (and long-term) undertakings for the development of its energy and raw materials resources. These joint development projects are agreed to at the *governmental level* and are based on negotiated compensation arrangements. Hence, they are not inter-firm co-operation agreements, as commonly understood. According to the UN, over 100 long-term agreements and contracts are said to be in operation covering more than 60 large compensation projects. Among the most important of these are the agreements concluded with Italian, German, French, British and Austrian firms, designed to develop the Soviet Union's energy (notably gas) and raw materials resources, as well, as the chemical industry[115].

Despite current emphasis placed on the need to achieve greater independence from the West (as far as reliance on Western technology is concerned), the intent to maintain (and further) mutually beneficial economic relations with the Western countries continues to be stressed – see Chapter 4. The desirability of expanding and diversifying industrial co-operation, notably co-operative forms which establish long-term and close contacts with the Western partner, is likewise underlined. Like its Eastern partners, the Soviet Union views industrial co-operation as a means to raise the technical level of indigenous production, speeding up the introduction of modern technology and know-how, expanding domestic production at lower cost and, increasing export potential – notably of engineering products[116].

Current developments indicate that, in future, particular attention will be paid to seeking turnkey type arrangements which not only involve the supply of Western machinery and equipment, but for which the Western partner is entirely responsible for construction[117]. Long delays in meeting construction deadlines for large-scale industrial projects undertaken by the Soviet Union (hence failure to meet production and export targets), appear to be behind this policy.

The official literature also suggests that the Soviet Union is interested in becoming more involved in co-production, since this form of co-operation "accelerates the production of modern technologies, raises the efficiency of production and creates additional export resources"[118]. Furthermore, it is a "higher level form of co-operation which presupposes, along with the engineering and technical interaction of partners, unification of the applied technologies and arrangements for production, co-ordination of sales and after-sales service"[119]. This interest in furthering co-production can be directly linked to the Soviet Union's inability to quickly, and efficiently, absorb and diffuse acquired technology. The close contact with Western partners implicit in co-production arrangements, clearly provides crucial assistance at the installation stage: one of the weakest points in the Soviet Union's production system.

Nevertheless, despite official support of industrial co-operation (with the aim of eliciting Western participation in those areas singled out as being priorities for the economic development and industrial policy to be pursued during the 1986-1990 plan period), it is evident that the Soviet Union is interested primarily in pursuing those types of EWIC agreements which encompass countertrade. Indeed, official discussion indicates that from the Soviet viewpoint, industrial co-operation encompasses buy-back or large-scale compensation[120].

## 2. Preferred Sectors and Partners in Industrial Co-operation

According to the UN, the chemical industry continues to account for the highest share of industrial co-operation agreements, followed by mechanical engineering (including machine tools) – see Table 3 of Chapter 7. The agreements listed in Table 14 of the chapter corroborate the findings of the UN survey. As the Table indicates, the chemical industry predominates by far. The majority of the multisector agreements (the second largest category) provide primarily for co-operation in the engineering field – notably, the production of machinery, equipment, machine tools and so forth. The same applies to the agreements which fall into the category of agriculture and equipment, construction and mining and metallurgy.

Based on Table 14, Italian firms are most prominently involved in industrial co-operation with the Soviet Union, followed by firms from Germany, France and Finland – the latter two having concluded the same number of agreements.

Concerning Italy (whose agreements are most prevalent in the chemical industry, though construction is also quite well represented), ENI (Ente Nazionale Idrocarburi) is said to be the Soviet Union's major Italian partner. ENI is a public holding which controls mainly oil companies, but also such enterprises as Snam Progretti (an engineering firm) and Nuovo Pignone (production of various machinery and, in particular, equipment for the gas and oil industries)[121]. According to Soviet sources, in 1983 some 46 per cent (volume terms) of trade between the Soviet Union and Italy ensued from co-operation with ENI: a volume turnover not matched by any other Western firm[122]. Whether (or to what extent) this figure is exaggerated, could not be verified.

Under the terms of the agreements concluded with Italian firms (notably ENI), the Soviet Union received machinery and equipment for construction of the Urengoy-Uzhgorod gas pipeline and for enterprises in the chemical, metallurgical, pulp-paper and constructions materials industries[123]. Co-operation with FIAT and its affiliate UTS (Ufficio Tecnico Studi) has also been instrumental in developing the Soviet Union's automotive industry.

Table 14 indicates that co-operation with German firms spans virtually every major industrial branch, although those agreements concerning mining and metallurgy, construction, the automotive branch and the chemical industry are somewhat in the lead. As far as the chemical industry is concerned, co-operation is said to have become wide-scale in nature and to have resulted in the construction of numerous complete installations – on the basis of compensation payment[124].

Relative to French firms, the chemical industry accounts for the largest number of agreements, followed by agriculture and equipment. According to the UN, the agreements signed with French firms (notably those concluded prior to 1980) involved primarily the provision of equipment to be used in transporting Siberian gas, for the manufacture of platforms for the Caspian Sea, and for the metallurgical, chemical and petro-chemical industries. It is also noted, that since co-operation agreements are becoming increasingly compensatory in nature, the participation of small and medium-sized French firms is steadily decreasing[125]. This would imply that, in future, co-operation will involve large-scale industrial projects – a preference which, in any event, is underlined by Soviet commentators.

Based on Table 14 (of the chapter), no clear-cut pattern emerges as far as co-operation with Finnish firms is concerned, although power and energy and the transport sector (notably railways) appear to be somewhat in the lead. According to the UN, industrial co-operation between Finland and the Soviet Union has been primarily concerned with joint planning, manufacturing and marketing of products[126]. The official literature claims that some 80 draft proposals on co-operation are presently under elaboration. The most important of these concern co-production of special freight cars, industrial robots and equipment for exploiting the sea-shelf. Furthermore, for the 1986-1990 period, co-production is to be expanded in order to increase mutual deliveries of machinery and equipment, as well as, exports to third countries[127]. In 1982, shipments of co-operative products are said to have accounted for some 30 per cent of the total volume of Soviet plant and machinery exports to Finland[128].

## 3. Forms of Co-operation

### a) *Co-production and Specialisation*

According to the UN survey, co-production is the most prevalent form of co-operation and the Soviet Union's share (in the total number of contracts) is also the highest – see Tables 2 and 5 of Chapter 7. The Soviet Union's preference for this type of arrangement is also substantiated by Table 14 (of the chapter), which indicates that the vast majority of the agreements entail joint production, co-operation in production or development.

### b) *Joint Ventures*

Joint ventures with foreign equity participation are not allowed in the Soviet Union – although there is some indication that a change of policy might occur. The number of joint companies operating in the West increased modestly between 1976-1980 (see Table 1 of Chapter 7). At the end of 1981, their total number had risen to 111. Most are located in the United Kingdom(14), closely followed by France and Germany (12 each) and Belgium (10). The majority of these mixed companies are involved in marketing and services[129]. Based on the UN survey, participation in this form of co-operation seems to be declining – see Table 2 of the chapter.

### c) *Delivery of Plant or Equipment*

According to the UN, delivery of plant or equipment remains the second most important category, although a decline – in terms of its percentage share – has occurred since 1982. Nevertheless, compared to its Eastern partners, the Soviet Union commands some 50 per cent of this type of agreement – see Tables 2 and 5 of Chapter 7. Table 14 of the chapter, indicates that plant construction, or agreements which involve co-operation in plant construction, are rather well represented, particularly in the chemical industry.

### d) *Licensing Agreements*

The latest UN survey indicates that licencing agreements account for a small share of the total number of contracts concluded by the Soviet Union and, compared to most of its Eastern partners, participation in such agreements is also modest (see Tables 2 and 5 of Chapter 7).

UN data suggest that licencing agreements have been primarily concluded for the chemical industry, followed by transport equipment and the light industry. Table 14 of the chapter indicates that, while the number of licencing agreements concluded are modest (30 over the 1980-1984 period), they are, nevertheless, the second most prevalent form. In contrast to the findings of the UN, most such agreements correspond to the UN's category of "light industry", while the chemical industry is very modestly represented.

It is evident from the official literature, that access to Western licences is considered crucial for internal domestic (and intra-CEMA) economic development. It is, after all the Soviet Union, which has advocated the need to articulate a common strategy towards Western licence acquisitions.

The Soviet Union's attitude towards licence agreements is very explicit. That is, "the objectives of licence agreements are... to obtain the right to use patented technical inventions, to use trademarks, industrial samples, transfer the technological innovations to the licensee together with know-how, [as well as] the technical assistance and equipment necessary for the utilisation of the process transferred". It is these elements which are said to yield greater economic benefit (than straight forward licence purchases) and to be also propitious for promoting exports[130]. Problems related to assimilating know-how, as well as marked reluctance on the part of Western firms to sell their latest know-how, explain primarily the Soviet Union's interest in acquiring licences within the framework of agreements which entail assistance in applying the licence as well as in updating the technology.

The official literature suggests that, in future, licence agreements are to be sought primarily for the engineering and metallurgy branches with payment on the basis of compensation.

## 4. Other Technology Transfer Channels

Statistical data concerning Soviet purchases of *licences* is scanty and ambiguous. Information as to exports ensuing from such purchases is unavailable. This is surprising, particularly in view of the official importance attached to Western licences: in terms of reducing the cost of developing a new product, raising production efficiency, facilitating market penetration and increasing export potential.

The significance attached by the Soviet Union to licence trade is also illustrated by the fact that, since 1970, licence purchases are incorporated in the annual and five-year economic plans. Current Soviet policy (like that of its Eastern partners) is to purchase the most up-to-date foreign licence and to aim for export promotion: as opposed to past practice, where licences were intended for import substitution.

According to official data, some 700 licences have been purchased over the past two decades[131]. The major suppliers of Western licences are said to be German, French, British, Austrian and Italian firms. According to the source cited, licences have been primarily purchased for the engineering and automotive industries. It is noted that on the basis of these licences, the Soviet Union is producing such equipment and products as medium-speed diesel engines, railway track repair machines, thyristors, torque d.c. electric motors, sprinkling machines, gas analysing machinery and chloroprene from butadiene.

Licence purchases have also been prevalent in the chemical industry, the iron and steel branch and for the production of electrical equipment. Examples of "output" cited include: manufacture of electric arc steel melting furnaces, regulated plate pumps, spherical graphite cast-iron gas pipes and mechanical treatment of graphite-coated electrodes[132].

In view of the fact that the Soviet Union frequently purchases licences on the basis of credit or compensation payment, it is not possible to quantify the value of licencing transactions.

*Turnkey projects* (which include licence and know-how purchases) have always been an attractive form of technology transfer. The Soviet Union's interest in turnkey projects is primarily related to the fact that such projects provide support at the design and construction stages: also among the weakest points of its production system.

These aspects may be illustrated by the very comprehensive specifications which form part of turnkey contracts. For intance, such contracts include: "the design of a technical documentation, manufacture and shipment of complete sets of equipment and materials, construction and assembly, site testing of the equipment, technical assistance in commissioning the completed project, servicing the equipment during the guarantee period and training local personnel to operate the constructed enterprise". Furthermore, turnkey contracts "usually entail the conclusion of subsequent and separate contracts which provide technical assistance in running the project (by sending specialists), shipping spare parts or by the transfer of production experience and know-how. Moreover, such contracts may be concluded on a compensation basis"[133]. The latter, is in fact, frequently the case. This also explains the Soviet Union's interest in turnkey projects.

A sample compilation of turnkey projects covering the 1971-1980 period (280 projects analysed), suggests that such projects have been most prevalent in the chemical industry and in the steel and non-ferrous metals branches. For the 1986-1980 period, France was the major supplier in the chemical industry (some 20 contracts) followed by Germany, Italy and Great Britain (roughly 10 each). In the steel and non-ferrous metals branches, France (5 contracts) slightly surpassed Germany (4 contracts)[134].

A similar attempt to classify turnkey projects by participant and end-use sector seems not to have been made for the more recent period. Those turnkey projects recorded in the general literature are too few to establish any pattern. However, those which have been listed, indicate that the majority pertain to the light industry, followed by the chemical industry and the metallurgical sector.

In the official literature particular significance seems to be attached to two recent turnkey projects. One is the contract with VOEST-Alpine (Austria) signed in 1982. This agreement (for which VOEST is general contractor) calls for the supply of 750 000 tons per annum turnkey steel works and a rolling mill. The project is reportedly the first Soviet order for a steel mill to be built on a turnkey basis. Exports ensuing from co-operation with VOEST-Alpine (iron and steel products, plates and pipes) appear to be a major factor behind the Soviet Union's interest in continuing its relationship with VOEST[135]. The second is the contract concluded with Germany (Krupp-Kautex Maschinenbau) for delivery of a complete plant (worth over DM10 million) for production of polyethylene fuel tanks for passenger cars; notably for a new Lada. The contract also provides for the supply of tools, various equipment and training of local personnel.

*Non-commercial* channels are also an important source of Western technology. These include scholarly and scientific-technical exchange programmes, information gathered and analysed from the Western technical press, analysis of patent data and so forth[136].

## *Section IV.* CONCLUDING OBSERVATIONS

### 1. General Trends and Considerations

Notwithstanding the conceptual and definitional problems involved in evaluating the evolution of EWIC (quantitatively or in value terms), certain general trends can be discerned.

The available data indicate, first of all, that EWIC growth seems to have been most active during the period of rapid expansion of East-West trade (1970-1975). Since the late 1970s, a general downward trend in EWIC growth has occurred due, in part, to changes in the economic climate: recession in a number of Western countries; declining demand on Western markets for products ensuing from co-operation (notably from co-production agreements); adoption by some Western countries of voluntary restraint agreements; financial problems on the side of the East (notably lack of hard currency needed to fulfill purchasing obligations); cut-backs in investment due to less easily available Western credits, as well as, the need to pay back debts incurred, and so forth. This suggests that industrial co-operation which, in essence, has developed as an extension of foreign trade, is as vulnerable to changing external (economic as well as political) conditions, as is East-West trade, in general.

Nevertheless, it must also be noted that despite additional hard currency outlays which usually accompany an industrial co-operation agreement, balance of payments problems will continue to exert considerable pressure on Eastern enterprises to obtain needed Western technology and know-how through industrial co-operation. However, current trends and discussions in the Eastern literature suggest that those forms of industrial co-operation which minimise hard currency expenditures and which include a "self-financing" element – that is, agreements which entail countertrade – will be increasingly sought.

Co-production agreements (including agreements which involve co-operation in production, development and marketing) are clearly predominant – at least as indicated by the UN surveys and by the agreements listed in Tables 8-14 of the chapter. According to the UN, delivery of plant or equipment is the second most important category of agreements, although commanding a much smaller (and evidently declining) percentage share than co-production arrangements.

However, it is difficult to assess which form is potentially most advantageous – as seen from the Eastern perspective. As has been noted, both provide for on-going participation of the Western partner in production processes and both usually lead to the further provision of Western technology beyond the terms of the original agreement. The latter consideration also holds true for licencing and joint venture agreements; an aspect which renders such agreements more important than their general "ranking" would suggest. The problem of assessing which form of co-operation is potentially more advantageous for the Eastern partner can also be illustrated if one takes into account the financing aspect, notably the element of buy-back. As mentioned, buy-back is an implicit component of agreements involving delivery of plant or equipment and licencing.

Relative to technology, it is evident that all EWIC arrangements entail the transfer of Western technology, in the broadest sense of the term. However, there are vastly different types of technology involved and whether advanced technology is actually transferred within the framework of an EWIC agreement, is difficult to assess and opinions about it are clearly divergent.

It may be noted in this regard, that the more complex forms of technology transfer in which the Western partner supplies the drawings, specifications, know-how and equipment to manufacture a complex/sophisticated product, is obviously preferred by the Eastern partner because of perceived long-term economic benefits. Nevertheless, expected economic benefits frequently do not materialise because of the Eastern partner's inability to handle the scale of the technology transferred and to turn it into a marketable product (for instance, Citroen's joint venture for the production of small passenger cars).

While available case studies and surveys suggest that EWIC agreements have increased the Eastern countries' technical level of production and that the quality of Eastern output has been improved, it also appears that, *in general*, the majority of the Eastern countries have been unable to quickly assimilate and replicate the acquired technology. Whether or not advanced technology is transferred may, therefore, not be the relevant policy issue. Rather, from the Eastern perspective, the issues are whether or not the acquired technology meets individual production needs or the objectives of the general development strategy articulated for the rest of the decade. Discussion in the Eastern literature suggests that, despite shortcomings, industrial co-operation does, on the whole, meet these objectives – although sharp differences in opinion on this point also emerge.

As far as trade is concerned, it is difficult to quantify trade flows stemming from EWIC agreements. Nevertheless, it is important to recall the UN's observation that despite continued, albeit slower growth of EWIC (as measured by the number of agreements), the share of exports ensuing from industrial co-operation is still almost negligible (see I B point 2 above).

Concerning the sectoral distribution of EWIC agreements, the producer's goods sector dominates in EWIC arrangements. Taking the Eastern countries as a whole, it is the chemical industry which accounts for the largest number of agreements concluded. According to the UN, mechanical engineering (including machine tools) is the second most important sector, followed by the electrical and electronics industries (see Table 3 of Chapter 7). Based on the agreements listed in Tables 8-14 of the chapter, one can conclude that the chemical industry is certainly predominant, while mechanical engineering and the electrical and electronics industries, appear to be less well represented than is indicated by the UN survey.

The most active Eastern participants in industrial co-operation are Hungary and the Soviet Union. On the Western side, German firms are predominently involved in industrial co-operation. In general, the role of the Western partner is to supply the technology, plant and equipment, as well as components and parts. The Eastern partner's contribution is, essentially, provision of manpower and the industrial site.

## 2. Assessment of EWIC: Some Implications

Given the variety and complexity of motivations involved (different from country to country and from firm to firm), no clear-cut verdict can be given as to the success or failure of EWIC, as a whole. Nevertheless, analytical surveys of actual experience in industrial co-operation, as well as discussion in the Western and Eastern literature, suggest that Western and Eastern expectations have, on the whole, been only partially met. The primary reasons for this, as well as problems pertaining to industrial co-operation, are summed up below.

### a) *Western Views*

On the Western side, one of the problems most frequently cited by firms is that the process of negotiating an industrial co-operation agreement is usually difficult and excessively long.

In part, this is the natural outcome of diverging motivations (between the Eastern and Western partner) as to the objectives set forth by an agreement. Conflict most often arises over the content of an agreement: that is, over the selection of components for the final product to be manufactured by the respective partners and over output. Choice of technology is also a critical problem. The Eastern partner usually wishes to obtain the latest technology, whereas the Western firm is interested in transferring technology which does not affect its competitive position or, at least, protects it. Conflicts in negotiating an agreement also arise over prices, the duration of an agreement, quality control and marketing arrangements.

Difficulties in negotiating an agreement are also the outcome of the Eastern management system itself. That is, while the principle of furthering industrial co-operation is supported at all levels, the interests and motives of the various actors involved frequently conflict. As noted, export promotion and meeting debt obligations are a primary concern – particularly for the East European countries. This implies that co-operation proposals are often finally not approved "at the top", since they would require greater financial outlay for Western imports than they would assure immediate exports.

For the Western firm, protracted (and sometimes fruitless) negotiations have resulted in higher costs, than originally foreseen. Relative to cost, it has also been pointed out by some Western company executives that co-operation has, at best, contributed only a negligible amount to their company's net income in recent years, and in some years, the company actually lost money.

Implementing an industrial co-operation agreement is also fraught with difficulties since it is at this stage, that the differing (and often conflicting) motivations for entering into an industrial co-operation agreement become apparent. It is also at the implementation stage that the inherent differences between the Eastern and Western economic and political systems – in terms of flexibility, management style, incentives for pursuing industrial co-operation, and so forth – are highlighted. As is noted, these differences make joint decision-making and co-ordination more difficult and frequently lead to problems in implementing an agreement[137].

The complaint most frequently voiced is that the Eastern partner often fails to meet delivery obligations: in terms of quantity or quality. Inflexible and rigid countertrade requirements of some Eastern countries (notably, in the case of Romania) have also proven to be stumbling blocks for the fulfilment of an agreement.

Furthermore, either because of hard currency constraints or because of the nature of a co-operation agreement, Western firms are subjected to increasing pressure to augment the proportion of counter-purchase or compensation (product-payback) arrangements within the framework of an industrial co-operation agreement. In essence, this shifts the burden of financing to the Western firm. While some Western firms have been able to conclude a co-operation agreement only as a result of their assent to such arrangements (or have done so because of their desire to retain or expand their share of Eastern markets), this development is not propitious for the future evolution of EWIC.

For one thing, current economic problems encountered by some Western countries (for instance, slow-down in economic growth and inflation), has reduced the absorptive capacity of many Western firms to cope with Eastern counter deliveries. Also, many small and medium-sized firms are simply unable to absorb countertrade deliveries. Since Western enterprises often build the cost of countertrade into their final price, this would also make an agreement more expensive for the Eastern partner.

Practical experience also indicates that if a co-operation agreement is too complex or ambitious, its parameters not clearly defined and the Western partner not sufficiently familiar with the Eastern partner's actual production capacities, its chances of meeting the Western firm's expectations are minimal. Indeed, such agreements frequently collapse.

Apart from the above-mentioned, fear of eventual market competition for certain products where Western quality standards or Eastern marketing skills are not a key issue (for instance cars or chemicals), has served to dampen a Western firm's interest in pursuing industrial co-operation. Expansion of EWIC has also been hampered by lack of information (on the Western side) about the technological and organisational possibilities of an Eastern enterprise and their lack of flexibility in handling the modalities of an agreement – particularly if the agreement is complex. Based on the experience of Hungarian and German firms (the most active participants in EWIC), it has been concluded that co-operation has not achieved the degree of market expansion (for either side) anticipated by industrial co-operation – a factor which has also served to dampen interest in pursuing EWIC[138].

Finally, it must also be mentioned that some Western governments, because of the technology transfer element, have discouraged or stopped industrial co-operation: for political/strategic and economic reasons.

b) *Eastern Views*

As noted, gaining access to Western technology and export promotion are the dominant motivations for the Eastern countries in seeking industrial co-operation and, EWIC continues to be considered as the best means of achieving these objectives. Nevertheless, it is also evident that these expectations have either not been, or only partially, met. Generalising the Eastern experience in EWIC is particularly hazardous. Systemic factors, as well as differing approaches toward industrial co-operation, obviously affect and nuance both expectations and outcome. Nevertheless, some common themes do emerge.

Inability to acquire advanced or "proven" industrial technology is frequently cited as a major point of dissatisfaction. It is linked to the observation that Western firms often exclude the Eastern partner from the more complex aspects of co-operation. This is regarded as relegating the Eastern partner to a passive position (as far as technology transfer is concerned) and, seen therefore, as an inherent constraint on the ability of an enterprise to "catch up" technologically or to diversify its product/output structure. For instance, in the case of Polish experience in the construction-machinery industry, it is noted that the Western partner frequently chose the simplest form of co-operation. That is, the Polish partner was reduced to playing the role of a sub-contractor for the production of simple (and minor) elements and was excluded from participating in the entire production cycle: starting from R&D, through supply and production, to joint sales.

Some Hungarian commentators have pointed out that many Western firms only transfer second or third-rate technology in order to prevent the Eastern partner from becoming competitors on world markets[139]. It is also said that, since East-West industrial co-operation has served as an alternative to direct foreign investment, this does not encourage Western firms to conclude agreements for the development of "advanced products". On the contrary, it leads them to transfer products which are already in the declining stage of their "life cycle"[140].

Marketing possibilities are another source of disappointment with EWIC. The major complaint voiced is that restrictive clauses attached to co-operation agreements severely curtail export possibilities to Western markets (hence ability to earn hard currency). It is also said that Western firms have frequently not honoured purchasing commitments or, having met the initial obligation, discontinued to buy. This has placed additional export burdens on the Eastern enterprise since, inability to gain access to Western markets (hence limiting the possibility of establishing sales contacts), provides an Eastern enterprise with little opportunity to find other purchasers.

Inability to increase exports to any significant extent, is also attributed to the fact that co-operation agreements are frequently small-scale in nature and that their average lifespan is short. While it is generally agreed that the volume of exports ensuing from EWIC has not increased to the extent anticipated, it is also pointed out that, at least in the case of Hungary, industrial co-operation "has fulfilled the expectation of increasing exports of marketable products. It has also made a modest contribution to redressing the imbalance of trade"[141]. It might also be mentioned in this regard, that the expectation of large export potential often fails to materialise because of domestic demand and the need to satisfy CMEA markets – leaving therefore, less products available for export to hard currency markets.

EWIC has also proven to be costly. That is, implementing an agreement has called for additional investment outlays or foreign currency expenditures to cover licence fees or purchase of additional inputs (components or parts) needed to meet output commitments. Given the East European countries' balance of payments difficulties, cost related considerations may not only become a serious constraint on the future development of EWIC, but has undoubtedly led to the increased emphasis placed on pursuing various forms of countertrade – both within the framework of industrial co-operation and in other commercial/trading practices.

Finally, industrial co-operation has not changed or improved production structures to the extent that was evidently anticipated, also compromising thereby, expectations of expanding export capability. For instance, some commentators have pointed out that co-operation has had only a marginal influence on the macro-structure of the economy. That is, co-operation has eased the production problems of a given enterprise, but has only rarely contributed "to the general development or stimulation of the domestic basic industries"[142]. Others have gone so far as to say that "the structure of exports has not only *not* become more modern, but has actually become inflexible"[143].

### 3. Future Prospects

In view of the fact that East-West industrial co-operation entails (at least in principle) the establishment of complex and long-term relationships and is a channel for the transfer of Western technology, its evolution is, as previously mentioned, particularly sensitive to the political and economic climate. However, difficulties encountered in industrial co-operation, as well as mutual disappointment relative to expectations, have evidently also led to the downswing in EWIC growth (or, in some instances, stagnation) that has occurred during the first half of the 1980s – at least as measured by the reduced number of new agreements which appear to have been concluded.

Nevertheless, despite economic/political constraints, operational problems and conflicting interests, it can be assumed that EWIC will remain as a permanent feature of East-West relations. It is, after all, unrealistic to assume that the motivations which have been the driving force behind the establishment of co-operation agreements, have become entirely invalidated. The fact that agreements do continue to be concluded, despite their frequently limited nature, would testify to the fact that the parties involved are "getting something out of the deal"!

While the manner in which industrial co-operation develops in future depends, of course, on assessments made (by both sides) as to the potential long-term economic benefits of EWIC, a primary determinant will remain the ability of the Eastern systems to better adapt to the requirements of industrial co-operation. Based on the Eastern literature, this factor is increasingly recognised. Recent writings (particularly in Eastern Europe) underline that basic systemic changes are needed if industrial co-operation is to be effective: notably, evolve beyond the "simple forms", which presently predominate.

The need to be more realistic about what the Eastern partner can, in fact, deliver (hence also better assessment of Western firms' objectives and requirements) as well as what can be achieved by means of industrial co-operation, is also stressed. Hungarian commentators, in particular, call attention to the fact that while industrial co-operation is superior to other technology transfer mechanisms, its results should not be over-estimated: as far as modernisation of indigenous production structures is concerned, or in terms of necessarily reducing the technological gap. This observation is all the more interesting, since Hungary is the most active participant in industrial co-operation and its participation generally considered to be most successful.

The fact that the Eastern countries continue to be interested in pursuing – and expanding – the scope of industrial co-operation has been noted. However, based on the Eastern literature, it appears that increased emphasis is being placed on seeking co-operation primarily with "proven partners". While agreements for developing the machine-building and engineering branches continue to be given priority, it is also interesting to note that almost uniform stress is placed on the desirability of increasing the co-production and joint venture form of co-operation and arrangements which extend co-operation on third markets. Agreements for such new areas as plant construction, development of the agricultural sector (notably for the production of agro-industrial equipment), environmental protection, non-polluting and renewable energy sources, are also being sought.

For some countries, the compensatory aspect of co-operation is to be favoured. In the case of the Soviet Union, for instance, the intent to expand the sphere of compensation deals to cover not only production of raw materials and semi-manufactured goods, but manufacturing industries as well, has been made explicit. Whether this strategy can be realised, is another matter. While some Western countries (in the interest of meeting energy and raw materials needs) have accepted compensation as a *sine qua non*, mounting concern over market disruption (and distortion) can be expected to lead to greater opposition on the part of Western firms toward such co-operation arrangements.

A number of East European countries are also placing greater emphasis on the fact that, in future, agreements which contain a larger "R&D element" and which increase production efficiency will be sought. Growing awareness that problems in effectively assimilating technology and, subsequent failure to meet Western delivery requirements (compromising, thereby, promising business possibilities), seem to be one reason for this change. Improving production in order to enhance export capability is, clearly, also a major consideration.

The Eastern countries' intent to either improve or change the modalities of co-operation, is made very explicit. Nevertheless, the question remains as to whether EWIC will, or can, evolve into more complex and long-lasting forms. The most realistic answer would seem to be "it depends". That is, the future development of EWIC is greatly dependent on the willingness of both sides to adapt to differences in their respective policy pursuits. Financial considerations and the manner in which East-West trade develops, will also have a decisive influence. Here, judgements made as to the economic and political desirability and/or profitability of forging closer links, will be crucial.

Table 1 (Chapter 7)

EAST-WEST INDUSTRIAL CO-OPERATION IN THE 1970S
(estimates of numbers of instances at mid- and end-decade) (1)

| Category of agreement | As of End-1976 | | As of End-1980 | |
|---|---|---|---|---|
| | Estimated numbers of agreements concluded | As % of subtotals | Estimated numbers of agreements concluded | As % of subtotals |
| I. General, framework agreements with Western firms | | | | |
| Bulgaria | 17 | 5 | 46 | 7 |
| CSSR | 4 | 1 | 14 | 2 |
| German Democratic Republic | 5 | 1 | 16 | 3 |
| Hungary | 12 | 3 | 48 | 8 |
| Poland | 21 | 6 | 57 | 9 |
| Romania | 23 | 6 | 38 | 6 |
| USSR | 279 | 77 | 406 | 64 |
| Total | 361 | 100 | 625 | 100 |
| II. Substantive, non-equity agreements with Western firms | | | | |
| Bulgaria | 50 | 6 | 63 | 5 |
| CSSR | 53 | 6 | 77 | 6 |
| German Democratic Republic | 32 | 3 | 48 | 4 |
| Hungary | 244 | 29 | 455 | 33 |
| Poland | 226 | 26 | 324 | 24 |
| Romania | 96 | 11 | 108 | 8 |
| USSR | 156 | 18 | 292 | 21 |
| Total | 857 | 100 | 1 367 | 100 |

Table 1 (Chapter 7) contd

| Category of agreement | As of End-1976 | | As of End-1980 | |
|---|---|---|---|---|
| | Estimated numbers of agreements concluded | As % of subtotals | Estimated numbers of agreements concluded | As % of subtotals |
| III. Agreements with Western firms for joint projects in third countries (mostly developing countries) | | | | |
| Bulgaria | 3 | 5 | 12 | 5 |
| CSSR | 7 | 13 | 19 | 7 |
| German Democratic Republic | 7 | 13 | 25 | 10 |
| Hungary | 19 | 34 | 76 | 30 |
| Poland | 5 | 9 | 71 | 28 |
| Romania | 7 | 13 | 28 | 11 |
| USSR | 7 | 13 | 26 | 10 |
| Total | 55 | 100 | 257 | 100 |
| IV. Joint East-West companies (by location) | | | | |
| A. In those Eastern countries permitting foreign equity investment | | | | |
| Bulgaria | n.a. | n.a. | 0 | 0 |
| Hungary | 3 | 38 | 8 | 47 |
| Poland | 0 | 0 | 1 | 6 |
| Romania | 5 | 62 | 8 | 47 |
| Total | 8 | 100 | 17 | 100 |

Table 1 (Chapter 7) contd

| Category of agreement | As of End-1976: Estimated numbers of agreements concluded | As of End-1976: As % of subtotals | As of End-1980: Estimated numbers of agreements concluded | As of End-1980: As % of subtotals |
|---|---|---|---|---|
| B. In the industrialised West | | | | |
| Bulgaria | 20 | 10 | 29 | 9 |
| CSSR | 11 | 5 | 24 | 8 |
| German Democratic Republic | 8 | 4 | 20 | 7 |
| Hungary | 38 | 18 | 51 | 16 |
| Poland | 48 | 23 | 75 | 25 |
| Romania | 25 | 12 | 32 | 11 |
| USSR | 60 | 29 | 73 | 24 |
| Total | 210 | 100 | 304 | 100 |
| C. In the developing countries | | | | |
| Bulgaria | 1 | 20 | 1 | 20 |
| CSSR | 0 | 0 | 0 | 0 |
| German Democratic Republic | 0 | 0 | 0 | 0 |
| Hungary | 2 | 40 | 2 | 40 |
| Poland | 1 | 20 | 1 | 20 |
| Romania | 1 | 20 | 1 | 20 |
| USSR | 0 | 0 | 0 | 0 |
| Total | 5 | 100 | 5 | 100 |

Table 1 (Chapter 7) contd

Note:

1. As noted by the author, the figures shown are cumulative totals and not (with the exception of category IV) agreements necessarily in force as of the dates indicated. Especially for the most recent period, the data are based heavily on secondary sources and are thus susceptible to variations in coverage by the specialised press. The information is based on information provided by 316 Western firms.

The Table has been compiled from data in the files of the East-West Project, Institute of Soviet and East European Studies, Carleton University, Ottawa. Documentation of these cases is based on: information obtained directly from the parties involved, lists of agreements provided by official and non-official organisations, in East and West, and published reports. All instances are individually documented, with the exception of the framework agreements concluded by the USSR, which are based in part on figures released by the State Committee for Science and Technology. The joint companies reported under category IV B are listed in the Directory of Soviet and East European Companies in the West published by the Institute. The files of the East-West Project do not include comprehensive coverage of category II agreements directly involving US-based firms. The earlier figures for category II therefore incorporate the findings of a survey of US participation reported by Marer and Miller*. In the absence of an update of this survey, the numbers of documented category II agreements involving non-US-based, Western firms have been adjusted upward by the shares which US-based firms occupied (by Eastern country) in 1975. The assumption is that the relative participation of US firms (averaging 22.8 per cent in 1976) did not change subsequently. This may have introduced an upward bias in the category II figures for end-1980; the adjustment factors were substantially higher than the 12.2 per cent share of US firms reported in the ECE's 1980 sample. [16th September, 1980, p. 5]

* See P. Marer and J. Miller, "American Industry in the USSR and Eastern Europe: US Participation in East-West Industrial Co-operation", Journal of International Business Studies, Fall-Winter, 1977, pp. 17-29.

Definition of categories:

I. General agreements or protocols with Western firms, establishing the intent to co-operate and the broad areas of co-operation but not containing details of specific co-operative projects or programs. Thus these are "umbrella" or "framework" agreements under which substantive, follow-up agreements are concluded between the partners. Included here are general agreements for co-operation between the partners in third countries. Most of the Soviet agreements in this category are protocols for scientific and technical co-operation signed by the Soviet State Committee for Science and Technology with individual Western firms.

Table 1 (Chapter 7) contd

II. Substantive agreements for co-operation on a non-equity basis in a broad range of activities and sectors, but for the most part directed towards the utilization, modernization, and/or expansion of industry in the Eastern partner countries. Agreements in this category fall within two broad types: a) Western sales of plant, equipment, or technology with subsequent payment in the resulting, or related, products (so-called "compensation" agreements) and b) agreements for specialization in production and trade between the partners in components and/or end-products (so-called "coproduction", "production sharing", or "product specialisation" agreements, with "sub-contracting" the simplest variant). Market sharing or joint marketing and co-operation in research and development are common secondary features of category II agreements, especially of type b).

III. Projects in third countries (mostly in the developing countries) carried out jointly by Eastern enterprises and Western firms. Most of these are turnkey and other capital projects in which one party is the principal and the other a sub-contractor.

IV. Joint equity ventures, generally in the form of jointly owned companies, established in the locations indicated, for a wide-range of economic purposes. Includes fishing ventures, mining companies, banks, insurance companies, shipping agencies, manufacturing firms, engineering-consulting firms, and so on. In contrast to the debtor-creditor relationships characteristic of the co-operation agreements of categories II and III, the partners in the joint venture companies of category IV share profits and risks.

Categories I through IV span the forms of co-operation included in the standard United Nations (ECE) definition of East-West Industrial co-operation.

All agreements included in the Table are with Western firms located in the industrialised economies of Western Europe, North America and Japan.

Source: Carl H. McMillan, "Trends in East-West Industrial Co-operation", Journal of International Business Studies, Fall, 1981, pp. 57-59.

Table 2 (Chapter 7)

CLASSIFICATION OF EAST-WEST INDUSTRIAL CO-OPERATION AGREEMENTS
(percentages)

| Country | Total | Supply of licence (a) | Delivery of plant | Special-isation | Sub-contracting | Joint venturing | Joint tendering or joint projects | Tripartite co-operation (b) |
|---|---|---|---|---|---|---|---|---|
| Total CMEA Countries | | | | | | | | |
| 1972 | 100.0 | 28.2 | 11.9 | 37.1 | 7.9 | 14.9 | - | - |
| 1975 | 100.0 | 26.1 | 21.7 | 33.3 | 6.8 | 2.9 | 9.2 | - |
| 1st June, 1976 | 100.0 | 17.1 | 20.5 | 39.3 | 7.4 | 10.5 | 5.1 | 1.1 |
| End September, 1978 | 100.0 | 6.1 | 17.4 | 45.2 | 3.8 | 16.9 | 4.2 | 6.4 |
| End September, 1979 | 100.0 | 15.0 | 20.0 | 39.1 | 3.3 | 13.1 | 4.1 | 5.4 |
| End August, 1980 | 100.0 | 15.6 | 19.9 | 38.5 | 2.6 | 12.6 | 10.7 | - |
| End August, 1981 | 100.0 | 14.6 | 19.9 | 39.9 | 2.3 | 12.9 | 10.4 | - |
| End September, 1982 | 100.0 | 13.6 | 19.6 | 41.3 | 2.4 | 12.7 | 10.3 | - |
| End September, 1983 | 100.0 | 13.6 | 17.8 | 43.6 | 2.4 | 12.3 | 10.3 | - |
| End September, 1984 | 100.0 | 13.1 | 16.5 | 43.6 | 3.1 | 13.3 | 10.4 | - |
| By Country: | | | | | | | | |
| Bulgaria, 1975 | 100.0 | 62.5 | 25.0 | 12.5 | - | - | - | - |
| 1st June, 1976 | 100.0 | 17.1 | 25.7 | 31.4 | 11.4 | 3.0 | 11.4 | - |
| End September, 1978 | 100.0 | - | - | 63.7 | 9.1 | 27.3 | - | - |
| End September, 1979 | 100.0 | - | 21.4 | 50.0 | 7.1 | 21.4 | - | - |
| End August, 1980 | 100.0 | - | 31.6 | 47.4 | 5.3 | 15.8 | - | - |
| End August, 1981 | 100.0 | 4.2 | 25.0 | 45.8 | 4.2 | 12.5 | 8.3 | - |
| End September, 1982 | 100.0 | 3.6 | 21.4 | 42.9 | 3.6 | 17.8 | 10.7 | - |
| End September, 1983 | 100.0 | 2.4 | 14.3 | 52.4 | 2.4 | 19.0 | 9.5 | |
| End September, 1984 | 100.0 | 4.5 | 13.6 | 50.0 | 4.5 | 18.2 | 9.1 | - |
| Czechoslovakia, 1975 | 100.0 | 35.7 | - | 28.6 | 7.1 | - | 28.6 | - |
| 1st June, 1976 | 100.0 | 27.3 | - | 22.7 | 9.1 | 18.2 | 22.7 | - |
| End September, 1978 | 100.0 | 22.2 | - | 11.1 | - | 66.7 | - | - |
| End September, 1979 | 100.0 | 30.2 | - | 53.5 | - | 14.0 | 2.3 | - |
| End August, 1980 | 100.0 | 30.0 | - | 57.5 | - | 10.0 | 2.5 | - |
| End August, 1981 | 100.0 | 27.9 | 2.3 | 58.1 | - | 9.3 | 2.3 | - |
| End September, 1982 | 100.0 | 18.2 | 1.5 | 71.2 | - | 7.6 | 1.5 | - |
| End September, 1983 | 100.0 | 20.3 | 1.4 | 69.6 | - | 7.2 | 1.4 | |
| End September, 1984 | 100.0 | 20.3 | 1.4 | 69.6 | - | 7.2 | 1.4 | - |

Table 2 (Chapter 7) contd

| Country | Total | Supply of licence (a) | Delivery of plant | Special-isation | Sub-contracting | Joint venturing | Joint tendering or joint projects | Tripartite co-operation (b) |
|---|---|---|---|---|---|---|---|---|
| German Democratic Republic, 1975 | 100.0 | - | - | 66.7 | - | - | 33.3 | - |
| 1st June, 1976 | 100.0 | - | 23.5 | 14.2 | 7.1 | 24.6 | 9.2 | - |
| End September, 1978 | 100.0 | - | 71.4 | 28.6 | - | - | - | - |
| End September, 1979 | 100.0 | 6.2 | 62.5 | 18.8 | - | 6.2 | 6.2 | - |
| End August, 1980 | 100.0 | 9.2 | 59.1 | 13.6 | - | 4.6 | 13.6 | - |
| End August, 1981 | 100.0 | 7.7 | 65.4 | 11.5 | - | 3.8 | 11.5 | - |
| End September, 1982 | 100.0 | 6.0 | 54.5 | 12.1 | 3.0 | 6.0 | 18.2 | - |
| End September, 1983 | 100.0 | 6.1 | 54.5 | 12.1 | 3.0 | 6.1 | 18.2 | - |
| End September, 1984 | 100.0 | 6.1 | 54.5 | 12.1 | 3.0 | 6.1 | 18.2 | - |
| Hungary, 1975 | 100.0 | 32.3 | 13.2 | 44.1 | 7.4 | - | 3.0 | - |
| 1st June, 1976 | 100.0 | 29.5 | 16.3 | 32.6 | 9.6 | 10.2 | 1.2 | 0.6 |
| End September, 1978 | 100.0 | 2.6 | 1.3 | 44.7 | 9.2 | 17.1 | 5.2 | 19.7 |
| End September, 1979 | 100.0 | 21.2 | 10.9 | 35.3 | 5.9 | 12.5 | 3.8 | 10.3 |
| End August, 1980 | 100.0 | 23.1 | 10.8 | 34.5 | 5.4 | 11.8 | 14.3 | - |
| End August, 1981 | 100.0 | 22.1 | 10.3 | 35.2 | 5.2 | 13.1 | 14.1 | - |
| End September, 1982 | 100.0 | 21.7 | 9.9 | 35.7 | 5.4 | 13.6 | 13.6 | - |
| End September, 1983 | 100.0 | 21.7 | 8.7 | 38.0 | 5.5 | 13.4 | 12.7 | - |
| End September, 1984 | 100.0 | 20.2 | 7.6 | 36.4 | 6.2 | 17.2 | 12.3 | - |
| Poland, 1975 | 100.0 | 25.0 | 16.1 | 37.5 | 3.6 | 1.8 | 16.0 | - |
| 1s June, 1976 | 100.0 | 21.7 | 24.2 | 32.3 | 6.4 | 6.4 | 8.0 | 1.0 |
| End September, 1978 | 100.0 | 22.2 | 24.1 | 37.0 | 3.8 | 1.9 | 7.4 | 3.7 |
| End September, 1979 | 100.0 | 22.5 | 27.0 | 30.3 | 3.3 | 2.2 | 7.8 | 6.7 |
| End August, 1980 | 100.0 | 20.4 | 25.8 | 33.3 | 1.1 | 3.2 | 16.1 | - |
| End August, 1981 | 100.0 | 20.0 | 24.2 | 33.7 | 1.0 | 5.3 | 15.8 | - |
| End September, 1982 | 100.0 | 20.0 | 24.2 | 33.7 | 1.1 | 5.3 | 15.8 | - |
| End September, 1983 | 100.0 | 19.8 | 26.0 | 33.3 | 1.0 | 5.2 | 15.6 | - |
| End September, 1984 | 100.0 | 19.8 | 25.0 | 33.3 | 1.0 | 5.2 | 15.6 | - |

Table 2 (Chapter 7) contd

| Country | Total | Supply of licence (a) | Delivery of plant | Special-isation | Sub-contracting | Joint venturing | Joint tendering or joint projects | Tripartite co-operation (b) |
|---|---|---|---|---|---|---|---|---|
| Romania, 1975 | 100.0 | 22.8 | 34.3 | 8.6 | 14.3 | 14.3 | 5.7 | - |
| 1st June, 1976 | 100.0 | 19.4 | 25.5 | 14.2 | 7.1 | 24.6 | 9.2 | - |
| End September, 1978 | 100.0 | - | 3.6 | 14.3 | 7.2 | 67.9 | - | 7.1 |
| End September, 1979 | 100.0 | 5.7 | 8.6 | 14.3 | 5.7 | 60.0 | - | 5.7 |
| End August, 1980 | 100.0 | 5.6 | 8.3 | 13.9 | 2.8 | 58.3 | 11.1 | - |
| End August, 1981 | 100.0 | 5.2 | 7.9 | 15.8 | 2.6 | 57.9 | 10.5 | - |
| End September, 1982 | 100.0 | 5.2 | 7.9 | 15.8 | 2.6 | 57.9 | 10.5 | - |
| End September, 1983 | 100.0 | 5.3 | 7.9 | 15.8 | 2.6 | 57.9 | 10.5 | - |
| End September, 1984 | 100.0 | 5.3 | 7.9 | 15.8 | 2.6 | 57.9 | 10.5 | - |
| USSR, 1975 | 100.0 | - | 56.6 | 34.8 | 4.3 | - | 4.4 | - |
| 1st June, 1976 | 100.0 | 3.2 | 20.4 | 61.5 | 4.7 | 7.1 | 1.6 | 1.5 |
| End September, 1978 | 100.0 | 2.4 | 27.2 | 57.4 | - | 8.6 | 3.9 | 0.8 |
| End September, 1979 | 100.0 | 3.8 | 30.2 | 50.9 | 0.6 | 9.4 | 3.8 | 1.3 |
| End August, 1980 | 100.0 | 4.5 | 28.8 | 50.0 | 0.6 | 10.2 | 5.8 | - |
| End August, 1981 | 100.0 | 4.0 | 28.5 | 53.1 | - | 9.1 | 5.1 | - |
| End September, 1982 | 100.0 | 3.7 | 31.0 | 51.4 | - | 8.5 | 5.3 | - |
| End September, 1983 | 100.0 | 4.1 | 27.1 | 54.6 | - | 7.3 | 6.9 | |
| End September, 1984 | 100.0 | 3.7 | 24.7 | 56.4 | 0.8 | 6.6 | 7.8 | - |

Notes: Because of rounding, percentage shares may not add up to exactly 100.

a) Supply of licence and/or know-how in exchange (in part at least) for products or components including parts in varying percentages of the final product.

b) As of 1980, the ECE no longer treats tripartite co-operation as a separate category. They have been included in the category "joint tendering" or "joint projects" (customer located in a third country).

Table 2 (Chapter 7) contd

Sources: 1972: Analytical Report on Industrial Co-operation among ECE Countries, United Nations, Economic Commission for Europe, 1973. Based on 202 cases of industrial co-operation.

1975: Proceedings of the UN/ECE Seminar on the Management of the Transfer of Technology with Industrial Co-operation, 14th-17th July, 1975, United Nations, Economic Commission for Europe, ECE/SC/TECH./10, 16th February, 1976. Based on 275 cases of industrial co-operation.

1st June, 1976: Promotion of Trade Through Industrial Co-operation, Statistical Outline of Recent Trends in Industrial Co-operation, Note by the Secretariat, United Nations Economic and Social Council, Economic Commission for Europe, Committee on the Development of Trade, TRADE/R.355/Add. 2, 14th November, 1977, Table 3. Based on 658 contracts.

End September, 1978: Promotion of Trade Through Industrial Co-operation, Statistical Outline of Recent Trends in Industrial Co-operation, Note by the Secretariat, United Nations Economic and Social Council, Economic Commission for Europe, Committee on the Development of Trade, TRADE/R.373/Add. 5, 19th October, 1978, Table 3. Based on 314 contacts.

End September, 1979: Promotion of Trade Through Industrial Co-operation, Statistical Outline of Recent Trends in Industrial Co-operation, Note by the Secretariat, United Nations Economic and Social Council, Economic Commission for Europe, Committee on the Development of Trade, TRADE/R.392, 9th October, 1979, Table 3. Based on 540 contracts.

End August, 1980: Promotion of Trade Through Industrial Co-operation, Statistical Survey of Recent Trends in Industrial Co-operation, Note by the Secretariat, United Nations Economic and Social Council, Economic Commission for Europe, Committee on the Development of Trade, TRADE/R.407, 16th September, 1980, Table 3. Based on 569 contracts.

End August, 1981: Promotion of Trade Through Industrial Co-operation, Statistical Survey of Recent Trends in Industrial Co-operation, Note by the Secretariat, United Nations Economic and Social Council, Economic Commission for Europe, Committee on the Development of Trade, TRADE/R.429, 19th October, 1981, Table 3. Based on 614 contracts.

End September, 1982: Promotion of Trade Through Industrial Co-operation, Statistical Survey of Recent Trends in Industrial Co-operation, Note by the Secretariat, United Nations Economic and Social Council, Economic Commission for Europe, Committee on the Development of Trade, TRADE/R.450, 29th October, 1982, Table 3. Based on 668 contracts.

End September, 1983: Promotion of Trade Through Industrial Co-operation, Statistical Survey of Recent Trends in Industrial Co-operation, Note by the Secretariat, United Nations Economic and Social Council, Economic Commission for Europe, Committee on the Development of Trade, TRADE/R.468, 10th November, 1983, Table 3. Based on 749 contracts.

End September, 1984: Promotion of Trade Through Industrial Co-operation, Statistical Survey of Recent Trends in Industrial Co-operation, Note by the Secretariat, United Nations Economic and Social Council, Economic Commission for Europe, Committee on the Development of Trade, TRADE/R.487, 19th October, 1984, Table 3. Based on 814 contracts.

Table 3 (Chapter 7)

INDUSTRIES' SHARE IN THE TOTAL NUMBER OF INDUSTRIAL CO-OPERATION CONTRACTS, BY COUNTRY
(percentages)

| Country | Total | Chemicals industry | Metal-lurgy | Transport equipment | Machine tools | Mechanical engineering | Total | Elec-tronics | Electrical equipment | Total | Food and agriculture | Light industry | Total | Other branches |
|---|---|---|---|---|---|---|---|---|---|---|---|---|---|---|
| | | 2 | 3 | 4 | 5 | 6 | 5+6 | 7 | 8 | 7+8 | 1 | 9 | 1+9 | 10 |
| Bulgaria | | | | | | | | | | | | | | |
| 1st June, 1976 | 100.0 | 25.0 | 5.7 | 14.3 | 2.9 | 28.5 | 31.4 | 2.9 | 5.7 | 8.6 | 17.3 | - | 17.1 | 2.0 |
| End September, 1978 | 100.0 | - | - | 9.1 | - | 45.5 | 45.5 | 9.1 | 9.1 | 18.2 | 27.3 | - | 27.3 | - |
| End September, 1979 | 100.0 | 21.4 | - | 7.1 | - | 35.7 | 35.7 | 7.1 | 7.1 | 14.2 | 21.4 | - | 21.4 | - |
| End August, 1980 | 100.0 | 21.0 | - | 10.5 | - | 31.6 | 31.6 | 5.3 | 5.3 | 10.6 | 15.8 | 5.3 | 21.1 | 5.3 |
| End August, 1981 | 100.0 | 16.7 | - | 8.3 | - | 41.7 | 41.7 | 4.2 | 12.5 | 16.7 | 12.5 | 4.2 | 16.7 | - |
| End September, 1982 | 100.0 | 14.3 | 3.6 | 7.1 | - | 39.3 | 39.3 | 3.6 | 10.6 | 14.3 | 14.3 | 7.1 | 21.4 | - |
| End September, 1983 | 100.0 | 9.5 | 4.8 | 7.1 | 4.8 | 26.2 | 31.0 | 9.5 | 7.1 | 16.6 | 9.5 | 19.1 | 28.6 | 2.4 |
| End September, 1984 | 100.0 | 9.1 | 4.5 | 6.8 | 4.5 | 25.0 | 29.5 | 11.4 | 6.8 | 18.2 | 9.1 | 20.4 | 29.5 | 2.3 |
| Czechoslovakia | | | | | | | | | | | | | | |
| 1st June, 1976 | 100.0 | 22.7 | - | 22.7 | 9.1 | 36.4 | 45.5 | - | 9.1 | 9.1 | - | - | - | - |
| End September, 1978 | 100.0 | 33.3 | - | 11.1 | 22.2 | 22.2 | 44.4 | - | 11.1 | 11.1 | - | - | - | - |
| End September, 1979 | 100.0 | 9.3 | - | 18.6 | 7.0 | 20.9 | 27.9 | 9.3 | 18.6 | 27.9 | 9.3 | 4.7 | 14.0 | 2.3 |
| End August, 1980 | 100.0 | 12.5 | - | 20.0 | 5.0 | 20.0 | 25.0 | 5.0 | 20.0 | 25.0 | 10.0 | 5.0 | 15.0 | 2.5 |
| End August, 1981 | 100.0 | 13.9 | - | 18.6 | 4.6 | 20.9 | 25.5 | 4.6 | 18.6 | 23.3 | 9.3 | 7.0 | 16.3 | 2.3 |
| End September, 1982 | 100.0 | 9.1 | - | 12.1 | 15.2 | 22.7 | 37.9 | 3.0 | 15.2 | 18.2 | 7.6 | 12.1 | 19.7 | 3.0 |
| End September, 1983 | 100.0 | 8.7 | - | 11.6 | 16.0 | 21.7 | 37.7 | 2.9 | 14.5 | 17.4 | 10.1 | 11.6 | 21.7 | 2.9 |
| End September, 1984 | 100.0 | 8.7 | - | 11.6 | 16.0 | 21.7 | 37.7 | 2.9 | 14.5 | 17.4 | 10.1 | 11.6 | 21.7 | 2.9 |
| German Democratic Republic | | | | | | | | | | | | | | |
| 1st June, 1976 | 100.0 | 23.5 | 23.5 | 6.0 | 17.6 | 17.6 | 35.2 | - | 5.9 | 5.9 | 5.9 | - | 5.9 | - |
| End September, 1978 | 100.0 | 14.3 | - | 14.3 | 28.6 | 28.6 | 57.2 | - | - | - | 14.3 | - | 14.3 | - |
| End September, 1979 | 100.0 | 25.0 | 12.5 | 6.2 | 12.5 | 18.8 | 31.3 | 6.2 | - | 6.2 | 6.2 | 6.2 | 12.4 | 6.2 |
| End August, 1980 | 100.0 | 22.7 | 9.1 | 22.7 | 9.1 | 13.6 | 22.7 | 4.5 | - | 4.5 | 4.5 | 9.1 | 13.6 | 4.5 |
| End August, 1981 | 100.0 | 19.2 | 7.7 | 30.8 | 7.7 | 11.5 | 19.2 | 3.8 | - | 3.8 | 3.8 | 11.5 | 15.4 | 3.8 |
| End September, 1982 | 100.0 | 15.1 | 6.1 | 27.3 | 9.1 | 18.2 | 27.3 | 9.1 | - | 9.1 | 3.0 | 9.1 | 12.1 | 3.0 |
| End September, 1983 | 100.0 | 15.1 | 6.1 | 27.3 | 9.1 | 18.2 | 27.3 | 9.1 | - | 9.1 | 3.0 | 9.1 | 12.1 | 3.0 |
| End September, 1984 | 100.0 | 15.1 | 6.1 | 27.3 | 9.1 | 18.2 | 27.3 | 9.1 | - | 9.1 | 3.0 | 9.1 | 12.1 | 3.0 |

Table 3 (Chapter 7) contd

| Country | Total | Chemicals industry | Metal-lurgy | Transport equipment | Machine tools | Mechanical engineering | Total | Elec-tronics | Electrical equipment | Total | Food and agriculture | Light industry | Total | Other branches |
|---|---|---|---|---|---|---|---|---|---|---|---|---|---|---|
| | | 2 | 3 | 4 | 5 | 6 | 5+6 | 7 | 8 | 7+8 | 1 | 9 | 1+9 | 10 |
| Hungary | | | | | | | | | | | | | | |
| 1st June, 1976 | 100.0 | 17.5 | 3.6 | 13.3 | 4.8 | 20.5 | 25.3 | 7.2 | 10.8 | 18.0 | 4.2 | 12.1 | 16.3 | 6.0 |
| End September, 1978 | 100.0 | 14.5 | 2.6 | 11.8 | 3.9 | 13.2 | 17.1 | 11.8 | 21.1 | 32.9 | 5.3 | 11.8 | 17.1 | 3.9 |
| End September, 1979 | 100.0 | 17.4 | 3.3 | 9.8 | 6.0 | 11.9 | 17.9 | 8.1 | 20.1 | 28.2 | 6.5 | 13.0 | 19.5 | 3.8 |
| End August, 1980 | 100.0 | 18.2 | 2.5 | 9.8 | 5.9 | 10.8 | 16.7 | 8.9 | 17.2 | 26.1 | 9.4 | 13.3 | 22.7 | 3.9 |
| End August, 1981 | 100.0 | 18.3 | 2.8 | 9.9 | 5.2 | 10.8 | 16.0 | 8.4 | 17.9 | 25.3 | 8.9 | 14.6 | 23.5 | 4.2 |
| End September, 1982 | 100.0 | 17.6 | 3.2 | 9.5 | 5.4 | 10.9 | 16.3 | 8.1 | 16.3 | 24.4 | 9.5 | 14.9 | 24.4 | 4.5 |
| End September, 1983 | 100.0 | 18.2 | 3.2 | 9.9 | 5.1 | 12.3 | 17.4 | 7.9 | 14.6 | 22.5 | 8.7 | 14.2 | 22.9 | 5.9 |
| End September, 1984 | 100.0 | 18.2 | 3.1 | 8.6 | 4.8 | 11.7 | 16.5 | 7.6 | 14.1 | 21.7 | 8.9 | 15.8 | 24.7 | 7.2 |
| Poland | | | | | | | | | | | | | | |
| 1st June, 1976 | 100.0 | 23.4 | 11.3 | 10.5 | 4.8 | 23.4 | 28.3 | 9.0 | 5.6 | 14.6 | 1.6 | 4.0 | 5.6 | 6.4 |
| End September, 1978 | 100.0 | 16.7 | 7.4 | 9.3 | 1.9 | 25.9 | 27.8 | 13.0 | 1.9 | 14.9 | 5.6 | 13.0 | 18.6 | 5.6 |
| End September, 1979 | 100.0 | 18.0 | 9.0 | 15.7 | 2.2 | 18.0 | 20.2 | 13.5 | 6.7 | 20.2 | 3.4 | 7.9 | 11.3 | 5.6 |
| End August, 1980 | 100.0 | 18.3 | 14.0 | 16.1 | 2.1 | 14.0 | 16.1 | 12.9 | 5.4 | 18.3 | 2.1 | 7.5 | 9.6 | 7.5 |
| End August, 1981 | 100.0 | 17.9 | 11.6 | 15.8 | 2.1 | 18.9 | 21.0 | 12.6 | 5.3 | 17.9 | 4.2 | 6.3 | 10.5 | 5.3 |
| End September, 1982 | 100.0 | 17.9 | 11.6 | 15.8 | 2.1 | 18.9 | 21.0 | 12.6 | 5.3 | 17.9 | 4.2 | 6.3 | 10.5 | 5.3 |
| End September, 1983 | 100.0 | 17.7 | 12.5 | 15.6 | 2.1 | 18.7 | 20.8 | 12.5 | 5.2 | 17.7 | 4.2 | 6.2 | 10.4 | 5.2 |
| End September, 1984 | 100.0 | 17.7 | 12.5 | 15.6 | 2.1 | 18.7 | 20.8 | 12.5 | 5.2 | 17.7 | 4.2 | 6.2 | 10.4 | 5.2 |
| Romania | | | | | | | | | | | | | | |
| 1st June, 1976 | 100.0 | 20.5 | 7.1 | 22.5 | - | 30.6 | 30.6 | 6.1 | 2.0 | 8.1 | 4.1 | 3.0 | 7.1 | 4.1 |
| End September, 1978 | 100.0 | 39.3 | 10.7 | 21.4 | - | 10.7 | 10.7 | 10.7 | - | 10.7 | - | 7.1 | 7.1 | - |
| End September, 1979 | 100.0 | 37.2 | 11.4 | 25.7 | - | 8.6 | 8.6 | 8.6 | 2.8 | 11.4 | - | 5.7 | 5.7 | - |
| End August, 1980 | 100.0 | 33.3 | 13.9 | 30.6 | - | - | - | 2.8 | 2.8 | 5.6 | 2.8 | 8.3 | 11.1 | 5.6 |
| End August, 1981 | 100.0 | 34.2 | 10.5 | 34.2 | - | - | - | 5.3 | 2.6 | 7.9 | 2.6 | 5.3 | 7.9 | 5.3 |
| End September, 1982 | 100.0 | 33.3 | 10.3 | 33.3 | - | - | - | 5.1 | 2.6 | 7.7 | 2.6 | 5.1 | 7.7 | 7.7 |
| End September, 1983 | 100.0 | 34.2 | 10.5 | 34.2 | - | - | - | 5.3 | 2.6 | 7.9 | 2.6 | 5.3 | 7.9 | 5.3 |
| End September, 1984 | 100.0 | 34.2 | 10.5 | 34.2 | - | - | - | 5.3 | 2.6 | 7.9 | 2.6 | 5.3 | 7.9 | 5.3 |
| USSR | | | | | | | | | | | | | | |
| 1st June, 1976 | 100.0 | 12.1 | 9.7 | 7.7 | 1.5 | 15.8 | 17.3 | 9.7 | 3.6 | 13.3 | 7.2 | 7.6 | 14.8 | 5.1 |
| End September, 1978 | 100.0 | 36.4 | 13.2 | 5.4 | 3.9 | 16.3 | 20.2 | 9.3 | 3.1 | 12.4 | 2.3 | 6.2 | 8.5 | 3.9 |
| End September, 1979 | 100.0 | 38.4 | 13.2 | 6.3 | 4.4 | 13.8 | 18.2 | 7.5 | 2.5 | 10.0 | 2.5 | 6.3 | 8.8 | 5.0 |
| End August, 1980 | 100.0 | 37.8 | 10.9 | 7.7 | 4.5 | 10.9 | 15.4 | 7.7 | 3.8 | 11.6 | 3.2 | 6.4 | 9.6 | 7.0 |
| End August, 1981 | 100.0 | 36.0 | 10.9 | 6.9 | 4.6 | 14.3 | 18.9 | 7.4 | 3.4 | 10.9 | 4.0 | 6.9 | 10.9 | 5.7 |
| End September, 1982 | 100.0 | 38.2 | 10.2 | 6.4 | 4.3 | 14.5 | 18.8 | 7.0 | 3.2 | 10.2 | 4.3 | 6.4 | 10.7 | 5.4 |
| End September, 1983 | 100.0 | 36.7 | 10.1 | 6.9 | 5.5 | 14.7 | 20.2 | 6.0 | 2.7 | 8.7 | 4.1 | 6.4 | 10.5 | 6.9 |
| End September, 1984 | 100.0 | 35.0 | 9.9 | 7.8 | 5.3 | 13.6 | 18.9 | 5.8 | 3.3 | 9.1 | 4.9 | 5.8 | 10.7 | 8.6 |

Table 3 (Chapter 7) contd

| Country | Total | Chemicals industry | Metal-lurgy | Transport equipment | Machine tools | Mechanical engineering | Total | Elec-tronics | Electrical equipment | Total | Food and agriculture | Light industry | Total | Other branches |
|---|---|---|---|---|---|---|---|---|---|---|---|---|---|---|
| | | 2 | 3 | 4 | 5 | 6 | 5+6 | 7 | 8 | 7+8 | 1 | 9 | 1+9 | 10 |
| Total | | | | | | | | | | | | | | |
| 1st June, 1976 | 100.0 | 23.8 | 7.9 | 12.6 | 3.5 | 22.0 | 25.5 | 7.5 | 6.0 | 13.5 | 5.2 | 6.5 | 11.7 | 5.0 |
| End September, 1978 | 100.0 | 26.1 | 8.3 | 9.6 | 4.1 | 18.2 | 22.3 | 10.2 | 7.3 | 17.5 | 4.5 | 8.3 | 12.8 | 3.5 |
| End September, 1979 | 100.0 | 24.6 | 7.6 | 11.3 | 4.6 | 14.8 | 19.4 | 8.9 | 10.5 | 19.4 | 5.0 | 8.5 | 13.5 | 4.1 |
| End August, 1980 | 100.0 | 24.4 | 7.4 | 12.8 | 4.4 | 12.1 | 16.5 | 8.3 | 9.8 | 18.1 | 6.1 | 9.1 | 15.2 | 5.4 |
| End August, 1981 | 100.0 | 23.9 | 6.8 | 12.9 | 4.1 | 14.3 | 18.4 | 8.0 | 9.6 | 17.6 | 6.4 | 9.4 | 15.8 | 4.6 |
| End September, 1982 | 100.0 | 23.2 | 6.6 | 12.0 | 5.2 | 15.1 | 20.3 | 7.6 | 9.1 | 16.7 | 6.6 | 9.9 | 16.5 | 4.6 |
| End September, 1983 | 100.0 | 22.8 | 6.7 | 11.7 | 5.7 | 15.1 | 20.8 | 7.5 | 8.3 | 15.8 | 6.4 | 10.3 | 16.7 | 5.5 |
| End September, 1984 | 100.0 | 22.5 | 6.5 | 11.3 | 5.5 | 14.4 | 19.9 | 7.4 | 8.3 | 15.7 | 6.8 | 10.8 | 17.6 | 6.5 |

LEGEND:

1. Food and agriculture industry (including beverages);
2. Chemicals industry (including pharmaceuticals);
3. Metallurgy (including mining);
4. Transport equipment: includes aircraft, automobiles, lorries, tractors (even for agriculture), rolling stock, earth-moving equipment, diesel engines (even stationary);
5. Machine-tools;
6. Mechanical engineering (all other non-electrical engineering);
7. Electronics (computers and other office equipment, radio and television sets, communication equipment);
8. Electrical equipment (all other equipment including electric locomotives and household appliances);
9. Light industry (textiles, footwear, rubber, glass, furniture, consumer goods);
10. Other, such as construction, hotel management, tourism, etc.

Note: Because of rounding, percentage shares may not add up to exactly 100.

Sources:
- 1st June, 1976: Promotion of Trade Through Industrial Co-operation, Statistical Outline of Recent Trends in Industrial Co-operation, TRADE/R.355/Add. 2, op.cit., Table 2.
- End September 1978: Promotion of Trade Through Industrial Co-operation, Statistical Outline of Recent Trends in Industrial Co-operation, TRADE/R.373/Add. 5, op.cit., Table 2.
- End September 1979: Promotion of Trade Through Industrial Co-operation, Statistical Outline of Recent Trends in Industrial Co-operation, TRADE/R.392, op.cit., Table 2.
- End August, 1980: Promotion of Trade Through Industrial Co-operation, Statistical Survey of Recent Trends in Industrial Co-operation, TRADE/R.407, op.cit., Table 2.
- End August, 1981: Promotion of Trade Through Industrial Co-operation, Statistical Survey of Recent Trends in Industrial Co-operation, TRADE/R.429, op.cit., Table 2.
- End September 1982: Promotion of Trade Through Industrial Co-operation, Statistical Survey of Recent Trends in Industrial Co-operation, TRADE/R.450, op.cit., Table 2.
- End September, 1983: Promotion of Trade Through Industrial Co-operation, Statistical Survey of Recent Trends in Industrial Co-operation, TRADE/R.468, op.cit., Table 2.
- End September, 1984: Promotion of Trade Through Industrial Co-operation, Statistical Survey of Recent Trends in Industrial Co-operation, TRADE/R.487, op.cit., Table 2.

Table 4 (Chapter 7)

SHARE OF THE VARIOUS TYPES OF CONTRACT IN THE TOTAL NUMBER OF INDUSTRIAL CO-OPERATION CONTRACTS, BY COUNTRY
(percentages)

| Country | Total | of which: A.1 | A.2 | A* | B.1 | B.2 | B.3 | B* | C.1 | C.2 | C.3 | C* | D.1 | D.2 | D* | E.1 | E.2 | E* | F.1 | F.2 | F* |
|---|---|---|---|---|---|---|---|---|---|---|---|---|---|---|---|---|---|---|---|---|---|
| Bulgaria | 100.0 | 4.5 | - | 4.5 | 13.6 | - | - | 13.6 | 40.9 | 4.6 | 4.5 | 50.0 | 4.5 | - | 4.5 | 4.6 | 13.6 | 18 | .8 | 2.3 | 9.1 |
| Czechoslovakia | 100.0 | 18.8 | 1.5 | 20.3 | 1.4 | - | - | 1.4 | 63.8 | 4.4 | 1.4 | 69.6 | - | - | - | 5.8 | 1.4 | 7.2 | 1.4 | - | 1.4 |
| German Democratic Republic | 100.0 | 3.0 | 3.0 | 6.1 | 42.4 | 12.1 | - | 54.5 | 12.1 | - | - | 12.1 | - | 3.0 | 3.0 | 3.0 | 3.0 | 6.1 | 18.2 | - | 18.2 |
| Hungary | 100.0 | 16.8 | 3.4 | 20.2 | 7.6 | - | - | 7.6 | 27.8 | 4.1 | 4.5 | 36.4 | 3.8 | 2.4 | 6.2 | 6.2 | 11.0 | 17.2 | 10.6 | 1.7 | 12.3 |
| Poland | 100.0 | 15.6 | 4.2 | 19.8 | 17.7 | 7.3 | - | 25.0 | 28.1 | 4.2 | 1.0 | 33.3 | 1.0 | - | 1.0 | 2.1 | 3.1 | 5.2 | 13.5 | 2.1 | 15.6 |
| Romania | 100.0 | 2.6 | 2.6 | 5.3 | 7.9 | - | - | 7.9 | 7.9 | 2.6 | 5.3 | 15.8 | 2.6 | - | 2.6 | 18.4 | 39.5 | 57.9 | 10.5 | - | 10.5 |
| USSR | 100.0 | 2.9 | 0.8 | 3.7 | 20.2 | 4.5 | - | 24.7 | 17.7 | 3.3 | 35.4 | 56.4 | 0.8 | - | 0.8 | 4.1 | 2.5 | 6.6 | 5.3 | 2.5 | 7.8 |
| Total | 100.0 | 10.8 | 2.3 | 13.1 | 13.8 | 2.7 | - | 16.5 | 27.0 | 3.7 | 12.9 | 43.6 | 2.1 | 1.0 | 3.1 | 5.4 | 7.9 | 13.3 | 8.7 | 1.7 | 10.4 |

LEGEND:

Licensing:

A.1 -- Supply of licences and/or know-how (sometimes also some special equipment) in exchange -- at least partially -- for products or components;
A.2 -- Same as A.1, but the supply also includes parts in varying percentages of the final product in exchange for products or components;
A* -- A.1 + A.2.

Delivery of plant or equipment:

B.1 -- Supply of plant or equipment, including the corresponding technology, in exchange for products or components, at least partially;
B.2 -- For the exploitation of natural resources: same as B.1, plus studies of the availability and accessibility of the resources and research connected with the application of technology to particular circumstances;
B.3 -- Supply of plant or equipment on a leasing basis in exchange for products;
B* -- B.1 + B.2 + B.3.

Co-production and specialisation:

C.1 -- Co-operation, including or not sales, in which each party manufactures parts or components of a final product, the technology being supplied by one of the parties or both;
C.2 -- Co-operation in which each party specialises in part of the manufacturing programme and the parties then exchange units in order to complete each other's range of products;
C.3 -- Co-production and specialisation involving R&D only;
C* -- C.1 + C.2 + C.3.

Table 4 (Chapter 7) contd

<u>Sub-contracting</u>:

D.1 -- Short-term agreements providing for the delivery of an agreed quantity of finished or semi-finished goods produced through the use of documentation and know-how (and sometimes parts, machinery and equipment) provided by the contractor;
D.2 -- Long-term agreements providing for the delivery on a continuing basis, same as above;
D* -- D.1. + D.2.

<u>Joint ventures</u>:

E.1 -- Joint ventures involving marketing only;
E.2 -- Joint ventures involving production, marketing and R&D;
E* -- E.1 + E.2.

<u>Joint tendering or joint projects</u>:

F.1 -- Customer located in a third country;
F.2 -- Customer located in one of the countries;
F* -- F.1 + F.2.

<u>Note</u>: Because of rounding, percentage shares may not add up to exactly 100.

<u>Source</u>: <u>Promotion of Trade Through Industrial Co-operation, a Statistical Survey of Recent Trends in Industrial Co-operation</u>, TRADE/R.487, <u>op. cit.</u>, Table 3.

Table 5 (Chapter 7)

COUNTRIES' SHARES IN THE TOTAL NUMBER OF INDUSTRIAL CO-OPERATION CONTRACTS, WITH BREAKDOWN BY TYPE OF CONTRACT
(percentages)

| Type of contracts | Total for seven countries | Bulgaria | Czecho-slovakia | of which: German Democratic Republic | Hungary | Poland | Romania | USSR |
|---|---|---|---|---|---|---|---|---|
| A.1 | 100.0 | 2.3 | 14.8 | 1.1 | 55.7 | 17.0 | 1.1 | 7.9 |
| A.2 | 100.0 | - | 5.3 | 5.3 | 52.6 | 21.0 | 5.3 | 10.5 |
| A.1 + A.2 | 100.0 | 1.9 | 13.1 | 1.9 | 55.1 | 17.7 | 1.9 | 8.4 |
| B.1 | 100.0 | 5.4 | 0.9 | 12.5 | 19.6 | 15.2 | 2.7 | 43.7 |
| B.2 | 100.0 | - | - | 18.2 | - | 31.8 | - | 50.0 |
| B.3 | 100.0 | - | - | - | - | - | - | - |
| B.1 + B.2 + B.3 | 100.0 | 4.5 | 0.7 | 13.4 | 16.4 | 17.9 | 2.2 | 44.8 |
| C.1 | 100.0 | 8.2 | 20.0 | 1.8 | 36.8 | 12.3 | 1.4 | 19.5 |
| C.2 | 100.0 | 6.7 | 10.0 | - | 40.0 | 13.3 | 3.3 | 26.7 |
| C.3 | 100.0 | 1.9 | 0.9 | - | 12.4 | 0.9 | 1.9 | 81.9 |
| C.1 + C.2 + C.3 | 100.0 | 6.2 | 13.5 | 1.1 | 29.9 | 9.0 | 1.7 | 38.6 |
| D.1 | 100.0 | 11.7 | - | - | 64.7 | 5.9 | 5.9 | 11.7 |
| D.2 | 100.0 | - | - | 12.5 | 87.5 | - | - | - |
| D.1 + D.2 | 100.0 | 8.0 | - | 4.0 | 72.0 | 4.0 | 4.0 | 8.0 |
| E.1 | 100.0 | 4.5 | 9.1 | 2.3 | 40.9 | 4.5 | 15.9 | 22.7 |
| E.2 | 100.0 | 9.3 | 1.6 | 1.6 | 50.0 | 4.7 | 23.4 | 9.3 |
| E.1 + E.2 | 100.0 | 7.4 | 4.6 | 1.8 | 46.3 | 4.6 | 20.4 | 14.8 |
| F.1 | 100.0 | 4.2 | 1.4 | 8.4 | 43.7 | 18.3 | 5.6 | 18.3 |
| F.2 | 100.0 | 7.1 | | - | 35.7 | 14.3 | - | 42.9 |
| F.1 + F.2 | 100.0 | 4.7 | 1.2 | 7.1 | 42.3 | 17.6 | 4.7 | 22.3 |
| Total | 100.0 | 5.4 | 8.5 | 4.0 | 35.7 | 11.8 | 4.7 | 29.8 |

Note: Because of rounding, percentage shares may not add up to exactly 100. See Table 4 for the Legend.

Source: Ibid., Table 4.

Table 6 (Chapter 7)

INDUSTRIAL CO-OPERATION CONTRACTS,
BY WESTERN COUNTRY (NINE PRINCIPAL COUNTRIES), TYPE OF CONTRACT AND INDUSTRY
(percentages)

| Country / Type of contract and industry | Grand Total | Total A-F | of which: A | B | C | D | E | F | Total 1-10 | of which: 1 | 2 | 3 | 4 | 5 | 6 | 7 | 8 | 9 | 10 |
|---|---|---|---|---|---|---|---|---|---|---|---|---|---|---|---|---|---|---|---|
| Germany | 25.6 | 100.0 | 22.0 | 10.8 | 39.3 | 5.2 | 14.1 | 8.5 | 100.0 | 7.9 | 16.7 | 4.6 | 14.1 | 8.2 | 15.1 | 7.2 | 8.5 | 12.5 | 5.2 |
| Italy | 11.7 | 100.0 | 9.3 | 12.2 | 39.6 | 1.4 | 25.2 | 12.2 | 100.0 | 5.7 | 33.1 | 5.7 | 12.2 | 2.2 | 7.9 | 3.6 | 9.3 | 11.5 | 8.6 |
| United States | 11.6 | 100.0 | 26.1 | 6.5 | 32.6 | 5.1 | 26.1 | 3.6 | 100.0 | 6.5 | 22.5 | 7.2 | 11.6 | 0.7 | 12.3 | 9.4 | 11.6 | 10.1 | 8.0 |
| France | 9.7 | 100.0 | 14.6 | 20.7 | 38.8 | 0.9 | 17.2 | 7.8 | 100.0 | 6.9 | 22.4 | 5.2 | 19.8 | 6.9 | 12.1 | 6.9 | 6.0 | 6.9 | 6.9 |
| United Kingdom | 7.3 | 100.0 | 24.1 | 9.2 | 28.7 | 3.4 | 31.0 | 3.4 | 100.0 | 6.9 | 25.3 | 8.0 | 17.2 | 2.3 | 5.7 | 10.3 | 6.9 | 10.3 | 6.9 |
| Austria | 7.1 | 100.0 | 27.1 | 15.3 | 28.2 | 1.2 | 15.3 | 12.9 | 100.0 | 5.9 | 16.5 | 8.2 | 4.7 | 1.2 | 20.0 | 3.5 | 10.6 | 14.1 | 15.3 |
| Japan | 5.7 | 100.0 | 16.2 | 26.5 | 22.0 | - | 11.8 | 23.5 | 100.0 | 1.5 | 32.3 | 10.3 | 14.7 | 4.4 | 14.7 | 4.4 | 8.8 | 2.9 | 5.9 |
| Sweden | 5.2 | 100.0 | 14.5 | 9.7 | 50.0 | - | 22.6 | 3.2 | 100.0 | 4.8 | 11.3 | 8.1 | 16.1 | 4.8 | 25.8 | 14.5 | 6.4 | 6.4 | 1.6 |
| Switzerland | 4.3 | 100.0 | 13.7 | 3.9 | 33.3 | 2.0 | 27.4 | 19.6 | 100.0 | 2.0 | 23.5 | 5.9 | 3.9 | 5.9 | 5.9 | 7.8 | 13.7 | 19.6 | 11.8 |
| Other Western countries | 11.8 | 100.0 | 12.8 | 11.3 | 31.9 | - | 24.8 | 19.1 | 100.0 | 17.7 | 14.2 | 8.5 | 7.8 | 3.5 | 9.9 | 5.0 | 18.4 | 8.5 | 6.4 |
| Total | 100.0 | | | | | | | | | | | | | | | | | | |

LEGEND: A: Licensing; B: Delivery of plant or equipment; C: Co-production and specialisation; D: Sub-contracting; E: Joint ventures; F: Joint tendering or joint projects. For the industries (1-10) see the legend to Table 3.

Note: Because of rounding, percentage shares may not add up to exactly 100.

Source: Ibid., Table 6.

Table 7 (Chapter 7)

COUNTRIES' SHARES IN THE TOTAL NUMBER OF INDUSTRIAL CO-OPERATION CONTRACTS WITH BREAKDOWN BY INDUSTRY
(percentages)

| Industry | Total for seven countries | of which: Bulgaria | Czecho-slovakia | German Democratic Republic | Hungary | Poland | Romania | USSR |
|---|---|---|---|---|---|---|---|---|
| Chemicals industry (2) | 100.0 | 2.2 | 3.3 | 2.7 | 29.0 | 9.3 | 7.1 | 46.4 |
| Metallurgy (3) | 100.0 | 3.8 | - | 3.8 | 17.0 | 22.6 | 7.5 | 45.3 |
| Transport equipment (4) | 100.0 | 3.3 | 8.7 | 9.8 | 27.2 | 16.3 | 14.1 | 20.6 |
| Machine-tools (5) | 100.0 | 4.4 | 24.4 | 6.7 | 31.1 | 4.4 | - | 28.9 |
| Mechanical engineering (6) | 100.0 | 9.4 | 12.8 | 5.1 | 29.1 | 15.4 | - | 28.2 |
| (5) + (6) | 100.0 | 8.0 | 16.0 | 5.6 | 29.6 | 12.3 | - | 28.4 |
| Electronics (7) | 100.0 | 8.3 | 3.3 | 5.0 | 36.7 | 20.0 | 3.3 | 23.3 |
| Electrical equipment (8) | 100.0 | 4.4 | 14.7 | - | 60.3 | 7.3 | 1.5 | 11.8 |
| (7) + (8) | 100.0 | 6.2 | 9.4 | 2.3 | 49.4 | 13.3 | 2.3 | 17.2 |
| Food & agriculture (1) | 100.0 | 7.3 | 12.7 | 1.8 | 47.3 | 7.3 | 1.8 | 21.8 |
| Light industry (9) | 100.0 | 10.2 | 9.1 | 3.4 | 52.3 | 6.8 | 2.3 | 15.9 |
| (1) + (9) | 100.0 | 9.1 | 10.5 | 2.8 | 50.3 | 7.0 | 2.1 | 18.2 |
| Other branches (10) | 100.0 | 1.9 | 3.8 | 1.9 | 39.6 | 9.4 | 3.8 | 39.6 |
| Total | 100.0 | 5.4 | 8.5 | 4.0 | 35.7 | 11.8 | 4.7 | 29.8 |

LEGEND:

1. Food and agriculture industry (including beverages);
2. Chemicals industry (including pharmaceuticals);
3. Metallurgy (including mining);
4. Transport equipment: includes aircraft, automobiles,lorries, tractors (even for agriculture), rolling stock, earth-moving equipment, diesel engines (even stationary);
5. Machine-tools;
6. Mechanical engineering (all other non-electrical engineering);
7. Electronics (computers and other office equipment, radio and television sets, communication equipment);
8. Electrical equipment (all other equipment including electric locomotives and househould appliances);
9. Light industry (textiles, footwear, rubber, glass, furniture, consumer goods);
10. Other, such as construction, hotel management, tourism, etc.

Note: Because of rounding, percentage shares may not add up to exactly 100.

Source: Ibid., Table 1.

Tables 8-14 of Chapter 7

EWIC: AGREEMENTS CONCLUDED DURING 1980-1984

GLOSSARY

| | |
|---|---|
| C | Cooperation Agreement |
| JP-jp | Joint Production |
| JV | Joint Venture |
| L,l | Licencing |
| LS,ls | Leasing |
| PC,pc | Plant Construction |
| TK | Turn-key |
| d | development |
| e | equipment |
| k | know-how |
| m | marketing |
| p | production |
| r | research |
| t | technology |
| 1 | Counter-purchase (non resultant product) |
| 2 | Compensation-Buy-back (resultant product) |
| 3 | Barter-Switch (direct exchange of goods having offsetting values) |
| 4 | Marketing in the West |
| 5 | Marketing in Third Countries |
| 6 | Third Country Projects |

N.B.:

1. The figure in front of the agreement denotes the number of agreements signed in this category.
2. Tables 8-14 include only agreements said to be in effect. Furthermore, only those turn-key projects have been included which entail some form of co-operation. Those turn-key contracts which entail outright purchase are discussed, where applicable, under the heading "other technology transfer channels".

Sources:

1. Defense Intelligence Agency, Western Technology Transfer to the Warsaw Pact (DDB-1900-46-82), Washington, D.C. November 1982.
2. Business Eastern Europe, 1980-1984.
3. East-West Fortnightly Bulletin, 1982-1984.
4. East-European Markets, 1982-1984.
5. Foreign Trade (English edition), Moscow, 1980-1984.
6. Moscow Narodny Bank (Press Bulletin), 1983-1984.
7. Various East European publications.

Table 8 (Chapter 7)

Country: BULGARIA 1980-1984 (SUMMARY)   Agreements: 87

| Type & No. | Coop | Licence | JV | PC | TK | Other |
|---|---|---|---|---|---|---|
| | 57 | 18 | 9 | 1 | 2 | - |
| Percentage: | 65.5 | 20.6 | 10.3 | 1.1 | 2.2 | |
| Sector | | | No. | | % | |
| Multisector | | | 17 | | 19.5 | |
| Textiles | | | 14 | | 16.1 | |
| Chemicals | | | 9 | | 11.5 | |
| Food & Tobacco | | | 7 | | 8.7 | |
| Automotive | | | 6 | | 6.9 | |
| Electronics | | | 5 | | 5.7 | |
| Western Partner | | | | | | |
| Germany | | | 21 | | 24 | |
| Japan | | | 14 | | 16.0 | |
| Italy | | | 13 | | 14.9 | |
| UK | | | 12 | | 13.8 | |

Year : 1980

| | Aus. | Bel. | Can. | Den. | Fin. | Fra. | Ger. | It. | Jap. | Neth. | Nor. | Por. | Spa. | Swe. | Swi. | U.K. | U.S.A. |
|---|---|---|---|---|---|---|---|---|---|---|---|---|---|---|---|---|---|
| 1. Agriculture and equipment | | | | | | | Lpk | | | Cd | | | | | | | |
| 2. Automotive | | | | | | | | | | | | | | | | Cd | |
| 3. Chemicals and petrochemicals | | | | | | | | | | | | | | Crpm6 | | Crpm6 | |
| 4. Electronics | | | | | | | | | C | | | | | | | | |
| 5. Food and tobacco | | | | | | | | | Ct | | | | | | | | |
| 6. Iron and steel | | | | | | | | | JV | | | | | | | | |
| 7. Machinery | | | | | | | | | | Crp6 | | | | | | | |
| 8. Machine tools | | | | | | | | | Lp | | | | | | | | |
| 9. Mining and metallurgy | | | | | | | | Cpe | | | | | | | | | |
| 10. Multisector | | | | | | | | C6 | Cd6 C6 | | | | | | | | C |
| 11. Telecommunications | | | | | | C | | | | | | | | | | | |
| 12 Textiles, clothing and leather | | | | | | | C | | | | | | | | | | |

Table 8 (Chapter 7) contd
Bulgaria

Year : 1981

| | Aus. | Bel. | Can. | Den. | Fin. | Fra. | Ger. | It. | Jap. | Neth. | Nor. | Por. | Spa. | Swe. | Swi. | U.K. | U.S.A. |
|---|---|---|---|---|---|---|---|---|---|---|---|---|---|---|---|---|---|
| 1. Chemicals and petrochemicals | | | | | | | C | | | | | | | | | | C |
| 2. Construction | | | | | | | | Cp | | | | | | | | | |
| 3. Electronics | | | | | | | | | | | | | | C | | CD | |
| 4. Environment | | | | Cp | | | | | | | | | | | | | |
| 5. Food and tobacco | | | | | | | | | | | | | | | Cp | | |
| 6. Machine tools | | | | | | | | | JV | | | | | | | | |
| 7. Mat.Handling & Transport Eq. | | | | | | | | | Lp | | | | | | | | |
| 8. Multisector | TK6 | | | | | | | | C | | | | | | | C6 | |
| 9. Power and energy | | | | | | | Cp5 | | | | | | | | | | |

Year : 1982

| | Aus. | Bel. | Can. | Den. | Fin. | Fra. | Ger. | It. | Jap. | Neth. | Nor. | Por. | Spa. | Swe. | Swi. | U.K. | U.S.A. |
|---|---|---|---|---|---|---|---|---|---|---|---|---|---|---|---|---|---|
| 1. Automotive | | | | | | | | | | | | | | | | Cpm | |
| 2. Chemicals and petrochemicals | | | | | | | Lpc | | | | | | | | | Cpe | |
| 3. Consumer goods | | | | | | | | | JV5 | | | | | | | | |
| 4. Environment | | | | | | | C | | | | | | | | | | |
| 5. Food and tobacco | Lp | | | | | | Lk | 2C | | | | | | | | | |
| 6. Mat.Handling and Transport Eq. | | | | | C5 | | | | | | | | | | | | |
| 7. Multisector | | | | | | Cjd | | | | | | | | Crd | | JV | |
| 8. Shipping and shipbuilding | | | | | | | JV | | | | | | | | | | |
| 9. Textiles, clothing and leather | | | | | | | 2Lp | 2JV | | | | | | | | | |

Table 8 (Chapter 7) contd
Bulgaria

Year : 1983

| | Aus. | Bel. | Can. | Den. | Fin. | Fra. | Ger. | It. | Jap. | Neth. | Nor. | Por. | Spa. | Swe. | Swi. | U.K. | U.S.A. |
|---|---|---|---|---|---|---|---|---|---|---|---|---|---|---|---|---|---|
| 1. Automotive | Lp5<br>Cp | | | | | | | | | | | | | | | Cpm | |
| 2. Business Eq. | | | | | | | | | | | | | | | JV | | |
| 3. Chemicals and petrochemicals | | | | | | | Cp3 | | | | | | | | | | |
| 4. Consumer goods | | | | | | | | | | | | | | | | Cp | |
| 5. Electrical Eq. | | | | | | | Cjp | | | | | | | | | | |
| 6. Electronics | | | | | | C | | | | | | | | | | | |
| 7. Mat.Handling & Transport Eq. | | | | | | | JVm | | | | | | | | | | |
| 8. Multisector | C6<br>C | | | | | | Cd | | C | | | | | | | | |
| 9. Pharmaceuticals | | | | | | | Lp | | | | | | | | | | |
| 10. Shipping and shipbuilding | | | | Cjp | | | | | | | | | | | | | |
| 11. Textiles clothing and leather | | | | | | Lp<br>Cp | Lp | 2Lp<br>Cjp | Lp | Cp | Lp | | | | | | |

Year : 1984

| | | | | | | | | | | | | | | | | | |
|---|---|---|---|---|---|---|---|---|---|---|---|---|---|---|---|---|---|
| 1. Agriculture and equipment | | | | | | | | TK | | | | | | | | | |
| 2. Automotive | | | | | | | | | | | | | | | | | |
| 3. Business equipment | | | | | | | | | | | | | | | | Cp | |
| 4. Chemicals and petrochemicals | | | | | | | | PC1 | | | | | | | | Cd | |
| 5. Consumer Goods | Lp | | | | | | | | | | | | | | | | |
| 6. Electronics | | | | | | | Cp | | | | | | | | | | |
| 7. Food and Tobacco | | | | | | | | | | | | | | | C | | |
| 8. Machine Tools | | | | | | | Cp | | | | | | | | | | |
| 9. Medical Equipment | | | | | | | Lp<br>C | | | | | | | | | | |
| 10. Multisector | | | | | | | | C | 2C | | | | | | | | |
| 11. Precision instruments | | | | | | | C | | | | | | | | | | |

Table 9 (Chapter 7)

Country: CZECHOSLOVAKIA 1980-1984 (SUMMARY) Agreements: 40

| Type & No. | Coop | Licence | JV | PC | TK | Other |
|---|---|---|---|---|---|---|
| | 33 | 4 | - | 2 | 1 | - |
| Percentage: | 82.5 | 10.0 | - | 5.0 | 2.5 | |
| Sector | | No. | | % | | |
| Chemicals | | 6 | | 15 | | |
| Multisector | | 5 | | 11.2 | | |
| Agriculture | | 4 | | 10 | | |
| Automotive | | 3 | | 7.5 | | |
| Western Partner | | | | | | |
| Austria | | 10 | | 25 | | |
| France | | 6 | | 15 | | |
| Germany | | 4 | | 10 | | |
| Italy | | 4 | | 10 | | |

Year : 1980

| | Aus. | Bel. | Can. | Den. | Fin. | Fra. | Ger. | Gre. | It. | Jap. | Neth. | Nor. | Spa. | Swe. | Swi. | U.K. | U.S.A. |
|---|---|---|---|---|---|---|---|---|---|---|---|---|---|---|---|---|---|
| 1. Agriculture and Equipment | | | | | | Cjp | | | | | | | | | | | |
| 2. Chemicals and petrochemicals | | | | | | | | | Cm5 | | | | | | | | |
| 3. Consumer Goods | | | | | | Cjp | | | | | | | | | | | |
| 4. Mining and Metallurgy | | | | | | | | Cp | | | | | | | | | |
| 5. Multisector | C6 | | | | | | | | | | | | | | | | |
| 6. Packaging | | Lp | | | | | | | | | | | | | | | |
| 7. Wood, Pulp and Paper | PC | | | | | | | | | | | | | | | | |

Table 9 (Chapter 7) contd
Czechoslovakia

Year : 1981

| | Aus. | Bel. | Can. | Den. | Fin. | Fra. | Ger. | It. | Jap. | Neth. | Nor. | Por. | Spa. | Swe. | Swi. | U.K. | U.S.A. |
|---|---|---|---|---|---|---|---|---|---|---|---|---|---|---|---|---|---|
| 1. Agriculture | C | | | | | | | | | | | | | | | | |
| 2. Chemicals and petrochemicals | | | | | | | Crd | | | | | | | | | C | |
| 3. Glass and ceramics | | | | | | | | Cp | | | | | | | | | |
| 4. Machinery | | | | | | | Cp | | | | | | | | | | |
| 5. Medical Eq. | | | | | C&L | | | | | | | | | | | | |
| 6. Metalworking | | | | | | Cp | | | | | | | | | | | |
| 7. Multisector | | | | | | | | | 2C6 | | | | | | | | |
| 8. Pharmaceuticals | C | | | | | | | | | | | | | | | | |
| 9. Power and Energy | Cp5 | | | | | | | C6 | | | | | | | | | |
| 10. Telecommunications | | | | | | | | | Ct | | | | | | | | |

Year : 1982

| | | | | | | | | | | | | | | | | | |
|---|---|---|---|---|---|---|---|---|---|---|---|---|---|---|---|---|---|
| 1. Agriculture and Equipment | | | | Lp | | | | | | | | | | | | | |
| 2. Chemicals and Petrochemicals | | | | | | | Lp | | | | | | | | | | |
| 3. Machine tools | | | | | | | | | | | | | | | | Cjp | |
| 4. Power and Energy | Cp | | | | | | | | | | | | | | | | |

Year : 1983

| | | | | | | | | | | | | | | | | | |
|---|---|---|---|---|---|---|---|---|---|---|---|---|---|---|---|---|---|
| 1. Agriculture and Equipment | | | | | | | | | | Cp2 | | | | | | | |
| 2. Automotive | | | | | | Cp | | | | | | | | | | | |
| 3. Chemicals and Petrochemicals | C | | | | | | | | | | | | | | | | |
| 4. Electrical Equipment | | | | | | Ct | | | | | | | | | | | |
| 5. Glass and Ceramics | Lp | | | | | | | | | | | | | | | | |
| 6. Iron and Steel | C6 | | | | | | | | | | | | | | | | |
| 7. Mat. Handling and Transport Eq. | Cjp | | | | | | | | | | | | | | | | |
| 8. Multisector | | | | | | | | C | C | | | | | | | | |
| 9. Shipping and Shipbuilding | | | | | | | | | | | | | | Cp | | | |

Table 9 (Chapter 7) contd
Czechoslovakia

<u>Year : 1984</u>

| | Aus. | Bel. | Can. | Den. | Fin. | Fra. | Ger. | It. | Jap. | Neth. | Nor. | Por. | Spa. | Swe. | Swi. | U.K. | U.S.A. |
|---|---|---|---|---|---|---|---|---|---|---|---|---|---|---|---|---|---|
| 1. Automotive | | | | | | | Cle | C | | | | | | | | | |
| 2. Chemicals and petrochemicals | | | | | | PC | | | | | | | | | | | |
| 3. Construction | | | | | TK | | | | | | | | | | | | |
| 4. Multisector | | | | | | | | | | | | | | | | C | |
| 5. Textiles Clothing & Leather | | | | | | | | | | | | | | | | | C |

Table 10 (Chapter 7)

Country: GERMAN DEMOCRATIC REPUBLIC 1980-1984 (SUMMARY) Agreements: 31

| Type & No. | Coop | Licence | JV | PC | TK | Other |
|---|---|---|---|---|---|---|
| | 23 | 4 | 1 | 2 | - | - |
| Percentage: | 76.6 | 13.3 | 3.3 | 6.6 | | |

| Sector | No. | % |
|---|---|---|
| Chemicals | 9 | 29 |
| Multisector | 3 | 9.6 |
| Agriculture | 2 | 6.4 |
| Railway | 2 | 6.4 |
| Machine Tools | 2 | 6.4 |
| Machinery | 2 | 6.4 |
| Iron and Steel | 2 | 6.4 |
| **Western Partner** | | |
| France | 6 | 19.3 |
| Germany | 5 | 16.1 |
| Italy | 4 | 12.9 |
| Japan | 4 | 12.9 |
| Austria | 4 | 12.9 |
| UK | 3 | 9.6 |

Year : 1980

| | Aus. | Bel. | Can. | Den. | Fin. | Fra. | Ger. | It. | Jap. | Neth. | Nor. | Por. | Spa. | Swe. | Swi. | U.K. | U.S.A. |
|---|---|---|---|---|---|---|---|---|---|---|---|---|---|---|---|---|---|
| 1. Chemicals and petrochemicals | | | | | | C | | | | | | | | | | | |
| 2. Iron and Steel | Cl | | | | | | | | | | | | | | | Cjpd | |
| 3. Railway | | | | | | Cd<br>Cp | | | | | | | | | | | |
| 4. Services | | | | | | | | | | | | | | | | C | |

Table 10 (Chapter 7) contd
German Democratic Republic

Year : 1981

| | Aus. | Bel. | Can. | Den. | Fin. | Fra. | Ger. | It. | Jap. | Neth. | Nor. | Por. | Spa. | Swe. | Swi. | U.K. | U.S.A. |
|---|---|---|---|---|---|---|---|---|---|---|---|---|---|---|---|---|---|
| 1. Agriculture and Equipment | | | | | | | | | PC | | | | | | | | |
| 2. Chemicals and Petrochemicals | | | | | | | | Cd C3 | | | | | | | C3m5 | | |
| 3. Iron and Steel | | | | | | | | | PC | | | | | | | | |
| 4. Machinery | | | | | | | Lpm | | | | | | | | | | |
| 5. Metalworking | | | | | | | | Lp | | | | | | | | | |
| 6. Multisector | | | | | | | | C6 | | C6 | | | | | | | |

Year : 1982

| | Aus. | Bel. | Can. | Den. | Fin. | Fra. | Ger. | It. | Jap. | Neth. | Nor. | Por. | Spa. | Swe. | Swi. | U.K. | U.S.A. |
|---|---|---|---|---|---|---|---|---|---|---|---|---|---|---|---|---|---|
| 1. Agriculture and Equipment | | | | | | | | | C | | | | | | | | |
| 2. Chemicals and Petrochemicals | | | | | | | | | Cjp6 | | | | | | | | |
| 3. Machine Tools | | | | | JVm5 | | | | | | | | | | | | |
| 4. Multisector | | | | | | | | | | | | | | | | C | |
| 5. Power and Energy | | | | | | | C5 | | | | | | | | | | |

Year : 1983

| | Aus. | Bel. | Can. | Den. | Fin. | Fra. | Ger. | It. | Jap. | Neth. | Nor. | Por. | Spa. | Swe. | Swi. | U.K. | U.S.A. |
|---|---|---|---|---|---|---|---|---|---|---|---|---|---|---|---|---|---|
| 1. Chemicals and Petrochemicals | | | | | | | | | | | | | | C3 | | | |
| 2. Consumer Goods | | | | | | | Ltp | | | | | | | | | | |
| 3. Machine Tools | | | | | | | | | | | | | | Cd | | | |
| 4. Mining & Metallurgy | C | | | | | | | | | | | | | | | | |

Year : 1984

| | Aus. | Bel. | Can. | Den. | Fin. | Fra. | Ger. | It. | Jap. | Neth. | Nor. | Por. | Spa. | Swe. | Swi. | U.K. | U.S.A. |
|---|---|---|---|---|---|---|---|---|---|---|---|---|---|---|---|---|---|
| 1. Automotive | | | | | | | Lpt2 | | | | | | | | | | |
| 2. Chemicals and Petrochemicals | Cdp | | | | | 2C | | | | | | | | | | | |
| 3. Electronics | | | | | | C | | | | | | | | | | | |
| 4. Machinery | C | | | | | | | | | | | | | | | | |
| 5. Textiles, clothing and Leather | | | | | | | Cm5 | | | | | | | | | | |

Table 11 (Chapter 7)

Country: HUNGARY 1980-1984 (SUMMARY) Agreements: 294

| Type & No. | Coop 198 | Licence 52 | JV 33 | PC 5 | TK 1 | Other 5 |
|---|---|---|---|---|---|---|
| Percentage: | 67.4 | 17.8 | 10.9 | 1.7 | 0.3 | 1.7 |
| Sector | | No. | | % | | |
| Chemicals | | 35 | | 11.9 | | |
| Agriculture | | 26 | | 8.8 | | |
| Textiles | | 25 | | 8.5 | | |
| Food and Tobacco | | 22 | | 7.4 | | |
| Construction | | 21 | | 7.1 | | |
| Consumer goods | | 19 | | 6.4 | | |
| Automotive | | 19 | | 6.4 | | |
| Electronics | | 18 | | 6.1 | | |
| Electrical Eq. | | 12 | | 4 | | |
| Multisector | | 11 | | 3.7 | | |
| Machine Tools | | 9 | | 3 | | |
| Machinery | | 7 | | 2.4 | | |
| Materials Hand. | | 7 | | 2.4 | | |
| Pharmaceuticals | | 6 | | 2 | | |
| Western Partner | | | | | | |
| Germany | | 97 | | 33 | | |
| Austria | | 37 | | 12.5 | | |
| US | | 25 | | 8.5 | | |
| France | | 21 | | 7.1 | | |
| Italy | | 20 | | 6.8 | | |
| UK | | 19 | | 6.4 | | |
| Japan | | 15 | | 5.1 | | |
| Switzerland | | 15 | | 5.1 | | |
| Sweden | | 12 | | 4.1 | | |
| Finland | | 7 | | 1.3 | | |

Table 11 (Chapter 7) contd
Hungary

Year : 1980

| | Aus. | Bel. | Can. | Den. | Fin. | Fra. | Ger. | It. | Jap. | Neth. | Nor. | Por. | Spa. | Swe. | Swi. | U.K. | U.S.A. |
|---|---|---|---|---|---|---|---|---|---|---|---|---|---|---|---|---|---|
| 1. Agriculture and Equipment | | | | Cjp<br>Crd | | Cel<br>Cjp2 | Lp<br>Cel | | | | | | | | | | JV<br>Cl |
| 2. Automotive | Cd | | | | | | Cpk2<br>Cm5<br>Lp | | | | | | | | | Cd<br>Lek5 | |
| 3. Business Equipment | | | | | | | Cp | | | | | | | | | | |
| 4. Chemicals and Petrochemicals | | | | | | | Cp15<br>Lpl | | | | | | | | | Cl | |
| 5. Construction | Cp2 | | | | Cp | Cjp<br>Lp5 | Ckm5 | | | | | | | Cjp | | | |
| 6. Consumer Goods | Cjp | | | | | | Cp | Cpe | | | | | | Cjpm | | | |
| 7. Electrical Equipment | Cpk5 | | | | Cm | Cp | JV<br>2Lt | | | | | | | | | | |
| 8. Electronics | Cjp<br>Cpl | | | | Cp | Lt | JV | | Lk | | | | | | | | |
| 9. Food and Tobacco | | | | L | | | Cjp2 | | | | | | | | | | L2 |
| 10. Household Eq. | | | | | | | | | | | | | | Cp | | | |
| 11. Machinery | Cpl | | | | | | | | | | | | | | | | |
| 12. Machine Tools | Cdp2 | | | | | | Cp5<br>Lk | | Lp5 | | | | | | | | |
| 13. Mat.Handling & Transport Equipment | | | | | | | Lp<br>Cjp | | | | | | | | | | |
| 14. Medical Eq. | | | | | | | | | | | | | | | | | Lp |
| 15. Metalworking | | | | | | | 2Lp | | | | | | | | | | |
| 16. Mining and Metallurgy | | | | | | | | | | | | | | Cpt2 | | | |
| 17. Multisector | | | | | | | Clt5 | | Cm5 | | | | | | | | |
| 18. Packaging | | | | | | | | | | | | | | | Cl | | |
| 19. Petroleum and Gas | | Cp | | | | | | | | | | | | | | | |
| 20. Pharmaceuticals | | | | | | | | | | | | | | | Cp<br>Lp5 | | |
| 21. Precision Instruments | | | | | | | Lp | | | | | | | | | | |
| 22. R''bber | | | | | | | | Cm5 | | | | | | | | Cm5 | |
| 23. Textiles Clothing and Leather | Cpt | | | | | | Cpm<br>Cp | | | | | | | | | | Ck |
| 24. Wood, Pulp and Paper | | | | | Ce5 | | Ck | | | | | | | | | | |

Table 11 (Chapter 7) contd
Hungary

Year : 1981

| | Aus. | Bel. | Can. | Den. | Fin. | Fra. | Ger. | It. | Jap. | Neth. | Nor. | Por. | Spa. | Swe. | Swi. | U.K. | U.S.A. |
|---|---|---|---|---|---|---|---|---|---|---|---|---|---|---|---|---|---|
| 1. Agriculture and Equipment | JV<br>C | | | | | Cd6 | | C | | Cp | | | | | | | Cr |
| 2. Automotive | | | | | | | | | | | | | | Cjpm<br>Cjp | | | |
| 3. Business Equipment | | | | | | | | | | | | | | | Cm4 | | |
| 4. Chemicals and Petrochemicals | | | | | | | Lp<br>Lk<br>Clk<br>Lpe3 | | | | | | | | | Cp | |
| 5. Construction | | | | | | | | | | | | | | | | | |
| 6. Consumer Goods | LS | | | | | Cjp2 | | | | LS | | | | | | | |
| 7. Electronics | | | | | | | C3 | | C<br>Cl | | | | | | | | |
| 8. Environment | | | | Cjp | | | Cjp<br>Cjp5 | | | | | | | Cp | | C | |
| 9. Food and Tobacco | Lp | | | | | PC | | | | | | | | | Ckem | | |
| 10. Iron and Steel | | | | | | | | | | | | | | | | | C |
| 11. Machinery | | | | | Cpt2 | Cse2 | | | | | | | | | Cp | | |
| 12. Machine Tools | | | | | | | Cpkm | | | | | | | | | | |
| 13. Mat. Handling & Transport Eq. | | | | | | | Cjp | | | | | | | | Lp | | |
| 14. Multisector | | | | | | | 2C2 | Cjp | | | | | | | | | |
| 15. Petroleum and Gas | | | | | | | C1jp | Cd | | | | | | | | | |
| 16. Power and Energy | | | | | | | | | | | Cp | | | | | | |
| 17. Precision Instruments | | | | | | | Lp2 | | | | | | | | | | |
| 18. Rubber | | | | | | | | C | | | | | | | | | |
| 19. Textiles Clothing and Leather | Cp4 | | | | | | C14<br>Ck2 | Cp | | | | | | | Ctk | | |
| 20. Wood, Pulp and Paper | | | | | | | Cjp | | | | | | | | | | |

Table 11 (Chapter 7) contd
Hungary

Year : 1982

| | Aus. | Bel. | Can. | Den. | Fin. | Fra. | Ger. | It. | Jap. | Neth. | Nor. | Por. | Spa. | Swe. | Swi. | U.K. | U.S.A. |
|---|---|---|---|---|---|---|---|---|---|---|---|---|---|---|---|---|---|
| 1. Agriculture and Equipment | | | | | | | Cjpl<br>Cp | | | | | | | JV | | | Lt<br>2Cp |
| 2. Automotive | | | | | | Lp | Lk<br>Cjp<br>Cte | | | | | | | | | | |
| 3. Business Eq. | | | | | | | Cjp | | | | | | | | | | |
| 4. Chemicals and Petrochemicals | JV | Lt | | | | | Clp<br>Cp3<br>Ctd | Cp2<br>Ctrd | | | Ck<br>Cm | | | | Cp | | |
| 5. Construction | C&LS<br>Cjp | | | | | | Cjp<br>Cjp3<br>TK6 | Lk | | | Cm5 | | | | | | |
| 6. Consumer Goods | Cp | | | | | Lk | Cml | | | | | | | | | Cp | |
| 7. Electrical Eq. | | Cjp | | | | | | | | | | | | | | | |
| 8. Electronics | Cpke | | | | | | Cp2 | | Cp4 | | | | | | | | |
| 9. Food and Tobacco | Cjp | | | Cp | | | | | | | | | | | | | |
| 10. Iron and Steel | | | | | | | Cp2 | | | | | | | | | | |
| 11. Machine Tools | | | Cjp | | | | | | | | | | | | | | |
| 12. Mat. Handling & Transport Eq. | | | | | | | Lpm<br>Cjp | | | | | | | | | Cjp2 | |
| 13. Medical Eq. | | | | | | | | | | | | | | Cjp5 | | | |
| 14. Metallurgy & Mining | | | | | | | | | | | | | | | | Ce3 | |
| 15. Multisector | | | | | | | | C | Ctr | | | | | | | | Crd |
| 16. Non-ferrous metals | Cjp | | | | | | | | | | | | | | | | |
| 17. Power and Energy | Lp<br>Lk | | | | | | | | | | | | | | | | |
| 18. Precision Instruments | Cjp5 | | | | | | | | | | | | | | | | |
| 19. Rubber | | | | | | | | | | Cp | | | | | | | |
| 20. Services | | | | | | | JV<br>LS | | | | | | | | | | |
| 21. Telecommunications | | | | | | | Cp<br>Cr | | | | | | | | | | |
| 22. Textiles Clothing and Leather | | | | | | C5 | | | | | | | | | | | Lp |

Table 11 (Chapter 7) contd
Hungary

Year : 1983

| | Aus. | Bel. | Can. | Den. | Fin. | Fra. | Ger. | Gre. | It. | Jap. | Neth. | Nor. | Por. | Spa. | Swe. | Swi. | U.K. | U.S.A. |
|---|---|---|---|---|---|---|---|---|---|---|---|---|---|---|---|---|---|---|
| 1. Agriculture and Equipment | | | | Lp15 | | | Cp | | | | | | | | | | | Cm |
| 2. Automotive | Cm | | Cpm | | | | C | | | | | | | | | | | Cd |
| 3. Chemicals and Petrochemicals | | | | | | Cp2 | Cprd | | | Cp | | Crd<br>Cd | | | | | | Lk1 |
| 4. Construction | JV6<br>Lk1 | | | JV | | C1 | CTK5 | | | | | | | | | | | Cp |
| 5. Consumer Goods | | | | | | | Cm | | Cjp | | | | | | | JV | | Lp |
| 6. Electrical Equipment | | JVpm4 | | | | | Cd<br>C1kp5 | | | | | | | | | | Lke | |
| 7. Electronics | | | | | | | Cte2 | | | | | | | | | | | |
| 8. Food and Tobacco | Lk | | | Lp | | | C1k1<br>JV | JV | Cp1 | | | | | | | Lpt<br>C | | |
| 9. Machinery | | | | | | | Cpk3 | | | | | | | | | | | |
| 10. Machine Tools | JV | | | | | | C3 | | | | | | | | | | | |
| 11. Multisector | | | | | | | | | | | | | | | | Ck1 | | Cjp6 |
| 12. Non-ferrous metals | | | | | | | | | | | | | | | Cp | | | |
| 13. Packaging | | | | | | | JV | | | | | | | | | | | |
| 14. Petroleum and Gas | | | | | | Cr | | | | | | | | | | | | |
| 15. Pharmaceuticals | | | | | | | | | Crd | | | | | | JV | | Crd | C&L5 |
| 16. Power and Energy | | | | | | | Cp | | | | | | | | | | | |
| 17. Precision Instruments | | | | | | | | | | | | | Cjp35 | | | | | |
| 18. Railway | | | | | | Lp | | | | | | | | | | | | |
| 19. Rubber | | | | | | | | | | | | | | | | | | Crd |
| 20. Services | | | | | | | JV | | | | | | | | | | | JV |
| 21. Textiles Clothing and Leather | | | | | | | Cpk2<br>Cp5<br>JV | | C<br>LS | | | | | | | | Cp | Cp5 |

Table 11 (Chapter 7) contd
Hungary

Year : 1984

| | Aus. | Bel. | Can. | Den. | Fin. | Fra. | Ger. | It. | Jap. | Neth. | Nor. | Por. | Spa. | Swe. | Swi. | U.K. | U.S.A. |
|---|---|---|---|---|---|---|---|---|---|---|---|---|---|---|---|---|---|
| 1. Agriculture and Equipment | Cp2 | | | | | | C | | | JV<br>JV5 | | | | | | | |
| 2. Automotive | | | | | | | JV<br>Lp | | C | | | | | | | | |
| 3. Chemicals and Petrochemicals | C6<br>LS | | | | | PC | JV<br>Ck | Cd<br>C6 | C6<br>PC4<br>JV | | | | | Cp3 | | Cl | |
| 4. Construction | Lp | | | | | | Lke | | | | | | | | | | |
| 5. Consumer Goods | JVm4 | | | | Lp | | Cjpl | | | | | | | | | | JV |
| 6. Electrical Equipment | | | | | | | | | | | | | | | | | Cp1 |
| 7. Electronics | | | | | | JV | LS | | 2Cp | | | | | | | JV2<br>JV | |
| 8. Food and Tobacco | JV | | | | | | L&C<br>Cp3 | | | | | | | | Lp | Lp1 | Cdm |
| 9. Iron and Steel | JV | | | | | | | | | | | | | | | | |
| 10. Machinery | | | | | L&C | | Cd | | | | | | | | | | |
| 11. Machine Tools | | | | | | | Cjp | | | | | | | | | | |
| 12. Medical Eq. | | | | | | | | | | | | | | | | | Lp |
| 13. Mining and Metallurgy | | | | | | PC | | | | | | | | | | Cdp | |
| 14. Multisector | | | | | | | | Cm5 | | | | | | | | | |
| 15. Petroleum and Gas | | | | | | | | | | | | | | | C | Ctk | |
| 16. Pharmaceuticals | | | | | | | | | Cm | | | | | | | | |
| 17. Telecommunications | | | | | | | Cp | | | | | | | | | | |
| 18. Textiles, Clothing, and Leather | JV | | | | | Ltk2 | Cek<br>Cjp<br>Cke | Cp | | | | | | | | Cp | |
| 19. Wood, Pulp and Paper | | | | PC | | | | | | | | JV4 | | | | | |

Table 12 (Chapter 7)

Country: POLAND 1980-1984 (SUMMARY)   Agreements: 34

| Type & No. | Coop | Licence | JV | PC | TK | Other |
|---|---|---|---|---|---|---|
| | 17 | 9 | 3 | 5 | - | - |
| Percentage: | 51.5 | 27.2 | 6.0 | 15.1 | - | - |

| Sector | No. | % |
|---|---|---|
| Chemicals | 7 | 20.6 |
| Mining & Metallurgy | 4 | 11.7 |
| Construction | 3 | 8.8 |
| Power & Energy | 3 | 8.8 |
| Automotive | 2 | 5.8 |
| Machine Tools | 2 | 5.8 |
| **Western Partner** | | |
| Germany | 11 | 32.3 |
| US | 5 | 14.7 |
| Italy | 4 | 11.7 |
| Japan | 4 | 11.7 |
| UK | 4 | 11.7 |

Year : 1980

| | Aus. | Bel. | Can. | Den. | Fin. | Fra. | Ger. | It. | Jap. | Neth. | Nor. | Por. | Spa. | Swe. | Swi. | U.K. | U.S.A. |
|---|---|---|---|---|---|---|---|---|---|---|---|---|---|---|---|---|---|
| 1. Automotive | | | | | C | | | Cjp | | | | | | | | | |
| 2. Chemicals and Petrochemicals | | | | | | Lt | | | | | | | | | | Crd | |
| 3. Construction | | | | | | | | | PC | | | | | | | C | |
| 4. Electrical Eq. | | | | | | | Lp | | | | | | | | | | |
| 5. Food & Tobacco | | | | | | | | C | | | | | | | | | |
| 6. Machine Tools | | | | | | | | | Lp | | | | | | | | |
| 7. Mat. Handling & Transport Eq. | | | | | | | | | | | | | | | | Crpm | |
| 8. Mining & Metallurgy | | | | | | | | Crd | | | | | | | | | Cp4<br>Lp |
| 9. Multisector | | | | | | | C | | | | | | | | | | |
| 10. Petroleum and Gas | | | | | | | PC | | | | | | | | | | |
| 11. Power & Energy | PC | | | | | | | | | | | | | | | | |
| 12. Telecommunications | | | | | | | Cd5 | | | | | | | | | | |

Table 12 (Chapter 7) contd
Poland

Year : 1981

| | Aus. | Bel. | Can. | Den. | Fin. | Fra. | Ger. | It. | Jap. | Neth. | Nor. | Por. | Spa. | Swe. | Swi. | U.K. | U.S.A. |
|---|---|---|---|---|---|---|---|---|---|---|---|---|---|---|---|---|---|
| 1. Chemicals and Petrochemicals | | | | | | | | | | Lp | | | | | | | C |
| 2. Construction | | | | | | | | JV | | | | | | | | | JVm5 |
| 3. Electronics | | | | | | | Lp | | | | | | | | | | |
| 4. Mining & Metallurgy | | | | | | | Lp | | | | | | | | | | |
| 5. Power & Energy | | | | | | | Cr | | | | | | | | | | |

Year : 1982

| | Aus. | Bel. | Can. | Den. | Fin. | Fra. | Ger. | It. | Jap. | Neth. | Nor. | Por. | Spa. | Swe. | Swi. | U.K. | U.S.A. |
|---|---|---|---|---|---|---|---|---|---|---|---|---|---|---|---|---|---|
| 1. Chemicals and Petrochemicals | | | | | | | Cpke | | Lp | | | | | | | PC | |
| 2. Packaging | | JV | | | | | | | | | | | | | | | |
| 3. Textiles, Clothing and Leather | | | | | | | Cp2 | | | | | | | | | | |

Year : 1983

| | Aus. | Bel. | Can. | Den. | Fin. | Fra. | Ger. | It. | Jap. | Neth. | Nor. | Por. | Spa. | Swe. | Swi. | U.K. | U.S.A. |
|---|---|---|---|---|---|---|---|---|---|---|---|---|---|---|---|---|---|
| 1. Machine Tool | | | | | | | Cp | | | | | | | | | | |
| 2. Shipping & Shipbuilding | | | | | | | | | | | | | | | Lp | | |

Year : 1984

| | Aus. | Bel. | Can. | Den. | Fin. | Fra. | Ger. | It. | Jap. | Neth. | Nor. | Por. | Spa. | Swe. | Swi. | U.K. | U.S.A. |
|---|---|---|---|---|---|---|---|---|---|---|---|---|---|---|---|---|---|
| 1. Consumer Goods | | | | | | | | | | | | | | | | | Cp |
| 2. Metalworking | | | | | | | Cp | | | | | | | | | | |
| 3. Power & Energy | | | | | | | | | Pc | | | | | | | | |

Table 13 (Chapter 7)

Country: ROMANIA 1980-1984 (SUMMARY) Agreements: 18

| Type & No. | Coop | Licence | JV | PC | TK | Other |
|---|---|---|---|---|---|---|
| | 16 | 2 | - | - | - | - |
| Percentage: | 88.8 | 11.1 | - | - | - | - |

| Sector | No. | % |
|---|---|---|
| Chemicals | 4 | 22.2 |
| Multisector | 3 | 16.6 |
| Mining & Metallurgy | 2 | 11 |
| Iron & Steel | 2 | 11 |
| **Western Partner** | | |
| Italy | 8 | 44.4 |
| US | 5 | 27.7 |

Year : 1980

| | Aus. | Bel. | Can. | Den. | Fin. | Fra. | Ger. | It. | Jap. | Neth. | Nor. | Por. | Spa. | Swe. | Swi. | U.K. | U.S.A. |
|---|---|---|---|---|---|---|---|---|---|---|---|---|---|---|---|---|---|
| 1. Agriculture and Equipment | | | | | | | | | | | | | | | | | Ct |
| 2. Automotive | | | | | | Lp | | | | | | | | | | | |
| 3. Chemicals and Petrochemicals | | | | | | | Cp | | Lt | | | | | | | | Cm |
| 4. Mining and Metallurgy | Cjp | | | | | | | | | | | | | | | | Cd |
| 5. Multisector | | | | | | | | | | | | | | | | | Cm |
| 6. Petroleum & Gas | | | | | | | | | | | | | | | | | C6 |

Year : 1981

| | Aus. | Bel. | Can. | Den. | Fin. | Fra. | Ger. | It. | Jap. | Neth. | Nor. | Por. | Spa. | Swe. | Swi. | U.K. | U.S.A. |
|---|---|---|---|---|---|---|---|---|---|---|---|---|---|---|---|---|---|
| 1. Agriculture and Equipment | | | | | | | | Cpc | | | | | | | | | |
| 2. Multisector | | | | | | | | C<br>Cm | | | | | | | | | |
| 3. Power and Energy | | | | | | | | Cpc | | | | | | | | | |
| 4. Railway | | | | | | | | | | | | | | Cp | | | |

Table 13 (Chapter 7) contd
Romania

Year : 1982

| | Aus. | Bel. | Can. | Den. | Fin. | Fra. | Ger. | It. | Jap. | Neth. | Nor. | Por. | Spa. | Swe. | Swi. | U.K. | U.S.A. |
|---|---|---|---|---|---|---|---|---|---|---|---|---|---|---|---|---|---|
| 1. Chemicals and Petrochemicals | | | | | | | | Cd | | | | | | | | | |
| 2. Electrical Eq. | | | | | | | | Cp | | | | | | | | | |
| 3. Iron and Steel | | | | | | | | Cd | | | | | | | | | |

Year : 1983

| | Aus. | Bel. | Can. | Den. | Fin. | Fra. | Ger. | It. | Jap. | Neth. | Nor. | Por. | Spa. | Swe. | Swi. | U.K. | U.S.A. |
|---|---|---|---|---|---|---|---|---|---|---|---|---|---|---|---|---|---|
| 1. Iron and Steel | | | | | | | | Ct | | | | | | | | | |

**Year : 1984**

**No agreements were traced in 1984.**

Table 14 (Chapter 7)

Country: USSR 1980-1984 (SUMMARY)    Agreements: 254

| Type & No. | Coop<br>205 | Licence<br>30 | JV<br>- | PC<br>9 | TK<br>8 | Other<br>2 |
|---|---|---|---|---|---|---|
| Percentage: | 81.0 | 11.4 | - | 3.5 | 3.16 | 0.7 |
| Sector | | No. | | % | | |
| Chemicals | | 53 | | 20.9 | | |
| Multisector | | 22 | | 8.6 | | |
| Agriculture | | 18 | | 7.1 | | |
| Construction | | 15 | | 5.9 | | |
| Mining & Metallurgy | | 13 | | 5.1 | | |
| Textiles | | 10 | | 3.9 | | |
| Power & Energy | | 10 | | 3.9 | | |
| Rubber | | 9 | | 3.5 | | |
| Iron and Steel | | 9 | | 3.5 | | |
| Automotive | | 8 | | 3.1 | | |
| Petroleum & Gas | | 7 | | 2.7 | | |
| Western Partner | | | | | | |
| Italy | | 67 | | 26.4 | | |
| Germany | | 38 | | 15 | | |
| France | | 30 | | 11.8 | | |
| Finland | | 30 | | 11.8 | | |
| UK | | 16 | | 6.3 | | |
| Japan | | 15 | | 5.9 | | |
| Austria | | 12 | | 4.7 | | |
| US | | 11 | | 4.3 | | |
| Sweden | | 10 | | 3.9 | | |
| Switzerland | | 10 | | 3.9 | | |

Table 14 (Chapter 7) contd
USSR

Year : 1980

| | Aus. | Bel. | Can. | Den. | Fin. | Fra. | Ger. | It. | Jap. | Neth. | Nor. | Por. | Spa. | Swe. | Swi. | U.K. | U.S.A. |
|---|---|---|---|---|---|---|---|---|---|---|---|---|---|---|---|---|---|
| 1. Automotive | | | | | | | Cd | Cd | | | | | | | | | |
| 2. Aviation | | | | | | Cdpm | | | | | | | | | | | |
| 3. Chemicals and Petrochemicals | | | | | C6 | Cl<br>Lt<br>Cd | | Cr<br>Cpcl | | Crd | | | | | | | |
| 4. Construction | | | | | | | | Cd<br>2Cdp<br>2C | Cd | | | | | | | | |
| 5. Electrical Eq. | Cd | | | | | | | | | | | | | | | | |
| 6. Glass and Ceramics | | | | | | | | Cd<br>PC | | | | | | | | | |
| 7. Iron and Steel | | | | | | | Crd<br>Ctd | Ct | | | | | | | | | |
| 8. Medical Eq. | | | | | | Lp | | | | | | | | | | | |
| 9. Mining and Metallurgy | Lt | Cd | | | C6 | | Cd<br>Cdm | | | | | | | | Crd | | |
| 10. Multisector | | | | | | Cl<br>Cp | | C | Cl | | | | | | | Crd | |
| 11. Petroleum and Gas | | | Ctr | | | | Cd | Cr | | Ct | | | | | | | |
| 12. Pharmaceuticals | | | | | | | | | | Cp | | | | | Cd | | |
| 13. Power and Energy | | | | | | | | Cd | | | | | | | | | |
| 14. Precision Instruments | | | | | | | Lp | | | | | | | | | | |
| 15. Printing | | | | | | | Lp | | | | | | | | | | |
| 16. Railway | | | | | | | Lp | | | | | | | | | | |
| 17. Shipping and Shipbuilding | | | | | | | | | | | | | | | | | LS |
| 18. Textiles Clothing and Leather | | | | | | | Lp | Cp2<br>Lpt | | | | | | | | | |
| 19. Wood, Pulp and Paper | | | | | Cdp | | | | | | | | | | | | |

Table 14 (Chapter 7) contd
USSR

Year : 1981

| | Aus. | Bel. | Can. | Den. | Fin. | Fra. | Ger. | It. | Jap. | Neth. | Nor. | Lux. | Spa. | Swe. | Swi. | U.K. | U.S.A. |
|---|---|---|---|---|---|---|---|---|---|---|---|---|---|---|---|---|---|
| 1. Agriculture and Equipment | | | | | | | | | | | | | | Cpd | | | Ckt |
| 2. Automotive | | | | | | | | | | | | | | | | Lk | |
| 3. Chemicals and Petrochemicals | | | | | | | Cd<br>PC | Ctr<br>Cdk<br>Cjp<br>C | C<br>Crd | | | | | | Cdp | | Cd |
| 4. Construction | | | | | | | Ce | Cr<br>Cd | | | | | | | | | |
| 5. Consumer Goods | | | | | Lk | | | Cp | | | | | | | | | |
| 6. Electronics | | | | | | Cdm5<br>Cd | | | | | | | | | | | |
| 7. Environment | | | | | | | Ld | | | | | | | | | | |
| 8. Food and Tobacco | Cd1 | Lpc<br>Lp | | | | | | | | | | | | | | | |
| 9. Iron and Steel | | | | | | | | | | | | Cd | | | | | |
| 10. Machine Tools | | | | | | | Lp | | | | | | | Ce | | | |
| 11. Mat.Handling & Transport Eq. | | | | | | | | | | | | | | | LS | | |
| 12. Metalworking | | | | | | Cdp | | | | | | | | | | | |
| 13. Mining and Metallurgy | | | | | | | | | | | | | | | | Cjp<br>Cd | |
| 14. Multisector | | | | | | | C123<br>Cdp | Cm5 | 2C | | | | | | | | C |
| 15. Packaging | | | | | | | | | | | Cd | | | | | | |
| 16. Petroleum and Gas | | | | | | | | | Cd | | | | | | | | |
| 17. Power and Energy | Cjp | | | | Cdp | | | C6<br>Cd | | | | | | | | | |
| 18. Railway | Cd | | | | Cd | | | | | | | | | | | | |
| 19. Telecommunications | | | | | Cjp2 | | | | | | | | | | | | |
| 20. Textiles Clothing and Leather | | | | | | Lk | | | | | | | | | | | |
| 21. Wood, Pulp and Paper | | | | | | | | | Cd | | | | | | | | |

Table 14 (Chapter 7) contd
USSR

Year : 1982

| | Aus. | Bel. | Can. | Den. | Fin. | Fra. | Ger. | It. | Jap. | Neth. | Nor. | Por. | Spa. | Swe. | Swi. | U.K. | U.S.A. |
|---|---|---|---|---|---|---|---|---|---|---|---|---|---|---|---|---|---|
| 1. Agriculture and Equipment | | | | | | Cd | | Lp | | | | | | Cd | | | |
| 2. Automotive | | | | | | | Cd | | | | | | | | | | |
| 3. Chemicals and Petrochemicals | | | | | | | | Cdp | | | | | | C3 | | | Cl |
| 4. Construction | | | | | | | Cjp2 | | | | | | | | | | |
| 5. Consumer Goods | | | | | | | Lp | | | | | | | | | | |
| 6. Electrical Equipment | | | | | | | | Cd | | | | | | | | | |
| 7. Electronics | Cd | | | | | | Lp | | | | | | | | | | |
| 8. Environment | | | | | | | | | Cdr | | | | | | | | |
| 9. Food and Tobacco | | | | | | | | Crdk | | | | | | | | | C |
| 10. Iron and Steel | Cpc | | | | PC | | | | | | | | | | | | |
| 11. Machinery | | | | | | | | | | | | | | | | | Cd |
| 12. Machine Tools | | | | | | | Cjp | | | | | | | | | | |
| 13. Metalworking | | | | | | | | | | | | | | | | Lp | |
| 14. Mining and Metallurgy | | | | | | Cd | | | TK | | | | | | | | |
| 15. Multisector | | | | | Cjp | Cjp | | Cd<br>C | Cd | | | | | | | | |
| 16. Pharmaceuticals | | | | | | | | | | | | | | | Ct | | |
| 17. Precision Instruments | | | | | | | Cpm5 | | | | | | | | | | |
| 18. Rubber | | | | | | | | Cd | | | | | | | | | |
| 19. Shipping and Shipbuilding | | | | | Cd | | | | | | | | | | | | |
| 20. Textiles Clothing and Leather | | | | | | | | Cd | | | | | | Lp2 | | | |
| 21. Wood, Pulp and Paper | | | | | Cjp | | | | | | | | | | | | |

Table 14 (Chapter 7) contd
USSR

Year : 1983

| | Aus. | Bel. | Can. | Den. | Fin. | Fra. | Ger. | It. | Jap. | Neth. | Nor. | Por. | Spa. | Swe. | Swi. | U.K. | U.S.A. |
|---|---|---|---|---|---|---|---|---|---|---|---|---|---|---|---|---|---|
| 1. Agriculture and and Equipment | | | | | | Cdt | | 2TK | | | | | | | TK | Cte | C |
| 2. Automotive | | | | | | | | | | | | | | | | | |
| 3. Chemicals and Petrochemicals | Ckl | PC | | Lpc | Cjpr | Crd<br>Cd<br>Cdp | Lp<br>Cjpd<br>Cp | 2Cjp<br>Cjp5<br>2Cd<br>Crd<br>Cd<br>Cd | PC | | | | | Crd | | | |
| 4. Construction | | | | | | | Cpm2 | | | | | | | | | | |
| 5. Consumer Goods | | | | | | | Crdk | | | | | | | | | | |
| 6. Food and Tobacco | | | | | TK | | | | | | | | | | | | |
| 7. Glass and Ceramics | Crd | | | | | | | | | | | | | | | | |
| 8. Iron and Steel | | | | | | | Ck | | Ct | | | | | TK | | | |
| 9. Machinery | | | | | Cpl | | | | | | | | | | | | |
| 10. Machine Tools | | | | | | | Cd | | | | | | | | | | |
| 11. Mat. Handling & Transport Eq. | | | | | Cjp | | | | | | | | | | | | |
| 12. Mining and Metallurgy | | | | | | | Cd2<br>Cd<br>Cdp | | | | | | | | | | |
| 13. Multisector | | | | | | | | | | | | | | | | | |
| 14. Petroleum and Gas | | | | | | | | | | | Crd | | | | | C | C |
| 15. Pharmaceuticals | | | | | | C | | | | | | | | | | | |
| 16. Power and Energy | | | | | C | | Cd | | | | | | | | Crd | | |
| 17. Railway | | | | | Cp | | | | | | | | | | | | |
| 18. Rubber | | | | | | | C | Cdjp | | | | | | | | | |
| 19. Shipping and Shipbuilding | | | | | | Lp | | | | | | | | | | | |
| 20. Textiles Clothing and Leather | | | | | TK | | | 2Lp | | | | | | | | | |
| 21. Wood, Pulp and Paper | | | | | | | | | | | | | | Ce2 | | | |

Table 14 (Chapter 7) contd
USSR

Year : 1984

| | Aus. | Bel. | Can. | Den. | Fin. | Fra. | Ger. | It. | Jap. | Neth. | Nor. | Por. | Spa. | Swe. | Swi. | U.K. | U.S.A. |
|---|---|---|---|---|---|---|---|---|---|---|---|---|---|---|---|---|---|
| 1. Agriculture and Equipment | | | | | Cp<br>Cd | Cp<br>2Cd<br>Cdp | | Cjp | | | | | | | | | Lk |
| 2. Automotive | | | | | | | | 3C | | | | | | | | | |
| 3. Chemicals and Petrochemicals | CD<br>C | C | | | | PC<br>C6 | Cd | Cdp<br>Ce<br>3PC<br>TK<br>Cp2 | | | | | | | Lp | 2Cd<br>Ct | |
| 4. Construction | | | | | | | Cjpm5<br>Cjp | | | | | | | | | | |
| 5. Consumer Goods | Lp | | | | | | | | | | | | | | | | |
| 6. Electrical Eq. | | Cpc | | | | | | | | | | | | | | | Cp2 |
| 7. Electronics | | | | | Crd | | | C | | | | | | | | Cd<br>Ck | |
| 8. Food and Tobacco | | | | | Cd<br>Lpm4 | | | | | | | | | | C | | |
| 9. Iron and Steel | | | | | | Cte | | | | | | | | | | | |
| 10. Machine Tools | | | | | C | | | | | | | | | | | | |
| 11. Mat. Handling & Transport Eq. | | | | | | Cp | | Cd | | | | | | | | | |
| 12. Medical Eq. | | | | | Cd | | | | | | | | | | | Cp | |
| 13. Mining and Metallurgy | | | | | Cjp | | | Cpc6 | C | | | | | | | | |
| 14. Multisector | | | | | | | | C | C | | | | | | C&TK | | |
| 15. Non-ferrous metals | | | | | | | | | | | | | | | | Cl<br>C6 | |
| 16. Petroleum and Gas | | | | | | | | | | | | | | | | | |
| 17. Power and Energy | | | | | Cjp<br>Cjpm2 | Cpc6 | | | | | | | | | | | |
| 18. Precision Instruments | | | | | | Cdm | | | | | | | | Crd | | | |
| 19. Railway | | | | | Cd | | | | | | | | | Cd | | | |
| 20. Rubber | | | | | Cd | | | 3Cd<br>Cd5<br>Cp | | | | | | | | | |
| 21. Shipping and Shipbuilding | | | | | | | | | C | | | | | | | | |
| 22. Textiles, Clothing and Leather | | | | | | | | | | | | | | | | Lk | |

## NOTES AND REFERENCES

1. *Practical Measures to Remove Obstacles to Intra-Regional Trade and to Promote and Diversify Trade: Register of Long-Term Agreements on Economic Co-operation and Trade* (TRADE/R.485), Note by the Secretariat, Committee on the Development of Trade, Economic Commission for Europe, Geneva, 3rd October, 1984.
2. *Promotion of Trade Through Industrial Co-operation: Recent Developments In and Prospects for East-West Industrial Co-operation* (TRADE/R.488), Note by the Secretariat, Committee on the Development of Trade, Economic Commission for Europe, Geneva, 21st September, 1984, paragraphs 16-22.
3. M. Maximova (Head of Department, Institute for World Economy and International Relations, USSR Academy of Sciences), "Industrial Co-operation between Socialist and Capitalist Countries: Forms, Trends and Problems", as given in C.T. Saunders (editor), *East-West Co-operation in Business: Inter-firm Studies*, Wiener Institut für Internationale Wirtschaftsvergleiche, Vol. 2, Springer Verlag, Vienna-New York, 1977, pp. 21-22. It should be noted that the viewpoint cited continues to be reiterated in the Eastern literature.
4. Definitional and conceptual difficulties are discussed in detail by Carl H. McMillan, "East-West Industrial Co-operation", in *East-European Economies Post Helsinki-A Compendium of Papers*, Joint Economic Committee, Congress of the United States, Washington, D.C., August 25, 1977. See also F. Levcik and J. Stankovsky, *Industrielle Kooperation zwischen Ost und West*, Springer Verlag, Vienna-New York, 1977.
5. See Carl H. McMillan, "Trends in East-West Industrial Co-operation", *Journal of International Business Studies*, Fall, 1981.
6. *Legal Forms of Industrial Co-operation Practised by Countries having Different Economic and Social Systems with Particular Reference to Joint Ventures* (TRADE/AC.3/R.10), Note by the Secretariat, Committee on the Development of Trade, Economic Commission for Europe, Geneva, 15th October, 1976, p. 5. See also Annex Table 1 of Chapter 7.
7. The concept of effectiveness is discussed, *inter alia*, by: Eric W. Hayden and Henry R. Nau, "East-West Technology Transfer: Theoretical Models and Practical Experiences", *Columbia Journal of World Business*, Fall, 1975; *An Analysis of Export Control of US Technology-A DOD Perspective*, A report of the Defense Science Board Task Force on Export of US Technology, Washington, D.C., February 4, 1975 and, Gary Bertsch, Richard Cupitt, John R. McIntyre, Miriam Steiner, "East-West Technology Transfer and Export Controls", *Osteuropa Wirtschaft*, February 1981.
8. *A Statistical Outline of Recent Trends in Industrial Co-operation* (TRADE/AC.3/R.8), Note by the Secretariat, Committee on the Development of Trade, Economic Commission for Europe, Geneva, 3rd August, 1976.
9. *Ways and Means of Expanding Trade and Economic Relations between Countries having Different Economic and Social systems* (TD/B/AC.38/2), Study by the UNCTAD Secretariat, Trade and Development Board, Geneva, 12th April, 1984, p. 26.
10. *Trade Relations Among Countries having different Economic and Social Systems and all Trade Flows Resulting Therefrom, Review of Trends and Policies in Trade between Countries having Different Economic and Social Systems* (TD/B/859), Report by the UNCTAD Secretariat, Trade and Development Board, Geneva, 7th August, 1981, p. 29. A later estimate has not been given by the UN.

11. *East-West Industrial Co-operation* (ECE/TRADE/132), United Nations, Economic Commission for Europe, New York, 1979, pp. 3-4.

12. It is the opinion of some Swedish experts that real specialisation seldom occurs and "that the Western firm usually continues to manufacture the co-operation product itself". As the author points out, there are exceptions and the successful cases tend, over time, to develop into genuine specialisation between the partners. B. Högberg, "East-West Industrial Co-operation", *Eastern Business Magazine*, Vol. 3, No. 5, 1978, p. 2. Based on the UN survey, real specialisation is seemingly still rare: accounting for only 4 per cent. See Table 4 (column C.2) of this chapter.

13. According to the survey, 44 per cent of the sample cases involved the supply of complete plants, whereas co-production and specialisation accounted for 26 per cent. See B. Högberg, *art.cit.*, p. 3 and B. Högberg, A. Adahl, *Industrial Co-operation with the Soviet Union and Eastern Europe, the Experiences of Swedish Industry* (a summary of the main report), Styrelsen för Teknisk Utveckling, Stockholm, 1979. It is noted by the analysts that the types of agreements classified as industrial co-operation (hence included in the survey) correspond fairly well to the definition used by the ECE.

14. Malcolm R. Hill, *East-West Trade, Industrial Co-operation and Technology Transfer, the British Experience*, Gower Publishing Company Ltd., 1983, p. 167.

15. N. Leise, *Transfermechanismen des technischen Fortschritts in und zwischen verschiedenen Wirtschaftssystemen*, Forschungsbericht Nr. 4 (AZ:12 2686, n.d.), Institut für Aussenhandel und Überseewirtschaft der Universität Hamburg. See also M. R. Hill, *op. cit.*

16. *East-West Industrial Co-operation*, (ECE/TRADE/132), *op. cit.*, pp. 5 and 6.

17. See *The Transfer of Disembodied Technology with Special Regard to Licencing* (SC.TECH./SEM.8/R.3/Rev.1, TRADE/SEM.5/R.3./Rev. 1), Seminar on Trends in Different Forms of Technology Transfer and their Impact on the Development of Trade and Economic Co-operation, Note by the Secretariat, Senior Advisers to ECE Governments on Science and Technology, Economic Commission for Europe, Geneva, 2nd April, 1982, p. 5.

18. It might be noted that even the Western statistics are skewed since the item under which such payments appear ("incomes from and expenditures on invisibles") also includes fees for consulting and engineering services, sales and purchases of films, copyright payments and so forth. See *Ibid.*

19. See *Licencing and Leasing*, Committee on the Development of Trade, Economic Commission for Europe, Geneva, 1976, p. 15. Though falling outside the immediate framework of this discussion, brief mention must be made of patents *per se*, considered to be one of the most important channels for the transfer of disembodied technology. One exhaustive study of this topic concluded that patents are the first step in the international technology transfer process in that a patent sale is invariably followed by licence sales, sales of goods and by direct investment: the latter, only modestly developed in the East-West context. The study also found that a correlation existed between patent application and weakness in the CMEA area's R&D structure. Thus, Western patent applications in Eastern Europe were particularly pronounced in the chemicals sector (organic and synthetic) which is insufficiently developed. The "pilot function" of patenting was found to be particularly noticeable in the agricultural and foodstuffs sector (Bulgaria) and in the mining industry (Soviet Union and Poland). See Jiri Slama, *Technologietransfer durch den Lizenz-und Patentverkehr in den Ost-West Wirtschaftsbeziehungen unter besonderer Berücksichtigung der Bundesrepublik Deutschland*, prepared for the Bundesminister für Wirtschaft, Osteuropa Institut, Munich, 1981.

20. Malcolm R. Hill, *op. cit.*, p. 165.

21. Since licencing agreements generally provide for the marketing of products manufactured under the licence, the agreements include arrangements for a strict division of markets. The basic arrangement is that the East European partner is granted unlimited but not exclusive sales rights in CMEA markets, while the Western partner retains all other markets for itself. See, for instance: *Marketing Aspects of East-West Industrial Co-operation Contracts*

(TRADE/SEM.4/R.1), Note by the Secretariat, Committee on the Development of Trade, Economic Commission for Europe, Geneva, 16th February, 1979, and *Marketing of Industrial Products (Machinery, Instruments and Chemicals), in East-West Trade* (TRADE/INF.5), Note by the Secretariat, Committee on the Development of Trade, Economic Commission for Europe, Geneva, 8th December, 1980.

22. Stanislav Simanovsky, "Conditions and Prospects for Licence Exchanges of the CMEA Countries", *Foreign Trade*, (English edition), No. 2, Moscow, 1980, p. 18. As an example, the author cites the licencing package concluded with FIAT, Italy which "serves as a technological basis for car manufacture at the Volga Motor Works" On the basis of the technology provided by FIAT, the general efficiency of the automotive industry is said to have been raised, making the VAZ car models competitive on world markets.

23. *Ibid.*, p. 19. Examples mentioned by the author are the Ikarus buses, developed on the basis of Western licences and which enabled Hungary to organise large-scale production of these buses (subsequently exported to CMEA and third countries), and Polish specialisation in shipbuilding.

24. *East-West Industrial Co-operation* (ECE/TRADE/132), *op. cit.*, p. 13.

25. C. H. McMillan and D. P. St. Charles, *Joint Ventures in Eastern Europe: A Three-Country Comparison*, The Canadian Economic Policy Committee and the C.D. Howe Research Institute, 1974, p. 1.

26. *Ibid.*, p. 2.

27. See *Promotion of Trade Through Industrial Co-operation: Recent Developments In and Prospects For East-West Industrial Co-operation* (TRADE/R.488), *op.cit.*, paragraphs 49 and 52.

28. See *Co-operation in Third Markets of East-West Joint Ventures Domiciled in Eastern Europe*, (TRADE/R.469), Note by the Secretariat, Committee on the Development of Trade, Economic Commission for Europe, Geneva, 18th October, 1983.

29. *East-West Industrial Co-operation* (ECE/TRADE/132), *op. cit.*, p. 15.

30. See Patrick Gutman, "Tripartite Industrial Co-operation and East Europe", in *East European Economic Assessment: Part 2-Regional Assessments, A Compendium of Papers, op.cit.*, pp. 833 and 849.

31. See *Promotion of Trade Through Industrial Co-operation: Recent Developments In and Prospects For East-West Industrial Co-operation* (TRADE/R.488), *op.cit.*, paragraphs 68-79.

32. For an assessment of the significance of tripartite co-operation see, for instance, Gérard Ballot and Patrick Gutman, "Political Economy of East-West-South Industrial Co-operation", in *East European Economies: Slow Growth in the 1980s*, Volume 2, Joint Economic Committee, Congress of the United States, Washington, D.C., forthcoming, and Michel Gomez, "La coopération industrielle tripartite: Mythe ou réalité?", in *Le Courrier des pays de l'Est*, No. 284, May 1984.

33. It is interesting to note that the Swedish survey found that almost half of the co-operation agreements analysed concerned the machine-building sector, and almost four-fifths the engineering sector as a whole. See B. Högberg, Andreas Adahl, *Industrial Co-operation with the Soviet Union and Eastern Europe, op. cit.*, p. 4.

34. See *inter alia*, L. Rodina, "Economic Integration of the CMEA Countries and East-West Business Co-operation", *Obshchestvennye Nauki*, No. 6, November-December, 1983, as given in *Joint Publications Research Service* (UIE-84-003), *USSR Report*, 24th January 1984, p. 39, and E. Tabaczynski, *Industrial Co-operation and Transfer of Technology: the Polish Experience* (SC.TECH./SEM.8/R.13/Rev.1; TRADE/SEM.5/R.13/Rev.1), Seminar on Trends in Different Forms of Technology Transfer and their Impact on the Development of Trade and Economic Co-operation, Senior Advisers to ECE Governments on Science and Technology, Economic Commission for Europe, Geneva, 22nd March, 1982, p. 3.

35. See *Ibid.* and G. N. Prokhorov "The Role of Foreign Economic Relations in Providing for the Economic Interests of the Socialist Countries", *Izvestiva Akademii Nauk SSSR, Seriya Ekonomicheskaya*, No. 1, January-February, 1983, as given *Joint Publications Research Service* (83815), *USSR Report*, 1st July, 1983, p. 39.

36. Malcolm R. Hill, *op.cit.*, p. 98.

37. L. Rodina, *art. cit.*, and O. Rybakov, "The Problems of the Co-ordination of the National Economic Plans of the CMEA Member Countries", *art. cit.*

38. See *inter alia*:
    - Malcolm R. Hill, *op. cit.*;
    - Paul Marer, E. Tabacznski (editors), *Polish-US Industrial Co-operation in the 1980s, Findings of a Joint Research Project*, Indiana University Press, Bloomington, 1981;
    - Klaus Bolz, Peter Pissulla, *Die Erfahrungen deutscher Unternehmen aus der Kooperation mit polnischen Wirtschaftsorganisationen,* Weltarchiv GmbH, Hamburg, 1981;
    - P. Gutman, "East-West Industrial Co-operation in the Automative Industry and the International Division of Labour", *The ACES Bulletin*, Vol. XXII, No. 3-4, Tempe Arizona, Fall-Winter, 1980;
    - B. Högbert, Andreas Adahl, *op. cit.*;
    - K. E. Schenk and A. Wass von Czege, *Technologietransfer und Strukturwandel durch Industrielle Ost-West Kooperation,* Forschungsbericht No. 18, Institut für Aussenhandel und Überseewirtschaft der Universität Hamburg, Hamburg 1979;
    - A. Wass von Czege, *Erfahrungen Deutscher Firmen beim Technologietransfer im Rahmen von Ost-West-Kooperation-Ergebnisse einer Befragung,* Institut für Aussenhandel und Ubersee-wirtschaft der Universität Hamburg, Hamburg, 1977 and,
    - E.W. Hayden, *Technology Transfer to East Europe, US Corporate Experience*, Praeger Publishers, New York, 1976.

39. V. Kartsev and Alexandr Sokol, "Bulgaria's Foreign Trade: Main Aspects of a New and Economic Mechanism", *Foreign Trade* (English edition), No. 8, Moscow, 1983, pp. 26-27.

40. I. Savyolova, "CMEA Member States and Western Countries' Co-operation in Manufacturing Parts and Components", *Foreign Trade* (English edition), No. 7, Moscow, 1980, p. 32.

41. *Bolgarskaya Vunshna Torgoviya*, No. 4, 1982, pp. 3-6, as given in *Presseschau Ostwirtschaft, No. 8/9*, Österreichisches Ost-und Südosteuropa Institut, Vienna, 1982, p. 62.

42. See *Co-operation in Third Markets of East-West Joint Ventures Domiciled in Eastern Europe* (TRADE/R.469), *op. cit.*, paragraphs 10-37 and *Promotion of Trade Through Industrial co-operation, East-West Joint Ventures Domiciled in Eastern Europe and Yugoslavia: Current Status of Legislation and Enterprise Response* (TRADE/R.489), Note by the Secretariat, Committee on the Development of Trade, Economic Commission for Europe, Geneva, 9th October 1984, paragraphs 16 and 24.

43. See *Trade Relations Among Countries Having Different Economic and Social Systems and All Trade Flows Resulting Therefrom: Review of Trends and Policies in Trade between Countries Having Different Economic and Social Systems* (TD/B/965), Report by the UNCTAD Secretariat, Trade and Development Board, Geneva, 1st September 1983, paragraph 108; *East-West Fortnightly Bulletin*, No. 326, 3rd November, 1983; and, *Vunshna Torgoviya* No. 6, 1981, pp. 3-5, as given in *Presseschau Ostwirtschaft, Österreichisches Ost-und Südosteuropa Institut*, No. 1, Vienna, 1982, p. 44.

    The analysis of joint investments by McMillan, indicates that at the end of 1981, Bulgaria had invested in 44 Western firms (Finland was not included). Investments were primarily for marketing of products from the CMEA countries. See, C.H. McMillan, "Soviet and East European Participation in Business Firms and Banks Established in the West: Policy Issues", in *External Economic Relations of CMEA Countries: Their Significance and Impact in a Global Perspective*, Colloquium, NATO, Economics Directorate, Brussels, 6th-8th April, 1983, Tables 1 and 2.

44. *East-West Fortnightly Bulletin*, No. 326, 3rd November, 1983, p. 5.

45. F.L. Altmann, Jiri Slama, *Die Aussenwirtschaftsbeziehungen der CSSR under dem Gesichtspunkt der Ost-West-Zusammenarbeit*, Stiftung Wissenschaft und Politik, Ebenhausen, July, 1980, pp. 26-27. The authors cite official sources. Levcik and Skolka (also quoting official data) give a figure of 22 co-operation agreements by the end of 1976. See, F. Levcik and J. Skolka, *East-West Technology Transfer, Study of Czechoslovakia*, OECD, Paris 1984, p. 61.

46. L. Strougal, introducing the Government Report on "The State of Work and Main Problems in Preparing the 7th Five-Year Plan for 1981-1985" to the Central Committee in October, 1980. Cited in F. Levcik and J. Skolka, *op.cit.*, p. 76 (note 66).

47. The relevant Decrees are: No. 42/1980, concerning economic relations with foreign countries; Decree No. 284/1980, on the "More Effective Utilisation of Co-operation with Non-Socialist Countries and Yugoslavia" and the Decrees on Hard Currency Incentives, Nos. 4/1982 and 26/1982.

48. See F. Levcik and J. Skolka, *op. cit.*, pp. 60-62 for a more detailed discussion.

49. *Svet Hospodarstvi*, No. 16, 5th February 1985, p. 1, as given *Presseschau Ostwirtschaft*, No. 2, Österreichisches Ost-und Südosteuropa-Institut, 1985, p. 13.

50. *Review of Recent and Prospective Trends, Policies and Problems in Intra-Regional Trade* (TRADE/R.479), Note by the Secretariat, Committee on the Development Trade, Economic Commission for Europe, Geneva, 17th October 1984, paragraph 41.

51. F. Levcik and J. Skolka, *op. cit.*, Table 30. The source of the Table is the Czechoslovak *Review of Trade, Industry and Economy*, No. 2, 1983, p. 35.

52. *Svet Hospodarstvi*, No. 16, *op.cit.*, p. 13.

53. J. Terek and V. Kroc, "Development of Co-operation in Production with Companies of Non-Socialist Countries", *Zahranicni Obchod*, No. 9, 1982, as given in *Joint Publications Research Service* (82532), *East Europe Report*, 27th December 1982, p. 3.

54. *Svet Hospodarstvi*, No. 16, *op.cit.*, p. 13.

55. C. H. McMillan, "Soviet and East European Participation in Business Firms and Banks Established in the West: Policy Issues", *art. cit.*

56. J. Blisak, "Potential for Technology Transfer between the CSSR and Advanced Capitalist States", *Ekonomicky Casopis*, No. 6, 1983, as given in *Joint Publications Research Service* (L/11678), *East Europe Report*, 20th October, 1983, p. 3.

57. *Ibid.*, p. 5.

58. F. Levcik and J. Skolka, *op. cit.*, p. 38.

59. See F. Levcik and J. Skolka, *op. cit.*, p. 39 and J. Kozehuba, "Licence Utilisation in the Slovak Socialist Republic", *Zahranicni Obchod*, No. 9, 1982, as given in *Joint Publications Research Service* (82456), *East Europe Report*, 14th December, 1982, pp. 8-9.

60. See *East European Markets*, Vol. 4, No. 7, 2nd April, 1984, p. 7 and F.L. Altmann and H. Clement, *Die Kompensation als Instrument im Ost-West Handel*, G. Olzog, Munich, 1979.

61. C.H.McMillan, "Soviet and East European Participation in Business Firms and Banks Established in the West: *Policy Issues*", *art. cit.*

62. *The Transfer of Disembodied Technology with Special Regard to Licencing* (SC.TECH./SEM.8/R.3/Rev.1; TRADE/SEM.5/R.3/Rev.1), *op. cit.*, p. 10.

63. *Review of Recent and Prospective Trends, Policies and Problems in Intra-Regional Trade, A Profile of the East-West Trade of Hungary* (TRADE/R.402/Add.2), Note by the Secretariat, Committe on the Development of Trade, Economic Commission for Europe, Geneva, 12th November, 1980, p. 22.

64. I. Kristof, "The Situation of the Basic Industries and Certain Problems of their Development", *Statisztikai Szemle*, August-September, 1981, as given in *Joint Publications Research Service* (79368), *East Europe Report*, 4th November, 1981, p. 42.

65. K. Ravasz, *The Role of Technology Transfer in Co-operation Agreements of Hungarian Enterprises,* (SC.TECH./SEM.8/R.15); TRADE/SEM.5/R.15), Seminar on Technology Transfer as it Affects the Development of Trade and Economic Co-operation, Senior Advisers to ECE Governments on Science and Technology, Economic Commission for Europe, Geneva, 19th March 1981, p. 3.

66. See, for instance, K. Kovats, "Timely Issues of East-West Co-operation", *Marketing in Hungary*, No. 1, 1984, pp.30-32.

67. *Business Eastern Europe*, 2nd September, 1983, p. 275.

68. *Ibid.*, p. 276.

69. See, respectively, *Co-operation in Third Markets of East-West Joint Ventures Domiciled in Eastern Europe* (TRADE/R.469), *op. cit.*, *Review of Recent and Prospective Trends, Policies and Problems in Intra-Regional Trade* (TRADE/R.460), Note by the Secretariat, Committee on the Development of Trade, Economic Commission for Europe, Geneva, 17th November 1983, paragraph 42, and *Promotion of Trade Through Industrial Co-operation, East-West Joint Ventures Domiciled in Eastern Europe and Yugoslavia: Current Status of Legislation and Enterprise Response* (TRADE/R.489), *op. cit.*, paragraphs 17, 29 and 70.

70. E. Kortvelyes, "Mixed Doubles", *Heti Vilaggazdasag*, 26th June, 1982, as given in *Joint Publications Research Service* (81429), *East Europe Report*, 3rd August, 1982, pp. 34-36.

71. C.H. McMillan, "Soviet and East European Participation in Business Firms and Banks Established in the West: Policy Issues", *art. cit.*

72. K. Ravasz, *op. cit.*, p. 14.

73. Paul Marer, *East-West Technology Transfer: Study of Hungary, 1968-1984, op.cit.*, Chapter IV.

74. "To Buy or Not to Be", *Figyelo*, No. 1, 1985, as given in *Joint Publications Research Service* (EEI-85-017), *East Europe Report*, 13th February 1985, p. 23.

75. Paul Marer, *East-West Technology Transfer: Study of Hungary, 1968-1984, op.cit.*, Chapter IV. See also *The Transfer of Disembodied Technology with Special Regard to Licencing*, (SC.TECH/SEM.8/R.3/Rev. 1; TRADE/SEM.5/R.3/Rev. 1), *op. cit.*, and *Hungary's International Trade in Licences and Technical Designs* (SC.TECH./SEM.8/R.4; TRADE/SEM.5/R.4), Seminar on Technology Transfer as it Affects the Development of Trade and Economic Co-operation, Note by the Secretariat, Senior Advisers to ECE Governments on Science and Technology, Economic Commission for Europe, Geneva, 10th March, 1981.

76. "To Buy or Not to Be", *art.cit.*, p. 24.

77. E.B. Somorjai, "The Experiences of Capitalist Co-operation", *Figyelo*, 5th August, 1982, as given in *Joint Publications Research Service* (81819), *East Europe Report*, 22nd September, 1982, p. 41. For a discussion of systemic difficulties which have hindered the purchase and application of licences, see also Paul Marer, *East-West Technology Transfer: Study of Hungary, 1968-1984, op.cit.*

78. J. Cieslik, "The Regulation of Technology Transfer in Poland", *Osteuropa Wirtschaft*, No. 3, 1983, p. 223.

79. A. Wieczorkiewicz, "Poland's Co-production Ties with the West", *Sprawy Miedzynarodow*, No. 1, 1983, as given in *Joint Publications Research Service* (83599), *East Europe Report*, 2nd June, 1983, pp. 41 and 42 and, Z. Fallenbuchl, *East-West Technology Transfer: Study of Poland, 1971-1980, op. cit.*, pp. 64 and 65.

80. *Review of Recent and Prospective Trends, Policies and Problems in Intraregional Trade* (TRADE/R/460), *op.cit.*, p. 10.

81. *Policies and Methods for Facilitating the Participation of Small and Medium-Sized Enterprises in Intraregional Trade* (TRADE/R. 446/Add.1), Committee on the Development of Trade, Economic Commission for Europe, Geneva, 3rd November, 1982, pp. 10-11.

82. Z. Fallenbuchl, *East-West Technology Transfer: Study of Poland, 1971-1980, op. cit.*, p. 65.

83. See, for instance, E. Tabaczynski, "Manufacturing Ventures Involving Foreign Capital in Poland", *Polish Foreign Trade*, No. 4, 1984, p. 8 and E. Sadowska-Cieslak, "Compensation Transactions – A Controversial Phenomenon in East-West Trade", *Polish Foreign Trade*, No. 11, 1984, pp. 21-22.

84. Z. Fallenbuchl, *East-West Technology Transfer: Study of Poland, 1971-1980, op.cit.*, p. 65.

85. *Ibid.*

86. *Business Eastern Europe*, 16th March, 1984, p. 81.

87. A. Wieczorkiewicz, *art. cit.*, p. 34.

88. R. Pac, "Remarks on Industrial Co-operation between PPR and FRG", *Polish Foreign Trade*, No. 1, 1984, p. 11.

89. A. Wieczorkiewics, *art.cit.*, p. 34.

90. Interview with E. Tabaczynski in *Preglad Techniczny*, No. 1, 13th June, 1982, as given in *Joint Publications Research Service* (81633), East Europe Report, 26th August 1982, p. 77.

91. *Joint Venture Legislation in Eastern Europe and Yugoslavia: A Survey of Recent Developments, 1977-1979* (TRADE/AC.14/R.2), Committee on the Development of Trade, Economic Commission for Europe, Geneva, 29th April, 1980, paragraghs 35-80, and *Promotion of Trade Through Industrial Co-operation, East-West Joint Ventures Domiciled in Eastern Europe and Yugoslavia: Current Status of Legislation and Enterprise Response* (TRADE/R/489), *op.cit.*, paragraphs 18, 19, 30 and 37.

92. *Zycie Gospodarcze*, No. 14, 4th March, 1983, as given in *Presseschau Ostwirtschaft*, No. 4, 1983, p. 34.

93. *Rzeczpospolita*, No. 7, 29th July, 1983, as given in *Presseschau Ostwirtschaft*, No. 8/9, 1983, p. 50. See also C. H. McMillan, "Soviet and East European Participation in Business Firms and Banks Established in the West: *Policy Issues*", *art. cit.* His breakdown – though referring to 1981 – is somewhat different.

94. J. Cieslik and R. Rapacki, *Restrictive Contract Clauses in East-West Licencing Trade* (SC.TECH./SEM.8/R.10/Rev.1; TRADE/SEM.5/R.10/Rev.1), Seminar on Trends in Different Forms of Technology Transfer and their Impact on the Development of Trade and Economic Co-operation, Senior Advisers to ECE Governments on Science and Technology, Economic Commission for Europe, Geneva, 23rd March, 1982. The discussion of licencing agreements is primarily based on this paper.

95. H. A. Janiszewski, "Transfer of Technology: Present Situation and Prospects", *Przeglad Techniczny*, 13th February, 1983, as given in *Joint Publications Research Service* (83839), *East Europe Report*, 7th July, 1983, p. 25.

96. See, for instance, J. Mujzel and K. Poznanski, *System-Related Determinants of a Country with Regard to the International Transfer of Technology* (SC.TECH./SEM.8/R.14; TRADE/SEM.5/R.14), Seminar on Technology Transfer as it Affects the Development of Trade and Economic Co-operation, Senior Advisers to ECE Governments on Science and Technology, Economic Commission for Europe, Geneva, 19th March, 1981.

97. Z. Fallenbuchl, *East-West Technology Transfer: Study of Poland, 1971-1980, op. cit.*, p. 45.

98. W. Brzost, *The Role of Technology Transfer in Economic Development Strategy* (SC.TECH./SEM.8/R.12/Add.1; TRADE/SEM.5/R.12/Add.1), Seminar on Trends in Different Forms of Technology Transfer and their Impact on the Development of Trade and Economic Co-operation, Senior Advisers to ECE Governments on Science and Technology, Economic Commission for Europe, Geneva, 23rd March, 1982, p. 1.

99. *Ibid.*, pp. 1 and 3.

100. Z. Cieslik, "The Regulation of Technology Transfer in Poland", *art. cit.*, p. 220.

101. H. A. Janiszewski, "Technology Imports into Poland", *Les Nouvelles*, June, 1982, p. 130.

102. *Review of Recent and Prospective Trends, Policies and Problems in Intraregional Trade: A Profile of the Trade of Romania with ECE Countries in the 1970s* (TRADE/R.442/Add.4), Note by the Secretariat, Committe on the Development of Trade, Economic Commission for Europe, Geneva, 18th October, 1982, paragraph 79.

103. S. Arsene and S. Manoliu, *Revista Economica*, No. 30, 29th July, 1983, as given in *Joint Publications Research Service* (84368), *East Europe Report*, 20th September, 1983, p. 32.

104. See, for instance, "Rumänian sucht mehr Westkooperation", *Nachrichten für den Aussenhandel*, Köln, 15th October 1984 and A. Constantinescu and V. Babiuc, *Revista Economica*, No. 36, 7th September 1984, as given in *Joint Publications Research Service* (EEI-84-114), *East Europe Report*, 15th October 1984, pp. 6-9.

105. *Review of Recent and Prospective Trends, Policies and Problems in Intraregional Trade: A Profile of the Trade of Romania with ECE Countries in the 1970s* (TRADE/R/442/ Add.4), *op. cit.*, paragraphs 95-103.

106. See *Co-operation in Third Markets of East-West Joint Ventures Domiciled in Eastern Europe* (TRADE/R/469), *op.cit.*, paragraph 97 (Note 62).

107. See *Review of Recent and Prospective Trends, Policies and Problems in Intraregional Trade: A Profile of the Trade of Romania with ECE Countries in the 1970s* (TRADE/R.442/Add.4), *op. cit.*, paragraphs 84-103; *Romanian Preoccupation for a More Active Participation in International Technology Trade* (SC.TECH./SEM.8/R.16, TRADE/SEM.5/R.16), submitted by the Government of Romania, Seminar on Technology Transfer as it Affects the Development of Trade and Economic Co-operation, Senior Advisers to ECE Governments on Science and Technology, Economic Commission for Europe, Geneva, 20th March, 1981; *Co-operation in Third Markets of East-West Joint Ventures Domiciled in Eastern Europe* (TRADE/R. 469), *op. cit.*, and, *Promotion of Trade Through Industrial Co-operation, East-West Joint Ventures Domiciled in Eastern Europe and Yugoslavia: Current Status of Legislation and Enterprise Response* (TRADE/R.489), *op.cit.*, paragraphs 26, 28, 58, 69 and 80.

108. *Co-operation in Third Markets of East-West Joint Ventures Domiciled in Eastern Europe* (TRADE/R.469), *op. cit.*, paragraph 59, (note 59).

109. C.H. McMillan, "Soviet and East European Participation in Business Firms and Banks Established in the West: Policy Issues", *art. cit.*

110. J. Monkiewicz, "International Technical Connections of the Socialist Countries", *Handel Zagraniczny*, No. 11, 1979, as given *Joint Publications Research Service* (75765), *East Europe Report*, 23rd May 1980, pp. 20-21.

111. *Review of Recent and Prospective Trends, Policies and Problems in Intraregional Trade: A Profile of the Trade of Romania with ECE Countries in the 1970s* (TRADE/R/ 442/Add.4), *op. cit.*, paragraph 98.

112. *Svet Hospodarstvi*, No. 100, 23rd August, 1983, p. 3, as given in *Presseschau Ostwirtschaft*, No. 8/9, 1983, p. 54.

113. *Romanian Preoccupation for a More Active Participation in International Technology Trade* (SC.TECH./SEM.8/R.16, TRADE/SEM.5/R.16), *op. cit.*, pp. 3-4.

114. M. Hill, *op. cit.*, p. 96.

115. See *Review of Recent and Prospective Trends, Policies and Problems in Intra-Regional Trade: A Profile of the East-West Trade of the USSR* (TRADE/R.442/Add.5), Note by the Secretariat, Committee on the Development of Trade, Economic Commission for Europe, Geneva, 11th November, 1983. The role of Western technology in developing the Soviet chemical industry and compensation trade is discussed, *inter alia*, by P. Gutman, in F. Geze, P. Gutman, A. Valladao, *Les modes d'imbrication de la division internationale du travail occidentale et de la division internationale socialiste du travail, études sectorielles*, study for the DGRST, January,

1981 and V. Sobeslavsky and P. Beazley, *The Transfer of Technology to Socialist Countries: The Case of the Soviet Chemical Industry*, The Royal Institute of International Affairs, Gower, 1980.

116. See V. Malkevich, "Industrial Co-operation as a New Form of Economic Collaboration between the USSR and Industrial Capitalist Countries", *Foreign Trade* (English edition), No. 1, Moscow, 1980; V. Klochek, "Soviet Foreign Trade on the Threshold of the Eleventh Five-Year Plan Period", *Foreign Trade* (English edition), No. 5, Moscow, 1981 and, V. Chekurov and N. Chekurova, "Production-Technical Ties in the Economic Relations between the CMEA Member and the Capitalist Countries", *Ekonomicheskiye Nauki,* No. 9, 1983, as given in *Joint Publications Research Service* (UIE-84-001), *East Europe Report,* 3rd January 1984, pp. 54-63.

117. See, for instance, "USSR Plans First IIdPE Plant at Budyennovsk", *European Chemical News,* 17th September 1984, p. 28 and "USSR Invites Bid for $1bn. Polyester Complex", *European Chemical News,* 3rd December 1984, p. 30.

118. V. Malkevich, "USSR Industrial Co-operation with Western Countries", *Foreign Trade* (English edition), No. 9, Moscow, 1983, p. 3.

119. *Ibid.*

120. For a recent review of future trends in EWIC as well as those areas deemed most propitious for industrial co-operation relative to the 1986-1990 Plan period, see J.M. Gvishiani, "Economic, Scientific and Technological Co-operation and Trade between the USSR and Industrially Developed Western Countries in 1985-1990, Prospects of Development", presentation at the Conference sponsored by Business International, Geneva, March 1985. Countertrade in the Soviet Union and its role in technology transfer is discussed concisely in M. Bornstein, *East-West Technology Transfer, The Transfer of Western Technology to the USSR,* Chapter 2, OECD, Paris, 1985.

121. *Review of Recent and Prospective Trends, Policies and Problems in Intra-Regional Trade: A Profile of the East-West Trade of Italy in the 1970s* (TRADE/R.422/Add.2), Note by the Secretariat, Committee on the Development of Trade, Economic Commission for Europe, Geneva, 28th October 1982, paragraph 39.

122. See *Ecotass*, No. 18, 30th April 1984, p. 12 and *Ekonomicheskaya Gazeta*, No. 15, April, 1983, p. 20.

123. *Ekonomicheskaya Gazeta*, No. 1, January, 1984, p. 21.

124. See V. Chistov, "USSR-FRG: Development of Economic Ties Serving Both States' Interests", *Foreign Trade* (English edition), No. 11, Moscow, 1984, p. 21.

125. *Review of Recent and Prospective Trends, Policies and Problems in Intraregional Trade: A Profile of the East-West Trade of France in the Nineteen-Seventies* (TRADE/R.462/Add.2), Note by the Secretariat, Committee on the Development of Trade, Economic Commission for Europe, Geneva, 25th November 1983, paragraph 74.

126. *Review of Recent and Prospective Trends, Policies and Problems in Intra-Regional Trade: A Profile of the East-West Trade of Finland in the 1970s* (TRADE/R.442/Add.1), Note by the Secretariat, Committee on the Development of Trade, Economic Commission for Europe, Geneva, 6th October, 1982, paragraph 106.

127. A. Manzhulo, "USSR-Finland: New Prospects for Trade and Economic Co-operation", *Foreign Trade* (English edition), No. 12, Moscow, 1984, pp. 7-9.

128. V. Klochek, "Soviet Foreign Trade in 1982", *Foreign Trade* (English edition), No. 5, Moscow, 1983, p. 10.

129. C.H. McMillan, "Soviet and East European Participation in Business Firms and Banks Established in the West: Policy Issues", *art. cit.*

130. See V. Malkevich, "Industrial Co-operation as a New Form of Economic Collaboration between the USSR and Industrial Capitalist Countries", *art. cit.*, p. 8 and A.I. Danilenko, *et al.*, *International Economic Relations of the CMEA Countries with the West*, Moscow, 1983, as given in *Joint Publications Research Service* (UIE-85-003-L), *East Europe Report*, 8th February 1985, p. 69.

131. V. Malkevich, "USSR Industrial Co-operation with the Western Countries", *art. cit.*, p. 3.

132. *Ibid.*

133. V. Gribanov, "A Contract for Construction of Industrial and Other Projects in Soviet Foreign Economic Relations", *Foreign Trade* (English edition), No. 9, Moscow, 1982, p. 45.

134. L. Hess, "Les achats soviétiques d'usines clés en main à l'Occident", *Le courrier des pays de l'Est*, No. 257, December, 1981, pp. 34-52.

135. For a discussion of Austria's relations with the Soviet Union and the role of VOEST-Alpine, see Sandor Richter, *Some Aspects of Economic Relations between Austria and the Soviet Union*, Forschungsberichte, No. 101, Wiener Institut für Internationale Wirtschaftsvergleiche, November 1984.

136. See M. Bornstein, *East-West Technology Transfer, The Transfer of Western Technology to the USSR, op.cit.*, Chapter 3.

137. See B. Högberg and C. Wahlbin, "East-West Industrial Co-operation: The Swedish Case", *Journal of International Business Studies*, Spring/Summer 1984. The authors analyse in detail the various motivations and problems pertaining to industrial co-operation – based on the experience of Swedish firms.

138. K. Bolz, "Entrepreneurial Co-operation between East and West in the First Half of the 1980s", *Intereconomics*, September/October 1984, pp. 245-246.

139. B. Kadar, "Patterns and Motivation for Industrial Co-operation", *Marketing in Hungary*, No. 4, Budapest, 1983, p. 16.

140. Z. Krasznai, "Multinational Division of Labour: Dead End or a Chance to Catch Up", *Valosag*, No.3, 1983, as given in *Joint Publications Research Service* (83759), *East Europe Report*, 24th June 1983, p. 12.

141. K. Ravasz, *op. cit.*, p. 7.

142. See I. Kristof *art. cit.*, p. 43 and K. Ravasz, *op.cit.* p. 11.

143. Z. Krasznai, *art. cit.*, p. 13.

*Part IV*

# THE IMPACT OF EAST-WEST TECHNOLOGY TRANSFER

*Chapter 8*

# THE IMPACT OF EAST-WEST TECHNOLOGY TRANSFER ON EASTERN COUNTRIES[1]

Whereas in Chapter 6, the strategies and trends in Eastern countries' imports of capital goods were described, as suggested by their actual size, commodity and sector/branch structure of these flows, section 1 of the present chapter focuses mainly on the relationship between Eastern capital goods imports from the West and the development of Eastern exports to the West – paying attention to the growth of Eastern exports to other areas of the world. Progress in augmenting domestic production to achieve import savings is also reviewed. In section 2, the impact of embodied technology imports on domestic output is evaluated.

## 1. The Impact of Western Capital Goods on East-West Trade

### a) *Eastern Embodied Technology Imports and Exports to the West: General Aspects*

In order to investigate further the role played by technology imports from the West, embodied technology imports are first examined in the context of overall Eastern export performance, taking into account the sector and branch basis of such flows [sub-section *a)*]. In sub-section *b),* Eastern export performance in 1970-1983 is reviewed, and the relative performance of Eastern exports to the OECD relative to other suppliers is examined in sub-section *c).* Finally in sub-section *d),* Eastern imports of Western technology are reviewed in the light of import savings on a sector and branch basis.

It is not possible to separate out the share of Western capital goods directed to export activities. This is true no less at sector/branch end-user level than in aggregate. It is of course possible to compare capital goods imports by end-user branch, presented already in the previous Part (Chapter 6), to the exports of these branches to the West. The problem is that there is, in practice, no way of showing whether any of these exports derive directly from the use of Western technology. First, because of lags between the shipment of capital goods and its commissioning, the exports shown derive, if at all, from Eastern purchases of capital goods made in a previous year. Lags may range from a short period to up to three or four years, and there is no way of knowing what these lags were in the case of a given Eastern export flow. Second, the exports of a given sector or branch, even if its past technology imports have been large may, in fact, be derived either fully or in part from the use of indigenous technology or of technology imported from other areas. The mix of these with Western technology can vary both between sectors or branches and, for the exports of a given sector or branch, it may vary from year to year. Such problems affect all countries though of course, the more highly dependent an economy is on Western technology, the more chance there is that a given export to the West does derive from Western technology. Nevertheless, the problem remains.

Most of the discussion in Chapter 5 was based on data in current prices. As background to the present discussion, it is perhaps useful to examine first changes in the volume of West-to-East embodied technology sales relative to changes in the volume of total East-West trade. Table 1 of this chapter shows that, in volume terms, total Western exports of technology to Eastern Europe between 1970 and 1975 nearly doubled, and more than doubled to the Soviet Union, with intermediate (technology) imports rising faster than capital goods to the Soviet Union and slower to Eastern Europe. Between 1975 and 1980, and especially between 1980 and 1983, by contrast, exports of embodied technology items to Eastern Europe declined by almost a half (and by more than half for capital goods), whereas those to the Soviet Union, (especially capital goods) resumed faster growth after cutbacks in the late 1970s.

The growth of total OECD import volumes from the East European countries fell far behind the growth of OECD's total exports to these countries in 1970-1975. In the second half of the decade, the volume growth of OECD's total imports from Eastern Europe exceeded the export volume growth to those countries, but only because Eastern Europe cut its technology imports. The volume growth of OECD's total imports from Eastern Europe was in fact slower than in 1970-1975. Comparing 1983 with 1980, this trend intensified, with further declines in OECD's total imports from the area. It should perhaps be stressed that only the German Democratic Republic succeeded in increasing the volume of its exports to the OECD area between 1980 and 1983. This finding is significant. Even though virtually all the technology exported by the OECD countries to Eastern Europe in the 1970s was in place by the end of the period, it does not seem to have helped Eastern Europe in making the massive resource shift towards the foreign balances which was necessitated by unfavourable internal and external developments. Though the resource shift did occur, it was achieved by import cuts rather than by any rapid export expansion.

The Soviet case in the 1970s was, in general, similar. But cutbacks in capital goods imports from the OECD area were much less marked in 1975-1980 and the volume of Soviet imports of intermediate (technology) goods rose much more strongly than in Eastern Europe. In 1980-1983, however, and in contrast to Eastern Europe, the volume of capital goods imports again turned sharply upwards. Moreover, unlike Eastern Europe, the Soviet Union continued to increase the volume of its total exports to the OECD countries. Thus, again unlike Eastern Europe, the rise in the volume of Soviet imports of embodied technology over 1970-1983 exceeded by a considerable margin the rise in its total export volumes to the OECD countries. In fact, there was a considerable overall real resource shift *away* from the foreign balances, made possible by improvement in the Soviet terms of total trade due, first, to fuel price increases. This is also why, when the volume of total Soviet exports to the OECD countries rose more slowly than those of Eastern Europe during 1970-1980, the volume of technology purchases needed to be cut back by much less. It was also able to offset the fall in fuel prices in 1982 and 1983 by a higher volume of fuel shipments.

Despite these contrasts, the underlying general export positions of the Soviet Union and Eastern Europe with regard to goods other than fuel was, nonetheless, very similar. As will be shown below, with the partial exception of the German Democratic Republic, the performance of non-fuel exports has been sluggish throughout the Eastern region. Most of the technology imported over the period reviewed was in place by the beginning of the 1980s. Yet, with the exception of pipeline-related fuel exports, the most that can be said of the expanded purchases of embodied technology during the 1970s is that it was associated with some, but not very, dynamic export growth. This occurred under admittedly difficult external demand conditions due to the Western recession.

Sector and branch linkages between imports of Western capital goods and export performance are shown in Table 2 of this chapter. It should be noted that the coefficients presented relate, for each Eastern country, exports of a given sector or branch and its imports of Western technology in a given year and hence do not imply any causal linkage: the problem of lags between technology imports and exports of resultant products has already been mentioned. The Table does no more than illustrate the size and changes in the size of past relationships between individual sector and branch imports of capital goods and simultaneous export performance. Thus a high coefficient in one sector or branch relative to others indicates that a given industry in a given year imports only a small amount of technology in relation to its export effort and *vice versa*. A rise in the coefficient over time thus means either that, where exports have risen, capital goods imported have not kept pace, or that where exports have fallen, capital goods have been maintained or fallen by less – and *vice versa*. The changes recorded thus need to be interpreted on a case by case basis. For instance, in the first case, a rise in the coefficient could indicate a successful past effort to harness Western capital goods to production for export. But, a rise in the coefficient could also signify a reduction (loss in priority) in imports of capital goods for the industry concerned.

At aggregate level, the coefficients during 1970-1980 varied, with some exception between 3 and 5 in the material sphere and industry as a whole. During 1981-1983, the coefficient rose in all countries except Bulgaria and the German Democratic Republic – sometimes, as in Poland and Romania, very steeply. Given poor overall export performance (except in the Soviet Union), this betokened a noticeable shift in priorities, with export earnings being shifted very markedly towards expenditures for other purposes (other types of imports, foreign debt servicing, etc.). Relationships *between* branches have remained fairly constant over time since 1970, although there has been a more rapid relative rise in the coefficient for the fuel industry, reflecting substantial expansion in the volume of fuel exports in several countries, and also in the textile, other light and food industries in a number of countries – notably in Czechoslovakia, Poland and Romania. In these three countries, this represented cuts in capital goods imports for what are usually considered relatively low priority sectors. The chemical industry coefficient has also risen relatively fast, and in some countries very fast. Chemical industry export performance suggests that a substantial creation of export capacity in the industry has already taken place – a point which is confirmed in the following sub-section. A similar movement of the coefficient for engineering in all countries but the German Democratic Republic since 1980, in contrast, reflects not rising exports but cuts in capital goods imports for the industry (Czechoslovakia, Hungary and the Soviet Union). In other countries, coefficients have not changed greatly. As is also shown below, efforts to develop engineering exports on the basis of imported Western capital goods have apparently been rather limited and, in any, case unsuccessful.

### b) *The "Boomerang" Effect*

The purpose of this sub-section is to explore further the relationship between imports of Western technology and exports to OECD countries by the sectors or industrial branches which received this technology (the "boomerang" effect). Table 3 shows the degree of similarity between the growth rates of imports of Western capital goods by main sector/branch end-users in individual East European countries on the one hand, and export performance of the corresponding sectors and branches in these countries on the other. Both growth rate series were obtained from OECD-reported data on Eastern capital goods imports by sector/branch end-users, and on Eastern exports to the OECD by sector/branch of origin, contained in the OECD technology transfer data base. The growth rates refer to changes between the first and last years of the two periods examined. The periods covered are 1970-1977 and 1977-1983. The reason for this particular division is that there was a significant discontinuity in the growth of machinery imports of East European countries from the West after 1977. The decision taken by several East European countries to cut back capital goods imports in 1977/1978 (in some countries even earlier), means that the East-West trade flows and their structure were affected by a deliberate action of the Eastern planners.

The degree of similarity between the two indicators was estimated with the help of the Spearman Rank correlation coefficient. The value of the coefficient will be equal to +1 if there exists a perfect similarity (identity) between sectoral export performance and sectoral and branch imports of Western capital goods by the Eastern countries concerned. The value of the coefficient will be equal to –1 if the two indicators are perfectly dissimilar and it will be zero if no relationship exists between them. The problem of lags betwen the imports of capital goods and subsequent exports is mitigated by calculating Table 3 of this chapter on the basis of averages for the period as a whole. It is assumed here that Western technology was assimilated during the period under consideration.

It may be useful to stress again that the Table does not show in quantitative terms the contribution of Western technology to export performance of Eastern Europe. The results do not show, for example, that Western technology increased competitiveness of East European exports by, say 20 per cent resulting in an increase of exports by 20 per cent. What the Table does show, however, is whether Western capital goods were allocated primarily to sectors with dynamic export performance rather than to sectors whose exports were sluggish, or to those which did not produce for export at all. In other words, the Spearman Rank correlation coefficients can be interpreted as measures of the success or failure of East European planners to adjust the import growth rate of Western capital goods with the intention of augmenting existing levels of exports.

Turning now to the findings, we see that each coefficient is very considerably less than one. This is not entirely surprising, since some Western technology will always be used in production of goods which are either non-tradeables or which do not represent the most dynamic exports to the West, or which are used for production destined to internal use or, for exports to destinations other than the West. What is rather unexpected, however, are the low relative values of the coefficients – with the limited exceptions of the Soviet Union in 1970-1977 and in Romania in 1970-1977 and in 1977-1983. Only the coefficients for the Soviet Union are, in fact, significant at the 5 per cent level (Olds test). The extraordinarily low value of the coefficients for the German Democratic Republic could result from the exclusion of figures on trade with the Federal Republic of Germany. But the data for Hungary, usually considered to be the most export-oriented of all the Eastern countries, are negative in both periods. For the region as a whole, therefore, there appears to be no evidence of a strong relationship between imports of Western technology on the one hand and successful export performance on the other. In other words, successful export performance tended to be in sectors which received relatively less priority to purchase Western technology. In addition, as noted in Chapter 6, Western capital goods were quite extensively used in sectors which produced non-traded goods or for the domestic market. For the Soviet Union, at least in 1970-1977, the case is rather different; for example, the dynamic export sectors include fuel, chemicals, and raw-material producing sectors. This conforms quite closely to the nature of countertrade agreements concluded by OECD firms with that country. They involved mainly natural gas and chemicals, and in this respect, Western capital goods contributed significantly to export growth of the Soviet Union. On the other hand the engineering sectors, among the major users of Western technology, have performed relatively poorly as exporters. According to the above estimates countries can be grouped into three classes: *i)* Romania and the Soviet Union, representing countries with a certain degree of export orientation based on Western technology – at least in the 1970s; *ii)* countries with a barely traceable export orientation of export capacity from Western technology: Czechoslovakia throughout, the German Democratic Republic in 1977-1983 and Poland in 1970-1977 only; and *iii)* countries with a weak or even negative export orientation of Western technology: Bulgaria and, surprisingly, Hungary.

This is a simple and rough test aimed to assess the contribution of Western technology to East European exports. There is very little evidence of any significant "boomerang" effect in East-West trade. This is obviously not to deny the possibilities of Western technology making important contributions to the sale of particular East European products and evidence of such individual cases indeed exists – notably chemicals and, in some countries, fuels. However, even at the macro-economic, and even more so at the sector and branch level, the individual contribution of Western technology would, had it been detectable, have been reflected in the tests.

In practice, there are some reasons to believe that the pattern of expenditures on imported Western capital goods will be to a large extent correlated with the pattern of investment expenditures. This policy can be observed particularly in the development of 'priority' sectors whereby those sectors receiving larger shares of investment spending were also more likely to obtain hard currency allocations. But it would be foolhardy to argue that investment policy has followed this rule in Eastern Europe at all times. Thus, for this reason or any one of the other reasons mentioned above, the branch-user share of imports of Western capital goods, not tabulated here, is in part significantly different from the branch share of investment spending. In such a case a slow export growth of branch X may be due to the small contribution of Western capital input (or due to low return on domestic investment).

Given the "leakages" which can weaken the relationship between Eastern imports of Western capital goods and Western exports noted above, it is not surprising that attempts made to relate technology imports with exports on a year-to-year basis by sector and branch do not show any systematic relationship – with a few rare exceptions such as Soviet imports of Western pipeline equipment and export of fuel to the West. (Even in that case, much of this equipment is used to facilitate oil and gas transport for domestic uses – especially in Eastern Europe – or for export to Soviet CMEA trade partners.) Nevertheless, even inconclusive results have their own significance and the preliminary conclusion that, at macro-economic (sector, branch and aggregate) level the import of Western technology has not been significantly correlated with export performance on exports to the West, is a relevant finding, and is consistent with the analysis of Eastern import strategies presented earlier[2].

Table 4 shows the fastest-growing (i.e. average or above-average) export commodity groups for Eastern Europe and the Soviet Union between 1975 and the years 1980 or 1981. The list of products which grew at average or above average rates is a short one for both – especially for the Soviet Union – and shows the very narrow range of commodities upon which Eastern export growth has depended since 1975. In fact, most of the commodities grouped – though accounting for 42.6 per cent and 33.1 per cent of total exports to the West (in current prices) in the East European and Soviet cases respectively in 1975 – are, with certain exceptions, individually rather small in relation to total exports. As the Table shows, the list narrowed even further between 1980 and 1983. Moreover, the bulk of the East European group in 1980/1981 consisted of five groupings only: clothing, basic industrial chemicals, refined petroleum, iron and steel and non-ferrous metals (which alone accounted for over half of the fastest growing commodity groups in the 1980s). In the Soviet case, the fastest growing export products are even smaller in number and two of them – petroleum and gas and basic industrial chemicals – accounted for no less than 72 per cent of the whole group of above-average export growth commodities in 1983.

This exercise already indicates that, so far as export potential is concerned, the import of chemical equipment by the East European countries appears to be one of the main success stories. Not only were imports of chemical equipment the principal or second most important item of Western technology for all of them, but chemical products have been one of the main constituents of real export growth.

A second area of relative success has been iron and steel. This branch – third or fourth place in total imports of Western industrial capital goods over the decade – was among the most buoyant export producers. The third item is textiles – especially clothing. Here, growth rates since 1975 have been less buoyant than in the case of iron and steel, or of refined petroleum. The share of capital goods for the textile industry in total capital goods imports from the West was high in relation to investment in most East European countries.

For the Soviet Union, the position is more simple and more striking. In addition to oil, gas and related products, industrial goods export performance to the West has relied predominantly on basic industrial and some other chemicals. The high share of pipeline and chemical equipment in total OECD capital goods imports from the East confirms that good export performance in fuels and chemicals is, in fact, strongly related to imports of Western capital goods.

Perhaps the most striking feature of Eastern export performance is the almost complete absence of high growth performance commodities from the engineering sector – the first or second most important purchaser of industrial capital goods of virtually all countries in the area throughout the 1970s. In fact, only three engineering product groups appear in the list of Soviet leading export products, and though more appear for Eastern Europe, their combined shares amount to less than that of processed wood. With regard to technology items, it is recalled from Chapter 5 (Table 6) that the share of capital goods accounted for only 3-5 per cent of imports of the OECD countries from the East in recent years – less than in 1970. Their low level was mainly due to the Soviet Union, where they account for only 1 per cent of OECD imports from that country. In contrast, all East European countries except Czechoslovakia and the German Democratic Republic, recorded rises which brought the share of capital goods to nearly 10 per cent of OECD's imports from the six countries as a whole in 1983. Eastern motor car exports, though they have roughly doubled since 1975 (starting however at very low levels), still account for only a small share of total exports and hence only a small percentage of total export growth.

### c) *The Competitivity of Eastern Exports in Western Markets*

Clearly, the sluggish export performance described above cannot be interpreted solely from the export supply side: that is, without reference to demand in OECD markets. The market share approach presented in the following paragraphs enables the above direct examination of Eastern countries' export performance to be assessed in the light of performance on Western markets relative to other suppliers, and hence also, in effect, independently of the effects of depressed market conditions in OECD countries in the early 1980s.

An overview of the Eastern countries' total, non-industrial and industrial exports (including and excluding fuels) to the OECD on this basis is first presented (Table 5). Given that industrial goods (including fuels) account for roughly 90 per cent of the Eastern countries' exports to the West, a more detailed analysis of the performance of various commodity groups produced by industry is then set out. The period covered is 1970-1983. The evidence of relative stagnation of Eastern exports to the West in 1984 suggests that no big changes have occurred since the final year covered.

The market share for total Eastern countries' exports show, first, that the seven countries together accounted for 3.2 per cent of the OECD imports from the world in 1983. This share has grown from 2.9 per cent in 1975, and from 2.5 per cent in 1970. However, the increases are due to the German Democratic Republic and the Soviet Union only. These two countries' shares in the OECD market rose respectively by 11 and 81 per cent since 1970. Eastern Europe as a whole saw its share of the OECD market decline by 15 per cent.

Additionally, virtually all of the increase for the seven countries together is due to one major commodity grouping only: fuel (crude and refined). If these flows are excluded, not only did the German Democratic Republic and Soviet market shares decline, but the seven-country share falls to a little above 2 per cent in 1983, substantially less than in 1970 and even further below 1975. (The Eastern countries together in fact supply some 15 per cent of OECD's total imports of crude oil, natural gas and other fuels, and some 16 per cent of its refined petroleum.)

The seven countries' non-industrial (i.e. agricultural) goods account for a smaller share (2.7 per cent) of OECD countries' agricultural imports from the world in 1983 than industrial commodities as a whole. This represents a considerable reversal since 1970, when they accounted for 4¼ per cent (i.e. a bigger share than industrial and total goods). The fall in share has been most marked in Eastern Europe, where it has diminished in all countries.

The position of Eastern agricultural goods as a whole on OECD markets owes much to a rather small number of products: wood and timber, of which 11 per cent was supplied by the Soviet Union and 4 per cent by Eastern Europe in 1983; textile fibres, of which the Soviet Union has supplied 5-8 per cent in 1980-1983 and, live animals (of which Eastern Europe still supplies about 2-3 per cent of OECD's needs).

The narrow commodity base on which the share of the Eastern countries' agriculture in OECD markets rests is echoed within industry. As a starting point for the analysis of industrial goods performance, a 61-group commodity breakdown has been used. This is based on OECD 5-digit SITC trade returns with each Eastern country, re-aggregated into conveniently recognizable groupings[3]. In order to assess Eastern countries' export performance, all commodity groups having registered increases in the share of the OECD market have been extracted and their size in total Eastern countries' industrial exports calculated (Table 6). Those commodities accounting for more than 2 per cent of each Eastern country's industrial exports to the OECD area in 1983 have also been noted in the Table. The share of both groups as a whole in OECD's imports from each country of the area in 1970 and in 1983 is also shown. The broader group of all products which have increased their share in total OECD imports from the world is referred to in the Table as the "competing group". The narrower grouping of products which constituted more than 2 per cent of a country's exports to the OECD in 1983 is referred to as the "top group". Given that both groups concerned consist of goods whose share of OECD markets rose over time, an increased share of the groups in exports shows how far the export structure has moved in favour of competitive products: the bigger the change between the two reference years, the more competitive the structure has become.

The absolute size of the competitive group is also important. The bigger it is in the initial year, the greater the relative advantage possessed by the country at the outset in the light of its actual export competitivity – as revealed over time. In addition, the size of the two groups in relation to the number of products contained in them indicates the degree of commodity concentration of competitive export performance. In particular, the size of the 'top group', selected on the basis of the much smaller number of the biggest commodity contributions to exports in relation to the other broader competitive group and to total exports, illustrates the narrow (and, since 1980, narrowing) commodity concentration of competitive exports with particular clarity.

The competing group in fact accounted for just over one-quarter (by value) of the OECD's industrial imports from the seven countries in 1970. It rose to just over two-thirds by 1983. Some 21 commodity groups were involved. The top group, however, consisted of only four: crude petroleum and gas, clothing, basic industrial chemicals and refined petroleum. These items alone, which accounted for one-fifth of the Eastern countries' exports to the West in 1970, rose to just under two-thirds of the total in 1983. Thus, although a broader range of Eastern commodities also increased their market share, a high degree (over 90 per cent) of increased penetration depended on only the four named commodity groups.

The Soviet expansion was, moreover, particularly narrowly based. Only 11 Soviet commodity groups increased their share of the OECD market, rising from 36 to 81 per cent of all Soviet exports to that market between 1970 and 1983. The three top commodities alone (oil and gas, refined petroleum and basic industrial chemicals) amounted to over three-quarters of Soviet exports to the OECD by 1983, compared with a third in 1970.

A similar, though not so pronounced, pattern is apparent for Bulgaria, the German Democratic Republic, Poland and Romania – though the competing group, wich contains a greater number of products than in the Soviet Union, accounted for a rather smaller share of these countries' exports to the OECD in both 1970 and in 1983 (except in Bulgaria). Both Bulgaria and Romania relied heavily on exports (mainly re-exports) of refined petroleum to stimulate their hard currency-earning efforts. The top group of commodities, the number of which ranged from four only in Bulgaria, the German Democratic Republic and in Poland, and six in Romania accounted, as in the Soviet Union, for a big share of the competing group (also over 90 per cent except in Poland). The situation was different in Czechoslovakia and Hungary – though for contrasting reasons and with contrasting effects. Both countries also lost market shares in overall and industrial exports. However, whereas the number of commodities whose competivity and share in exports to the OECD increased, was low in Czechoslovakia both at the beginning and the end of the period, in Hungary they were not only more numerous but they accounted for a much bigger share of the country's exports to the OECD. In other words, Czechoslovakia earmarked no specific groups of products for export expansion, and its loss of competivity was widely spread over the major part of its exports. In Hungary, some successes were, on the contrary, achieved over a much wider range of products. But, they were not enough to avoid a general loss of competivity, and, by extension, a rather big fall in the competivity of those products which did not increase their OECD market share.

A study of the Economic Commission for Europe also notes the drop in the market share of Eastern exports in East-West trade[4]. It confirms the close link between competitivity and the fall of Eastern products in the industrialized countries' total imports. Only in Poland was the "product" effect – the loss in share due to the low weighting in Polish exports to the West of products for which Western demand was buoyant – found to be significant. However, the study shows that Eastern exports won an increasing share of OECD markets between 1965 – before the expansion of Eastern capital goods imports began – and about 1977. Falls thereafter reflected a number of factors. For instance, the constraints on Eastern export industries due to shortages of materials resulting from the slow growth of domestic output, and balance of payments problems which dictated import cuts. Both played an important role. The need to direct an increasing volume of Eastern European exports to the Soviet Union to finance fuel and raw material imports reduced the range of East European exportables.

Finally, the sharp competition from the newly industrialized countries' export-led growth policies also led to losses in market shares. The trend of Eastern market share gains was thus reversed after 1977, even though Western equipment imported earlier in the decade was presumably coming on stream. The prior existence of an upward trend before the mid-1970s provides another reason for the weak relationship between Eastern imports of Western capital goods and Eastern exports to the West noted in Table 3.

It should be stressed that there is no sign that these losses in competitivity have been reversed in recent years. Non-fuel industrial goods have registered further declines in all countries but the German Democratic Republic since 1980 or 1981. Moreover, the range of competing goods has narrowed further. A comparison of the number of competing goods as defined for the purpose of Table 6 shows that their number declined from 27 to 7 in Bulgaria between 1981 and 1983. For the other countries, the numbers fell as follows: from 14 to 13 in Czechoslovakia; from 20 to 10 in the German Democratic Republic; from 39 to 27 in Hungary; from 32 to 17 in Poland; from 20 to 14 in Romania; and from 21 to 11 in the Soviet Union. Thus, the commodity basis of competitive export performance continues to decline.

d) *Import Savings*

To the extent that Eastern imports of Western technology result in import saving (and the reduction of adverse foreign trade deficits on industrial inputs is now a central preoccupation in most Eastern countries and particularly in Eastern Europe), successful use of Western technology for this purpose could have important foreign trade consequences. Up to 1980, there was little sign that the Eastern countries had registered successes in this area. Indeed, there is strong evidence that the expanded import of Western capital goods in the 1970s worsened the problem. As can be derived from Table 6 of Chapter 6, the value of intermediate (technology) goods, which had been slightly lower (98 per cent) than the value of capital goods imported from the West by the East in 1970, rose so fast that by 1980 their value actually exceeded that of capital goods by 30 per cent. As a result, the balance of trade deficit on these goods approached the deficit on capital goods, whereas it had been less than half the former in 1970. This imbalance persisted up to 1983 in Eastern Europe but not in the Soviet Union, where capital goods imports from the West climbed fast after 1981.

As these figures imply, the demand for intermediate (technology) goods accompanied or even antedated the expansion of capital goods imports. But the timing suggests that the expansion of capital goods imports from the West during the 1970s was one factor which disproportionately increased demand for intermediate (technology) goods. While some increases could be expected, the fact that the increase in intermediate (technology) goods took place during a period in which efforts were being made to improve the technological level of the economy – and hence, presumably, the output of intermediate (technology) goods from capital goods imports of different kinds – suggests that the difficulty in doing so was underestimated by the Eastern planners. It seems likely that more of these products of advanced technology were needed as inputs for the capital goods imported – and they, too, had to be imported. Since 1980, intermediate (technology) goods could not be cut back as sharply as capital goods imports, but the effects of the fall in absolute terms which nonetheless took place on industrial output, were a major, if unquantifiable, element in the economic slowdown during this period.

As shown in Table 7, the deficits *in real terms* on production by all main trading sectors and branches of the economy increased sharply between 1970 and 1975 in all Eastern countries. They continued to do so, though at a much lower rate, in some sectors in some countries even after 1975 – until the more severe cutbacks in imports began to be implemented after 1978 and again in 1981. For 1976-1980, the Soviet Union, where import cutbacks have been much less severe (and less necessary) shows an almost unchanged – and in some cases faster – growth in its various branch deficits in real terms compared with 1970-1975. Apart of course from fuels, where the big Soviet surplus continued to grow even in constant price terms, only the engineering sector showed a diminution in the West's surplus in trade with the Soviet Union up to 1980 but this rose strongly rose up to 1983. This reflected primarily the cuts in the volume of Soviet imports of capital goods, which are the main constituents of engineering branch output (since 1975). Similarly, engineering imports have been one of the principal branches subjected to volume cutbacks in all the East European countries but the German Democratic Republic (excluding imports from the Federal Republic of Germany) and Hungary. Despite the turnaround in real deficits in Eastern Europe since 1980, developments up to that year suggest that little import substitution had taken place. By 1983, cuts in imports had transformed previous deficits into surpluses, but the continuing high level of the deficits up to 1980 suggests that this reflected less the substitution of imports by domestic or other non-OECD suppliers than the simple necessity of doing without.

## 2. The Domestic Impact of Technology Transfer

### a) *Macroeconomic Estimates*[5]

Although the macroeconomic approach is abstract and prone to a number of well-known methodological limitations[6], it is nevertheless an essential part of any assessment of the economic impact of imported technologies. Disaggregated microeconomic approaches rest upon detailed studies of particular industrial sectors, and are thus liable to be unrepresentative of industry as a whole. They may also fail to capture the full impact of technologies whose effects carry far beyond the confines of one industrial branch. At one end of the spectrum is the innovative Green-Levine study which concluded that at the margin, one ruble's worth of machinery imported by the Soviet Union in the period 1961-1974 was as productive as 8.1 rubles of non-Western capital equipment in chemicals, petrochemicals and petroleum products, as 21.9 rubles in machine-building and metal-working and as 14.0 rubles in industry as a whole. It was also implied that a 10 per cent growth of machinery imports from the West (mean annual growth for 1961-1974 was 10.5 per cent) added 1.1 per cent to the annual growth rate of Soviet industrial output[7]. These benefits are very substantial indeed: so substantial that, intuitively, they are difficult to accept. Weitzman, using the same data but specifying the relationships in a different way in order to take greater account of indigenous technical progress, remained undecided about the impact of Western technology because the results of calculations depended so much on the initial assumptions[8]. In his study of Soviet technical progress from 1950 to 1973 Rosefielde concluded that domestic R&D was the prime mover, with imported capital playing a negligible role[9]. Stanislaw Gomulka, in a more refined version of this conclusion, arrived at by different methods, argues that Western technology has been important but that machinery imports are a relatively unimportant channel of that transfer in the case of countries like the Soviet Union which have a huge R&D establishment. Here, and in contrast to many of the other East European countries, disembodied diffusion and induced innovation are thought to be relatively more important[10]. Yasushi Toda, critizing some of the assumptions underlying previous attempts by Weitzman and Green-Levine, found that there was no statistically significant difference between the marginal productivity of imported and indigenous capital goods[11]. At the opposite end of the spectrum from Green-Levine, Kaser in his study of Romania has even hinted that the importation of CMEA capital equipment may have been relatively *more* effective than Western equipment because it was easier to assimilate. This conclusion stems from an analysis of the relationship between Western equipment imports and net increments to industrial output, lagged for various periods[12]. No significant correlations could be established.

A later study by Brada and Hoffman on Soviet imports of Western capital goods offers a further reformulation of the functions employed in the above studies[13]. Arguing first that the varying answers found reflect conflicting conceptions of the specification of the production function, the authors point out that three of the studies on the Soviet Union depend on restrictions of the production function that are inconsistent with our knowledge of the production relationship in Soviet industry. In particular, this applies first to the assumption retained in all three studies that Soviet industry and its branches are characterized by linear homogenous production functions. The authors also criticize the Green-Levine and Toda assumption that the elasticity of substitution between capital and labour is equal to one. Finally, they also point out that since imports consist only of equipment while indigenous capital consists of both structure and equipment, the necessary conditions for the aggregation of the capital inputs may not be satisfied in the Soviet case.

The functions so derived show differences between the productivity of Western capital inputs and those deriving from indigenous (or other CMEA country) sources – but for some branches and for industry as a whole they indicate that indigenous equipment is, surprisingly, more productive than that imported from the West. The question of why, if this is so, the Soviet Union continues to import Western capital goods is answered in the following terms. First, it may be that the result stems from difficulties in using Western plant at full capacity. Second, and assuming that Western plant is, in fact, *potentially* more productive than the indigenous product, enterprises using them may find that it enables certain inputs to be used more economically, and as such it may serve as a performance model. Third, the copying of techniques embodied in imported equipment could also help to raise the productivity of Soviet indigenous capital. Finally, an indirect but potentially large contribution could be made by complementarity between Western supplied and indigenous technology. Case studies have shown numerous instances of this : for example, mineral fertilizer production based on Western technology for producing ammonia or submersible pumps for secondary oil recovery. Moreover, the econometric tests carried out by Brada and Hoffman to identify such complementarity proved positive for the chemical and petroleum industries, though not for engineering or total industry.

The wide variations in the results of econometric analyses of the impact of imported Western technology on the East European economies appear to be the result of a number of inherent limitations:

- the slender data base upon which elaborate calculations are often based;
- difficulties in specifying the most appropriate production function (especially, in making adequate allowances for technical progress and bringing into agreement different assumptions about the degree of substitutability between domestic capital and imported machinery);
- the difficulty in isolating significant relationships among the variables or interpreting the precise meaning of residuals,
- difficulty in estimating the exact contribution of complementary Soviet inputs and governmental priorities, which accompany the importation of Western plant and equipment[14].

The problem of the slenderness of the data bases remains. It is reinforced by the fact that, in most of the studies quoted, official data have been adjusted to remove presumed inaccuracies. There is no way of verifying the accuracy of these adjusted data. It is perhaps regrettable that no results are presented on the basis of the official data as they stand, or for the larger number of industrial branches available from most Eastern countries' statistics, or for more of the countries of Eastern Europe.

Problems of specifying the production function are also real. The successive refinements and corrections made to the underlying assumptions affect the results very considerably (contrast Green-Levine and Brada-Hoffman in the Soviet case). In fact, however, most of the earlier studies referred to, including the Brada-Hoffman paper which explicitly takes these into account, suggest that the direct effects of Western technology may be rather limited in terms of differential marginal productivities. The fact of continuing imports of Western technology, despite this several times recurring indication from the econometric work, does suggest that benefits may accrue mainly in terms of opportunities to emulate Western techniques, to duplicate and replicate processes and, perhaps, to copy equipment. That is, the effects are indirect and educative, rather than directly production or export-enhancing. Also, in some instances Western equipment fills gaps in areas where indigenous technology is unavailable – the "bottleneck" removing function. In these cases it plays an important complementary role. These differences in the role played by Western technology can, moreover, be expected to have different kinds of economic effects. Hence, quite apart from the difficulties inherent in measuring overall relationships, the importance of these individual effects may contribute (to a varying degree over time) to either cost saving or output enhancement. They also could, where emulation is of importance, have an effect lagged over a certain number of years, rather than having an immediate direct effect of any kind, and so on. Moreover, such effects could, and almost certainly do, vary in magnitude from one sector or industrial branch to another. The fact that the Eastern countries continue to import capital goods [as well as intermediate (technology) goods] is thus not inconsistent with the somewhat inconclusive nature of the macro-economic investigations. This may simply reflect the hard fact that technology transfer is no simple matter. It may contribute in different ways and at different times in different countries. Hence, no general approach to measuring its effects is likely to be valid.

### b) *Eastern Imports of Western Technology and Internal Developments*

The inconclusive nature of econometric measures of the economic impact of technology transfer does not preclude a simpler statistical approach designed to shed light on a few general questions. It is clear however, that assessing the impact of technology imports from the West on the economies of the Eastern countries necessarily involves many elements of judgement. The statistical information available can, at most, only serve to set broad limits within which different interpretations appear equally plausible. The reasons are several. First, our data refer to the import of various categories of goods. What kind of technology and how much of it is embodied in these goods or accompanies them, and with what speed and success this technology is put to use in the destination countries, are not directly reflected in the figures. Furthermore, imports of technology interact with other features and trends of the recipient economies, and at the same time these features and trends also interact with each other. The following among them are likely to have influenced the process of absorption of Western technology in the 1970s:

- the pre-existing technological capacity of each Eastern country. The term is meant to be understood in a broad sense. It includes not only the number of scientists, engineers, managers, technicians and skilled workers available, but also their qualifications, experience and self-confidence. Beyond this, technological capacity is also affected by the work traditions and entire culture of the recipient society;

- the supply of labour and its change over time in terms of numbers, age and sex distribution, geographical location, level, and structure of qualifications. The evolution of qualifications depends, in turn, strongly – though not exclusively – on the resources which each country devotes to investment in human capital. This is another aspect which our statistics do not cover;
- investment in physical capital. In this field, beyond the global figures and proportions, structure is important – both in terms of distribution among the sectors and branches of the economy as well as between buildings and machinery, between existing and new enterprises, etc. The gestation period of investment projects exerts an influence: the long delays in completion are a constant source of concern in all East European countries;
- economic efficiency, incentives and motivation. It seems that in Eastern Europe today the practically important issue is not the point of principle whether technological innovators, enterprises and persons, should be rewarded. The debate between "material" and "moral" incentives has been decided in favour of material incentives long ago. Rather, the crucial questions are the following. How far does the planning and management system encourage, or make possible at all, the development or adoption of new technology? Do the people capable of such initiatives occupy the positions that command the resources required by the initiatives? How far do the rewards of successful innovation reach the persons actually responsible for the achievement? How does the size of the attainable rewards relate to the personal risks attending unsuccessful attempts at technical innovation – and attending successful attempts?,
- attitudes and policies shaping economic develoments on the national level. For example, even though promotion of the "technical-scientific revolution" is a declared official aim, in specific cases this is likely to clash with such explicit and implicit goals as the avoidance of open unemployment and protection of the positions and quiet life of the bureaucracy. In such cases the primacy of technical progress can by no means be taken for granted.

In view of the number and complexity of the influences mentioned, one cannot expect to find a clear correlation between imports of technological goods from the West and some index of technical progress in the East. A number of relevant variables are shown in Tables 8A-8G, expressed in the form of compound annual growth rates at current values from 1970 to 1975, 1975 to 1980 and 1980 to 1983. The first observation, recalling the movements described in Chapter 5, is that in all the Eastern countries the growth of total capital goods imports in value terms from the OEDC area was slower, and often negative, in the second half of the 1970s. They declined further during 1980-1983. The same can be said of imports of capital goods for use in industry, with the sole exception of Czechoslovakia where this flow grew faster in the second half of the decade. In most of the countries concerned, imports of capital goods from within the CMEA also decelerated, even if in this case, the difference between the two halves of the 1970s was less marked. Thus, even if a clear-cut influence of capital goods imports on some other variables could be discerned in the Tables, it would be difficult to disentangle the impact of capital goods imports from the West on one hand and of such imports from the East on the other hand.

The growth of total fixed assets was also slower in 1975-1983 than in 1970-1975 in most of the Eastern countries, including the Soviet Union. However, in three countries it was faster.

The greatest acceleration was registered in Poland where the growth rate of capital goods imports, both from the West and from the world as whole, fell the most. However, if the change in investment rather than the change in fixed assets, is considered, the slowdown in imports of capital goods by Poland is seen to be associated with a sharp reduction in the growth rate of investment. These began to fall in absolute terms in the second half of the 1970s. This finding fits in with the known tendency of Polish economic policy in this period to use foreign borrowing to fill not only a "trade gap" in the economy – a shortage of convertible currency earnings to finance imports from the West – but a "savings gap" as well. Thus, when in 1975-1980 circumstances made a deceleration of capital goods imports from the West inevitable (amounting for most countries to a reduction in real terms), and the growth of trade in capital goods within the CMEA also slowed down, the other Eastern countries were able to maintain the rate of investment by shifting its composition towards domestically produced buildings and machinery to a much greater extent than could Poland, where the necessary savings capacity was also lacking. The difficulties associated with the sharp change in the direction of Polish economic policy are also reflected in a negative rate of change of total employment in 1975-1980. In 1970-1975, of the six East European countries, employment had risen fastest in Poland, – equalling the Soviet Union in this respect. For this reason, the rate of growth of capital intensity per worker in Poland rose by half in the latter period, but even so was less than the rate of increase of this variable registered in Romania and Bulgaria.

With respect to production, labour productivity and capital productivity, the Tables do not permit comparison of the total economies of the European CMEA countries. However, comparisons of the performance of the industrial sectors are feasible. The growth of production slowed down in all of these countries – by most in Poland and by least in the German Democratic Republic. In the first half of the decade, labour productivity in industry rose fastest in Poland; in the second half it increased there most slowly. However, the growth rate of industrial labour productivity was reduced in the other Eastern countries too, except in Bulgaria, the deceleration lying between 3 and 1.5 percentage points in the Soviet Union, Czechoslovakia, Hungary and Romania. The rate of change of capital productivity in industry was negative in 1970-1975 in all countries but Poland. In 1975-1983, the capital productivity decline accelerated in all countries and was highest in Poland.

Turning to the individual sectors and branches of the Eastern economies, it is a known fact that, in relation to their share in total output, some activities are allocated much more investment than others. In relation to their share in total investment some absorb more in the way of imported capital goods than others and, within the total of imported capital goods, for some sectors and branches the proportion of imports of Western origin is significantly higher than for others. In terms of total investment, in the Soviet Union, for instance, it is construction and the food and light industries which consistently receive much less than their share in total output, and in this respect no change over the 1970s can be observed. Between 1970 and 1975, Soviet industry as a whole lost a few percentage points in the distribution of investment and agriculture gained a little. In the second half of the decade and thereafter, the new proportions seem to have stabilized, but whether this development had any connection with capital goods imports from the West is not clear. What appears, on the other hand evident, is that the more plentiful availability of imported capital goods in the mid-1970s made it possible to reduce the share allotted to industry from Soviet imports of both Eastern and Western origin, and that by the end of the decade these shares rebounded to approximately their previous level. Within industry, the easier economic circumstances up to 1975, showed up in a significant increase in the share of light industries in imported capital goods. Between 1975 and 1983, this share rose marginally further for imports from the

CMEA but was reduced drastically to less than half of its 1970 level in imports from the rest of the world. In metallurgy, the figures show a shift from Eastern to Western imports and then about half way back, whereas the proportion of Western capital goods imports absorbed by engineering and metal-working showed a consistently falling trend and that of chemicals a consistently rising one. The increased attention paid to the Soviet energy sector is reflected in the rising share of fuel in total investment (presumably spurred on by the gradual decline of its share in output) and by the approximate doubling of the share of this branch in imports of capital goods of both origins over the decade. For imports from the West, the jump occurred between 1970 and 1975, and for imports from the East between 1975 and 1983. While electrical energy did not increase its share either in total Soviet investment or in Soviet imports of capital goods from the CMEA, the branch's share in capital goods imports from the West was over twice as high in 1983 as in 1970, and almost three times as high as in 1975.

For the six countries of Eastern Europe, the following Table shows the correlation coefficients (r) between the distribution of their capital goods imports from the West in 1980 and the corresponding distribution of the Soviet Union:

| | Bu | Cz | GDR | Hu | Po | Ro |
|---|---|---|---|---|---|---|
| Distribution among 4 sectors | 0.960 | 0.890 | 0.972 | 0.913 | 0.999 | 0.966 |
| Distribution among 9 industrial branches | 0.783 | 0.004 | 0.204 | 0.173 | 0.859 | 0.883 |

Whereas the coefficients referring to the sectoral distributions are very high, reflecting the basic similarity of long-term development strategies, within the industrial sector on the branch level two groups of countries appear to have emerged. Those which are still in the relatively early stage of industrialisation – Bulgaria, Poland and Romania – show structures similar to that of the Soviet Union. By contrast, the main concern of industrial policy in Czechoslovakia, the German Democratic Republic and Hungary is to find ways of productivity improvement and profitable international specialisation. The branch distribution of their imports of capital goods from the West for use in industry differs widely from the Soviet pattern. Nevertheless, inspection of the Table suggests that in all the CMEA countries (with the possible exception of Romania) the concern with energy supplies – which can be said to be more a sectoral than an industrial branch matter – has manifested itself in the choice of capital goods imports from the world outside the CMEA.

Tables 8A-8G also make possible an analysis of the impact of capital goods imports from the West on other important indicators of the Eastern economies. However, in its sectoral and branch detail, the picture presented by the Tables is too complex to permit the drawing of any obvious inferences. To provide greater insight Table 9 has been prepared, using Tables 8A-8G as its source. It shows correlation coefficients between changes from 1970-1975 to 1975-1980 in the compound growth rates of various variables, first for each of the seven Eastern countries, using observations from all branches of industry, and then for each of the sectors and industrial branches using observations from all the countries taken together. The calculations were not extended to 1981-1983 because of the sharp breaks in the trend of both domestic and capital goods import developments. A serious shortcoming here is that, while the domestic indicators are shown in constant prices, the two series on capital goods imports, not shown in Tables 8A-8G, are in current values. Although a Western-reported series at constant prices has been used in other sections of this chapter for imports, no constant price series is available for intra-CMEA capital goods trade. The Eastern-reported data at current prices have thus been retained for reasons of comparability. This limitation on the basic data, together with the small number of observations contained in each series, already explains why probably none of the coefficients is statistically significant at any conventionally accepted level. Nevertheless, in the aggregate they suggest – though they do not prove – certain relationships among the data.

In the countries under strongest balance of payments pressure (Poland and Romania), imports of capital goods from East and West tended to vary together. Curiously, this appears to have been the case in the Soviet Union too. By contrast Czechoslovakia and Hungary were more successful in compensating the fall in the growth rate of capital goods imports from one source with increased reliance on the other source. Taken together, the countries tended to practice such offsetting for the total economy and for industry as a whole, for agriculture, and for the priority industrial branches (fuel, electrical energy and chemicals). In food and tobacco processing, light industries and engineering, capital goods imports from the two regions of origin were more often allowed to slow down together. The results for engineering perhaps suggest that, for many capital goods imported for this branch from the West, no technically satisfactory substitute was available from Eastern suppliers or, at least no substitute that would be preferred to the equipment produced by each country for domestic use.

Capital goods imports from the world and investment activity in industry slowed down together in Poland, but to a lesser extent in Hungary and Bulgaria. In contrast the German Democratic Republic (which has the most advanced capital goods producing industries in the area) managed to accelerate industrial investment in the face of a slowdown of capital goods imports. In the Soviet Union, Romania and Czechoslovakia virtually no connection between the two variables was registered. Over the whole group of Eastern countries, a strong connection can be discerned between changes in the rate of growth of capital goods imports and in investment, in the total economy and in industry as a whole, as well as in most industrial branches. However, in electrical energy, and in the food industry the relationship appears to be weaker, and virtually non-existent in agriculture, construction and fuel.

Changes in the growth rate of investment were, on the whole, associated with changes in the same direction in the productivity of labour, albeit very weakly, in Bulgaria, Hungary and Poland. The correlation was weakly in the opposite direction in Czechoslovakia and the German Democratic Republic. This is a surprising finding, especially as one would expect the tight labour situation of these two countries to encourage "capital deepening" (productivity-raising) rather than "capital widening" (capacity-increasing) investment. The negative coefficients of the Soviet Union and Romania perhaps reflect economy-wide co-ordination and management problems in these countries. The sectors and branches show positive relationships between investment and labour productivity though of widely varying strength, and there are four exceptions. Agricultural output per head is affected by the weather; the negative cases among the industrial branches are unexplained.

For five of the seven Eastern countries, changes in the growth rate of labour productivity are also found to have been correlated positively, though nowhere very strongly, with changes in the growth of capital goods imports – both from the world and from the rest of the world considered separately. Why Bulgaria and Czechoslovakia constitute exceptions is not clear. As between capital goods imports from the World and from the West alone, in some countries productivity growth appears to have been more closely associated with one and in other countries with the other. The relatively high correlation with Western imports shown by the German Democratic Republic may reflect a practice of buying capital goods from the OECD area very selectively: only if they are indispensable to eliminate a bottleneck or if they are technically much superior to any substitute produced in the German Democratic Republic or elsewhere in the European CMEA. Among the sectors and industrial branches the coefficients are also mostly positive, in some cases higher for total imports and in others for imports from the West. Nevertheless, wood and paper constitutes an exception with respect to capital goods imports from the world. The cause is not known but it may be noted that in this branch, changes in the growth rate of labour productivity have been negatively correlated with changes in the growth rate of investment as well. Finally, electrical energy and chemicals stand out as exceptions with respect to both total and Western capital goods imported by the Eastern countries. The grounds can only be guessed at. For electrical energy they may include the long gestation period of projects, so that much expensive equipment (e.g. for nuclear power stations) begins to yield output only years after it had been imported. Chemicals may have experienced problems of capacity utilisation and diversification in several countries. In this branch the association between changes in the growth rates of investment and of labour productivity was also found to be extremely weak.

Can any general conclusion be drawn about the impact of technology imports from the West on the Eastern economies? The inferences suggested by the information at our disposal appear to be mostly negative.

Neither the very fast growth of technology imports from the West in the first half of the 1970s nor their slowdown since can be regarded as the underlying cause of the internal economic difficulties of the CMEA countries. Rather these fluctuations were themselves sysmptoms of more deep-seated problems. The latter arise partly from the vicissitudes of the world economy and from the place and role of the Eastern countries in it, and partly from the economic and social system of these countries in its totality. Judging by the Polish experience it may nevertheless be suspected that technology imports from the West financed by large-scale borrowing contributed to the explosive emergence of these problems into the open (for a while they seemed to offer a promising way ahead and thereby distracted attention from the need to introduce reforms). This leads to the conclusion that the Eastern countries cannot realistically expect to find the solution to their current problems in the field of technology imports from the West: whether they seek it in attempting to restore the growth rate of these imports, or at least their volume, to the level reached in the mid-1970s, or whether they choose to compress the imports to the tolerable minimum.

## 3. Conclusions

### *Impact of Western Technology on the Eastern Economies*

The conclusions to Chapter 6 noted that Eastern technology import strategies consisted of a number of different elements. Hence the impact of the various strategies was, at the outset, likely to be diffused among several different outcomes (for instance, between efforts on exports, on import saving, on domestic output and efficiency indicators), any of which could also be diffused differently among different sectors and branches of the economy.

In order to clarify the findings suggested by the study, the question of impact will be discussed under two broad headings: impact on export and effects on the domestic economy. The impact with regard to individual sectors and branches will also be discussed under the same two headings.

#### a) *Export*

For most of the 1970s, Eastern exports to the West failed to match import growth. The Eastern export failure did not only reflect sluggish demand in the West. The competitiveness of Eastern exports did, to some extent, increase between 1970 and 1977; though, as indicated by their share of OECD's total imports, it did not rise spectacularly. However, these modest gains were wiped out after 1977. By 1983, the Eastern share of total OECD imports of non-fuel commodities had fallen below the 1970 level. Thus while market conditions in the Western countries certainly contributed to the growth of Eastern deficits, the main cause of the growth of trade deficits was the Eastern failure to export goods of the kind, quality and in the quantities necessary to maintain their previous shares of a more slowly growing Western market.

To the extent that a strategy of incurring medium-to-long term debt to pay for current imports of capital goods was adopted by the Eastern countries, a degree of short-term imbalance in East-West trade was obviously foreseen. What was perhaps not foreseen, was the extent to which the deficits in capital goods (some $68 billion over 1971-1980) would be supplemented by further deficits in other kinds of technology imports – notably intermediate goods with a significant technology content and also agriculture. For these categories the deficits built up to $49 billion and $14 billion respectively over the same period.

This failure is reflected in the difficulty of isolating a single "strategy" toward Eastern imports of Western technology noted in the conclusions to Chapter 6. There are clearly elements of export promotion – notably towards the chemical industry (the Soviet Union, the German Democratic Republic and perhaps also in Hungary), motor cars (Poland), and clothing and processed foods (Hungary). But the industries producing tradeable goods, let alone enterprises producing mainly for export, have not – as far as can be traced – been the main recipients of Western technology. This lack of emphasis on export production, despite (or because of) a degree of success in individual cases noted in section 2*c)* above, is the key to the lack of impact of Western technology on aggregate Eastern export performance on Western markets.

### b) *The Domestic Impact and Import Saving*

As noted above, the greater part of total imports of Western technology has been devoted to sectors and industries producing goods or providing services especially for domestic markets or for export outside the OECD area. Recent economic approaches to measuring the impact of Eastern technology imports from the West on the economy as a whole, though not always consistent with previous work, show some measure of agreement that differences between the productivity of imported Western and indigenous capital goods are rather small. The rationale underlying such imports is, therefore, to be found in the marginal improvements in domestic performance which they bring about, together with an educative effect which cannot be measured.

These somewhat imprecise findings are echoed in attempts made to find correlations between the growth rates of imports of Western capital goods in total, and on a sector branch basis, with the variations in the rate of output and factor productivity changes [section 2*b)* of this chapter]. While due to certain inconsistencies in the data totally conclusive results are not possible, some of the relationships among the data are suggestive. In five of the seven Eastern countries – the exceptions being Bulgaria and Czechoslovakia – a positive, though not a strong, correlation was found between sector and branch rates of growth of capital goods imports as a whole and labour productivity growth. A relatively high correlation is shown with regard to industry as a whole in only one country – the German Democratic Republic – but, a number of individual sectors and branches also gave a similar result.

The lack of consistency in these results suggests that imports of Western capital goods cannot provide the sole or even the main contribution to low or only slowly rising factor productivity. Nevertheless, the findings confirm what the continuing import of these items already suggest: Western capital goods make an appreciable contribution to Eastern economic performance, though it is not always possible to pinpoint it. The bottoming out of such imports following the recent period of import cuts suggests that a tolerable minimum level had been reached – though Romania is a striking exception. Given their general correlation with fluctuations in the growth of total investment, it is likely that such imports will resume their growth if, and when, overall economic growth resumes or investment ratios begin to rise.

Table 1 (Chapter 8)

WESTERN TECHNOLOGY EXPORTS TO THE EAST AND TOTAL EAST-WEST TRADE
(Indices (1970 = 100) based on data in US dollar prices of 1970)

| | 1975 | 1980 | 1982 | 1983 |
|---|---|---|---|---|
| A. Western technology exports to the East | | | | |
| OECD capital goods exports to the East: | | | | |
| Eastern Europe | 209 | 137 | 107 | 104 |
| Soviet Union | 210 | 167 | 237 | 270 |
| Total | 210 | 153 | 173 | 188 |
| OECD intermediate (technology) goods exports to the East | | | | |
| Eastern Europe | 183 | 189 | 130 | 135 |
| Soviet Union | 252 | 372 | 361 | 394 |
| Total | 208 | 257 | 215 | 230 |
| B. Total East-West trade | | | | |
| Total OECD exports to the East | | | | |
| Eastern Europe | 179 | 191 | 141 | 136 |
| Soviet Union | 265 | 346 | 428 | 462 |
| Total | 215 | 256 | 260 | 271 |
| Total OECD imports from the East | | | | |
| Eastern Europe | 128 | 162 | 146 | 148 |
| Soviet Union | 126 | 147 | 156 | 157 |
| Total | 127 | 156 | 150 | 152 |

Source: OECD East-West technology transfer data base (Western statistics).

Table 2 (Chapter 8)

RATIOS OF WESTERN IMPORTS FROM THE EAST AND WESTERN EXPORTS OF CAPITAL GOODS TO THE EAST, BY SECTOR AND BRANCH
(coefficients based on absolute value at current prices)

| | Bulgaria | | | | | Czechoslovakia | | | | | German Democratic Republic | | | | | Hungary | | | | |
|---|---|---|---|---|---|---|---|---|---|---|---|---|---|---|---|---|---|---|---|---|
| | 1970 | 1975 | 1980 | 1982 | 1983 | 1970 | 1975 | 1980 | 1982 | 1983 | 1970 | 1975 | 1980 | 1982 | 1983 | 1970 | 1975 | 1980 | 1982 | 1983 |
| Energy | - | - | - | 2.00 | 1.93 | - | 0.12 | 0.24 | 0.27 | 0.27 | 1.32 | 0.37 | 5.91 | 13.68 | 11.19 | - | 0.01 | 0.27 | 0.03 | - |
| Fuel | 0.31 | 0.07 | 8.68 | 4.81 | 9.50 | 4.40 | 5.95 | 10.96 | 11.53 | 11.31 | - | - | - | - | - | 2.76 | 0.75 | 2.06 | 1.66 | 1.22 |
| Metallurgy | 10.66 | 3.27 | 6.74 | 4.26 | 4.27 | 8.39 | 4.56 | 7.44 | 11.12 | 8.53 | 13.82 | 16.95 | 8.45 | 8.26 | 2.88 | 17.94 | 10.54 | 8.75 | 11.21 | 15.10 |
| Engineering and | | | | | | | | | | | | | | | | | | | | |
| metal-working | 4.17 | 1.24 | 3.62 | 0.95 | 0.46 | 10.04 | 6.06 | 6.08 | 7.75 | 6.94 | 12.98 | 20.70 | 8.94 | 9.82 | 7.26 | 9.46 | 8.13 | 8.74 | 14.24 | 9.86 |
| Chemicals | 1.24 | 1.12 | 10.44 | 3.41 | 3.97 | 2.09 | 3.28 | 4.89 | 4.66 | 5.66 | 4.10 | 10.24 | 15.51 | 10.83 | 12.40 | 4.19 | 2.75 | 5.22 | 4.25 | 6.46 |
| Construction | | | | | | | | | | | | | | | | | | | | |
| materials | - | - | - | - | - | - | - | - | - | - | - | - | - | - | - | 6.07 | 8.73 | 10.30 | 9.14 | 18.40 |
| Wood and paper | 0.92 | 0.38 | 4.51 | 1.46 | 0.99 | 2.51 | 3.07 | 2.63 | 5.47 | 11.89 | .. | .. | .. | .. | .. | .. | .. | .. | .. | .. |
| Textiles | 0.35 | 0.12 | 1.06 | 0.27 | 0.63 | 1.13 | 2.20 | 4.05 | 3.77 | 5.81 | 1.22 | 1.23 | 2.52 | 2.55 | 5.15 | 3.24 | 1.92 | 2.95 | 1.81 | 2.66 |
| Other light industry | 11.29 | 1.47 | 5.52 | 4.66 | 3.55 | 8.13 | 11.49 | 15.96 | 13.73 | 22.22 | 4.38 | 5.78 | 7.09 | 5.89 | 4.67 | 7.39 | 9.39 | 17.80 | 6.62 | 16.53 |
| Food | 20.95 | 2.76 | 7.37 | 6.13 | 5.68 | 4.30 | 4.19 | 4.18 | 7.99 | 6.41 | 4.62 | 13.29 | 5.97 | 4.89 | 8.69 | 20.93 | 23.63 | 17.35 | 14.04 | 17.11 |
| Other | 0.21 | 0.25 | 0.55 | 0.31 | 0.24 | 1.05 | 0.92 | 1.64 | 1.29 | 1.53 | 0.83 | 2.54 | 1.17 | 0.95 | 0.77 | 1.02 | 0.78 | 1.21 | 0.99 | 1.42 |
| Total industry | 2.36 | 0.96 | 3.58 | 1.91 | 1.71 | 3.28 | 3.42 | 4.64 | 5.13 | 6.05 | 3.50 | 5.75 | 5.54 | 5.76 | 4.74 | 5.08 | 4.01 | 5.05 | 4.08 | 5.44 |
| Agriculture | 94.46 | 8.68 | 8.74 | 5.72 | 6.14 | 16.35 | 5.46 | 20.21 | 12.56 | 7.96 | 61.34 | 18.07 | 9.22 | 9.02 | 48.04 | 57.28 | 5.50 | 4.69 | 4.38 | 3.87 |
| Total material sphere | 3.65 | 1.29 | 4.07 | 2.20 | 1.94 | 3.54 | 3.51 | 4.91 | 5.36 | 6.17 | 3.75 | 6.18 | 5.65 | 5.82 | 4.85 | 6.76 | 4.21 | 5.05 | 5.04 | 5.27 |

| | Poland | | | | | Romania | | | | | Soviet Union | | | | |
|---|---|---|---|---|---|---|---|---|---|---|---|---|---|---|---|
| Energy | - | 0.03 | 0.23 | 0.78 | 1.52 | - | 0.16 | - | - | - | 0.20 | 0.05 | 0.49 | 0.36 | 0.21 |
| Fuel | 31.82 | 13.24 | 15.28 | 18.87 | 23.27 | 5.40 | 11.76 | 32.47 | 103.2 | 198.2 | 23.34 | 27.43 | 85.46 | 31.45 | 41.75 |
| Metallurgy | 11.09 | 1.62 | 9.15 | 13.94 | 18.94 | 4.11 | 3.80 | 6.86 | 33.35 | 31.63 | 2.02 | 0.71 | 1.11 | 1.09 | 1.50 |
| Engineering and | | | | | | | | | | | | | | | |
| metal-working | 3.43 | 2.43 | 7.12 | 5.02 | 9.96 | 1.08 | 1.65 | 2.42 | 19.90 | 9.01 | 0.62 | 0.76 | 0.93 | 0.87 | 0.64 |
| Chemicals | 4.69 | 1.47 | 4.28 | 8.67 | 7.33 | 2.31 | 2.51 | 3.94 | 17.13 | 45.53 | 1.72 | 0.75 | 2.41 | 1.57 | 1.55 |
| Construction | | | | | | | | | | | | | | | |
| materials | - | - | - | - | - | - | - | - | - | - | - | - | - | - | - |
| Wood and paper | 10.16 | 2.39 | 7.10 | 22.31 | 15.89 | 13.21 | 10.94 | 40.35 | 161.3 | 174.9 | 7.03 | 4.17 | 13.28 | 6.31 | 5.17 |
| Textiles | 0.95 | 0.48 | 3.74 | 6.63 | 2.12 | 0.73 | 1.52 | 2.15 | 7.82 | 13.69 | 1.01 | 0.27 | 0.84 | 0.47 | 0.51 |
| Other light industry | 5.65 | 5.19 | 11.86 | 28.61 | 13.35 | 6.46 | 23.76 | 31.90 | 167.7 | 222.2 | 1.14 | 0.79 | 1.65 | 0.48 | 0.28 |
| Food | 29.54 | 5.39 | 16.49 | 25.42 | 24.02 | 14.12 | 25.13 | 20.46 | 73.69 | 195.5 | 8.42 | 2.77 | 1.59 | 1.73 | 1.39 |
| Other | 1.32 | 0.33 | 0.63 | 1.65 | 1.57 | 0.26 | 0.36 | 0.53 | 0.92 | 2.47 | 2.54 | 1.41 | 3.08 | 2.59 | 2.27 |
| Total industry | 6.00 | 2.34 | 5.44 | 8.84 | 8.84 | 2.52 | 3.85 | 6.15 | 18.53 | 33.28 | 3.30 | 2.47 | 6.65 | 6.54 | 5.87 |
| Agriculture | 168.5 | 4.79 | 18.45 | 44.56 | 66.07 | 100.3 | 68.32 | 72.63 | 169.6 | 119.9 | 18.91 | 9.70 | 13.33 | 10.92 | 27.70 |
| Total material sphere | 6.85 | 2.43 | 5.77 | 9.55 | 9.73 | 3.13 | 4.08 | 6.33 | 19.07 | 34.04 | 3.91 | 2.78 | 6.88 | 6.69 | 6.09 |

Note: Total material sphere includes industry and agriculture only.

Source: OECD East-West technology transfer data base (Western statistics).

Table 3 (Chapter 8)

DEGREE OF SIMILARITY BETWEEN SECTORAL PERFORMANCE OF EAST EUROPE EXPORTS TO THE WEST AND SECTORAL PURCHASES OF WESTERN TECHNOLOGY IN EASTERN EUROPE 1970-1977 and 1977-1983
(Spearman Rank correlation coefficients)

| | 1970-1977 | 1977-1983 |
|---|---|---|
| Current prices | | |
| Bulgaria | 0.14 | 0.03 |
| Czechoslovakia | 0.23 | 0.20 |
| German Democratic Republic | -0.01 | 0.18 |
| Hungary | -0.45 | -0.35 |
| Poland | 0.20 | -0.38 |
| Romania | 0.47 | 0.34 |
| Soviet Union | 0.56* | -0.29 |
| Constant prices | | |
| Bulgaria | 0.13 | 0.36 |
| Czechoslovakia | -0.12 | 0.20 |
| German Democratic Republic | -0.33 | 0.41 |
| Hungary | -0.28 | -0.28 |
| Poland | 0.09 | -0.01 |
| Romania | 0.38 | 0.24 |
| Soviet Union | 0.41 | -0.20 |

Note: Based on agriculture plus 11 individual industrial branches (ten for the German Democratic Republic).

* Significant at the 5 per cent level (Olds test).

Source: OECD East-West technology transfer data base (Western statistics).

Table 4 (Chapter 8)

EASTERN COUNTRIES' INDUSTRIAL EXPORT COMMODITIES WHICH HAVE GROWN AS FAST OR FASTER THAN TOTAL INDUSTRIAL EXPORTS BETWEEN 1975 AND 1980 OR 1983
($million at 1970 prices)

| | Eastern Europe | | | | Soviet Union | | | |
|---|---|---|---|---|---|---|---|---|
| | 1975 | 1980 | 1982 | 1983 | 1975 | 1980 | 1982 | 1983 |
| Crude minerals | 52.2 | 59.1 | 49.4 | 44.6 | | | | |
| Crude petroleum, gas | 1.3 | 8.6 | 7.1 | 14.1 | 308.3 | 613.7 | 726.0 | 784.3 |
| Energy | 3.5 | 4.0 | 9.0 | 12.2 | 2.7 | 7.0 | 14.3 | 19.1 |
| Beverages | 38.4 | 65.3 | 52.4 | 51.4 | 5.1 | 4.0 | 4.7 | 6.4 |
| Knitted products | 6.6 | 12.5 | 8.3 | 7.7 | | | | |
| Other textiles | 4.3 | 5.7 | 4.2 | 6.5 | 4.5 | 8.0 | 6.0 | 7.[illegible] |
| Clothing | 333.6 | 423.3 | 416.5 | 416.7 | | | | |
| Leather, etc. | 97.4 | 127.5 | 113.0 | 114.1 | | | | |
| Furniture | 93.0 | 140.4 | 139.9 | 145.3 | | | | |
| Paper and products | 25.3 | 50.9 | 54.1 | 58.0 | 17.2 | 56.2 | 46.3 | 52.2 |
| Glass | 72.8 | 94.2 | 79.9 | 79.4 | | | | |
| China | 21.9 | 24.4 | 23.8 | 23.7 | | | | |
| Basic chemicals | 148.6 | 184.6 | 202.9 | 204.7 | 125.6 | 331.2 | 366.9 | 295.3 |
| Fertilizers | 45.4 | 96.2 | 69.4 | 95.2 | | | | |
| Synthetic resins | 52.5 | 122.3 | 138.4 | 149.2 | 6.2 | 25.2 | 56.9 | 50.7 |
| Refined petroleum | 153.5 | 254.6 | 196.9 | 254.1 | | | | |
| Rubber products | 19.9 | 42.8 | 36.3 | 45.9 | | | | |
| Other chemicals | | | | | 5.3 | 4.1 | 3.7 | 2.8 |
| Other non-metallic minerals | 19.5 | 21.9 | 17.6 | 20.6 | | | | |
| Iron and steel | 309.7 | 495.9 | 469.3 | 433.5 | | | | |
| Non-ferrous metals | 206.4 | 371.6 | 364.6 | 323.1 | | | | |
| Construction, mining machinery | 3.8 | 7.3 | 8.5 | 12.0 | 4.1 | 1.8 | 2.6 | 1.2 |
| Machinery for specialised industries | 4.5 | 3.8 | 6.9 | 10.6 | | | | |
| Office equipment, computers | 20.5 | 35.8 | 30.4 | 28.5 | | | | |
| Electrical machinery for industry | 63.0 | 88.1 | 74.1 | 71.8 | | | | |
| Railway equipment | 11.7 | 21.8 | 29.2 | 12.5 | | | | |
| Motor vehicles | 58.5 | 136.1 | 100.3 | 111.3 | | | | |
| Aircraft | 3.0 | 5.7 | 8.9 | 5.7 | 0.4 | 26.7 | 3.7 | 4.0 |
| Watches, clocks | | | | | 5.5 | 7.7 | 5.1 | 4.7 |
| Scientific and measuring equipment | 21.5 | 28.9 | 28.1 | 24.4 | | | | |
| Photo/optical equipment | 10.6 | 10.5 | 11.0 | 11.9 | | | | |
| Other electrical equipment | 36.3 | 67.0 | 63.3 | 64.5 | | | | |
| Structural parts | 17.4 | 28.2 | 21.7 | 24.1 | | | | |
| Miscellaneous other | 260.2 | 327.5 | 285.1 | 275.4 | 175.0 | 324.3 | 297.9 | 280.4 |
| Total | 2 216.8 | 3 366.5 | 3 120.5 | 3 152.7 | 659.9 | 1 409.9 | 1 534.1 | 1 508.2 |
| Percentage of total industrial goods exports | 55.1 | 65.0 | 67.6 | 67.2 | 26.2 | 44.0 | 45.1 | 43.5 |
| - at current prices | 42.6 | 69.9 | 72.4 | 72.7 | 33.1 | 55.6 | 55.0 | 54.2 |

Source: OECD East-West technology transfer data base (Western statistics).

Table 5 (Chapter 8)

SHARE OF OECD IMPORTS FROM EASTERN EUROPE AND THE SOVIET UNION IN OECD's INDUSTRIAL, AGRICULTURAL AND TOTAL IMPORTS FROM THE WORLD IN 1970, 1975, 1980 and 1983
(per cent)

| | Bulgaria | | | | Czechoslovakia | | | | German Democratic Republic | | | |
|---|---|---|---|---|---|---|---|---|---|---|---|---|
| | 1970 | 1975 | 1980 | 1983 | 1970 | 1975 | 1980 | 1983 | 1970 | 1975 | 1980 | 1983 |
| Total industry | 0.08 | 0.05 | 0.07 | 0.06 | 0.33 | 0.29 | 0.23 | 0.22 | 0.19 | 0.18 | 0.16 | 0.21 |
| - excluding fuel | 0.08 | 0.07 | 0.06 | 0.05 | 0.35 | 0.35 | 0.29 | 0.26 | 0.20 | 0.21 | 0.17 | 0.18 |
| Agriculture | 0.36 | 0.19 | 0.13 | 0.11 | 0.26 | 0.18 | 0.24 | 0.19 | 0.12 | 0.19 | 0.10 | 0.08 |
| Total trade | 0.11 | 0.07 | 0.07 | 0.06 | 0.32 | 0.28 | 0.23 | 0.22 | 0.18 | 0.18 | 0.15 | 0.20 |
| | Hungary | | | | Poland | | | | Romania | | | |
| Total industry | 0.19 | 0.20 | 0.20 | 0.19 | 0.46 | 0.55 | 0.39 | 0.26 | 0.22 | 0.29 | 0.26 | 0.24 |
| - excluding fuel | 0.21 | 0.25 | 0.26 | 0.22 | 0.49 | 0.69 | 0.51 | 0.33 | 0.21 | 0.29 | 0.22 | 0.18 |
| Agriculture | 0.61 | 0.39 | 0.30 | 0.30 | 0.57 | 0.42 | 0.47 | 0.32 | 0.47 | 0.18 | 0.11 | 0.09 |
| Total trade | 0.24 | 0.21 | 0.20 | 0.19 | 0.47 | 0.54 | 0.40 | 0.27 | 0.24 | 0.28 | 0.25 | 0.23 |
| | Eastern Europe | | | | Soviet Union | | | | Total Seven Countries | | | |
| Total industry | 1.46 | 1.56 | 1.31 | 1.17 | 1.02 | 1.34 | 1.72 | 2.08 | 2.48 | 2.90 | 3.04 | 3.24 |
| - excluding fuel | 1.55 | 1.86 | 1.50 | 1.22 | 0.80 | 0.84 | 0.84 | 0.71 | 2.35 | 2.70 | 2.34 | 2.04 |
| Agriculture | 2.39 | 1.56 | 1.34 | 1.19 | 2.05 | 2.55 | 1.78 | 1.45 | 4.43 | 4.11 | 3.12 | 2.64 |
| Total trade | 1.56 | 1.56 | 1.32 | 1.17 | 1.13 | 1.44 | 1.73 | 2.04 | 2.68 | 3.00 | 3.04 | 3.21 |

Source: OECD East-West technology transfer data base (Western statistics).

Table 6 (Chapter 8)

OECD IMPORTS OF INDUSTRIAL GOODS FROM EASTERN COUNTRIES:
TOTALS OF COMMODITY GROUPS WHOSE SHARE OF CORRESPONDING OECD IMPORTS FROM THE WORLD HAS GROWN DURING 1970-1983
(per cent of OECD industrial imports from the Eastern country concerned)

| | Bulgaria | | Czechoslovakia | | German Democratic Republic | | Hungary | |
|---|---|---|---|---|---|---|---|---|
| | 1970 | 1983 | 1970 | 1983 | 1970 | 1983 | 1970 | 1983 |
| Competitive group (a) | 11.1 | 48.3 | 19.3 | 30.6 | 22.9 | 51.1 | 52.9 | 70.3 |
| Number of product items | 7 | | 13 | | 10 | | 27 | |
| Top group (b) | 11.1 | 46.5 | 18.2 | 27.3 | 19.9 | 46.7 | 46.3 | 51.5 |
| Number of product items | 4 | | 6 | | 4 | | 6 | |
| Identified as: | 1. Refined petroleum<br>2. Iron and steel<br>3. Furniture<br>4. Energy | | 1. Coal<br>2. Meat<br>3. Manuf. textiles<br>4. Processed wood<br>5. Paper and prods.<br>6. Synth. resins and fibres | | 1. Furniture<br>2. Synth. resins and fibres<br>3. Refined petroleum<br>4. Iron and steel | | 1. Meat<br>2. 'Other' processed food<br>3. Clothing<br>4. Basic industrial chemicals<br>5. Agrochemicals<br>6. Refined petroleum<br>7. Electrical apparatus n.e.s. | |

| | Poland | | Romania | | Total Six | | Soviet Union | | Total Seven | |
|---|---|---|---|---|---|---|---|---|---|---|
| | 1970 | 1983 | 1970 | 1983 | 1970 | 1983 | 1970 | 1983 | 1970 | 1983 |
| Competitive group (a) | 12.4 | 27.6 | 31.6 | 74.3 | 22.0 | 43.2 | 35.9 | 80.5 | 28.7 | 67.9 |
| Number of product items | 17 | | 14 | | 23 | | 11 | | 21 | |
| Top group (b) | 10.0 | 20.3 | 28.1 | 70.1 | 16.1 | 35.3 | 32.4 | 78.6 | 21.0 | 62.5 |
| Number of product items | 4 | | 6 | | 5 | | 3 | | 4 | |
| Identified as: | 1. Crude minerals<br>2. Clothing<br>3. Non-ferrous metals<br>4. Motor vehicles | | 1. Manuf. textiles<br>2. Clothing<br>3. Leather, footwear and furs<br>4. Furniture<br>5. Agrochemicals<br>6. Refined petroleum | | 1. Clothing<br>2. Agrochemicals<br>3. Synth. resins and fibres<br>4. Refined petroleum<br>5. Non-ferrous metals | | 1. Petroleum and gas<br>2. Basic industrial chemicals<br>3. Refined petroleum | | 1. Petroleum and gas<br>2. Clothing<br>3. Basic industrial chemicals<br>4. Refined petroleum | |

Notes:

a) Group whose share in total OECD imports from the corresponding country has grown.
b) Group accounting for more than 2 per cent of imports by the OECD from the corresponding country at end of period.

Source: OECD East-West technology transfer data base (Western statistics).

Table 7 (Chapter 8)

EASTERN COUNTRIES' BALANCES ON TRADE IN INDUSTRIAL GOODS WITH THE WEST, IN TOTAL AND BY INDUSTRIAL BRANCH (a)
(Western exports f.o.b. minus Western imports f.o.b. at 1970 dollar prices)

| | Bulgaria | | | | | Czechoslovakia | | | | | German Democratic Republic | | | | |
|---|---|---|---|---|---|---|---|---|---|---|---|---|---|---|---|
| | 1970 | 1975 | 1980 | 1982 | 1983 | 1970 | 1975 | 1980 | 1982 | 1983 | 1970 | 1975 | 1980 | 1982 | 1983 |
| Energy | - | - | - | -4.1 | -5.5 | - | -0.7 | -1.2 | -1.5 | -1.0 | 8.1 | -3.2 | -21.7 | -34.9 | -50.5 |
| Fuel | 8.7 | 1.7 | -9.7 | -9.1 | -3.5 | -47.0 | -40.1 | -65.5 | -51.2 | -57.9 | .. | .. | .. | .. | .. |
| Metallurgy | 19.2 | 52.2 | 68.7 | 24.2 | 32.1 | -51.5 | -45.1 | -82.0 | -97.8 | -74.9 | 4.2 | 24.7 | 19.8 | -77.6 | -44.1 |
| Engineering | 95.6 | 244.7 | 122.5 | 212.8 | 237.4 | 165.8 | 214.7 | 172.4 | 171.2 | 149.2 | 63.5 | 26.1 | 32.4 | 58.9 | 113.4 |
| Chemicals | 37.4 | 51.4 | 64.2 | 69.2 | 79.0 | 63.5 | 113.5 | 96.4 | 45.7 | 56.0 | -46.4 | -2.1 | -17.5 | -111.0 | -112.5 |
| Construction materials | 0.9 | 2.9 | 3.9 | 4.5 | 6.5 | -6.1 | -4.0 | -4.3 | -2.9 | -1.4 | -1.4 | 0.3 | 1.3 | -0.3 | -3.2 |
| Wood and paper | 8.2 | 10.9 | 15.2 | 13.3 | 8.8 | -26.3 | -15.3 | -71.8 | -113.0 | -124.4 | .. | .. | .. | .. | .. |
| Textiles | 22.2 | 22.6 | 6.5 | 12.5 | 14.4 | -13.8 | -16.6 | -38.6 | -30.3 | -34.2 | -0.9 | 30.9 | 23.9 | 25.8 | 21.8 |
| Other light industry | -14.0 | -13.3 | -16.4 | -14.8 | -15.0 | -55.9 | -95.8 | -116.4 | -114.7 | -113.6 | -20.8 | -10.6 | -37.2 | -54.4 | -60.3 |
| Food | -29.4 | -16.7 | -0.8 | 0.7 | 2.7 | 8.7 | 0.4 | 9.5 | 18.4 | -10.4 | 14.9 | -20.9 | 57.5 | 52.9 | 69.5 |
| Other | 7.2 | 13.6 | 14.0 | 22.1 | 27.3 | -24.4 | -16.2 | -19.5 | -19.6 | -19.0 | -21.1 | -26.2 | -25.5 | -36.1 | -28.0 |
| Total East-West industrial goods trade | 156.1 | 370.0 | 268.1 | 331.2 | 384.1 | 13.0 | 94.8 | -120.8 | -195.6 | -231.5 | .. | 19.0 | 32.8 | -176.8 | -93.7 |
| - at current prices | 159.0 | 769.5 | 600.0 | 774.7 | 851.7 | 30.2 | 218.1 | -456.1 | -585.3 | -669.0 | 2.6 | 89.8 | -174.4 | -965.0 | -827.6 |
| All East-West trade | 84.9 | 337.4 | 269.2 | 328.8 | 384.3 | 43.9 | 120.1 | -38.2 | -170.9 | -244.5 | 26.0 | 21.4 | 226.1 | -47.4 | 54.1 |
| - at current prices | 87.8 | 707.1 | 612.9 | 747.4 | 828.4 | 61.1 | 236.1 | -249.7 | -520.2 | -675.6 | 28.6 | 95.0 | 387.5 | -680.1 | -442.0 |

Table 7 (Chapter 8) contd

EASTERN COUNTRIES' BALANCES ON TRADE IN INDUSTRIAL GOODS WITH THE WEST, IN TOTAL AND BY INDUSTRIAL BRANCH (a)
(Western exports f.o.b. minus Western imports f.o.b. at 1970 dollar prices)

| | Hungary | | | | | Poland | | | | |
|---|---|---|---|---|---|---|---|---|---|---|
| | 1970 | 1975 | 1980 | 1982 | 1983 | 1970 | 1975 | 1980 | 1982 | 1983 |
| Energy | - | -0.1 | -1.0 | -0.1 | .. | - | -0.8 | -1.9 | -3.2 | -5.7 |
| Fuel | -12.5 | -9.1 | -10.1 | -9.1 | -6.4 | -223.6 | -309.2 | -251.7 | -158.0 | -184.1 |
| Metallurgy | -8.7 | 16.3 | -20.7 | -34.9 | -23.3 | 18.7 | 337.0 | -107.6 | -185.5 | -131.7 |
| Engineering | 116.6 | 188.8 | 192.7 | 245.3 | 202.2 | 182.9 | 897.9 | 295.7 | 51.6 | 79.4 |
| Chemicals | 81.6 | 143.9 | 165.3 | 158.7 | 137.1 | 59.6 | 210.3 | 256.0 | 178.5 | 183.6 |
| Construction materials | 11.8 | 2.9 | -5.4 | -2.0 | -5.0 | -2.8 | 12.2 | 3.2 | - | -0.2 |
| Wood and paper | .. | .. | .. | .. | .. | -30.2 | 12.8 | -41.7 | -59.4 | -68.2 |
| Textiles | 22.0 | 88.2 | 64.7 | 71.5 | 67.6 | 6.5 | 84.3 | 38.4 | 24.6 | 55.2 |
| Other light industry | -3.5 | -68.3 | -96.9 | -89.1 | -106.4 | -23.3 | -82.3 | -110.9 | -88.0 | -83.2 |
| Food | -35.2 | -107.6 | -165.7 | -141.7 | -130.4 | -155.9 | -135.9 | -67.1 | 50.9 | -35.8 |
| Other | 5.1 | 19.9 | 9.3 | 20.4 | 20.3 | -4.9 | 60.3 | 17.7 | 11.0 | 41.9 |
| Total East-West industrial goods trade | 177.3 | 275.0 | 132.3 | 218.9 | 156.1 | -172.9 | 1086.7 | 30.1 | -177.6 | -148.7 |
| - at current prices | 187.9 | 725.6 | 550.7 | 752.6 | 405.5 | -158.0 | 1914.8 | -120.4 | -511.9 | -568.9 |
| All East-West trade | 78.7 | 192.8 | 69.6 | 118.6 | 70.3 | -194.0 | 1246.8 | 412.1 | 0.7 | -108.8 |
| - at current prices | 89.3 | 581.2 | 447.1 | 561.9 | 241.4 | -179.0 | 2325.6 | 1007.2 | -98.1 | -372.8 |

Table 7 (Chapter 8) contd

EASTERN COUNTRIES' BALANCES ON TRADE IN INDUSTRIAL GOODS WITH THE WEST, IN TOTAL AND BY INDUSTRIAL BRANCH (a)
(Western exports f.o.b. minus Western imports f.o.b. at 1970 dollar prices)

| | Romania | | | | | Soviet Union | | | | |
|---|---|---|---|---|---|---|---|---|---|---|
| | 1970 | 1975 | 1980 | 1982 | 1983 | 1970 | 1975 | 1980 | 1982 | 1983 |
| Energy | - | -1.8 | - | - | - | -1.4 | -2.7 | -7.0 | -14.3 | -19.1 |
| Fuel | 23.2 | -59.8 | -15.3 | -17.6 | -61.5 | -791.9 | -870.1 | -1124.0 | -1288.0 | -1441.0 |
| Metallurgy | 84.6 | 131.8 | 117.7 | 11.1 | -28.3 | 157.4 | 1051.5 | 1543.9 | 1485.2 | 1785.7 |
| Engineering | 253.4 | 311.2 | 230.4 | 37.9 | -0.2 | 982.5 | 2130.0 | 1482.0 | 2231.3 | 2509.9 |
| Chemicals | 30.4 | 57.4 | 100.1 | 24.9 | 3.2 | 192.0 | 306.3 | 547.2 | 588.2 | 784.7 |
| Construction materials | 2.6 | 5.4 | 15.3 | 8.1 | 5.8 | -1.3 | 1.5 | 16.0 | 7.1 | 4.6 |
| Wood and paper | -69.7 | -73.9 | -94.9 | -101.2 | -112.6 | -48.5 | 77.7 | 172.3 | 178.0 | 147.7 |
| Textiles | 9.1 | 18.6 | 1.1 | 6.7 | 9.2 | 162.8 | 233.7 | 306.0 | 280.9 | 368.5 |
| Other light industry | -33.5 | -109.7 | -153.1 | -180.5 | -170.8 | 149.5 | 173.6 | 171.1 | 245.3 | 179.6 |
| Food | -58.6 | -136.4 | 50.9 | 3.1 | -4.3 | -32.4 | 216.0 | 1046.0 | 1128.7 | 942.8 |
| Other | 21.7 | 26.1 | 22.1 | 4.1 | 6.2 | 5.6 | 6.7 | -121.6 | 58.0 | 86.7 |
| Total East-West industrial goods trade | 216.9 | 168.9 | 274.4 | -203.4 | -353.1 | 391.3 | 2744.0 | 3657.1 | 4556.6 | 4883.6 |
| - at current prices | 218.8 | 240.9 | -75.8 | -1092.0 | 1615.0 | 361.3 | 3366.9 | -4326.0 | -5216.0 | -5074.0 |
| All East-West trade | 147.8 | 201.8 | 436.3 | -142.3 | -318.2 | -271.7 | 3796.5 | 5380.9 | 7302.2 | 8182.7 |
| - at current prices (a) | 149.8 | 330.5 | 382.7 | -948.3 | -1527.0 | 89.6 | 4053.1 | -2202.0 | -2003.0 | -2073.0 |

Note: a) Imports cif. Hence Western surpluses understated (or deficits overstated). The symbol "-" indicates a Western deficit.

Source: OECD East-West technology transfer data base (Western statistics).

Table 8A (Chapter 8)

BULGARIA, AVERAGE ANNUAL PERCENTAGE CHANGES OF SELECTED VARIABLES, 1970-1983

| | | Capital goods imports from OECD | Exports to OECD | Fixed assets | Fixed invest-ment | Employ-ment | Capital intensity | Pro-duction | Labour produc-tivity | Capital produc-tivity |
|---|---|---|---|---|---|---|---|---|---|---|
| Energy | 1971-1975 | 10.5 | - | 9.4 | 6.2 | 2.8 | 6.4 | 5.4 | 2.5 | -3.7 |
| | 1976-1980 | -9.3 | - | 8.3 | 6.7 | 5.4 | 2.8 | 7.9 | 2.4 | -0.4 |
| | 1981-1983 | 32.2 | 0000 | 9.0 | 5.1 | 4.3 | 4.5 | 3.8 | -0.5 | -4.8 |
| Fuel | 1971-1975 | 20.6 | -20.9 | 3.6 | -1.5 | -2.0 | 5.7 | 8.4 | 10.7 | 4.7 |
| | 1976-1980 | -11.5 | 110.7 | 7.1 | 15.8 | 1.2 | 5.8 | 5.5 | 4.2 | -1.5 |
| | 1981-1983 | -2.9 | -2.9 | 10.2 | -0.5 | -2.7 | 13.2 | -11.9 | -9.5 | -20.1 |
| Metallurgy | 1971-1975 | 25.3 | 3.2 | 7.7 | 5.5 | 1.3 | 6.4 | 10.9 | 9.6 | 3.0 |
| | 1976-1980 | -12.9 | 3.8 | 1.9 | 14.4 | 2.1 | -0.2 | 6.2 | 4.0 | 4.2 |
| | 1981-1983 | 18.7 | 2.3 | 8.4 | 3.8 | 2.2 | 6.1 | 3.8 | 1.6 | -4.3 |
| Engineering and metal-working | 1971-1975 | 25.3 | 2.8 | 13.3 | 5.9 | 5.0 | 7.9 | 14.1 | 8.7 | 0.7 |
| | 1976-1980 | -11.4 | 11.4 | 6.5 | 11.5 | 2.0 | 4.4 | 9.1 | 7.0 | 2.4 |
| | 1981-1983 | 62.6 | -17.3 | 10.9 | 17.6 | 1.4 | 9.3 | 8.3 | 6.7 | -2.3 |
| Chemicals | 1971-1975 | 2.6 | -1.0 | 12.2 | 1.7 | 4.9 | 7.0 | 11.7 | 6.5 | -0.5 |
| | 1976-1980 | -4.7 | 21.7 | 7.2 | -1.2 | 2.1 | 5.0 | 9.4 | 7.2 | 2.1 |
| | 1981-1983 | 26.3 | -0.1 | 4.9 | 2.4 | 2.9 | 1.9 | 5.5 | 2.5 | 0.6 |
| Construction materials | 1971-1975 | - | 2.6 | 10.9 | 21.7 | 4.0 | 6.7 | 9.4 | 5.3 | -1.3 |
| | 1976-1980 | - | 21.9 | 11.6 | 6.8 | 1.1 | 10.4 | 7.5 | 6.3 | -3.7 |
| | 1981-1983 | - | -18.6 | 4.6 | -17.0 | 1.9 | 2.7 | 3.7 | 1.8 | -0.8 |
| Wood and paper | 1971-1975 | 4.7 | -9.8 | 11.2 | 2.2 | 0.5 | 10.6 | 7.4 | 6.9 | -3.4 |
| | 1976-1980 | -30.0 | 11.1 | 6.4 | -0.6 | -1.3 | 7.8 | 3.4 | 4.8 | -2.8 |
| | 1981-1983 | 66.5 | 11.2 | 9.5 | 4.5 | 1.0 | 8.4 | 3.3 | 2.3 | -5.7 |
| Textiles | 1971-1975 | 27.1 | 6.5 | 8.5 | 5.7 | 2.5 | 5.9 | 7.3 | 4.6 | -1.2 |
| | 1976-1980 | -25.5 | 18.0 | 6.6 | -15.0 | -0.4 | 7.1 | 5.0 | 5.4 | -1.6 |
| | 1981-1983 | 5.1 | -15.4 | 4.0 | 4.2 | 0.6 | 3.4 | 4.2 | 3.5 | 0.2 |
| Other light industry | 1971-1975 | 48.9 | 1.7 | 10.7 | 13.9 | 2.0 | 8.5 | 7.3 | 5.2 | -3.1 |
| | 1976-1980 | -19.5 | 5.5 | 7.8 | -13.9 | -0.1 | 7.9 | 3.4 | 3.5 | -4.1 |
| | 1981-1983 | 12.4 | -1.6 | 6.3 | 4.5 | 2.7 | 3.5 | 3.0 | 0.3 | -3.1 |
| Food | 1971-1975 | 40.0 | -1.4 | 8.2 | 9.5 | - | 8.1 | 6.0 | 6.0 | -2.0 |
| | 1976-1980 | -22.0 | -2.0 | 7.3 | -3.0 | -0.8 | 8.1 | 2.4 | 3.2 | -4.5 |
| | 1981-1983 | 9.6 | -0.8 | 5.3 | -5.7 | 0.4 | 4.9 | 4.5 | 4.1 | -0.7 |
| Other | 1971-1975 | 16.4 | 19.3 | 2.6 | 1.6 | 1.2 | 1.4 | 11.7 | 10.4 | 8.9 |
| | 1976-1980 | -4.5 | 8.6 | 18.1 | 11.9 | 2.3 | 15.4 | 6.2 | 3.8 | -10.0 |
| | 1981-1983 | 12.9 | -22.6 | 7.6 | 6.1 | - | 7.6 | 4.3 | 4.3 | -3.1 |
| All industry | 1971-1975 | 17.9 | 1.2 | 9.1 | 5.6 | 2.3 | 6.6 | 9.2 | 6.7 | 0.1 |
| | 1976-1980 | -12.8 | 8.0 | 7.6 | 5.7 | 1.0 | 6.5 | 5.9 | 4.8 | -1.6 |
| | 1981-1983 | 22.0 | -2.5 | 7.8 | 5.8 | 1.1 | 6.6 | 4.4 | 3.3 | -3.2 |
| Construction | 1971-1975 | 36.1 | - | 14.4 | 16.6 | -0.3 | 14.8 | 6.5 | 6.8 | -6.9 |
| | 1976-1980 | -42.6 | - | 9.2 | -3.4 | 0.8 | 8.3 | 4.5 | 3.7 | -4.3 |
| | 1981-1983 | 92.1 | - | 9.2 | 13.3 | 1.2 | 7.9 | 6.0 | 4.8 | -2.9 |
| Agriculture | 1971-1975 | 42.9 | -9.6 | 6.6 | 8.1 | -4.1 | 11.1 | 1.6 | 5.9 | -4.6 |
| | 1976-1980 | -8.2 | -2.2 | 5.4 | 1.3 | -2.7 | 8.3 | -5.1 | -2.5 | -10.0 |
| | 1981-1983 | 9.0 | -7.4 | 2.7 | -2.9 | -2.5 | 5.3 | -2.5 | -0.1 | -5.1 |
| Total material sphere | 1971-1975 | 21.0 | -2.1 | 8.9 | 8.4 | - | 8.9 | 6.8 | 6.7 | -2.0 |
| | 1976-1980 | -13.0 | 5.9 | 8.1 | 3.6 | - | 8.1 | 5.3 | 5.3 | -2.6 |
| | 1981-1983 | 19.6 | -3.2 | 7.0 | 4.5 | 0.7 | 6.3 | 3.7 | 3.0 | -3.1 |

Table 8B (Chapter 8)

CZECHOSLOVAKIA, AVERAGE ANNUAL PERCENTAGE CHANGES OF SELECTED VARIABLES, 1970-1983

| | | Capital goods imports from OECD | Exports to OECD | Fixed assets | Fixed invest-ment | Employ-ment | Capital intensity | Pro-duction | Labour produc-tivity | Capital produc-tivity |
|---|---|---|---|---|---|---|---|---|---|---|
| Energy | 1971-1975 | 7.1 | -1.0 | 6.4 | 14.8 | 2.1 | 4.2 | 6.1 | 3.9 | -0.3 |
| | 1976-1980 | 5.2 | 7.8 | 8.1 | 7.6 | 2.6 | 5.4 | 1.8 | 2.1 | -3.1 |
| | 1981-1983 | -7.9 | -3.5 | 7.3 | -2.2 | 1.7 | 5.6 | 1.5 | -0.2 | -5.4 |
| Fuel | 1971-1975 | 3.6 | -1.0 | 4.2 | 9.7 | -0.1 | 4.3 | 3.2 | 3.3 | -1.0 |
| | 1976-1980 | -1.9 | 7.8 | 6.9 | 13.1 | 0.9 | 5.9 | 2.6 | 1.7 | -3.9 |
| | 1981-1983 | -0.1 | -3.5 | 5.6 | 6.4 | 1.7 | 3.9 | -0.4 | -2.1 | -5.7 |
| Metallurgy | 1971-1975 | 9.2 | -0.6 | 3.2 | 5.1 | 0.6 | 2.6 | 5.4 | 4.8 | 2.1 |
| | 1976-1980 | -8.7 | 5.2 | 4.5 | 7.3 | 0.7 | 3.8 | 2.8 | 2.1 | -1.6 |
| | 1981-1983 | -6.7 | -1.8 | 5.3 | -5.0 | -0.2 | 5.5 | 0.2 | 0.3 | -4.9 |
| Engineering and metal-working | 1971-1975 | 10.1 | 2.7 | 6.0 | 11.0 | 0.7 | 5.3 | 8.4 | 7.7 | 2.3 |
| | 1976-1980 | -2.3 | -0.6 | 6.5 | 7.0 | 1.1 | 5.3 | 6.7 | 5.5 | 0.2 |
| | 1981-1983 | -9.2 | -5.4 | 5.9 | -3.4 | 0.7 | 5.2 | 4.2 | 3.4 | -1.7 |
| Chemicals | 1971-1975 | -3.7 | 4.5 | 7.8 | 0.9 | 1.7 | 6.1 | 10.0 | 8.2 | 2.0 |
| | 1976-1980 | -1.5 | 11.5 | 5.0 | -1.4 | 0.7 | 4.3 | 5.8 | 5.1 | 0.7 |
| | 1981-1983 | 1.4 | 3.2 | 4.3 | -0.5 | 0.7 | 3.6 | 2.1 | 1.4 | -2.1 |
| Construction materials | 1971-1975 | - | 1.0 | 8.3 | 5.8 | 0.4 | 7.8 | 6.6 | 6.2 | -1.5 |
| | 1976-1980 | - | -0.6 | 7.1 | -0.8 | 0.4 | 6.7 | 4.3 | 4.0 | -2.6 |
| | 1981-1983 | - | -5.7 | 4.8 | 12.7 | -0.6 | 5.4 | 0.1 | 0.7 | -4.4 |
| Wood and paper | 1971-1975 | -2.3 | -0.2 | 9.4 | 3.5 | 1.0 | 8.4 | 7.2 | 6.2 | -2.0 |
| | 1976-1980 | 12.0 | 15.9 | 7.2 | 16.0 | 0.8 | 6.4 | 5.2 | 4.4 | -1.9 |
| | 1981-1983 | -35.8 | 13.2 | 8.5 | -15.6 | - | 8.5 | 2.5 | 2.5 | -5.5 |
| Textiles | 1971-1975 | -6.8 | 9.0 | 6.9 | -1.7 | -0.1 | 7.0 | 5.9 | 6.0 | -1.0 |
| | 1976-1980 | -12.7 | 2.4 | 5.2 | 2.4 | -0.3 | 5.4 | 3.5 | 3.8 | -1.5 |
| | 1981-1983 | -8.1 | -2.3 | 4.7 | -12.1 | -0.3 | 5.1 | 1.9 | 2.2 | -2.7 |
| Other light industry | 1971-1975 | -2.4 | 7.9 | 6.6 | 6.5 | 0.1 | 6.5 | 5.8 | 5.6 | -0.8 |
| | 1976-1980 | -3.9 | 2.5 | 6.1 | 4.1 | -0.4 | 6.5 | 3.8 | 4.3 | -2.1 |
| | 1981-1983 | -11.8 | -1.8 | 4.3 | -1.6 | - | 4.3 | 1.5 | 1.5 | -2.7 |
| Food | 1971-1975 | 1.3 | 2.4 | 5.6 | 7.9 | 0.3 | 5.2 | 4.7 | 4.4 | -0.8 |
| | 1976-1980 | -5.6 | 0.1 | 5.9 | 3.9 | -0.1 | 6.0 | 2.7 | 2.8 | -3.0 |
| | 1981-1983 | -6.5 | 2.7 | 5.0 | -1.2 | 0.2 | 4.9 | 1.1 | 0.9 | -3.7 |
| Other | 1971-1975 | 5.6 | 2.7 | 4.3 | 6.8 | 1.0 | 3.3 | 8.6 | 7.6 | 4.2 |
| | 1976-1980 | -5.2 | 1.9 | 5.1 | -2.2 | 0.9 | 4.2 | 4.4 | 3.4 | -0.7 |
| | 1981-1983 | -1.4 | -5.1 | 4.7 | 6.3 | 0.7 | 4.0 | 0.1 | -0.5 | -4.4 |
| All industry | 1971-1975 | 1.9 | 2.9 | 5.6 | 7.3 | 0.6 | 5.0 | 6.6 | 6.0 | 1.0 |
| | 1976-1980 | -3.0 | 4.5 | 6.1 | 5.9 | 0.6 | 5.4 | 4.6 | 3.9 | -1.4 |
| | 1981-1983 | -7.8 | - | 5.6 | -2.3 | 0.4 | 5.2 | 2.0 | 1.6 | -3.4 |
| Construction | 1971-1975 | 12.7 | - | 9.2 | 11.1 | 2.3 | 6.7 | 7.7 | 5.3 | -1.4 |
| | 1976-1980 | -28.8 | - | 8.0 | 14.9 | 1.1 | 6.9 | 1.9 | 0.8 | -5.7 |
| | 1981-1983 | -1.3 | - | 4.3 | -2.7 | -0.7 | 5.0 | - | 0.6 | -4.2 |
| Agriculture | 1971-1975 | 16.9 | -7.2 | 5.9 | 11.3 | -2.7 | 8.8 | 1.4 | 4.2 | -4.2 |
| | 1976-1980 | -18.9 | 13.1 | 6.6 | 0.7 | -1.3 | 8.0 | 0.8 | 2.2 | -5.4 |
| | 1981-1983 | 29.1 | 2.9 | 5.4 | 7.3 | -0.7 | 6.2 | -1.6 | -0.8 | -6.6 |
| Total material sphere | 1971-1975 | 4.0 | 2.2 | 5.8 | 8.4 | 0.3 | 5.4 | 5.7 | 5.3 | -0.1 |
| | 1976-1980 | -3.8 | 5.0 | 6.2 | 4.7 | 0.4 | 5.8 | 3.6 | 3.2 | -2.5 |
| | 1981-1983 | -5.3 | 0.2 | 5.4 | -1.2 | 0.3 | 5.1 | 0.8 | 0.6 | -4.3 |

Table 8C (Chapter 8)

GERMAN DEMOCRATIC REPUBLIC, AVERAGE ANNUAL PERCENTAGE CHANGES OF SELECTED VARIABLES, 1970-1983

| | | Capital goods imports from OECD | Exports to OECD | Fixed assets | Fixed invest-ment | Employ-ment | Capital intensity | Pro-duction | Labour produc-tivity | Capital produc-tivity |
|---|---|---|---|---|---|---|---|---|---|---|
| Fuel and Energy | 1971-1975 | 33.9 | 17.1 | 6.0 | 8.8 | 2.2 | 3.7 | 2.7 | 0.4 | 3.2 |
| | 1976-1980 | -9.3 | 46.9 | .. | 4.3 | 1.5 | .. | 4.8 | 3.2 | .. |
| | 1981-1983 | 26.2 | 32.1 | .. | 1.0 | 1.8 | .. | 3.6 | 1.7 | .. |
| Metallurgy | 1971-1975 | -11.2 | -4.5 | 5.7 | 10.7 | 1.1 | 4.6 | 6.8 | 5.7 | 1.0 |
| | 1976-1980 | 25.8 | 13.3 | .. | 19.1 | 0.7 | .. | 4.0 | 3.3 | .. |
| | 1981-1983 | 67.1 | 19.2 | .. | 6.5 | 1.5 | .. | 4.0 | 2.4 | .. |
| Mechanical engineering | 1971-1975 | -7.6 (a) | 6.1 (a) | 6.8 | 4.0 | 1.5 | 5.2 | 5.7 | 4.1 | -1.0 |
| | 1976-1980 | 18.8 (a) | 3.0 (a) | .. | 6.4 | 1.2 | .. | 6.2 | 4.9 | .. |
| | 1981-1983 | 4.6 (a) | -1.3 (a) | .. | 2.2 | 0.9 | .. | 4.6 | 3.7 | .. |
| Electrical engineering | 1971-1975 | .. | .. | 9.1 | -7.2 | 3.4 | 5.6 | 9.2 | 5.6 | 0 |
| | 1976-1980 | .. | .. | .. | 11.4 | 0.1 | .. | 8.7 | 8.5 | .. |
| | 1981-1983 | .. | .. | .. | 4.9 | 1.4 | .. | 8.8 | 7.3 | .. |
| Chemicals | 1971-1975 | -11.6 | 3.1 | 7.4 | -1.0 | 0.7 | 6.6 | 8.3 | 7.5 | 0.8 |
| | 1976-1980 | 0.2 | 6.7 | .. | 7.6 | 0.3 | .. | 4.8 | 4.4 | .. |
| | 1981-1983 | 26.4 | 12.3 | .. | -1.6 | 0.1 | .. | 3.8 | 3.7 | .. |
| Construction materials | 1971-1975 | - | -7.7 | 6.1 | -1.3 | 0.4 | 5.7 | 7.0 | 6.5 | 0.8 |
| | 1976-1980 | - | 2.5 | .. | 2.8 | 0.4 | .. | 2.3 | 1.9 | .. |
| | 1981-1983 | - | 50.6 | .. | -20.0 | -0.7 | .. | -0.4 | 0.3 | .. |
| Textiles | 1971-1975 | 0.1 | 2.9 | 5.6 | 6.7 | -0.3 | 5.9 | 5.2 | 5.6 | -0.3 |
| | 1976-1980 | -10.6 | 6.9 | .. | -1.3 | -1.3 | .. | 3.7 | 5.1 | .. |
| | 1981-1983 | -24.1 | -8.9 | .. | -3.7 | -1.4 | .. | 2.3 | 3.8 | .. |
| Other light industry | 1971-1975 | -2.7 | 4.5 | 6.9 | 7.1 | 2.3 | 4.4 | 5.9 | 3.5 | -0.9 |
| | 1976-1980 | 2.9 | 7.7 | .. | -1.9 | -0.6 | .. | 4.2 | 4.8 | .. |
| | 1981-1983 | 15.7 | 2.0 | .. | 6.1 | 0.4 | .. | 3.2 | 2.8 | .. |
| Food | 1971-1975 | 0.1 | 28.8 | 6.0 | 9.7 | 2.3 | 3.6 | 5.5 | 3.2 | -0.4 |
| | 1976-1980 | -10.0 | -21.4 | .. | -1.9 | 2.4 | .. | 2.7 | 0.4 | .. |
| | 1981-1983 | -15.7 | -7.0 | .. | 1.2 | 0.6 | .. | 1.7 | 1.2 | .. |
| Other | 1971-1975 | -14.8 | 7.5 | 4.3 | 5.1 | 0.8 | 3.4 | 3.7 | 2.9 | -0.5 |
| | 1976-1980 | 22.3 | 3.6 | .. | 5.2 | 2.2 | .. | 3.4 | 1.2 | .. |
| | 1981-1983 | 16.6 | 3.7 | .. | -4.8 | 1.7 | .. | 2.7 | 1.0 | .. |
| All industry | 1971-1975 | -6.0 | 5.0 | 6.6 (b) | 4.0 | 1.7 | 6.2 (b) | 6.3 | 4.6 | -0.6 (b) |
| | 1976-1980 | 7.4 | 14.9 | 5.4 (b) | 5.7 | 0.6 | 4.8 (b) | 5.0 | 4.4 | -0.5 (b) |
| | 1981-1983 | 18.4 | 5.9 | 5.9 (b) | 0.9 | 0.6 | 5.5 (b) | 1.1 | 3.5 | -1.2 (b) |
| Construction | 1971-1975 | -99.4 | - | 7.2 | 6.1 | -0.9 | 8.2 | 4.8 | 5.8 | -2.2 |
| | 1976-1980 | - | - | 8.0 | -2.8 | 0.9 | 7.1 | 2.7 | 1.8 | -5.0 |
| | 1981-1983 | 0000 | - | 5.2 | -22.2 | -0.9 | 6.1 | 3.3 | 4.2 | -1.8 |
| Agriculture | 1971-1975 | 41.1 | 14.9 | 4.9 | 2.8 | -2.0 | 7.1 | 2.2 | 4.3 | -2.6 |
| | 1976-1980 | -0.7 | -6.9 | 5.2 | -0.4 | -0.2 | 5.4 | 0.1 | 0.3 | -4.8 |
| | 1981-1983 | -45.8 | -5.3 | 4.8 | -4.8 | 0.8 | 4.0 | 1.0 | 0.2 | -3.6 |
| Total material sphere | 1971-1975 | 1.1 | 5.9 | 5.9 | 4.2 | 0.1 | 3.7 | 5.4 | 5.3 | -0.5 |
| | 1976-1980 | 0.7 | 3.6 | 5.5 | 3.3 | 0.5 | 3.5 | 4.1 | 3.6 | -1.3 |
| | 1981-1983 | 13.2 | 6.6 | 5.4 | -0.8 | 0.3 | 3.6 | 3.9 | 3.7 | -1.4 |

Notes: a) Total for the engineering branch as a whole.
b) Based on net output national accounts definitions.

Table 8D (Chapter 8)

HUNGARY, AVERAGE ANNUAL PERCENTAGE CHANGES OF SELECTED VARIABLES, 1970-1983

| | | Capital goods imports from OECD | Exports to OECD | Fixed assets | Fixed invest-ment | Employ-ment | Capital intensity | Pro-duction | Labour produc-tivity | Capital produc-tivity |
|---|---|---|---|---|---|---|---|---|---|---|
| Energy | 1971-1975 | 13.4 | 0000 | 8.6 | 10.2 | 1.5 | 7.0 | 7.6 | 6.0 | -0.9 |
| | 1976-1980 | -5.1 | 60.6 | 7.1 | 10.8 | -1.2 | 8.4 | 6.6 | 8.0 | -0.4 |
| | 1981-1983 | 7.8 | -100.0 | 3.7 | -0.5 | 1.0 | 2.7 | 3.7 | 3.1 | 0.4 |
| Mining | 1971-1975 | 7.7 | -12.5 | 5.3 | 0.6 | -2.8 | 8.3 | 2.0 | 5.0 | -3.1 |
| | 1976-1980 | 8.4 | 20.4 | 5.9 | 9.8 | -1.6 | 7.5 | 0.4 | 2.0 | -5.2 |
| | 1981-1983 | -4.6 | -13.5 | 7.3 | 1.9 | -0.5 | 7.8 | -1.0 | 0.1 | -7.1 |
| Metallurgy | 1971-1975 | 8.2 | 1.0 | 5.9 | -3.1 | -0.1 | 6.0 | 5.0 | 5.1 | -0.9 |
| | 1976-1980 | 8.6 | 7.6 | 5.7 | 17.2 | -1.3 | 7.2 | 1.3 | 2.7 | -4.2 |
| | 1981-1983 | -20.8 | -2.9 | 6.4 | -14.9 | -5.5 | 12.6 | -2.3 | 3.0 | -8.5 |
| Engineering and metal-working | 1971-1975 | 11.0 | 13.3 | 7.7 | 3.2 | 0.3 | 7.3 | 7.8 | 7.5 | 0.1 |
| | 1976-1980 | 9.5 | 13.7 | 8.3 | 4.6 | -1.3 | 9.7 | 3.1 | 4.5 | -4.7 |
| | 1981-1983 | -7.9 | -4.1 | 4.8 | -6.1 | -2.4 | 7.4 | 3.9 | 7.8 | 0.4 |
| Chemicals | 1971-1975 | 16.2 | 5.6 | 11.6 | 12.6 | 1.3 | 10.2 | 10.5 | 9.1 | -1.0 |
| | 1976-1980 | 2.5 | 18.5 | 8.8 | -7.8 | -1.4 | 10.3 | 6.2 | 7.7 | -2.4 |
| | 1981-1983 | 2.7 | 3.6 | 3.5 | 6.9 | -1.5 | 5.0 | 1.8 | 3.5 | -1.5 |
| Construction materials | 1971-1975 | 2.7 | 15.7 | 11.8 | 3.3 | - | 11.7 | 5.1 | 5.1 | -5.9 |
| | 1976-1980 | 5.9 | 9.3 | 7.3 | -2.4 | -0.9 | 8.2 | 4.9 | 5.8 | -2.2 |
| | 1981-1983 | -19.1 | -4.4 | 2.5 | -11.7 | -3.5 | 6.2 | 1.0 | 4.5 | -1.6 |
| Wood and paper | 1971-1975 | .. | .. | 13.1 | 5.1 | 0.4 | 12.7 | 6.1 | 5.7 | -6.2 |
| | 1976-1980 | .. | .. | 7.9 | -14.3 | -2.5 | 10.7 | 2.2 | 4.8 | -5.3 |
| | 1981-1983 | .. | .. | 2.8 | 21.0 | -6.6 | 10.1 | - | 5.4 | -4.3 |
| Textiles | 1971-1975 | 11.2 | 2.6 | 6.3 | 9.4 | -1.5 | 7.9 | 2.1 | 3.7 | -3.9 |
| | 1976-1980 | -4.1 | 8.4 | 6.0 | -8.2 | -2.6 | 8.8 | 1.4 | 4.0 | -4.4 |
| | 1981-1983 | 0.7 | -7.8 | 1.8 | -6.7 | -3.0 | 4.9 | -1.4 | 0.5 | -4.2 |
| Other light industry | 1971-1975 | 7.6 } incl. | 15.5 } incl. | 9.0 | 11.9 | -0.8 | 9.9 | 9.7 | 10.7 | 0.7 |
| | 1976-1980 | -6.5 } wood & | 6.5 } wood & | 6.9 | 1.2 | -1.8 | 8.9 | 2.9 | 4.8 | -3.8 |
| | 1981-1983 | 1.7 } paper | 0.7 } paper | 2.4 | -12.0 | -3.8 | 6.5 | 4.5 | 11.1 | 4.4 |
| Food | 1971-1975 | 3.1 | 11.9 | 7.3 | 10.3 | 1.6 | 5.5 | 4.7 | 3.0 | -2.4 |
| | 1976-1980 | 7.6 | 4.6 | 9.5 | 3.6 | - | 9.5 | 3.4 | 3.4 | -5.6 |
| | 1981-1983 | 3.4 | -1.8 | 6.4 | -10.0 | -0.4 | 6.8 | 2.8 | 3.9 | -2.8 |
| Other | 1971-1975 | 11.7 | 7.8 | 9.2 | 2.3 | -1.1 | 10.4 | 4.5 | 5.7 | -4.3 |
| | 1976-1980 | 5.9 | 10.2 | 5.9 | 5.2 | -4.0 | 10.2 | 3.5 | 7.8 | -2.2 |
| | 1981-1983 | -2.9 | -7.1 | 4.8 | -13.1 | -5.7 | 11.1 | 2.4 | 13.0 | 1.7 |
| All industry | 1971-1975 | 10.9 | 8.9 | 8.2 | 6.2 | -0.1 | 8.3 | 6.3 | 6.4 | -1.8 |
| | 1976-1980 | 3.6 | 9.1 | 7.5 | 3.6 | -1.5 | 9.1 | 3.4 | 5.0 | -3.8 |
| | 1981-1983 | -2.1 | -1.8 | 4.6 | -3.5 | -2.6 | 7.4 | 1.9 | 5.2 | -2.0 |
| Construction | 1971-1975 | -8.5 | - | 13.5 | 2.0 | 1.8 | 11.5 | 6.6 | 4.7 | -6.1 |
| | 1976-1980 | 10.8 | - | 10.8 | 6.2 | -1.3 | 12.3 | 4.4 | 5.8 | -5.8 |
| | 1981-1983 | 14.0 | - | 2.7 | -17.2 | -2.6 | 5.4 | 0.9 | 3.5 | -1.8 |
| Agriculture | 1971-1975 | 45.5 | -4.4 | 8.1 | 1.3 | -2.9 | 11.3 | 2.4 | 5.4 | -5.3 |
| | 1976-1980 | -1.7 | 1.7 | 5.5 | -0.5 | -0.5 | 6.0 | 0.2 | 0.7 | -5.0 |
| | 1981-1983 | 9.9 | 0.2 | 4.3 | -3.3 | 1.3 | 1.9 | 3.1 | 0.7 | -1.2 |
| Total material sphere | 1971-1975 | 11.6 | 5.9 | 6.3 | 6.2 | 2.5 | 5.5 | 6.2 | 6.3 | -0.1 |
| | 1976-1980 | 3.0 | 8.0 | 5.7 | 6.2 | 1.7 | 5.7 | 2.8 | 3.6 | -2.7 |
| | 1981-1983 | -1.7 | -1.6 | 4.0 | -4.2 | 0.4 | 4.0 | 2.6 | 3.1 | -1.4 |

Table 8E (Chapter 8)

POLAND, AVERAGE ANNUAL PERCENTAGE CHANGES OF SELECTED VARIABLES, 1970-1983

| | | Capital goods imports from OECD | Exports to OECD | Fixed assets | Fixed invest-ment | Employ-ment | Capital intensity | Pro-duction | Labour produc-tivity | Capital produc-tivity |
|---|---|---|---|---|---|---|---|---|---|---|
| Energy | 1971-1975 | 25.7 | 0000 | 9.0 | 11.8 | 3.4 | 5.5 | 9.1 | 5.5 | 0.1 |
| | 1976-1980 | -14.5 | 17.7 | 7.1 | 5.7 | 0.9 | 6.2 | 5.4 | 4.5 | -1.6 |
| | 1981-1983 | -20.7 | 44.7 | 4.7 | -9.9 | 3.9 | 0.8 | 0.8 | -3.0 | -3.8 |
| Fuel | 1971-1975 | 40.6 | 6.6 | 5.4 | 15.0 | 0.8 | 4.6 | 6.4 | 5.6 | 0.9 |
| | 1976-1980 | -10.1 | -4.2 | 8.7 | 4.5 | 1.9 | 6.6 | 2.5 | 0.6 | -5.7 |
| | 1981-1983 | -19.2 | -9.9 | 2.3 | -15.0 | 2.4 | -0.1 | -0.9 | -3.2 | -3.1 |
| Metallurgy | 1971-1975 | 46.4 | 5.0 | 10.2 | 28.4 | 1.4 | 8.7 | 9.9 | 8.4 | -0.2 |
| | 1976-1980 | -18.3 | 18.6 | 11.6 | -14.5 | 0.3 | 11.3 | 3.3 | 3.1 | -7.4 |
| | 1981-1983 | -33.5 | -12.7 | 1.2 | -31.8 | -4.0 | 5.5 | -4.9 | -0.9 | -6.1 |
| Engineering and metal-working | 1971-1975 | 29.8 | 25.1 | 14.2 | 26.8 | 4.6 | 9.2 | 14.2 | 9.2 | -0.1 |
| | 1976-1980 | -16.9 | 5.6 | 10.4 | -8.3 | 1.0 | 9.2 | 6.9 | 5.8 | -3.1 |
| | 1981-1983 | -25.3 | -16.1 | 3.3 | -11.5 | -3.7 | 7.2 | -2.3 | 1.4 | -5.4 |
| Chemicals | 1971-1975 | 34.8 | 5.7 | 9.6 | 15.4 | 3.1 | 6.3 | 12.0 | 8.7 | 2.3 |
| | 1976-1980 | -14.6 | 6.4 | 7.0 | -6.2 | -0.5 | 7.6 | 4.3 | 4.8 | -2.6 |
| | 1981-1983 | -24.5 | -12.8 | 2.1 | -10.4 | -2.7 | 4.9 | -1.4 | 1.4 | -3.4 |
| Construction materials | 1971-1975 | - | -3.7 | 8.4 | 20.6 | 0.7 | 7.7 | 8.1 | 7.4 | -0.3 |
| | 1976-1980 | - | 5.2 | 6.1 | -21.1 | -2.3 | 8.6 | 1.2 | 3.6 | -4.6 |
| | 1981-1983 | - | -9.7 | 1.7 | -10.6 | -2.8 | 4.6 | -4.1 | -1.3 | -5.7 |
| Wood and paper | 1971-1975 | 29.6 | -5.5 | 9.3 | 27.1 | 2.7 | 6.4 | 9.3 | 6.4 | - |
| | 1976-1980 | -14.8 | 12.3 | 7.4 | -11.6 | -1.6 | 9.1 | 4.0 | 5.6 | -3.2 |
| | 1981-1983 | -29.9 | -4.4 | 4.2 | -15.9 | -3.2 | 7.6 | -1.1 | 2.2 | -5.1 |
| Textiles | 1971-1975 | 17.5 | 4.6 | 12.3 | 23.0 | 1.9 | 10.2 | 8.3 | 6.0 | -3.8 |
| | 1976-1980 | -31.9 | 6.7 | 5.8 | -18.5 | -1.6 | 7.6 | 3.0 | 4.7 | -2.7 |
| | 1981-1983 | -2.9 | -23.9 | 1.8 | -11.0 | -6.7 | 9.2 | -7.5 | -0.9 | -9.2 |
| Other light industry | 1971-1975 | 20.7 | 22.4 | 10.5 | 15.5 | 3.7 | 6.6 | 9.8 | 6.0 | -0.5 |
| | 1976-1980 | -11.3 | 4.4 | 6.7 | -8.0 | -0.2 | 7.0 | 5.0 | 5.3 | -1.6 |
| | 1981-1983 | -13.3 | -9.5 | 2.7 | -2.0 | -2.7 | 5.6 | 0.3 | 3.1 | -2.3 |
| Food | 1971-1975 | 33.3 | 0.7 | 10.6 | 28.8 | 3.3 | 7.0 | 8.7 | 5.2 | -1.7 |
| | 1976-1980 | -21.4 | -1.0 | 7.5 | -11.6 | -0.2 | 7.8 | 2.5 | 2.7 | -4.7 |
| | 1981-1983 | -19.8 | -11.0 | 2.2 | -2.1 | 0.1 | 2.2 | -3.7 | -3.8 | -5.8 |
| Other | 1971-1975 | 38.1 | 4.9 | 8.6 | 21.7 | 3.3 | 5.1 | 9.8 | 6.3 | 1.2 |
| | 1976-1980 | -4.5 | 2.8 | 3.7 | -9.1 | 3.5 | 0.2 | 10.4 | 6.6 | 6.4 |
| | 1981-1983 | -27.5 | -15.2 | 38.0 | 11.6 | 18.4 | 16.5 | 3.3 | -12.8 | -25.1 |
| All industry | 1971-1975 | 32.2 | 7.1 | 9.9 | 21.8 | 3.1 | 6.7 | 9.9 | 6.6 | -0.1 |
| | 1976-1980 | -14.6 | 4.5 | 8.5 | -6.9 | 0.2 | 8.3 | 4.0 | 3.9 | -4.1 |
| | 1981-1983 | -24.4 | -12.0 | 2.9 | -12.5 | -2.2 | 5.2 | -2.7 | -0.5 | -5.5 |
| Construction | 1971-1975 | 44.6 | - | 14.0 | 26.5 | 5.9 | 7.6 | 13.3 | 7.0 | -0.5 |
| | 1976-1980 | -27.4 | - | 11.4 | -5.5 | -1.1 | 12.6 | -5.4 | -4.3 | -15.1 |
| | 1981-1983 | -35.7 | - | 1.1 | -29.1 | -3.7 | 4.9 | -9.6 | -6.1 | -10.6 |
| Agriculture | 1971-1975 | 90.9 | -2.0 | 5.0 | 14.0 | -0.8 | 5.9 | 1.2 | 2.1 | -3.6 |
| | 1976-1980 | -21.3 | 7.4 | 6.7 | -0.5 | 0.5 | 6.1 | -2.8 | -3.3 | -8.9 |
| | 1981-1983 | -35.8 | 0.7 | 2.6 | -7.8 | - | 2.6 | 4.3 | 4.2 | 1.6 |
| Total material sphere | 1971-1975 | 37.7 | 6.1 | 7.7 | 19.8 | 1.9 | 5.7 | 9.3 | 7.2 | 1.5 |
| | 1976-1980 | -14.4 | 4.8 | 7.8 | -4.5 | 0.4 | 7.4 | 0.7 | 0.3 | -6.6 |
| | 1981-1983 | -28.1 | -10.6 | 2.6 | -11.2 | -1.0 | 3.6 | -4.1 | -3.2 | -6.6 |

Table 8F (Chapter 8)

ROMANIA, AVERAGE ANNUAL PERCENTAGE CHANGES OF SELECTED VARIABLES, 1970-1983

| | | Capital goods imports from OECD | Exports to OECD | Fixed assets | Fixed invest-ment | Employ-ment | Capital intensity | Pro-duction | Labour produc-tivity | Capital produc-tivity |
|---|---|---|---|---|---|---|---|---|---|---|
| Energy | 1971-1975 | 7.1 | 0000 | .. | 10.6 | 1.4 | 6.3 (a) | 9.7 | 8.2 | .. |
| | 1976-1980 | -6.7 | -99.8 | .. | 4.7 | 1.2 | 6.7 (a) | 4.7 | 3.5 | .. |
| | 1981-1983 | -46.6 | - | .. | 17.9 | 3.4 | 7.8 (a) | 1.8 | -1.6 | .. |
| Fuel | 1971-1975 | 5.7 | 9.1 | .. | 12.2 | 0.9 | .. | 5.2 | 4.3 | .. |
| | 1976-1980 | -4.0 | 7.1 | .. | 8.6 | 4.2 | .. | 3.9 | -0.3 | .. |
| | 1981-1983 | -43.8 | -3.5 | .. | 9.6 | 4.1 | .. | 1.3 | -2.7 | .. |
| Metallurgy | 1971-1975 | 3.1 | 5.4 | .. | 16.3 | 3.8 | .. | 10.8 | 6.7 | .. |
| | 1976-1980 | -9.1 | 5.6 | .. | 10.0 | 4.3 | .. | 7.7 | 3.2 | .. |
| | 1981-1983 | -43.6 | -4.7 | .. | -12.4 | 1.1 | .. | 0.8 | -0.3 | .. |
| Engineering and metal-working | 1971-1975 | 1.3 | 13.6 | .. | 12.0 | 10.8 | .. | 18.1 | 6.6 | .. |
| | 1976-1980 | 0.3 | 10.6 | .. | 19.8 | 5.4 | .. | 12.6 | 6.8 | .. |
| | 1981-1983 | -39.8 | -5.4 | .. | -14.3 | 2.4 | .. | 3.7 | 1.3 | .. |
| Chemicals | 1971-1975 | 6.0 | 6.7 | .. | 18.5 | 7.3 | .. | 15.8 | 7.9 | .. |
| | 1976-1980 | -4.5 | 7.5 | .. | 9.0 | 3.2 | .. | 9.6 | 6.2 | .. |
| | 1981-1983 | -53.5 | 4.1 | .. | -9.8 | 4.0 | .. | 3.6 | -0.4 | .. |
| Construction materials | 1971-1975 | - | 3.8 | .. | 7.1 | 2.1 | .. | 10.0 | 7.8 | .. |
| | 1976-1980 | - | -0.8 | .. | 8.4 | 1.5 | .. | 12.8 | 11.1 | .. |
| | 1981-1983 | - | -23.1 | .. | -23.1 | -1.9 | .. | 1.5 | 3.4 | .. |
| Wood and paper | 1971-1975 | 3.4 | -0.9 | .. | 7.8 | 1.6 | .. | 6.9 | 5.3 | .. |
| | 1976-1980 | -20.0 | 5.1 | .. | -0.7 | 0.2 | .. | 6.5 | 6.3 | .. |
| | 1981-1983 | -38.1 | 2.2 | .. | 5.4 | 1.5 | .. | 2.7 | 1.2 | .. |
| Textiles | 1971-1975 | -3.5 | 14.1 | .. | 11.5 | 7.3 | .. | 12.1 | 4.4 | .. |
| | 1976-1980 | -3.8 | 7.2 | .. | 13.9 | 4.0 | .. | 10.8 | 6.5 | .. |
| | 1981-1983 | -46.9 | -7.3 | .. | -23.7 | 2.5 | .. | 3.6 | 1.1 | .. |
| Other light industry | 1971-1975 | -5.3 | 25.9 | .. | -0.5 | 6.4 | .. | 12.9 | 6.1 | .. |
| | 1976-1980 | 0.3 | 7.0 | .. | 5.7 | 3.0 | .. | 8.7 | 5.6 | .. |
| | 1981-1983 | -46.4 | 2.6 | .. | -11.9 | 1.6 | .. | 4.8 | 3.2 | .. |
| Food | 1971-1975 | -0.3 | 15.4 | .. | 12.8 | 4.1 | .. | 7.4 | 3.2 | .. |
| | 1976-1980 | -14.5 | -14.9 | .. | 1.4 | 1.2 | .. | 6.1 | 4.8 | .. |
| | 1981-1983 | -60.8 | -16.8 | .. | 12.7 | -0.3 | .. | 0.7 | 1.0 | .. |
| Other | 1971-1975 | 2.8 | 15.1 | .. | 6.8 | 8.7 | .. | 22.0 | 12.2 | .. |
| | 1976-1980 | -1.2 | 5.5 | .. | 4.4 | 2.7 | .. | 9.1 | 6.3 | .. |
| | 1981-1983 | -41.4 | -3.2 | .. | -7.1 | 2.0 | .. | 1.2 | -0.7 | .. |
| All industry | 1971-1975 | 2.4 | 10.6 | 13.1 (a) | 12.2 | 6.3 | .. | 11.6 | 5.0 | 0.1 (a) |
| | 1976-1980 | -3.6 | 3.2 | 10.4 (a) | 10.0 | 3.5 | .. | 8.9 | 5.2 | -1.7 (a) |
| | 1981-1983 | -43.5 | -2.1 | 9.3 (a) | 3.6 | 2.0 | .. | 2.6 | 0.7 | -5.0 (a) |
| Construction | 1971-1975 | -47.9 | - | 15.0 | 15.3 | 1.4 | 13.3 | 7.6 | 6.0 | -6.5 |
| | 1976-1980 | -99.3 | - | 14.0 | 8.8 | 0.8 | 13.1 | 6.3 | 5.4 | -6.8 |
| | 1981-1983 | - | - | 7.4 | -8.1 | -2.7 | 10.3 | 2.2 | 5.0 | -4.8 |
| Agriculture | 1971-1975 | -10.9 | -13.9 | 9.6 | 7.2 | -4.5 | 14.8 | 5.4 | 10.4 | -3.8 |
| | 1976-1980 | -9.6 | -4.4 | 7.5 | 7.5 | -4.4 | 12.4 | 2.1 | 6.7 | -5.1 |
| | 1981-1983 | -20.9 | -7.6 | 9.5 | 5.4 | -0.3 | 10.0 | 0.5 | 0.8 | -8.4 |
| Total material sphere | 1971-1975 | 4.9 | 7.3 | 11.8 | 11.3 | 0.2 | 11.6 | 10.3 | 10.1 | -1.4 |
| | 1976-1980 | -4.3 | 2.8 | 10.2 | 9.3 | 0.2 | 10.0 | 6.6 | 6.4 | -3.3 |
| | 1981-1983 | -41.9 | -2.4 | 9.1 | -1.8 | 0.2 | 8.8 | 2.9 | 2.7 | -5.7 |

Table 8G (Chapter 8)

SOVIET UNION, AVERAGE ANNUAL PERCENTAGE CHANGES OF SELECTED VARIABLES, 1970-1983

| | | Capital goods imports from OECD | Exports to OECD | Fixed assets | Fixed invest-ment | Employ-ment | Capital intensity | Pro-duction | Labour produc-tivity | Capital produc-tivity |
|---|---|---|---|---|---|---|---|---|---|---|
| Energy | 1971-1975 | 52.8 | 14.7 | 7.3 | 3.8 | 1.6 | 5.6 | 7.1 | 5.4 | -0.1 |
| | 1976-1980 | -17.3 | 20.8 | 6.0 | 4.2 | 2.4 | 3.5 | 5.0 | 2.5 | -0.9 |
| | 1981-1983 | 88.5 | 40.0 | 5.3 | 0.7 | 2.2 | 3.1 | 2.7 | 0.5 | -2.5 |
| Fuel | 1971-1975 | 14.7 | 2.2 | 7.4 | 7.7 | -1.4 | 9.0 | 5.9 | 7.4 | -1.4 |
| | 1976-1980 | -7.2 | 5.4 | 8.2 | 8.4 | 1.5 | 6.6 | 3.0 | 1.4 | -4.8 |
| | 1981-1983 | 46.6 | 8.8 | 8.9 | 8.6 | 1.9 | 6.9 | 1.7 | -0.2 | -6.6 |
| Metallurgy | 1971-1975 | 13.8 | -6.4 | 7.4 | 6.8 | 0.1 | 7.3 | 5.1 | 4.9 | -2.2 |
| | 1976-1980 | -17.3 | -5.4 | 6.2 | 1.7 | 0.7 | 5.4 | 2.0 | 1.2 | -4.0 |
| | 1981-1983 | 2.2 | 16.7 | 5.7 | 5.6 | 1.8 | 3.9 | 1.9 | 0.1 | -3.6 |
| Engineering and metal-working | 1971-1975 | 7.6 | 17.3 | 10.0 | 9.6 | 2.8 | 7.0 | 11.6 | 8.5 | 1.4 |
| | 1976-1980 | -4.5 | 0.3 | 9.5 | 3.7 | 2.2 | 7.1 | 8.2 | 5.8 | -1.2 |
| | 1981-1983 | 3.7 | -8.2 | 7.6 | 3.8 | 1.0 | 6.6 | 5.7 | 4.7 | -1.8 |
| Chemicals | 1971-1975 | 34.9 | 16.6 | 9.6 | 9.6 | 2.3 | 7.2 | 10.5 | 8.1 | 0.9 |
| | 1976-1980 | -6.1 | 19.7 | 9.2 | 1.0 | 1.9 | 7.2 | 5.7 | 3.8 | -3.2 |
| | 1981-1983 | 8.8 | -2.1 | 7.2 | 0.8 | 0.9 | 6.3 | 4.8 | 3.9 | -2.2 |
| Construction materials | 1971-1975 | - | 13.4 | 9.0 | 2.2 | 1.5 | 7.4 | 7.3 | 5.7 | -1.6 |
| | 1976-1980 | - | -20.8 | 6.4 | 1.5 | 1.3 | 5.1 | 1.9 | 0.6 | -4.2 |
| | 1981-1983 | - | 55.2 | 5.3 | 1.7 | 0.9 | 4.3 | 2.9 | 2.0 | -2.2 |
| Wood and paper | 1971-1975 | 6.1 | -5.5 | 7.7 | 5.8 | -0.4 | 8.1 | 5.6 | 6.0 | -2.0 |
| | 1976-1980 | -14.9 | 8.2 | 6.5 | -0.1 | 1.3 | 5.2 | 2.4 | 1.1 | -3.9 |
| | 1981-1983 | 26.3 | -2.2 | 6.2 | 2.8 | 0.4 | 5.8 | 4.0 | 3.6 | -2.1 |
| Textiles | 1971-1975 | 34.8 | 2.3 | 7.3 | 5.8 | 0.3 | 7.0 | 4.7 | 4.4 | -2.4 |
| | 1976-1980 | -19.8 | 1.4 | 6.5 | 5.4 | 0.7 | 5.8 | 2.7 | 2.0 | -3.6 |
| | 1981-1983 | 6.4 | -12.2 | 6.1 | 2.7 | -0.1 | 6.2 | 1.1 | 1.3 | -4.7 |
| Other light industry | 1971-1975 | 5.2 | 6.7 | 8.7 | 4.5 | 0.4 | 8.3 | 4.5 | 4.0 | -3.9 |
| | 1976-1980 | -14.3 | -7.7 | 6.1 | -6.5 | 1.0 | 5.0 | 4.7 | 3.6 | -1.3 |
| | 1981-1983 | 43.4 | -11.1 | 5.4 | -3.8 | - | 5.4 | 1.5 | 1.5 | -3.7 |
| Food | 1971-1975 | 24.8 | -0.1 | 7.4 | 5.3 | 0.8 | 6.6 | 5.4 | 4.6 | -1.9 |
| | 1976-1980 | -3.4 | -11.6 | 5.6 | 1.6 | 1.0 | 4.6 | 1.5 | 0.5 | -3.9 |
| | 1981-1983 | 5.1 | -2.8 | 5.4 | 2.1 | 0.4 | 5.0 | 3.7 | 3.3 | -1.6 |
| Others | 1971-1975 | 17.3 | 17.8 | 8.3 | 4.9 | 2.4 | 5.8 | 8.7 | 6.1 | 0.3 |
| | 1976-1980 | -2.8 | 27.3 | 7.5 | 2.3 | 0.8 | 6.7 | 4.8 | 3.9 | -2.6 |
| | 1981-1983 | 9.3 | -6.7 | 6.7 | -0.5 | 1.6 | 5.1 | 3.3 | 1.7 | -3.2 |
| All industry | 1971-1975 | 17.7 | 4.5 | 8.3 | 6.9 | 1.5 | 6.7 | 7.3 | 5.7 | -1.0 |
| | 1976-1980 | -6.7 | 4.9 | 7.5 | 3.7 | 1.6 | 5.8 | 4.3 | 2.7 | -3.0 |
| | 1981-1983 | 13.7 | 2.7 | 6.7 | 4.1 | 0.8 | 5.9 | 3.7 | 2.8 | -2.9 |
| Construction | 1971-1975 | 0000 | - | 9.7 | 7.8 | 3.2 | 6.4 | 6.0 | 2.7 | -3.4 |
| | 1976-1980 | 31.7 | - | 9.5 | 4.6 | 1.2 | 8.1 | 3.2 | 1.9 | -5.7 |
| | 1981-1983 | -12.8 | - | 8.9 | 2.5 | 0.2 | 8.6 | 3.1 | 2.8 | -5.4 |
| Agriculture | 1971-1975 | 20.0 | 5.9 | 9.5 | 10.2 | -0.4 | 9.9 | -4.9 | -4.6 | -13.2 |
| | 1976-1980 | -12.4 | -5.8 | 7.3 | 2.9 | -0.6 | 8.0 | -0.1 | 0.5 | -6.9 |
| | 1981-1983 | -26.7 | -1.9 | 6.8 | 2.6 | - | 6.8 | 3.4 | 3.4 | -3.2 |
| Total material sphere | 1971-1975 | 17.3 | 4.8 | 8.7 | 8.1 | 1.5 | 7.1 | 4.8 | 3.3 | -3.6 |
| | 1976-1980 | -4.8 | 3.1 | 7.4 | 3.7 | 1.1 | 6.3 | 3.9 | 2.8 | -3.3 |
| | 1981-1983 | 16.7 | 2.1 | 6.9 | 4.0 | 0.6 | 6.2 | 3.2 | 2.5 | -3.4 |

NOTES AND SOURCES TO TABLES 8A-8G

1. Output and productivity series for industrial branches and total industry are based on gross production; those for construction, agriculture and the material sphere in total are based on net output. Industry data follow the "enterprise" classification; construction, agricultural and total material sphere follow the national accounting "activity" classification.

2. The trade export data presented in the second column do not include material sphere services (transport and communications, trade etc.), since they are not commodity exporters. They are, however, included in the total for the material sphere in all other cases.

3. Sector/branch coverage is similar to the breakdown described in the Note to Table 4 of Chapter 5: Chemicals: includes oil refining in Bulgaria, the German Democratic Republic and Hungary. Other light industry: includes clothing, leather and footwear plus (except in the Soviet Union) glass, ceramics and printing. In the German Democratic Republic and Hungary it also includes glass and, in the former only, wood and paper. In Hungary, fuels are combined with all types of mining. In all other countries, ore mining is included in metallurgy. In Bulgaria and the Soviet Union, metallurgy includes ferrous metallurgy only, non-ferrous metals being included in "other" industry.

4. The symbol "0000" (certain entries in foreign trade columns only) indicates a very large rise due to a very small base figure.

5. In Table 8F, (a) indicates a figure based on net output using national accounts definition.

Sources: Eastern investment goods imports from the OECD and Eastern exports to the OECD (in constant prices): OECD East-West technology transfer data base (Western statistics); domestic indicators: UN Economic Commission for Europe.

Table 9 (Chapter 8)

CORRELATION COEFFICIENTS (r) OF CHANGES IN THE GROWTH
RATES OF RELEVANT VARIABLES FOR INDUSTRY BETWEEN 1970-1975 AND 1975-1980 (a)

| | Capital goods imports from CMEA with capital goods imports from R of W | Capital goods imports from world with investments | Investments with labour productivity | Capital goods imports from world with labour productivity | Capital goods imports from R of W with labour productivity |
|---|---|---|---|---|---|
| | All industrial branches (b) | | | | |
| Bulgaria | 0.180 | 0.446 | 0.160 | -0.160 | -0.384 |
| Czechoslovakia | -0.137 | 0.196 | -0.140 | -0.075 | -0.474 |
| German Democratic Republic | 0.454 | -0.719 | -0.305 | 0.325 | 0.607 |
| Hungary | -0.056 | 0.598 | 0.134 | 0.339 | 0.077 |
| Poland | 0.722 | 0.803 | 0.116 | 0.487 | 0.518 |
| Romania | 0.641 | -0.058 | -0.558 | 0.159 | 0.319 |
| Soviet Union | 0.694 | -0.107 | -0.400 | 0.353 | 0.138 |
| | All Eastern countries | | | | |
| Industry | -0.303 | 0.829 | 0.892 | 0.554 | 0.148 |
| Agriculture | -0.062 | 0.095 | -0.067 | 0.338 | 0.901 |
| Construction | 0.062 | 0.099 | 0.734 | 0.435 | 0.537 |
| Total economy | -0.613 | 0.863 | .. | .. | .. |
| Industrial branches: | | | | | |
| Electrical energy | -0.205 | 0.448 | 0.336 | -0.171 | -0.242 |
| Fuel | -0.218 | -0.137 | 0.397 | 0.021 | 0.497 |
| Metallurgy | 0.295 | 0.776 | 0.443 | 0.529 | 0.457 |
| Engineering, etc. | 0.609 | 0.712 | 0.204 | 0.681 | 0.378 |
| Chemicals | -0.418 | 0.773 | 0.056 | -0.411 | -0.717 |
| Construction materials | -0.139 | 0.892 | -0.032 | 0.220 | 0.178 |
| Wood and paper | 0.323 | 0.799 | -0.344 | -0.466 | 0.246 |
| Textiles | .. | .. | -0.720 | .. | .. |
| Other light industries | 0.628 | 0.853 | 0.530 | 0.351 | 0.332 |
| Food and tobacco processing | 0.874 | 0.456 | 0.295 | 0.226 | 0.237 |

Notes: a) Average compound rate of change in 1975-1980 minus average compound rate of change in 1970-1975. For capital goods imports in the first half of the decade, the rate of change of 1973-1975 has been used for Bulgaria and for Hungary that of 1971-1975.

b) Only the 10 branches listed under industry in the lower half of the Table; the non-industrial sectors and industry as a whole were not included in the calculations.

Source: See Sources to Tables 8A-8G.

# NOTES AND REFERENCES

1. Thanks are due Z. Drabek, R. Amann and P. Elkan who provided work on the boomerang effect, macro-economic effects and internal developments which was updated and expanded for sub-sections 1*b)*, (2)*a) and (2)b* respectively of this chapter. R. Amann's contribution has been published in "East-West Technology Transfer: A Concise Synthesis of the Issues", *CREES Discussion Papers*, Series RC/B 19, University of Birmingham, Centre for Russian and East European Studies, Spring, 1983.
2. See Chapter 6, last part of section 3*c)* of this study.
3. OECD East-West technology transfer data base.
4. *Economic Bulletin for Europe*, Vol. 35, N° 4, Pergamon Press for the United Nations, December 1983, pp. 441-534.
5. For a full comparative description of the macro-economic estimates see Stanislaw Gomulka and Alec Nove, *East-West Technology Transfer, Contribution to Eastern Growth: An Econometric Evaluation, Book I*, OECD, Paris, 1984. In the same publication Book II, by George D. Holliday, *Survey of Sectoral Case Studies*, the focus is placed on studies which have looked at the East-West technology transfer process by industry or by project; that is, he assesses the micro-economic impact. The author underlines that generalisations from findings of the studies surveyed should be made with caution. Nevetheless, he concludes that Western technology has made a significant contribution to the technological advancement of the Eastern industrial branches studied, but that domestic infrastructures and institutions tend to inhibit rapid and efficient assimilation of Western technology and the export of competitive manufactured goods to the West.
6. See R. Amann, J.M. Cooper and R.W. Davies, *The Technological Level of Soviet Industry*, Chapter 1, Yale University Press, 1977 and Philip Hanson, *Trade and Technology in Soviet-Western Relations*, Chapter 9, The Macmillan Press Ltd., 1981.
7. See S. Gomulka, A. Nove, *op. cit.*
8. *Ibid.*
9. *Ibid.*
10. *Ibid.*
11. Yasushi Toda, "Technology Transfer to the USSR: The Marginal Productivity Differential and the Elasticity of Inter-Capital Substitution in Soviet Industry", *Journal of Comparative Economics*, Vol. 3, N° 2, 1979, pp. 181-194.
12. Michael Kaser, "The Flow of Technology from the OECD Countries to Romania", Working Document, OECD, Paris, 1981, paras. 41 and 102.
13. Josef C. Brada and Dennis L. Hoffman, "The Productivity Differential between Soviet and Western Capital and the Benefits of Technology Imports to the Soviet Economy", *Quarterly Review of Economics and Business*, Vol. 25, N° 1, Spring 1985, pp. 6-18.
14. Martin Weitzman, "Technology Transfer to the USSR: An Econometric analysis", *Journal of Comparative Economics*, June 1979, p. 140.

*ANNEX TABLES*

Annex Table 1 (Chapter 4)

## MAJOR INVESTMENT PARTICIPATION OF THE GERMAN DEMOCRATIC REPUBLIC IN OTHER CMEA MEMBER COUNTRIES

| Project | Contract signed | Site country | Form of participation | Time of validity for supplies after completion |
|---|---|---|---|---|
| 1. Cellulose-combine Ust-Illimsk | 1972 | USSR | Multilateral | 12 years |
| 2. Asbestos combine Kiembai | 1973 | USSR | Multilateral | 12 years |
| 3. Ferro-alloys/ iron-ore | 1974 | USSR | Multilateral | 12 years |
| 4. Ammonphosphate Kingisepp | 1974 | USSR | Multilateral | 10 years |
| 5. Natural gas pipeline Orenburg-Uzhgorod | 1974 | USSR | Multilateral | 20 years |
| 6. 750 kV-line USSR-Hungary | 1974 | USSR | Multilateral | 30 years |
| 7. Nickel-combine | 1975 | Cuba | Multilateral | 30 years |
| 8. Feed-yeast factory Mosyr | 1979 | USSR | Multilateral | 15 years |
| 9. 750 kV-line USSR-Poland | 1979 | USSR | Multilateral | 30 years |
| 10. Natural gas pipeline Orenburg-Uzhgorod | 1981 | USSR | Bilateral | 5 years |

The amount of investment participation has developed as follows:

| Period | Amount (in billion Marks) | Proportion of total investments of the GDR economy (in %) |
|---|---|---|
| 1966-1970 | 1 995 | 1.4 |
| 1971-1975 | 3 160 | 1.6 |
| 1976-1980 | 8 000 | 3.3 |

*Source:* Christa Luft and Lutz Maier (German Democratic Republic), "International Investment Co-operation within the CMEA Area", paper presented at the *Seventh Congress of the International Economic Association,* Madrid, 5th-9th, September 1983, pp. 8 and 9.

Annex Table 2 (Chapter 4)

## EVOLUTION OF INTRA-CMEA SPECIALISATION AND CO-OPERATION IN SELECTED AREAS OF PRODUCTION: DIVISION OF TASKS

**Bulgaria**

- Electric elevators, computer equipment, electrical goods and equipment, agro-food products (canned fruit and vegetables)[1].
- Lift transport equipment (battery-powered trucks, electric hoists, loaders): 90 per cent for export; computers and peripheral devices for them, micro-electronic motors, individual electrical equipment subassemblies for motor vehicles[2].

*Export Specialisation:*

- Freight-handling and electrical engineering equipment, calculators, tractors and vessels[8].

**Czechoslovakia**

- Machinery and equipment for the metals and chemicals industries, cars and trucks, nuclear power station equipment, chemicals and industrially produced consumer goods (shoes, ceramics, etc.)[1].
- Single-item and small-series machine-building articles, nuclear reactors and equipment for nuclear power plants, rolling mills, equipment for the petroleum food and chemical industries, compressors, pumps, turbines and diesel engines; power transport and textile machine-building and the production of computer equipment; diesel locomotives and passenger cars: 40-50 per cent for export; motor scooters: 90 per cent for export; river ships and streetcars: 70-80 per cent for export; textiles, leather footwear: 45 per cent for export[2]
- By the end of 1980, Czechoslovakia was a treaty partner in more than 320 bi- and multilateral specialisation and production co-operation agreements, above all in the areas of machine-building, electrotechnology and electronics.
- Specialisation and production co-operation will be expanded particularly for the production of roller bearings, tractors, farm machinery and parts for Lada passenger cars[3].
- By 1983, Czechoslovakia had concluded 72 multilateral specialisation and co-operation agreements for the production of machinery and equipment[11].

**German Democratic Republic**

- Mechanical engineering products: machine tools, machinery and equipment for the textile industry[1].
- Precision mechanics and optics items, printing and publishing machinery and typewriters (80-85 per cent of all production exported).
- Textile machinery, ships, passenger cars and refrigerated cars: 70 per cent of output for export.
- Complete sets of equipment for open-pit mining, oil refining installations, chemical equipment, agricultural machinery, precision machine tools: 35-40 per cent of production for export[2].

*Export specialisation:*

- Machine tools, forging and pressing equipment, electrical engineering mining and metallurgical equipment, equipment for light and food industry, communication equipment, calculators, agricultural machinery, railroad rolling stock, vessels and various instruments[8].
- By the end of 1980, the German Democratic Republic had concluded some 379 bilateral and 101 multilateral co-operation agreements with its CMEA partners[12].

**Hungary**

- Aluminium industry products, pharmaceutical products, buses, equipment for the agro-food industry, agricultural machinery, light manufactures[1].
- Medical apparatus, weak-current devices and communications equipment, electronic automatic telephone exchanges: share of exports in their output – 60 per cent; electronics industry to be developed as a branch specialised for export. Buses (about 80 per cent being exported), equipment for food industry, poultry farms and feed lots and aluminium industry (about 50 per cent of output is exported)[2].
- The vast majority of output produced within the framework of production specialisation and co-operation is made up of machinery and equipment. Production specialisation and co-operation agreements are the basis for developing the production of buses, cranes, computers, communications equipment, agricultural machine-building, the "bauxite-alumina-aluminium" industrial chain and the chemical and pharmaceutical branches of industry[5].
- At present, Hungary has concluded more than 90 multilateral and 150 bilateral contracts for production specialisation and co-operation with its CMEA partners. Most co-operation arrangements are in engineering[6].

*Export Specialisation*:

- Communication equipment, buses, ships and floating cranes, rail transport products[8].

**Poland**

- Coal, non-ferrous metals (copper) excavators, machinery and equipment for the light and food industries, electrical products, ships[1].
- Industrial plant and assemblies for the automotive industry, agricultural machinery, equipment for the ship-building industry, ships, floating facilities, complete production lines for the chemical industry, equipment for the manufacture of ceramics and glassware, lathes, construction machines, mining equipment, electronic computers and anti-friction bearings.

Poland has signed bout 100 multilateral and some 220 bilateral agreements on specialised and cooperated production with the CMEA countries[10].

*Export Specialisation:*

- Power engineering and electrical engineering equipment, equipment for the food industry, railroad rolling stock, vessels and vessel equipment[8].

**Romania**

- Machinery and equipment for the chemical, petro-chemical, oil and food industries[1];
- Nitrogen fertilizers, sodium hydroxide, paints, rubber products, petroleum and prospecting and extraction equipment[2].
- Specialisation and co-operation in production of the machine-building industry is a priority for Romania. It will develop as follows:

Ships, electrical engineering equipment, tractors and farm machinery, in relationship with the *Soviet Union*: optical and fine mechanics equipment, electrical separators and power transformers, and diesel-

| | |
|---|---|
| | hydraulic locomotives, in relationship with the *German Democratic Republic*: machine-tools (heavy turning mills, jig borers), integrated circuits and semi-conductor devices, in relationship with *Czechoslovakia*; lifting and transportation equipment (16-ton cranes), machine-tools, sound boxes, in relationship with *Poland*; hydraulic units, assemblies and sub-assemblies for motor vehicles, medical equipment, in relationship with *Hungary*; trucking facilities (special trucks, tipping trucks, refrigerating vans, motor and electric stackers), machine tools for general purposes, in relationship with *Bulgaria*[7]. |
| | Romania has expressed its interest in furthering specialisation and increasing mutual deliveries in the motor vehicle, tractor and rolling traction material branches and in the production of machinery and equipment for the chemical, light and food industries[9]. |
| **USSR** | – Fuels, raw materials, cars, watches, cameras, aircraft and associated equipment, nuclear power station equipment, machine tools, agricultural machinery (tractors), computer equipment[1].<br>– Passenger cars, watches, cameras, cotton flax: 25-30 per cent of output for export; petroleum, iron ore, agricultural machinery, machine tools, computer and power equipment[2].<br>– In 1982, the Soviet Union had concluded 302 specialisation agreements with its CMEA partners[13]. |

*Sources:*


1. N.B. Boroch, C.N. Silvestrov, R. Schölf (ed.), "Proizvodstvennoye Sotrudnichestvo Stran SEV", *Ekonomika*, Moscow, 1982, pp. 11-12. Situation at the end of 1970.
2. I. Dudinskiy, "Socialist Integration as a Factor in Intensifying the Economy", *Voprosy Ekonomiki*, No. 8, 1982.
3. *Aussenwirtschaft*, No. 48, Berlin (East), 1982, as given in *Joint Publications Research Service* (82687), *East Europe Report*, 20th January, 1983, p. 38.
4. "Economic Co-operation of the CSSR with the Socialist Countries", *Dokumentacni Prehled*, No. 42, Prague, 1981, as given in *Joint Publications Research Service* (79925), *East Europe Report*, 22nd January, 1982, p. 2.
5. Peter Veresh, "The Prospects for the Development of Foreign Trade between the Hungarian People's Republic and the CMEA Member Countries in 1981-1985", *Ekonomicheskoye Sotrudnichestvo Stran-Chlenov SEV*, No. 2, 1982.
6. *Nepszabadsag*, No. 13, January 1983, p. 5, as given in *Joint Publications Research Service* (92787), *East Europe Report*, 2nd February, 1983, p. 1.
7. C. Lazar and I. Stan, *Revista Economica*, No. 33, 14th August, 1981, pp. 18-19, as given in *Joint Publications Research Service* (79014), *East Europe Report*, 18th September, 1981, p. 3.
8. N. Alekhin, "Production and Scientific-Technical Co-operation of CMEA Countries", *Voprosy Ekonomiki*, No. 10, 1982.
9. Ioan V. Totu, *Revista Economica*, No. 4, 26th January, 1984, pp. 21-24, as given in *Joint Publications Research Service* (EEI-84-025), *East Europe Report*, 24th February, 1984, pp. 3 and 7.
10. Janusz Kaczurba "Poland as a Member of the Council for Mutual Economic Assistance", *Foreign Trade* (English edition), No. 7, Moscow, 1980, p. 8.
11. "Greater Machinery Share in Exports", *Svet Hospodarstvi*, No. 5, 1985, as given in *Joint Publications Research Service* (EEI-84-098), *East Europe Report*, 30th August, 1984, p. 43.
12. H. Machowski, "Die Produktionsspezialisierung im Comecom: Das Beispiel der DDR", *Neue Zürcher Zeitung* (Foreign edition), No. 124, 30th May, 1984, p. 13.
13. *Soviet Export* (English edition), No. 6, 1983, as given in *Joint Publications Research Service* (UIE-84-006), *USSR Report*, 16th March, 1984, p. 31.


*Comment:* This repartition of tasks continues to be adhered to, though greater emphasis is being placed on co-operation directly related to developing energy resources. See, for instance, O. Ilin, "Investment Co-operation among the CMEA Member Countries and Formation of National Economic Structures", *Voprosy Ekonomiki*, No. 10, 1984.

Annex Table 3 (Chapter 4)

## SOME CO-OPERATION AND SPECIALISATION AGREEMENTS BETWEEN THE CMEA COUNTRIES IN THE ENGINEERING FIELD

### Multilateral agreements: 1970-1980

| Signatory countries | Date of signing | Implementation period | Scope of the agreement |
|---|---|---|---|
| Hungary, GDR, Poland, USSR, Czechoslovakia, Bulgaria, Romania | Sept. 1971 | 1971-1975 | Equipment for the manufacture of ceramic products. |
| Bulgaria, Hungary, GDR, Poland, Romania, USSR, Czechoslovakia | Sept. 1971 | 1971-1975 | Equipment for the manufacture of glassware. |
| Bulgaria, Hungary, GDR, Poland, USSR, Romania, Czechoslovakia | Sept. 1971 | 1971-1975 | Large capacity trucks. |
| Bulgaria, Hungary, Poland, USSR | 1972 | 1972-1975 | Weaving looms. |
| Bulgaria, Hungary, GDR, Romania, USSR, Poland, Czechoslovakia | Feb. 1972 | 1972-1975 | Certain types of numerical control machine tools. |
| Autoimpex (Bulgaria), Autoker (Hungary), Transportmaschinen (GDR), Autotraktor (Romania), Sojuzglavautozagranpostavaka (USSR), Motokov (Czechoslovakia) | Sept. 1971 | 1976-1980 | Multilateral specialisation and co-operation in motor vehicle production. |
| Bulgaria, Hungary, GDR, Poland | Feb. 1974 | 1978 | Vinnitsa (USSR) - Albertirsa (Hungary), high voltage power line. |
| Poland, Hungary, GDR, Romania, Czechoslovakia | Janv. 1974 | 1976-1980 | Specialisation and co-operation in the field of isotopes. |
| Bulgaria, Hungary, GDR, Poland, Romania, USSR, Czechoslovakia | Dec. 1974 | 1976-1980 | Specialisation and co-operation in production of ball-bearings. |
| Hungary, Poland, USSR, Czechoslovakia | March 1979 | 1984-1991 | Construction of a nuclear power station at Khmelnitskij (USSR). |
| Hungary, Poland, USSR, Czechoslovakia | March 1979 | 1983 | High voltage power line Khmelnitskij-Rzeszow (Poland). |
| Bulgaria, Hungary, Cuba, Poland | June 1979 | 1990 | Specialisation and co-operation in the field of energy-intensive chemical products in exchange for other chemical products. |
| Bulgaria, Hungary, GDR, Poland, Mongolia, Romania, USSR, Czechoslovakia | Feb. 1974 | 1990 | Plans for developing unified energy systems. |
| Bulgaria, Hungary, GDR, Poland, Romania, USSR, Czechoslovakia, Yugoslavia | June 1979 | 1990 | Production of nuclear power station equipment. |

Annex Table 3 (Chapter 4) *contd*

| Signatory countries | Date of signing | Implementation period | Scope of the agreement |
|---|---|---|---|
| GDR, Cuba, Poland, USSR, Czechoslovakia | | 1985 | Construction of a feed yeast plant at Mozyr (USSR). |
| GDR, Bulgaria, Hungary, Poland, Romania, USSR, Czechoslovakia | June 1980 | 1985 | Co-operation and specialisation in the production of computer equipment, components and assemblies. |

*Source: Mnogostoronnoe Ekonomicheskoe Sotrudnichestvo Socialisticheskikh Gosudarstv* (documents 1971-1980).

Annex Table 4 (Chapter 4)

## AGREEMENTS FOR SPECIALISATION AND CO-OPERATION IN PRODUCTION BETWEEN THE SOVIET UNION AND THE SIX: AREAS OF SPECIALISATION FORESEEN FOR THE EAST EUROPEAN COUNTRIES

| | |
|---|---|
| **USSR-Bulgaria** | – Co-operation is primarily foreseen for power engineering, ferrous and non-ferrous metallurgy, machine-building, the chemical and petroleum processing industry, agriculture and transportation. Co-operation in machine-building occupies a special place. Machine-building is involved in 25 of the 29 bilateral agreements in production specialisation. The exchange of components and parts is to be increased – particularly for electronic computer equipment.<br>Among the main specialised products of Bulgarian machine-building are electric telphers, automatic loaders, electric stackers, electric carts and electronic computer equipment.<br>About 80 per cent of exports of specialised products from Bulgaria to the USSR are from six branches: lifting and transport equipment, computer and organisational equipment, tractors and agricultural machines, power and electrotechnical equipment, communications equipment, radios and televisions and means of transportation[1].<br>– Some 170 contracts have been signed with Bulgaria in pursuance of the agreement signed between the two countries[4].<br>– The manufacture of engineering products, particularly electronics, is occupying an increasingly important position[10]. |
| **USSR-Czechoslovakia** | – Czechoslovak exports based on production specialisation include primarily Tatra trucks, shovel type loaders to be used with trucks, passenger and cargo ships of "river-sea" type, automated telephone stations, lathes and semi-automated machine bearings. Also delivered to the USSR are automated equipment, equipment for the textile and rubber industry, chemical preparations for plant protection and electro-insulating glass fabrics[1].<br>– Specialisation and co-operation in production is to be primarily developed in mechanical engineering; that is, machine-building[2].<br>– Here emphasis is placed on equipment for nuclear power plants, communication technology, technological units for the chemical industry, equipment for the food industry, machine tools and forming machinery, pumps, compressors and refrigeration technology[3].<br>– Czechoslovakia is also to specialise in the manufacture of numerical programme-controlled-machine tools, certain types of electric motors, and radar facilities[4].<br>– Twenty-five agreements have been signed of which 20 fall in the machine-building sector[11]. |
| **USSR-German Democratic Republic** | – The "co-operative programme" is devoted to developing power engineering and to reducing expenditures of fuel and raw materials. In the area of metallurgy, joint development and improvement of technological processes and equipment for producing steels and alloys are foreseen. It is intended to jointly develop and produce instruments made of synthetic diamonds and other ultra-hard materials which will contribute to the introduction of effective technologies in machine-buiding, the glass industry and metallurgy, among others. Importance is attached to furthering co-operation in the production of high-quality consumer goods[1]. |

| | |
|---|---|
| | – Nearly 160 inter-governmental and inter-departmental agreements have been concluded on the basis of the agreement for specialisation and co-operation in production[4].<br>– The emphasis of this agreement is on:<br>jointly solving important materials management tasks and enhanced refinement of available raw materials and developing microelectronics and robot technology[5];<br>further pooling of the two countries' material resources and scientific-technical potential for the period to 1990[12],<br>co-operation in manufacturing equipment for atomic power stations: high pressure accessories, centrifugal pumps, large lifting capacity electric overhead cranes, etc.[13].<br>– a total of 35 basic directions for specialisation and co-operation in production have been agreed upon[6, 14]. |
| **USSR-Hungary** | – The long-term agreement signed concerns 19 branches of industry. The automotive industry is the most important and Hungary is to expand its delivery of Ikarus buses. Production of rear axles is to double. In the area of machine tool building, reciprocal specialisation and production is foreseen for metal-cutting equipment, forge and press equipment. Co-operation in shipbuilding and power machine-building is to increase[1].<br>– Some 37 agreements on specialisation and co-operation in production had been signed by 1980 to develop the engineering and metallurgy industries[9].<br>– Items manufactured on the basis of specialisation and co-operation constitute 34 per cent of Hungary's exports to the USSR: of which, 85 per cent are machine-building products. Current emphasis is placed on furthering specialisation for the production of communication and electronic equipment. Micro-electronics is to become a priority area of co-operation[15]. |
| **USSR-Poland** | – Poland's specialisation includes shipbuilding, rolling stock for railroads, equipment for the light and food industry, chemical, petrochemical, metal processing and mining equipment and electronic computer equipment. In future, special attention is to be devoted to the production of standard parts, batching items and spare parts. Also earmarked for long-term co-operation are equipment for underground and open-pit mining, for electric and thermal stations and for extracting, transporting and processing petroleum and gas[1].<br>– Recently, "branch programmes" were signed for the development of specialisation and co-production through 1990 in the electronics industry, for the production of communications equipment, for the machinery and tools industry and the for production of road construction machinery[7]. Sixteen long-term specialisation agreements have been signed[16]. |
| **USSR-Romania** | – Romania is to specialise in the output of complex transformer substations, electric engines and automatic circuit breakers.<br>In *tractor and agricultural machine-building*, Romanian plants are to produce universal seeders, threshing machines and machines for cleaning corncobs.<br>In the area of *animal husbandry*, Romania is to produce trailer pickups. |

| | |
|---|---|
| | In *chemical and petroleum machine-building*, Romanian enterprises will have to deliver petroleum equipment for various technological purposes so that the USSR can construct installations for catalytic cracking and hydraulic cracking of vacum gas oil.<br>*In the electronic industry*, Romania will produce radio components, electronic optical systems and elements for colour picture tubes.<br>*In instrument building*, Romania will manufacture electromagnetic flow gauges, electrical and electro pneumatic positioners, analytical weights and balances and systems and devices for fire alarms.<br>In the area of *heavy and transportation machine-building*, Romania is to specialise in the production of coke and cinder equipment, equipment for continuous casting, brake equipment, hydraulic cylinders and hydraulic equipment.<br>*In the chemical industry*, specialisation and co-operation will be developed towards the production of various types of chemical plant protection products, mineral fertilizers and other chemical products[1].<br>– In future, specialisation and co-operation in production is to be expanded for metal-cutting lathes, petrochemical equipment, chemicals, electrical goods, tractors and farm machinery[8].<br>– Metallurgical equipment, agricultural machinery, oil extracting equipment and equipment for the light and food industries[17]. |

*Sources:*


1. I. Kareyev, "Bilateral Programmes for Production Specialisation and Co-operation", *Voprosy Ekonomiki*, No. 5, 1981.
2. "Main Trends in CSSR-USSR Economic Relations", *AW-DDR Aussenwirtschaft*, Vol. 11, No. 4, Berlin (East), 19th January, 1983, as given in *Joint Publications Research Service* (83160), *East Europe Report*, 29th March, 1983, p. 4.
3. "Economic Co-operation of the CSSR with the Socialist Countries", *Dokumentacni Prehled*, No. 42, 1981, as given in *Joint Publications Research Service* (79925), *East Europe Report*, 22nd January, 1982, p. 1.
4. N. Syomin, "On the Basis of Division of Labour", *APN Daily Review* (in English), Moscow, 12th May, 1983, as given in *Joint Publications Research Service* (84333), *USSR Report*, 15th September, 1983, p. 12-13.
5. Horst Tschanter (SED Central Committee), "The Vitality of CMEA", *Einheit*, Vol. 37, No. 11/12, Berlin (East), 1982, as given in *Joint Publications Research Service* (82828), *East Europe Report*, 9th February, 1983, p. 14.
6. N. Alekhin, "Production and Scientific Technical Co-operation of CMEA Countries", *Voprosy Ekonomiki*, No. 10, 1982.
7. Natalya Chudakova, "Soviet-Polish Trade and Economic Relations", *Gazeta Pomorska*, 26th May, 1983, as given in *Joint Publications Research Service* (84256), *East Europe Report*, 2nd September, 1983, p. 3.
8. V. Bondarchuk, "Soviet-Romanian Economic Co-operation", *Foreign Trade* (English edition), No. 11, Moscow, 1982, p. 23.
9. "USSR and Hungary", *Foreign Trade*, (English edition), No. 5, Moscow, 1980, p 16.
10. K. Khristov, "Bulgaria and the Soviet Union: Fraternal Co-operation", *Foreign Trade* (English edition), No. 8, Moscow, 1984.
11. V. Vladimirsky, "The USSR and Czechoslovakia: Today's Trade and Economic Co-operation", *Foreign Trade* (English edition), No. 8, Moscow, 1984.
12. Horst Sölle, "35 Years of the GDR's Successful Foreign Trade Policy", *Foreign Trade* (English edition), No. 10, Moscow, 1984.
13. I. Alexandrov, "USSR-GDR: By Joint Efforts", *Foreign Trade* (English edition), No. 6, Moscow, 1984.
14. *Die Wirtschaft* (East Berlin, Special Issue on the 1984 Leipzig Fair), as given in *Joint Publications Research Service* (ESA-84-041), *East Europe Report*, 20th November, 1984, p. 11.

15. I. Graham, "USSR-Hungarian People's Republic: Intensifying Production Co-operation", *Ekonomicheskaya Gazeta*, No. 51, 1984, p. 20.
16. "Stability - The Basis of Polish-Soviet Co-operation", *Foreign Trade* (English edition), No. 12, Moscow, 1984.
17. N. Mikhailenko, "USSR-Romania Ongoing Economic Co-operation", *Foreign Trade* (English edition) No. 10, Moscow, 1984.

*Comment*: As noted relative to Annex Table 2 (of Chapter 4), this repartition of tasks continues to be adhered to, although greater emphasis is being placed on the production of sophisticated machinery and equipment.

Annex Table 5 (Chapter 4)

SPECIALISED PRODUCTION SHARE OF INTRA-CMEA EXPORTS

(%)

| | 1975 | 1980 | 1981 |
|---|---|---|---|
| Bearings | 20.6 | 87.3 | – |
| Ships, boats, port installations | 81.6 | 81.3 | – |
| Lifting and handling equipment | 46.2 | 61.1 | 69.7 |
| Machinery for the textile industry | 25.2 | 53.1 | 59.4 |
| Road building equipment | 40.5 | 49.8 | – |
| Metal-cutting machine tools | 25.0 | 46.4 | – |
| Agricultural machinery | 31.9 | 44.6 | – |
| Tractors | 39.8 | 44.5 | – |
| Trucks and garages | 3.4 | 43.2 | 43.2 |
| Presses and forges | 11.2 | 43.2 | – |
| Equipment for the chemical industry | 25.5 | 34.0 | 39.1 |
| Telecommunications equipment | 15.3 | 24.9 | 35.7 |
| Comuter equipment | – | 42.0 | – |
| Energy and electrical engineering equipment | – | 19.0 | – |

*Source: Ekonomicheskoye Sotrudnichestvo Stran-Chlenov SEV,* No. 10, 1982, p. 26 and No.1, 1983, p. 18.

Annex Table 6 (Chapter 4)

## INTRA-CMEA EXPORTS OF SELECTED CATEGORIES OF MACHINERY AND EQUIPMENT AND PERCENTAGE OF CMEA IMPORT DEMAND MET BY THOSE EXPORTS

| | Unit | Intra-CMEA Exports 1980 | Percentage share in total imports of the products concerned |
|---|---|---|---|
| Metal-cuttins machine tools | 1 000 units | 28.4 | 64.0 |
| Energy and electrical engineering equipment | Million T.R. | 1 771 | 66.0 |
| Lifting and handling equipment | Million T.R. | 1 205 | 82.0 |
| Medical instruments and apparatus, laboratory equipment | Million T.R. | 1 128 | 60.0 |
| Equipment for the chemical, wood and paper industries | Million T.R. | 5 855 | 65.0 |
| Tractors and motor vehicle chassis | 1 000 units | 39.8 | 96.0 |
| Goods wagons | 1 000 units | 6.8 | 93.0 |
| Trucks | 1 000 units | 65.7 | 97.0 |
| Cars | 1 000 units | 319.0 | 95 |
| Buses | 1 000 units | 16.5 | 97.0 |
| Ships and associated equipment | Million T.R. | 922 | 63.0 |

*Source:* *Ekonomicheskoe Sotrudnichestvo Stran-Chlenov SEV,* No. 10, 1982, p. 27.

Annex Table 1 (Chapter 7)

## GLOSSARY OF TERMS FREQUENTLY USED IN EAST-WEST TRADE AND INDUSTRIAL CO-OPERATION

BARTER (TROC):*

Form of a short-term linked deal which provides for a direct exchange of goods with no money and no third party involved. (See *Buy-back*; *Compensation*; *Counter-trade*; *Counter-purchase*.)

BUY-BACK (RACHAT):*

Commitment of a seller of equipment, licences and/or know-how to accept in full or partial payment products manufactured with such equipment and/or licences and/or know-how. Large-scale and long-term *compensation projects* are based on buy-back commitments of resultant (or at least related) products.
(See *Compensation*; *Counter-purchase*; *Counter-trade*.)

COMPENSATION (COMPENSATION):*

Traditionally, compensation is a short-term operation whereby the supplier, through a single contract, accepts that the value of goods supplied be offset by a reverse flow of goods (counter-deliveries). These deliveries may be of equal value (full compensation) or only part of the transaction may be covered by the reverse flow of goods (partial compensation); in the latter case settlement of the remaining balance takes place in cash or through appropriate credit arrangements. In a broader sense, compensation is a form of industrial co-operation and may designate large-scale transactions where provision is made for long-term project financing and the purchase, through separate contracts, of resultant (or related) products. As such, it consists of deliveries on credit and *buy-back* commitments.

COMPLEX ENGINEERING (INGÉNIÉRIE COMPLEXE):*

A form of engineering encompassing: *consulting engineering* (intellectual services, management, technical assistance); *process engineering* (granting of know-how and process licences); and *contracting engineering* (supply of equipment and material, assembly, construction work). (See *Consulting engineering*; *Process engineering*; *Contracting engineering*.)

CONSULTING ENGINEERING (INGÉNIÉRIE-CONSEIL):*

A form of engineering relating essentially to intellectual services connected with the design of productive plant, the preparation of projects and blueprints, and the supervision of construction work. It can also comprise management and the provision of technical assistance. Technical assistance given by consulting engineers has often been considered part of the more general term "transfer of technology" and, in most cases, involves the provision of professional training. The terms "consulting engineering" and "pure engineering" are synonymous. (See *Complex engineering*; *Contracting engineering*.)

CONTRACTING ENGINEERING (INGÉNIÉRIE DE CONSTRUCTION ET/OU D'ENSEMBLE):*

A form of engineering related to the supply of equipment and material and/or the installation of productive plant on site. It includes, where necessary, civil engineering work. (See *Complex engineering*; *Consulting engineering*.)

CONTRACTUAL JOINT VENTURE (CO-ENTREPRISE CONTRACTUELLE):*

A synonym for the term non-equity joint venture: a flexible association lacking both corporate form and equity capital, whereby partners situated in different countries carry out together business activities without establishing any additional institution in the form of a separate joint venture enterprise. (See *Equity joint venture*.)

CO-PRODUCTION (CO-PRODUCTION):*

Apportionment of tasks under a contract for *industrial co-operation* under which each of the participating enterprises specialises, either in the manufacture of designated components or of a certain portion of a final product which is then assembled by one or both of the partners, or in the production of selected finished articles which are then exchanged, so that each partner acquires the full range of goods covered by the joint production programme.

COUNTER-PURCHASE (ACHAT DE CONTREPARTIE):*

Short- or medium-term *linked deal*, usually taking the form of several related contracts, whereby an exporter agrees to import goods and/or services from his customer in another country (or from a party designated by the latter) equal in value to a stipulated percentage of his own deliveries. (See *Barter*; *Buy-back*; *Compensation*; *Counter-trade*; *Parallel transaction.*)

COUNTER-TRADE (ÉCHANGES COMPENSÉS):*

Generic term which designates any linked deal according to which parties accept, as payment for the goods or services supplied, deliveries of other goods and/or services in place of or in addition to financial settlement.

EQUITY JOINT VENTURE (CO-ENTREPRISE AVEC CONSTITUTION D'UN CAPITAL SOCIAL PAR ACTIONS):*

In an international equity joint venture the share-owning members who are situated in different countries share the risk and the profits of their joint enterprise, which is formed as a separate legal entity. In the majority of East European countries the formation and operation of equity joint ventures with Western partners is regulated by special legislation. (See *Contractual joint venture*; *Joint venture.*)

EVIDENCE ACCOUNT (COMPTE TÉMOIN):*

Financial mechanism through which *counter-purchase* transactions and related short- and medium-term linked trading arrangements between Western firms and their partners in some East European countries are effected. The "evidence account" is maintained at the State foreign trade bank of the East European partner in an agreed convertible currency. All sales and purchases by the Western firms are registered in this account. The level of imports from the Western firm is usually linked to the amount of the positive balance on its account produced by its imports from the Eastern country. The "evidence account" should be distinguished from the *offsetting or balancing account* and from the *clearing* accounts.

INDUSTRIAL CO-OPERATION (COOPÉRATION INDUSTRIELLE):*

Relationship involving the creation of a long-term (in many cases initially running for ten years) community of interests between unrelated industrial enterprises of different nationalities. Industrial co-operation may comprise: supply of licences, plant or production lines; development of new technologies and sharing of information relating to these technologies; production, marketing, *joint tendering* or joint projects. The whole or part of the obligations arising from co-operation activities may be settled in kind (counter-deliveries of related products or services). (See *Co-production.*)

JOINT TENDERING (SOUMISSION EN ASSOCIATION):*

Form of *industrial co-operation* in which two or more enterprises submit a joint tender to a customer who may be located in either country or, more commonly, in a third country. (See *Tripartite industrial co-operation.*)

JOINT VENTURE (OPÉRATION EN ASSOCIATION):*

Form of business partnership involving joint management and the sharing of risks and profits. Joint ownership of capital may or may not be involved. If it is involved, the joint venture is referred to as an *equity joint venture*; if not, the joint venture is referred to as a *contractual (non-equity) joint venture.*

JUNKTIM TRANSACTION (OPÉRATION *JUNKTIM*):*

A short- or medium-term linked transaction, often extending over several years, under which a Western firm which purchases goods from an East European trading partner, and pays for them in convertible currency, acquires an "export credit" (or presumptive entitlement to sell) authorising exports of equivalent value to the Eastern country involved. The Western importer may use its export credits either for its own export sales to the country or transfer them to another exporting firm, usually against payment. (See *Barter*; *Buy-back*; *Counter-trade*; *Counter-purchase*.)

LICENCE AGREEMENT (ACCORD DE LICENCE):**

In general terms, a licence is a legal permit to do something. In relation to *technology transfer*, licence means the "consent given by the owner of an exclusive right (licensor) to another person (licensee) to perform certain acts which are covered by an exclusive right, or consent as to use of "*know-how*". Licence agreement refers to the *contract* between a licensor and a licensee on the granting of a licence. Such agreements, incorporated in one or more legal documents usually cover licence rights and may include as well the provision of various technical services and supply of materials.

CROSS LICENSING (CONCESSION RÉCIPROQUE DE LICENCES):*

An agreement whereby two enterprises exchange licences to manufacture some of each other's products.

KNOW-HOW LICENCE (LICENCE DE SAVOIR-FAIRE):***

Is granted for the disclosure of know-how which the licensor agrees to transfer to the licensee.

PACKAGE LICENCE (LICENCES GROUPÉES):**

The licensing of several of the licensor's *patents* imposed upon the licensee as a condition for obtaining a *licence.*

PATENT LICENCE (LICENCE DE BREVET):***

Provides for the use of a patented product or process.

TRADEMARK LICENCE (LICENCE DE MARQUE):***

Is an arrangement whereby the proprietor of a trademark allows someone else (usually in connection with a commercial advantage or for royalties) to use his trademark provided certain quality conditions are met.

LINKED OR LINKAGE DEAL (AFFAIRE LIÉE):*

A generic term designating any reciprocal trading arrangement. Often used to refer to: (i) the traditional concept of *compensation*; or (ii) two or more technically separate transactions which are concluded at the same time and where the conclusion and fulfilment of each of them is conditional on the conclusion and fulfilment of the other or others. (See *Counter-purchase*; *Counter-trade*; *Parallel transactions*.)

PARALLEL TRANSACTION (OPÉRATION OU TRANSACTION PARALLÈLE):*

A form of *counter-purchase* which consists of two separate contracts concluded simultaneously: in East-West trade these consist of (i) an export contract under which a Western firm sells goods or services to an East European buyer; and (ii) a *counter-purchase* contract under which the Western firm commits itself to import from the East European one, within a certain time-limit and subject to penalty, products worth a percentage of the sale specified in the export contract. Sometimes a third ("umbrella") contract is also concluded. A variant form of parallel transaction occurs when, at the time of the conclusion of the basic export contract, no formal counter-purchase contract is concluded but an acceptance obligation of the Western partner is recorded that permits later specification of the products to be imported by it from the Eastern enterprise. Parallel transaction differs from a *junktim transaction* in that the latter commits the Western participant to import first from the East European partner in anticipation of a future export. (See *Counter-trade.*)

PROCESS ENGINEERING (INGÉNIÉRIE DE PROCÉDÉ):*

Generally means: the development, scaling up and quantification of process steps used in a total process. (See *Complex engineering*; *Consulting engineering*; *Contracting engineering.*)

RECIPROCAL TRADING ARRANGEMENTS (ACCORDS DE COMMERCE RÉCIPROQUES):*

Generic term comprising large-scale and long-term compensation agreements and all short- and medium-term *linked deals.* (See *Compensation*; *Barter*; *Junktim*; *Counter-purchase*; *Counter-trade.*)

SUPPLIER'S CREDIT (CRÉDIT FOURNISSEUR):*

Short- or medium-term credit accorded generally at commercial interest rates by the seller of capital goods (or his bank) to the client for all or part of the purchase price. The repayment period is usually three to five years, starting once the plant comes on stream. It should be noted, however, that supplier's credit is often coupled with government-backed domestic export refinancing schemes.

TRIPARTITE INDUSTRIAL CO-OPERATION (COOPÉRATION INDUSTRIELLE TRIPARTITE):*

A type of co-operation in which an active part is taken by three or more enterprises located in two or three regions: an East European country, a Western country and – frequently – a Developing country. The subject of tripartite industrial co-operation is most frequently *joint tendering*, supply of complete plants or processing lines, or the execution of civil engineering projects. A tripartite industrial co-operation project may take one of two basic organisational forms: (a) the partners do not constitute a group as such but act independently; or (b) the partners form a group of one of three types: (i) simple consortium; (ii) a risk-and-profit sharing association; and (iii) a company having juridicial personality (*equity joint venture*). (See *Contractual joint venture.*)

TURNKEY CONTRACT (CONTRAT CLÉS EN MAIN):*

Contract under which the contractor assumes responsibility towards the client for the construction of productive installations and takes the client's place *vis-à-vis* the other participants in the project. In this way, complete responsibility is centered in one place and financing is arranged more easily for all the equipment in one package. The responsibility of the contractor ends when the completed installation is handed over to the client.

*Sources:*

* *Draft Glossary of Terms Frequently Used In East-West Trade and Industrial Co-operation* (TRADE/R.467), Note by the Secretariat, Committee on the Development of Trade, Economic Commission for Europe, Geneva, 15th November, 1983.

** *Proceedings of the UN/ECE Seminar on the Management of the Transfer of Technology within Industrial Co-operation* (ECE/SC.TECH./10), Senior Advisers to ECE Governments on Science and Technology, Economic Commission for Europe, Geneva, 16th February, 1976, pp. 62-63.

*** *Licencing and Leasing, op. cit.*, pp. 13 and 15.

Annex Table 2 (Chapter 7)

## TYPES OF CONTRACTUAL ARRANGEMENTS INCLUDED IN DIFFERENT DEFINITIONS OF EAST-WEST INDUSTRIAL (INTER FIRM) CO-OPERATION[a]

1. Sale of equipment for complete production systems, or "turnkey" plant sales (usually including technical assistance).
2. Licensing of patents, copyrights and production know-how.
3. Franchising of trademarks and marketing know-how.
4. Licensing or franchising with provision for market sharing and quality control.
5. Co-operative sourcing: long-term agreement for purchases and sales between partners, especially in the form of exchanges of industrial raw materials and intermediate products.
6. Sub-contracting: contractual agreement for provision of production services for a short-term and on the basis of existing capabilities, but often to design specifications furnished by the contractor. Some or all components also frequently supplied by contractor.
7. Sale of plant, equipment and/or technology (1-3 above) with provision for complete or partial payment in resulting or related products.
8. Production contracting: contractual agreement for production on a continuing basis, to partner specifications, of intermediate or final goods to be incorporated into the partner's product or to be marketed by him. In contrast to sub-contracting, production-contracting usually is on the basis of a partially transferred production capability, in the form of capital equipment and/or technology (on basis of a license or technical assistance contract).
9. Co-production: mutual agreement to narrow specialisation and exchange components so that each partner may produce and market the same end product in his respective market area. Usually on the basis of some shared technology.
10. Product specialisation: mutual agreement to narrow the range of end products produced by each partner and then to exchange them so that each commands a full line in his respective market area. In contrast to co-operative sourcing, product specialisation involves adjustment in existing product lines.
11. Co-marketing: agreement to divide market areas for some product(s) and/or to assume responsibilities for marketing and servicing each other's product(s) in respective areas. Joint marketing in third markets may be included. Does not in practice stand alone, unless in the form of a joint marketing company (i.e., combined with 14 below). Otherwise combined with various forms, especially 4, 8, 9, 10.
12. Project co-operation: joint tendering for development projects in third countries.
13. Joint Research and Development: joint planning, and the co-ordinated implementation of R&D programmes, with provision for joint commercial rights to all product or process technology developed under the agreement.
14. Any of the above in the framework of a specially formed mixed company or joint venture between the partner firms (on the basis of joint equity participation, profit and risk sharing, joint management).

*a)* The terminology used here is not standard, and the types are variously designated in the literature. For example, the term "production-sharing" is sometimes used to designate all or some of types 8 through 10.

*Source:* Carl H. McMillan, "East-West Industrial Co-operation", in *East European Economies Post Helsinki*, *op. cit.*, p. 1182.

Annex Table 3 (Chapter 7)

## CATEGORIES OF EAST-WEST INDUSTRIAL CO-OPERATION AND THEIR DEFINITION: A COMPARISON OF SCHEMES USED BY PAUL MARER AND JOSEPH C. MILLER AND BY THE UNITED NATIONS

| Category and Definition Used | ECE Definition |
|---|---|
| 1. Scientific-technical agreements (STA) are signed between a US corporation and the Soviet State Committee for Science and Technology (or its counterpart in an EE country) to explore what technology or know-how might be exchanged subsequently, on a commercial basis. While STAs are not commercial contracts, they are of interest because they identify industries and projects where the Easten partner intends to obtain Western technology and where US firms have both an interest and a capability to compete. | Category not included. |
| 2. Know-how (K) is transferred under a technical assistance contract. Only if such a contract was signed on its own, rather than as a component of a license or equipment sale or more complex forms of IC, was it included as a separate category[a]. | Category not included. |
| 3. Licensing (L), regardless of the means of payment. L is not shown separately if it occurred as part of a more complex transaction, such as turn-key, co-production of joint venture, to avoid double-counting the number of agreements or projects. | Licensing (L) with payment in resultant products. |
| 4. Turn-key (TK) includes all contracts where the supplier has significant on-site installation or supervision responsibilities, prime contractor; contractor (US firm is one of several contractors, but agreement is directly with the host country); and sub-contractor supplies machinery and equipment under sub-contract and has on-site installation/training responsibilities. | Turn-key (supply of complete plants or production lines) with payment in resultant products (Sub-contractors are not included). |
| 5. Sub-contracting (SC) in the Eastern country, with the Western partner usually providing the technical know-how and sometimes the machinery equipment, and parts. | Same |
| 6. Co-production (CP) each partner specialises either in the production of parts of a product assembled by one or both partners or in the production of a limited number of finished products exchanged to complete each partner's range of products. | Same |
| 7. Joint ventures (JV) co-ownership of capital, co-management, and sharing of risk and profit, if the JV is located on the territory of the Eastern partner. | Same, plus including joint marketing ventures typically located in the West which market products of the Eastern country. |

Annex Table 3 (Chapter 7) *contd*

| Category and Definition Used | ECE Definition |
|---|---|
| 8. Category not included (no known US cases). | Joint Tendering of Projects (JTP) supply of complete plants or production lines to a third Western party. |
| 9. Reverse Licensing (RL): US firm is importing a license. | Category not included. |

*a)* It is useful to distinguish know-how (1) to make: (2) to operate and maintain: and (3) to apply. Whereas a license and a patent normally cover what can be put on paper, know-how is less readily rendered on paper, rather it must be demonstrated.

*Source:* Paul Marer, Joseph C. Miller, "US Participation in East-West Industrial Co-operation Agreements", *Journal of International Business Studies,* Fall-Winter, 1977, pp. 28-29.

## SUMMARY GRAPHICS[1]

1. All graphics are based on value data used to construct Table 6 of Chapter 5.

Graphic 1. **OECD Exports to CMEA Countries**

Capital Goods

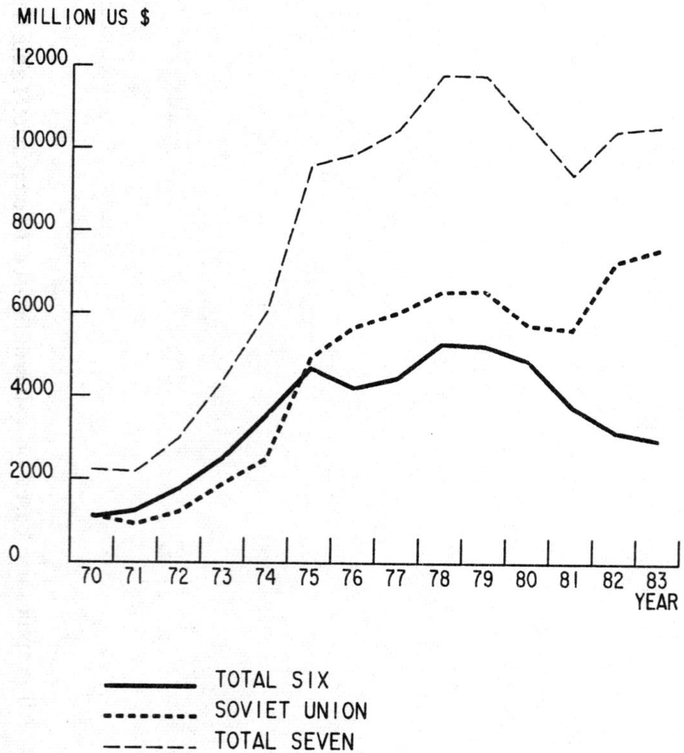


Intermediate Goods

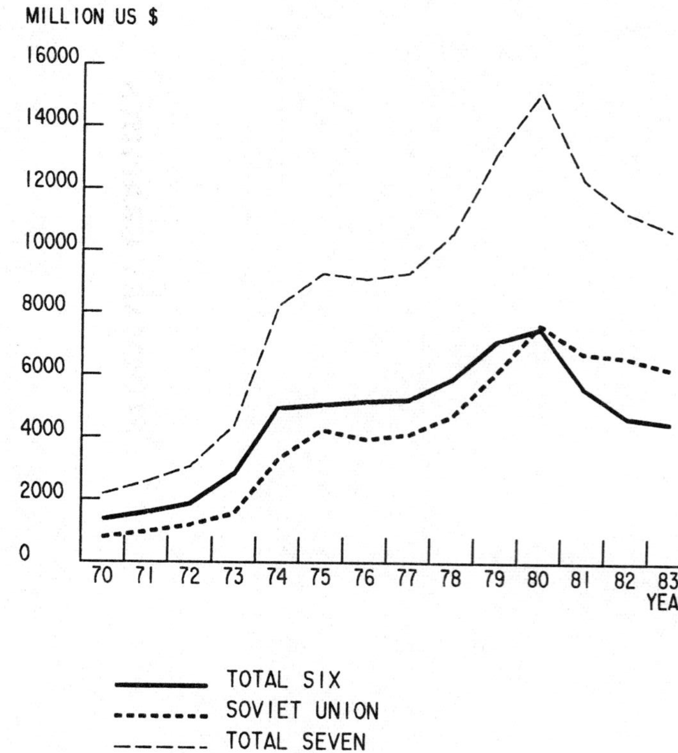

Graphic 2. **OECD Exports to CMEA Countries**

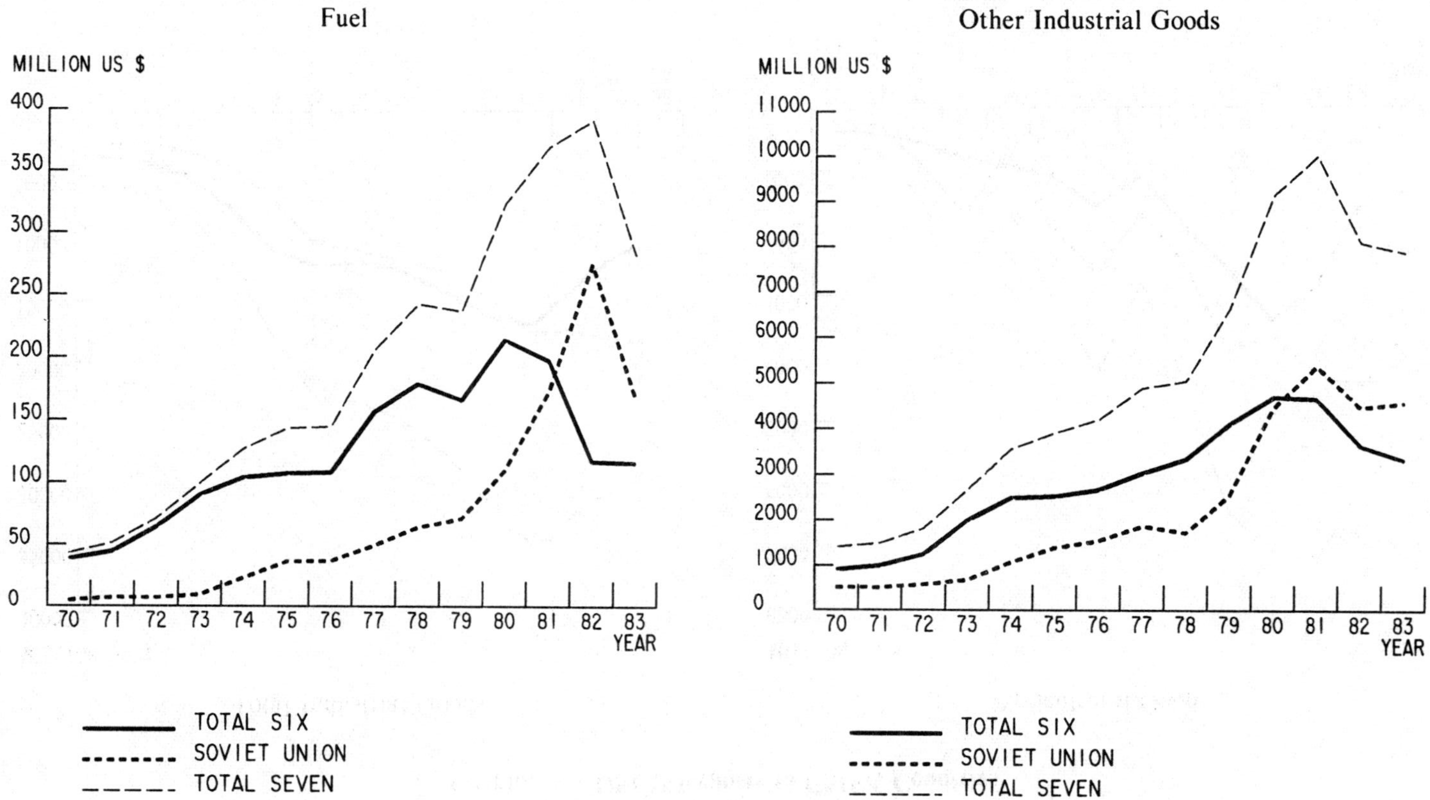

Graphic 3. **OECD Exports to CMEA Countries**

Total Industrial Goods

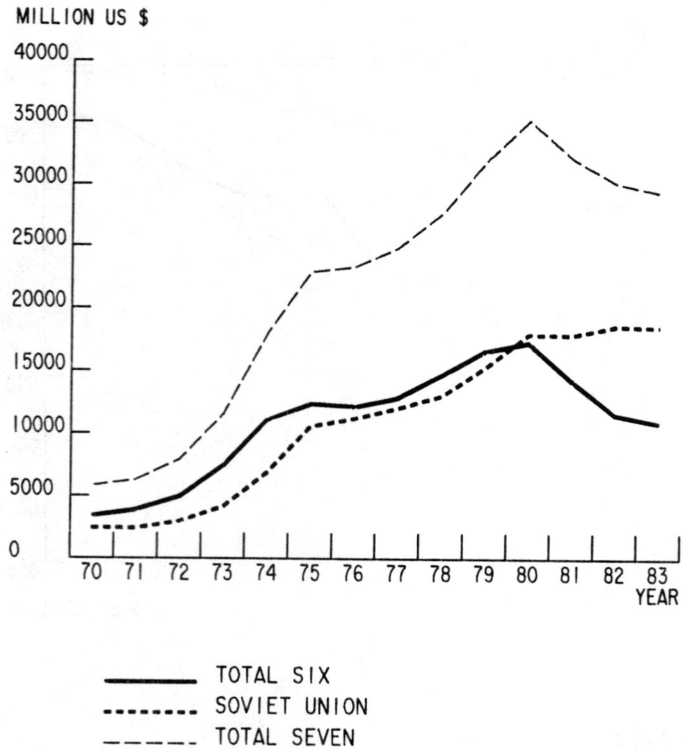


Agricultural Goods

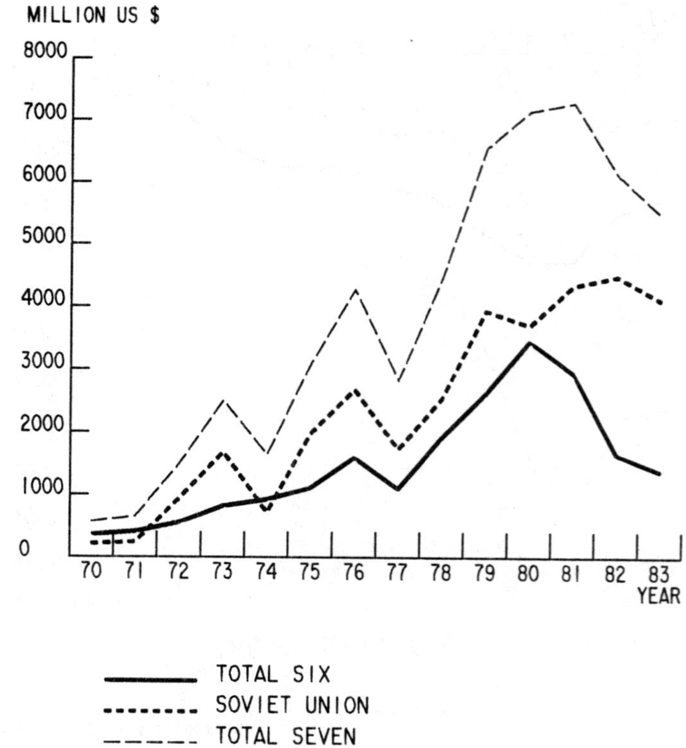

Graphic 4. **OECD Exports to CMEA Countries**

Total Trade

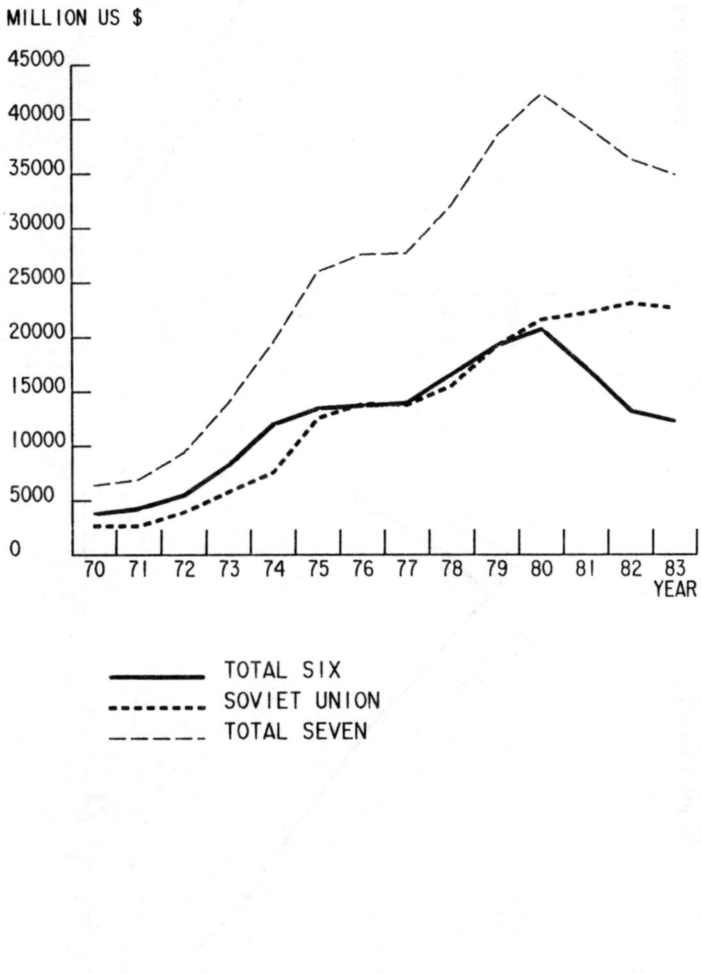

Graphic 5. **OECD Imports from CMEA Countries**

Capital Goods

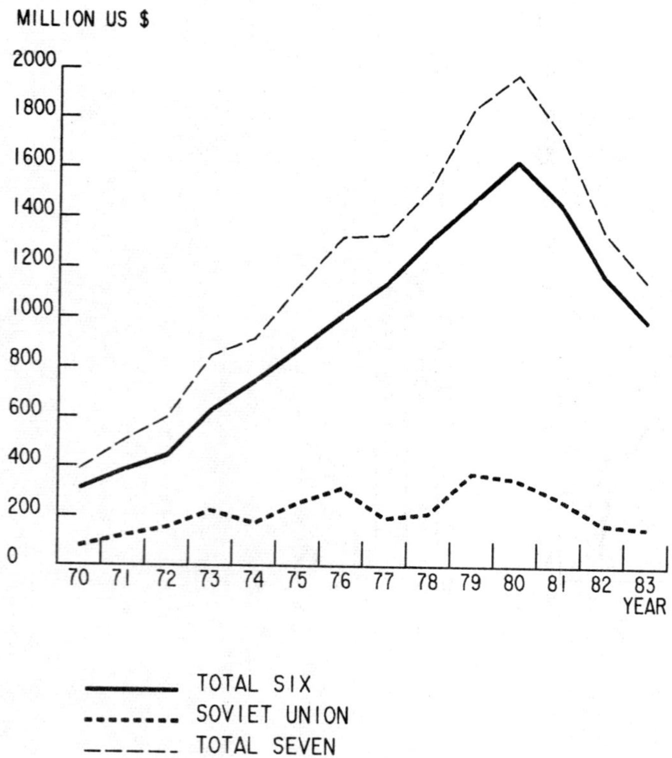


Intermediate Goods

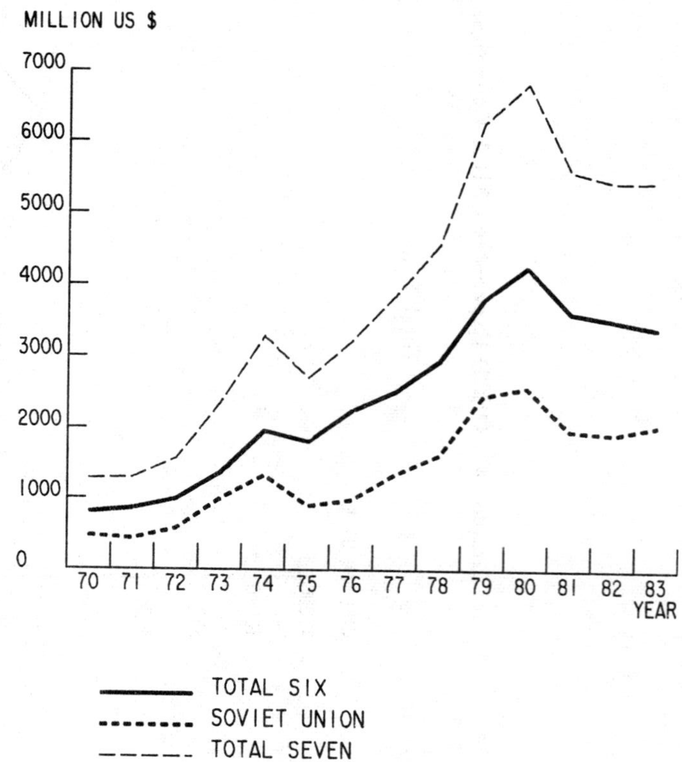

Graphic 6. **OECD Imports from CMEA Countries**

Fuel

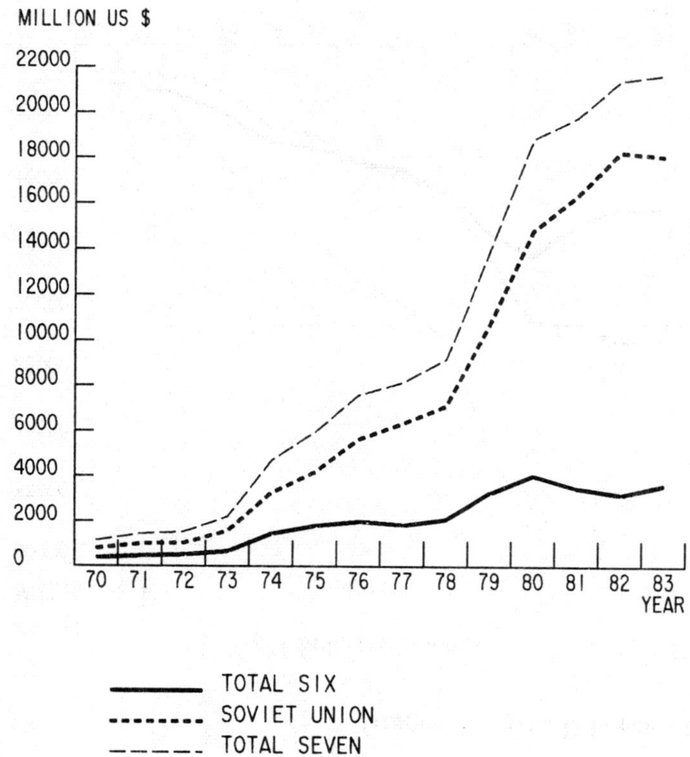


Other Industrial Goods

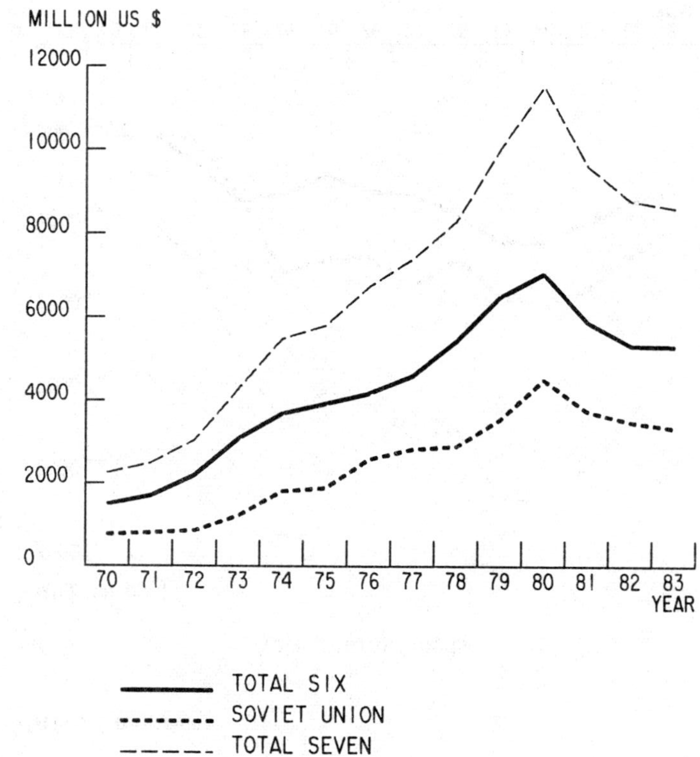

Graphic 7. **OECD Imports from CMEA Countries**

Total Industrial Goods

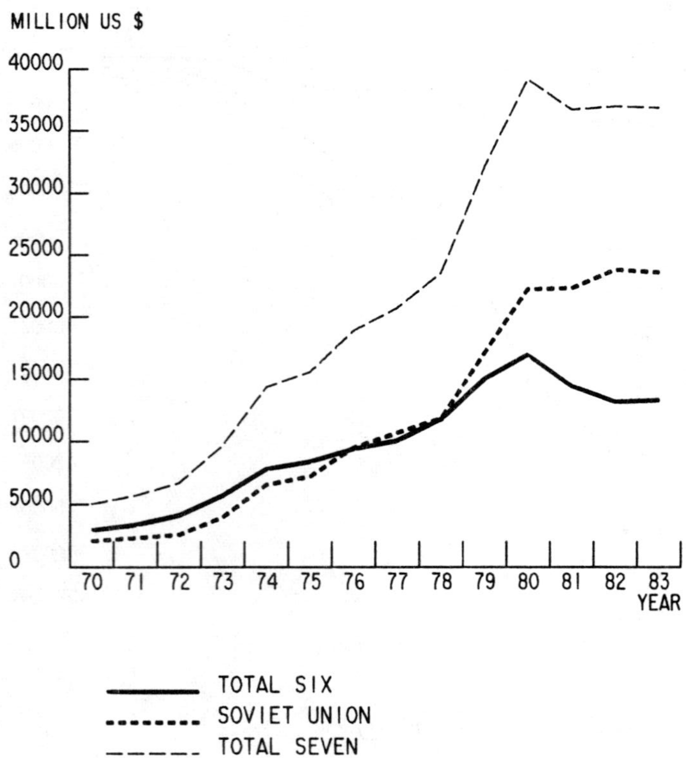


Agricultural Goods

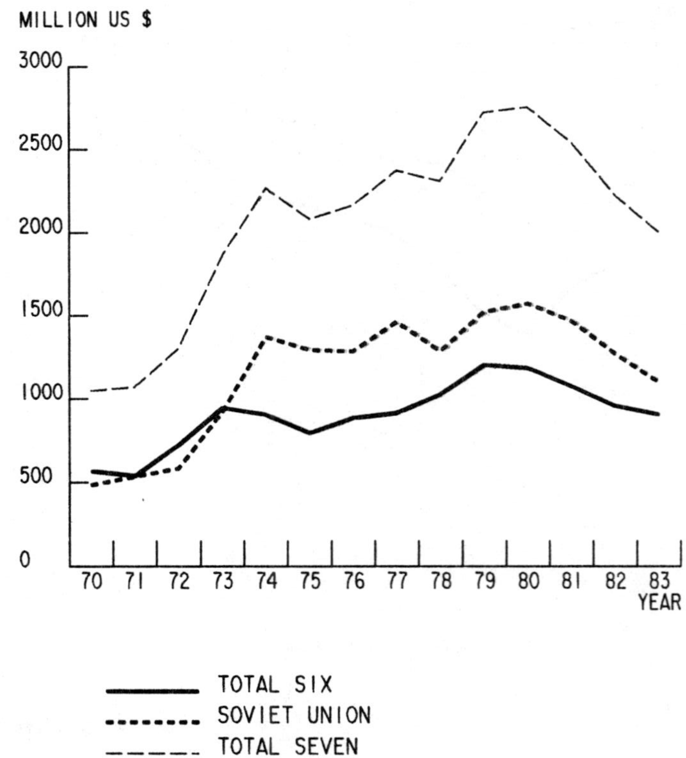

Graphic 8. **OECD Imports from CMEA Countries**

Total Trade

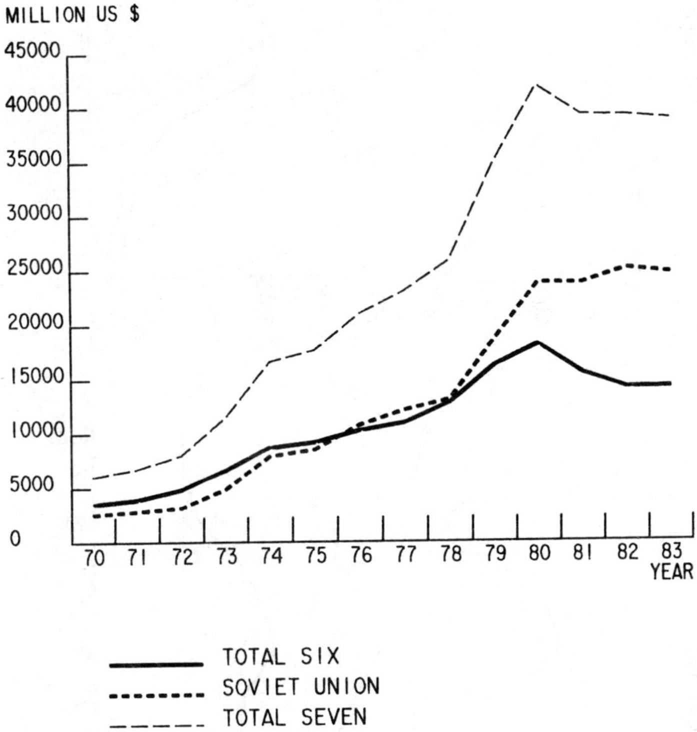

Graphic 9. **OECD Balance of Trade with CMEA Countries**

Capital Goods

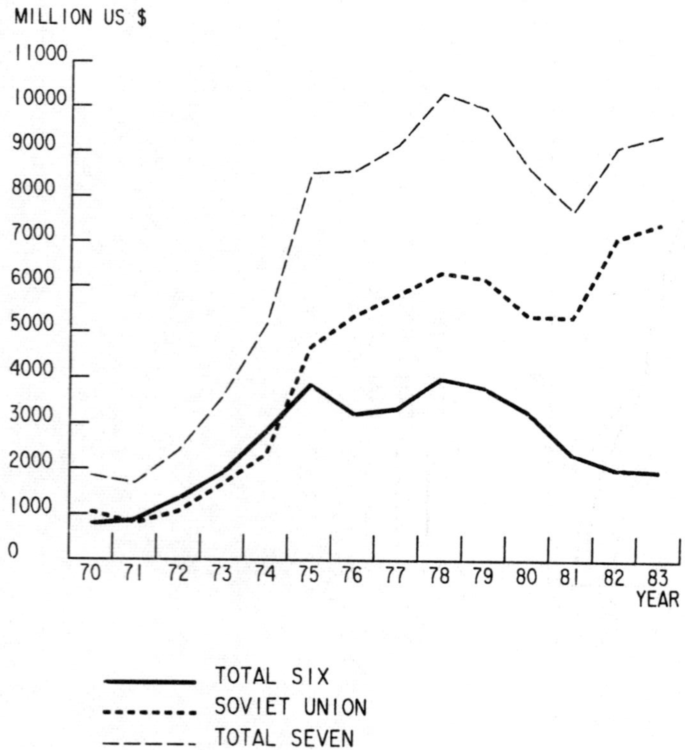


Intermediate Goods

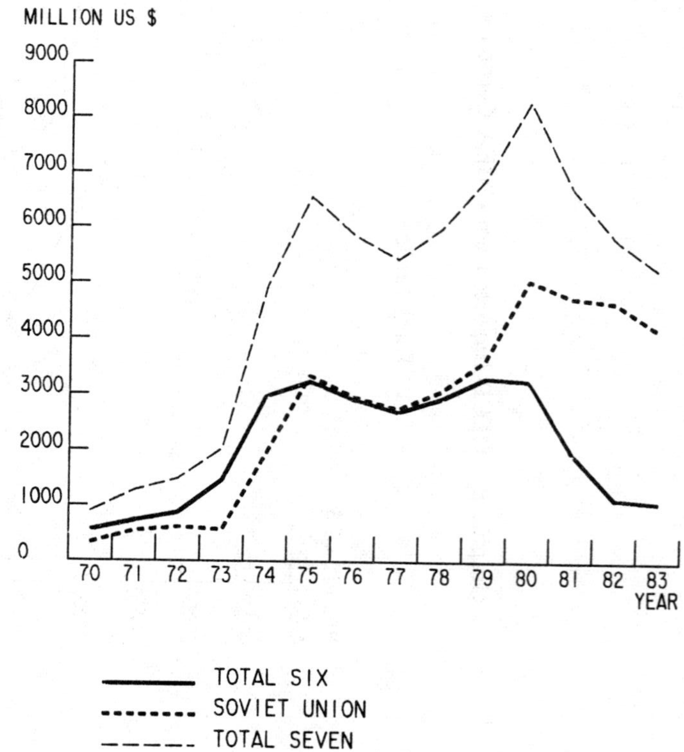

Graphic 10. **OECD Balance of Trade with CMEA Countries**

Fuel

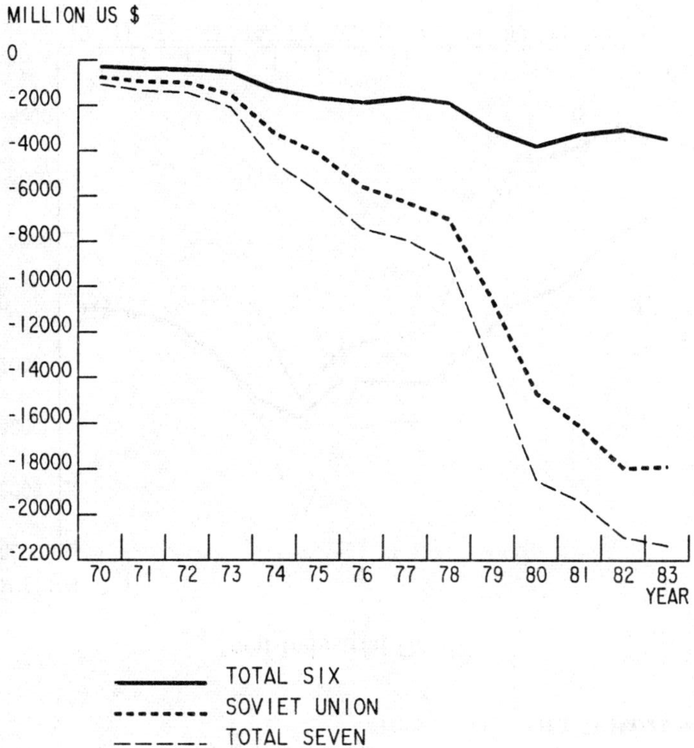


Other Industrial Goods

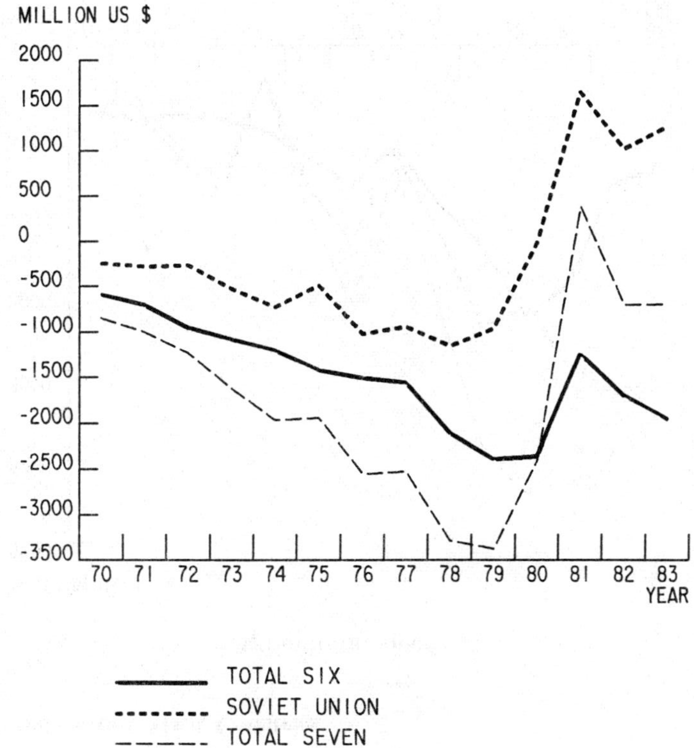

Graphic 11. **OECD Balance of Trade with CMEA Countries**

Total Industrial Goods

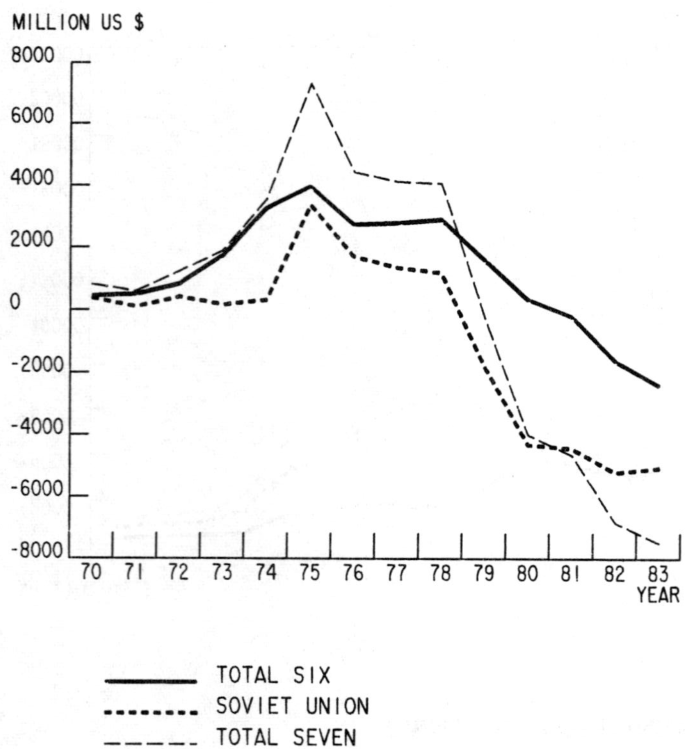


Agricultural Goods

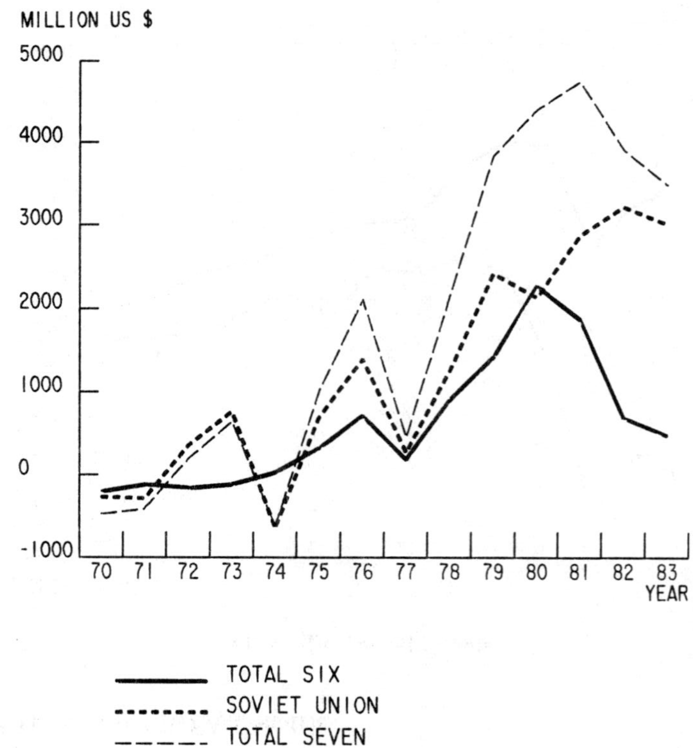

Graphic 12. **OECD Balance of Trade with CMEA Countries**

Total Trade

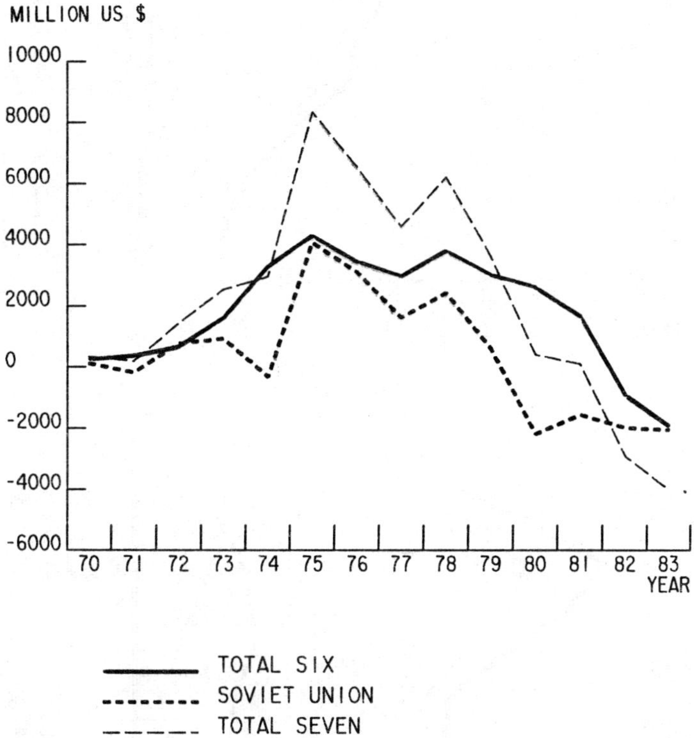

Graphic 13. **OECD Trade with Total Six**

Exports

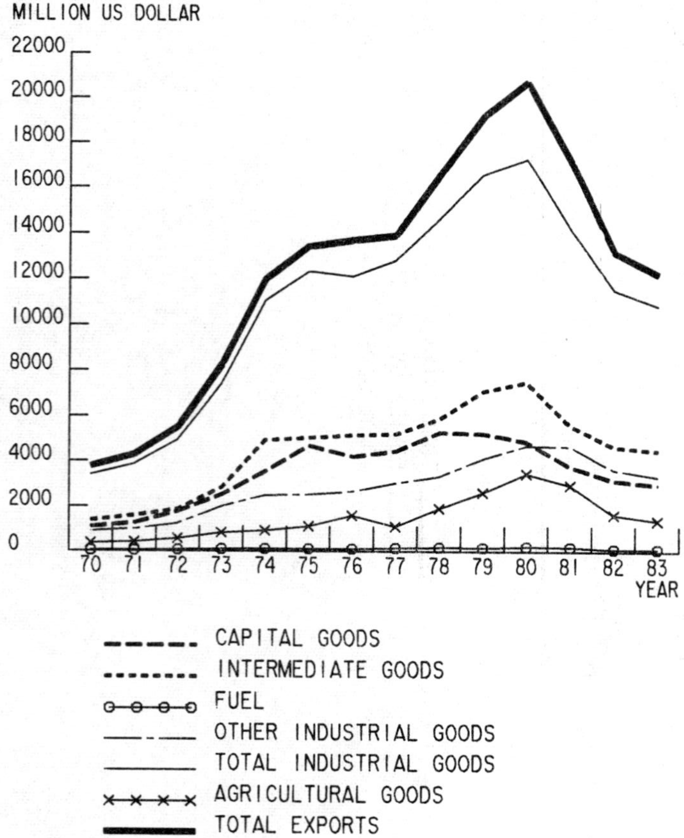


Imports

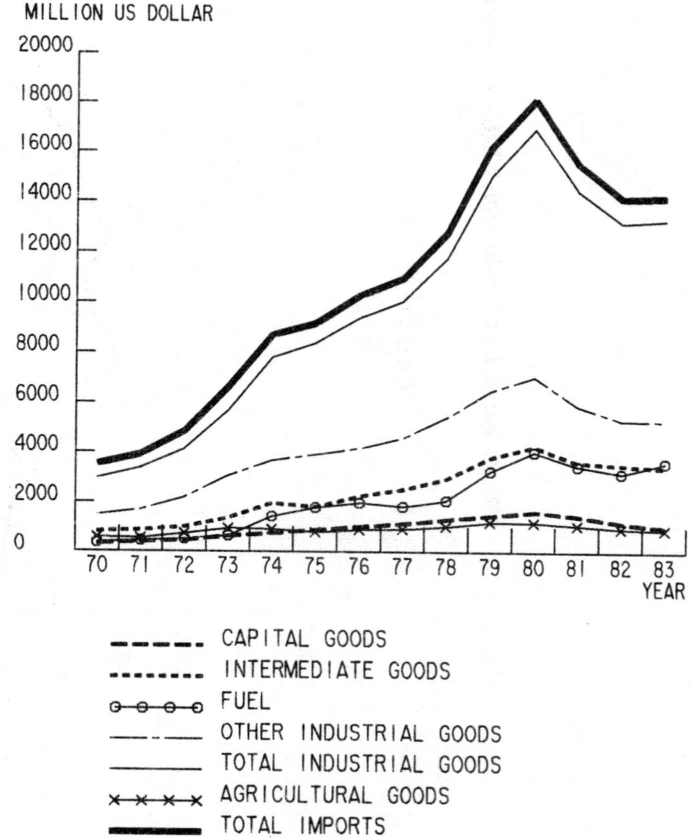

Graphic 14. **OECD Trade with Total Six**

Balance of Trade

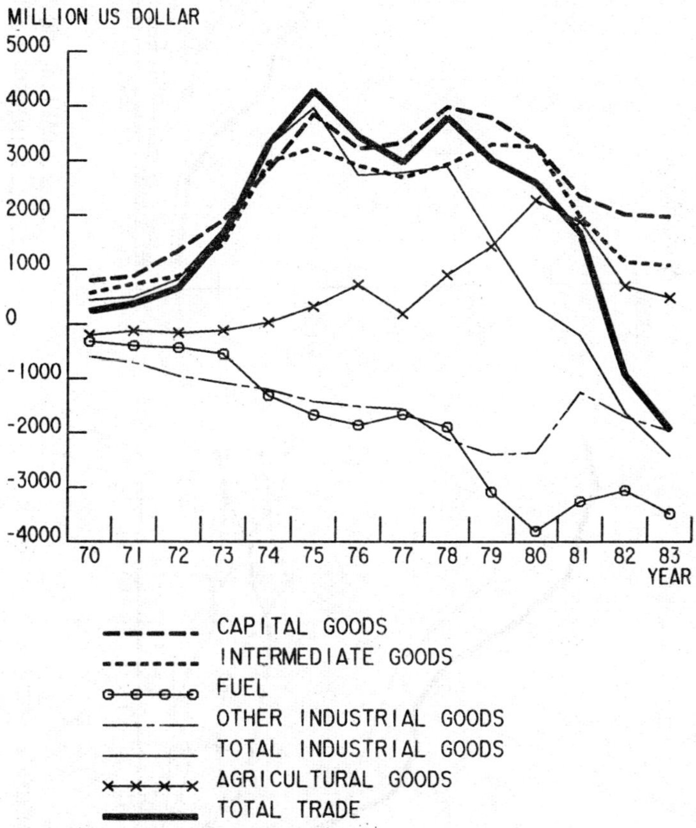

Graphic 15. **OECD Trade with USSR**

Exports

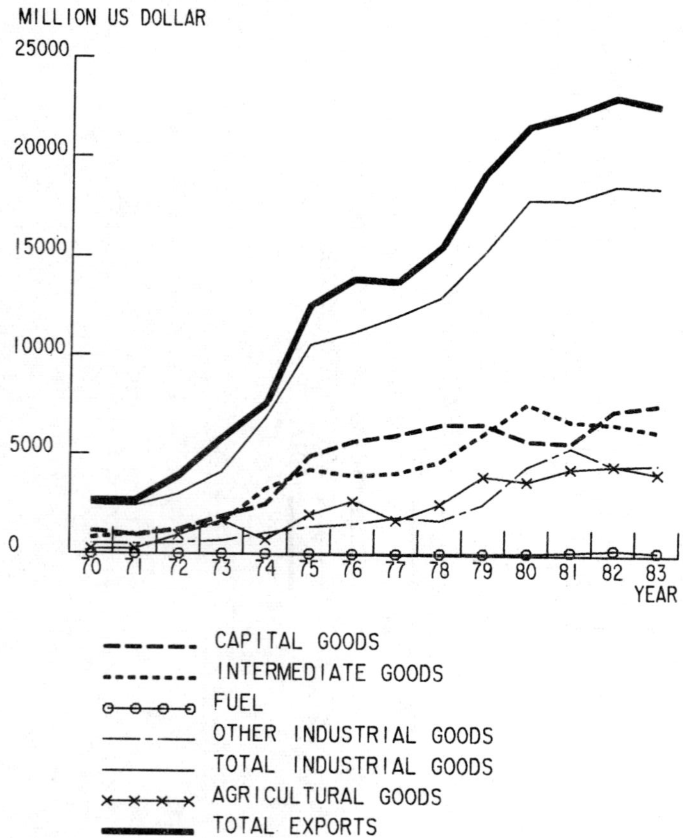


Imports

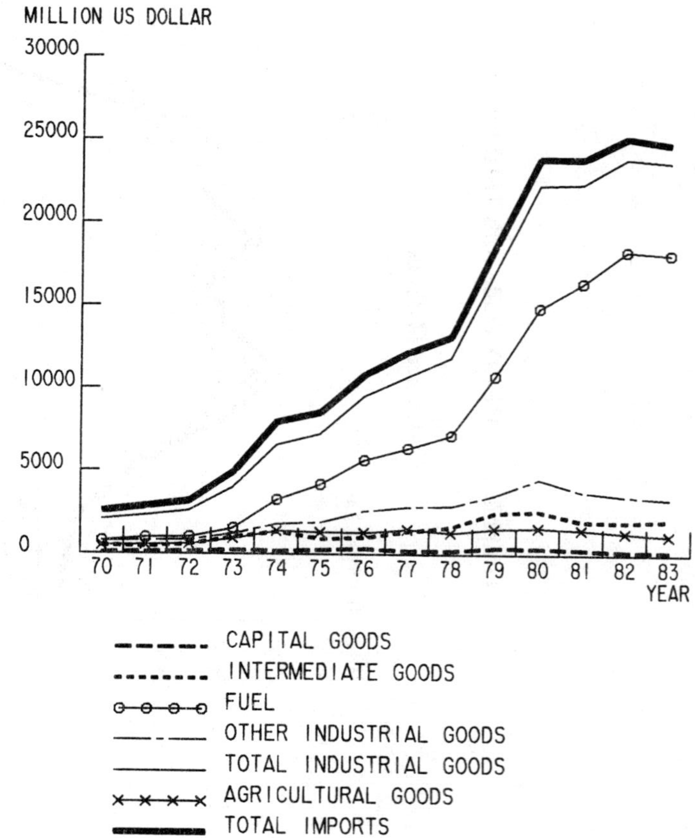

Graphic 16. **OECD Trade with USSR**

Balance of Trade

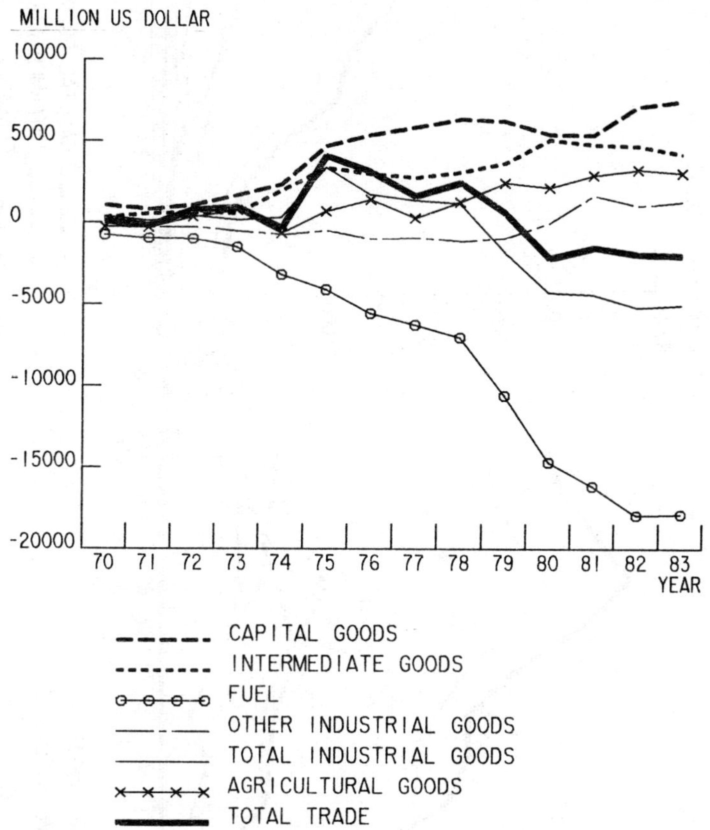

Graphic 17. **OECD Trade with Total Seven**

Exports

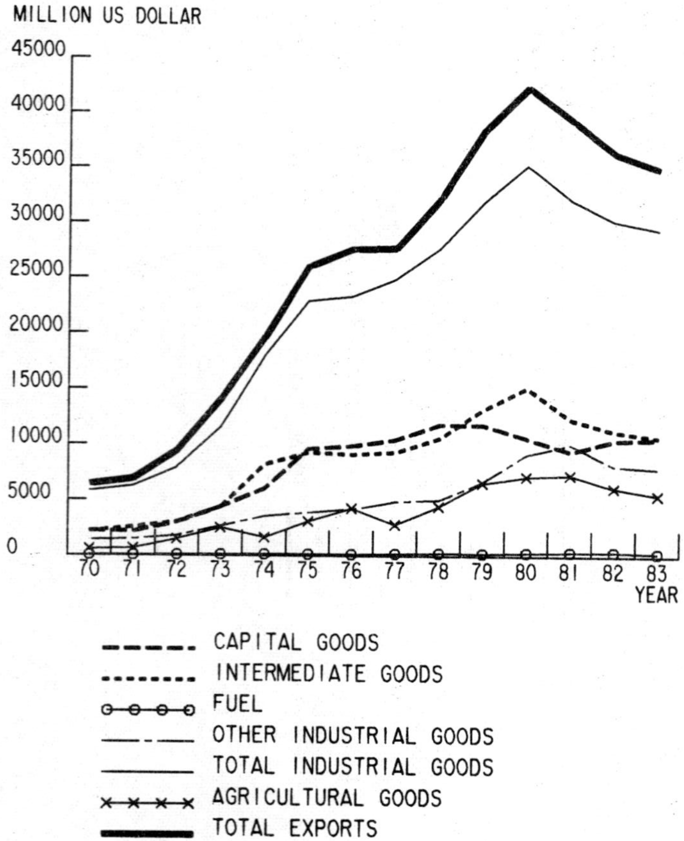


Imports

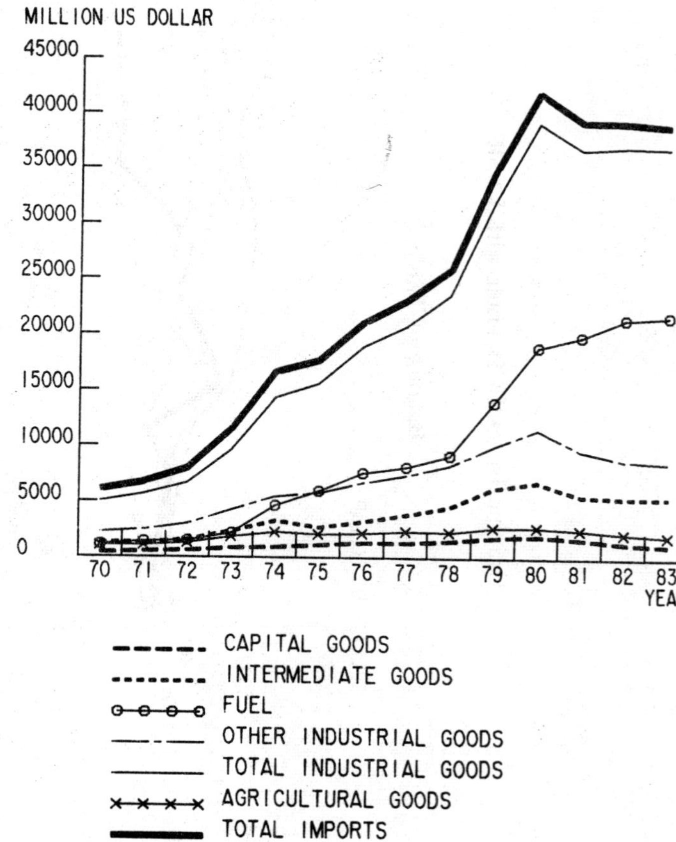

Graphic 18. **OECD Trade with Total Seven**

Balance of Trade

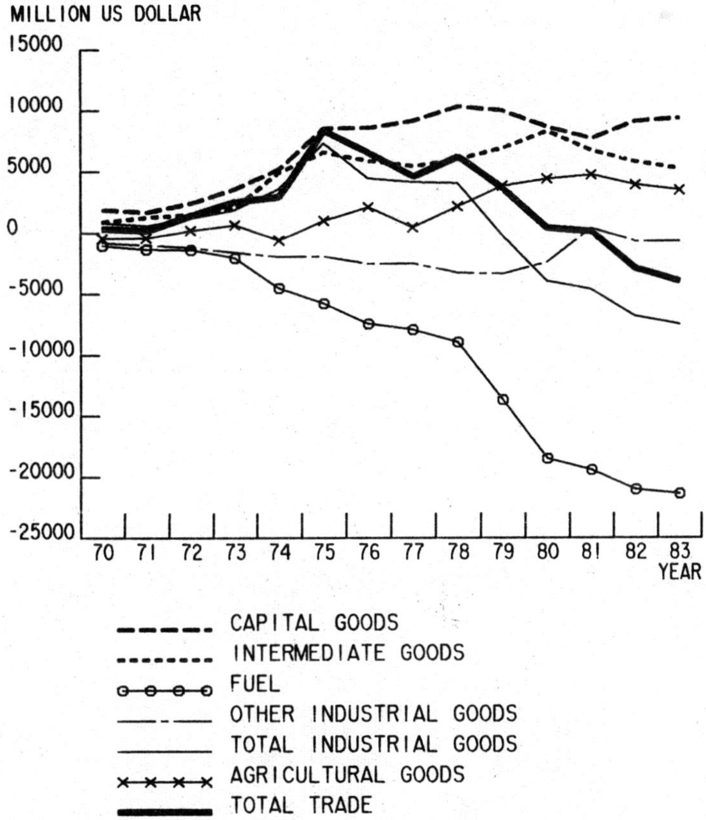

# OECD SALES AGENTS
# DÉPOSITAIRES DES PUBLICATIONS DE L'OCDE

ARGENTINA - ARGENTINE
Carlos Hirsch S.R.L.,
Florida 165, 4º Piso,
(Galeria Guemes) 1333 Buenos Aires
Tel. 33.1787.2391 y 30.7122

AUSTRALIA-AUSTRALIE
D.A. Book (Aust.) Pty. Ltd.
11-13 Station Street (P.O. Box 163)
Mitcham, Vic. 3132 Tel. (03) 873 4411

AUSTRIA - AUTRICHE
OECD Publications and Information Centre,
4 Simrockstrasse,
5300 Bonn (Germany) Tel. (0228) 21.60.45
Local Agent:
Gerold & Co., Graben 31, Wien 1 Tel. 52.22.35

BELGIUM - BELGIQUE
Jean de Lannoy, Service Publications OCDE,
avenue du Roi 202
B-1060 Bruxelles Tel. 02/538.51.69

CANADA
Renouf Publishing Company Limited/
Éditions Renouf Limitée Head Office/
Siège social – Store/Magasin :
61, rue Sparks Street,
Ottawa, Ontario K1P 5A6
Tel. (613)238-8985. 1-800-267-4164
Store/Magasin : 211, rue Yonge Street,
Toronto, Ontario M5B 1M4.
Tel. (416)363-3171
Regional Sales Office/
Bureau des Ventes régional :
7575 Trans-Canada Hwy., Suite 305,
Saint-Laurent, Quebec H4T 1V6
Tel. (514)335-9274

DENMARK - DANEMARK
Munksgaard Export and Subscription Service
35, Nørre Søgade, DK-1370 København K
Tel. +45.1.12.85.70

FINLAND - FINLANDE
Akateeminen Kirjakauppa,
Keskuskatu 1, 00100 Helsinki 10 Tel. 0.12141

FRANCE
OCDE/OECD
Mail Orders/Commandes par correspondance :
2, rue André-Pascal,
75775 Paris Cedex 16
Tel. (1) 45.24.82.00
Bookshop/Librairie : 33, rue Octave-Feuillet
75016 Paris
Tel. (1) 45.24.81.67 or/ou (1) 45.24.81.81
Principal correspondant :
Librairie de l'Université,
13602 Aix-en-Provence Tel. 42.26.18.08

GERMANY - ALLEMAGNE
OECD Publications and Information Centre,
4 Simrockstrasse,
5300 Bonn Tel. (0228) 21.60.45

GREECE - GRÈCE
Librairie Kauffmann,
28 rue du Stade, Athens 132 Tel. 322.21.60

HONG KONG
Government Information Services,
Publications (Sales) Office,
Beaconsfield House, 4/F.,
Queen's Road Central

ICELAND - ISLANDE
Snæbjörn Jónsson & Co., h.f.,
Hafnarstræti 4 & 9,
P.O.B. 1131 – Reykjavik
Tel. 13133/14281/11936

INDIA - INDE
Oxford Book and Stationery Co.,
Scindia House, New Delhi I Tel. 45896
17 Park St., Calcutta 700016 Tel. 240832

INDONESIA - INDONESIE
Pdin Lipi, P.O. Box 3065/JKT.Jakarta
Tel. 583467

IRELAND - IRLANDE
TDC Publishers – Library Suppliers
12 North Frederick Street, Dublin 1
Tel. 744835-749677

ITALY - ITALIE
Libreria Commissionaria Sansoni,
Via Lamarmora 45, 50121 Firenze
Tel. 579751/584468
Via Bartolini 29, 20155 Milano Tel. 365083
Sub-depositari :
Ugo Tassi, Via A. Farnese 28,
00192 Roma Tel. 310590
Editrice e Libreria Herder,
Piazza Montecitorio 120, 00186 Roma
Tel. 6794628
Agenzia Libraria Pegaso,
Via de Romita 5, 70121 Bari
Tel. 540.105/540.195
Agenzia Libraria Pegaso, Via S.Anna dei
Lombardi 16, 80134 Napoli. Tel. 314180
Libreria Hœpli,
Via Hœpli 5, 20121 Milano Tel. 865446
Libreria Scientifica
Dott. Lucio de Biasio "Aeiou"
Via Meravigli 16, 20123 Milano Tel. 807679
Libreria Zanichelli, Piazza Galvani 1/A,
40124 Bologna Tel. 237389
Libreria Lattes,
Via Garibaldi 3, 10122 Torino Tel. 519274
La diffusione delle edizioni OCSE è inoltre assicurata dalle migliori librerie nelle città più importanti.

JAPAN - JAPON
OECD Publications and Information Centre,
Landic Akasaka Bldg., 2-3-4 Akasaka,
Minato-ku, Tokyo 107 Tel. 586.2016

KOREA - CORÉE
Pan Korea Book Corporation
P.O.Box No. 101 Kwangwhamun, Seoul
Tel. 72.7369

LEBANON - LIBAN
Documenta Scientifica/Redico,
Edison Building, Bliss St.,
P.O.B. 5641, Beirut Tel. 354429-344425

MALAYSIA - MALAISIE
University of Malaya Co-operative Bookshop Ltd.,
P.O.Box 1127, Jalan Pantai Baru,
Kuala Lumpur Tel. 577701/577072

NETHERLANDS - PAYS-BAS
Staatsuitgeverij Verzendboekhandel
Chr. Plantijnstraat, 1 Postbus 20014
2500 EA S-Gravenhage Tel. 070-789911
Voor bestellingen: Tel. 070-789208

NEW ZEALAND - NOUVELLE-ZÉLANDE
Government Printing Office Bookshops:
Auckland: Retail Bookshop, 25 Rutland Street,
Mail Orders, 85 Beach Road
Private Bag C.P.O.
Hamilton: Retail: Ward Street,
Mail Orders, P.O. Box 857
Wellington: Retail, Mulgrave Street, (Head Office)
Cubacade World Trade Centre,
Mail Orders, Private Bag
Christchurch: Retail, 159 Hereford Street,
Mail Orders, Private Bag
Dunedin: Retail, Princes Street,
Mail Orders, P.O. Box 1104

NORWAY - NORVÈGE
Tanum-Karl Johan a.s
P.O. Box 1177 Sentrum, 0107 Oslo 1
Tel. (02) 801260

PAKISTAN
Mirza Book Agency
65 Shahrah Quaid-E-Azam, Lahore 3 Tel. 66839

PORTUGAL
Livraria Portugal,
Rua do Carmo 70-74, 1117 Lisboa Codex.
Tel. 360582/3

SINGAPORE - SINGAPOUR
Information Publications Pte Ltd
Pei-Fu Industrial Building,
24 New Industrial Road No. 02-06
Singapore 1953 Tel. 2831786, 2831798

SPAIN - ESPAGNE
Mundi-Prensa Libros, S.A.,
Castelló 37, Apartado 1223, Madrid-28001
Tel. 431.33.99
Libreria Bosch, Ronda Universidad 11,
Barcelona 7 Tel. 317.53.08/317.53.58

SWEDEN - SUÈDE
AB CE Fritzes Kungl. Hovbokhandel,
Box 16356, S 103 27 STH,
Regeringsgatan 12,
DS Stockholm Tel. (08) 23.89.00
Subscription Agency/Abonnements:
Wennergren-Williams AB,
Box 30004, S104 25 Stockholm. Tel. 08/54.12.00

SWITZERLAND - SUISSE
OECD Publications and Information Centre,
4 Simrockstrasse,
5300 Bonn (Germany) Tel. (0228) 21.60.45
Local Agent:
Librairie Payot,
6 rue Grenus, 1211 Genève 11
Tel. (022) 31.89.50

TAIWAN - FORMOSE
Good Faith Worldwide Int'l Co., Ltd.
9th floor, No. 118, Sec.2
Chung Hsiao E. Road
Taipei Tel. 391.7396/391.7397

THAILAND - THAILANDE
Suksit Siam Co., Ltd.,
1715 Rama IV Rd.,
Samyam Bangkok 5 Tel. 2511630

TURKEY - TURQUIE
Kültur Yayinlari Is-Türk Ltd. Sti.
Atatürk Bulvari No: 191/Kat. 21
Kavaklidere/Ankara Tel. 17.02.66
Dolmabahce Cad. No: 29
Besiktas/Istanbul Tel. 60.71.88

UNITED KINGDOM - ROYAUME UNI
H.M. Stationery Office,
Postal orders only:
P.O.B. 276, London SW8 5DT
Telephone orders: (01) 622.3316, or
Personal callers:
49 High Holborn, London WC1V 6HB
Branches at: Belfast, Birmingham,
Bristol, Edinburgh, Manchester

UNITED STATES - ÉTATS-UNIS
OECD Publications and Information Centre,
Suite 1207, 1750 Pennsylvania Ave., N.W.,
Washington, D.C. 20006 - 4582
Tel. (202) 724.1857

VENEZUELA
Libreria del Este,
Avda F. Miranda 52, Aptdo. 60337,
Edificio Galipan, Caracas 106
Tel. 32.23.01/33.26.04/31.58.38

YUGOSLAVIA - YOUGOSLAVIE
Jugoslovenska Knjiga, Knez Mihajlova 2,
P.O.B. 36, Beograd Tel. 621.992

Orders and inquiries from countries where Sales Agents have not yet been appointed should be sent to:
OECD, Publications Service, Sales and Distribution Division, 2, rue André-Pascal, 75775 PARIS CEDEX 16.

Les commandes provenant de pays où l'OCDE n'a pas encore désigné de dépositaire peuvent être adressées à :
OCDE, Service des Publications. Division des Ventes et Distribution. 2. rue André-Pascal. 75775 PARIS CEDEX 16.

69482-03-1986

OECD PUBLICATIONS, 2, rue André-Pascal, 75775 PARIS CEDEX 16 - No. 43571 1986
PRINTED IN FRANCE
(92 86 03 1) ISBN 92-64-12817-4